Lecture Notes in Mathematics

Edited by A. Dold and B. Eckmann

832

Representation Theory II

Proceedings of the Second International Conference on Representations of Algebras
Ottawa, Carleton University, August 13 - 25, 1979

Edited by V. Dlab and P. Gabriel

Springer-Verlag
Berlin Heidelberg New York 1980

Editors

Vlastimil Dlab
Department of Mathematics
Carleton University
Ottawa K1S 5B6
Canada

Peter Gabriel
Mathematisches Institut
Universität Zürich
Freiestrasse 36
8032 Zürich
Switzerland

AMS Subject Classifications (1980) 16A18, 16A52, 16A64, 20A20

ISBN 3-540-10264-7 Springer-Verlag Berlin Heidelberg New York
ISBN 0-387-10264-7 Springer-Verlag New York Heidelberg Berlin

Printing and binding: Beltz Offsetdruck, Hemsbach/Bergstr.
2141/3140-543210

PREFACE

The First International Conference on Representations of Algebras was held at Carleton University, Ottawa, on September 3-7, 1974; the Proceedings of the Conference appeared as Springer Lecture Notes #488 the following year.

Since then, the interest in the representation theory grew rapidly, and significant advances and contributions to the theory continued to take place. These were to be reflected in the meeting organized at Carleton University on August 13-25, 1979. The Organizing Committee consisted of Professors J. Alperin, M. Auslander, V. Dlab, P. Gabriel, I. Reiner, C.M. Ringel, A.V. Rojter and H. Tachikawa.

The developments of the five-year period preceding the meeting were the subject of five series of lectures in the WORKSHOP ON THE PRESENT TRENDS IN REPRESENTATION THEORY (August 13-18, 1979) given by J. Alperin (Block theory), P. Gabriel (Trends in representation theory), J.E. Humphreys (Highest weight modules for semi simple Lie algebras), C.M. Ringel (Algorithms for solving vector space problems) and V.A. Rojter (Differential graded categories). The first volume of these Proceedings contains reports from the WORKSHOP.

Recent advances in the representation theory were reported in 42 lectures during the second part of the meeting: THE SECOND INTERNATIONAL CONFERENCE ON REPRESENTATIONS OF ALGEBRAS; the lectures are listed on page VIII. Not all contributions to the CONFERENCE appear in these Proceedings; on the other hand, some papers which were not reported, are included. All published papers appear in the form submitted by the author; only very few technical alterations have been made. It should be pointed out that some of the contributions to these Proceedings have resulted out of discussions and conversations during the meeting. We should like to thank all referees for their assistance.

It has been suggested that the Proceedings contain a list of publications in the field of representation theory for the past 10 years. An attempt in this direction has been made; we wish to thank Mr. Ibrahim Assem for his assistance in preparing the Bibliography.

We should like to acknowledge financial assistance of the Natural Sciences and Engineering Research Council Canada and of Carleton University. The assistance of other national research bodies to support participants of the meeting has been also greatly appreciated. In particular, we wish to thank Carleton University for the generous support of the first part of the meeting; without its assistance the organization of the WORKSHOP would not have been possible. And, we wish to extend our thanks to the Secretary, Professor Luis Ribes for his unlimited help in the organization of the meeting and to Ms. Alejandra Leon for her efficient secretarial assistance.

Ottawa - Zurich, July 1980 Vlastimil Dlab and Peter Gabriel

TABLE OF CONTENTS

List of lectures VIII

List of registered participants XI

Volume I WORKSHOP

PETER GABRIEL

Auslander-Reiten sequences and representation-finite algebras 1

JAMES E. HUMPHREYS

Highest weight modules for semisimple Lie algebras 72

CLAUS MICHAEL RINGEL

Report on the Brauer-Thrall conjectures 104

CLAUS MICHAEL RINGEL

Tame algebras 137

V.A. ROJTER

Matrix problems and representations of BOCS's 288

Bibliography 1969 - 1979 325

Volume II CONFERENCE

M. AUSLANDER and IDUN REITEN

Uniserial functors 1

M. AUSLANDER and S.O. SMALØ

Preprojective modules: An introduction and some applications 48

R. BAUTISTA

Sections in Auslander-Reiten quivers 74

K. BONGARTZ

Zykellose Algebren sind nicht zügellos 97

SHEILA BRENNER and M.C.R. BUTLER

Generalizations of the Bernstein-Gelfand-Ponomarev reflection functors 103

H. BRUNE
On finite representation type and a theorem of Kulikov 170

CHARLES W. CURTIS
Homology representations of finite groups 177

EVERETT C. DADE
Algebraically rigid modules 195

VLASTIMIL DLAB and CLAUS MICHAEL RINGEL
The preprojective algebra of a modulated graph 216

P. DOWBOR, C.M. RINGEL and D. SIMSON
Hereditary artinian rings of finite representation type 232

JU. A. DROZD
Tame and wild matrix problems 242

EDWARD L. GREEN
Remarks on projective resolutions 259

DIETER HAPPEL, UDO PREISER and CLAUS MICHAEL RINGEL
Vinberg's characterization of Dynkin diagrams using subadditive functions with application to DTr-periodic modules 280

YASUO IWANAGA and TAKAYOSHI WAKAMATSU
Trivial extension of artin algebras 295

C.U. JENSEN and H. LENZING
Model theory and representations of algebras 302

VICTOR G. KAČ
Some remarks on representations of quivers and infinite root systems 311

HERBERT KUPISCH and EBERHARD SCHERZLER
Symmetric algebras of finite representation type 328

PETER LANDROCK
Some remarks on Loewy lengths of projective modules 369

NIKOLAOS MARMARIDIS
Reflection functors 382

ROBERTO MARTINEZ-VILLA

Algebras stably equivalent to ℓ-hereditary 396

FRANK OKOH

Hereditary algebras that are not pure-hereditary 432

WILHELM PLESKEN

Projective lattices over group orders as amalgamations of irreducible lattices 438

CHRISTINE RIEDTMANN

Representation-finite selfinjective algebras of class A_n 449

K.W. ROGGENKAMP

Representation theory of blocks of defect 1 521

EBERHARD SCHERZLER AND JOSEF WASCHBÜSCH

A class of self-injective algebras of finite representation type 545

DANIEL SIMSON

Right pure semisimple hereditary rings 573

HIROYUKI TACHIKAWA

Representations of trivial extensions of hereditary algebras 579

GORDANA TODOROV

Almost split sequences for TrD-periodic modules 600

JOSEF WASCHBÜSCH

A class of self-injective algebras and their indecomposable modules 632

KUNIO YAMAGATA

Hereditary artinian rings of right local representation type 648

LIST OF LECTURES

J.L. ALPERIN	Complexity of modules
M. AUSLANDER- I. REITEN*	Uniserial functors
M. AUSLANDER*- S.O. SMALØ	Subcategories of mod Λ over which mod Λ is functorially finite
M. AUSLANDER - S.O. SMALØ*	Preprojective modules
R. BAUTISTA	Sections in Auslander components
R.E. BLOCK	The algebraically irreducible representations of the Lie algebra $s\ell$ 2
K. BONGARTZ	Algebras of finite representation type without cycles
S. BRENNER	Some co-rank 2 quivers with relations and their null roots
D.W. BURRY	A module-oriented theory of blocks
M.C.R. BUTLER	Generalizations of the Bernstein-Gelfand-Ponomarev reflection functors
J.F. CARLSON	The dimension of modules and their restrictions over modular group algebras
Ch.W. CURTIS	Homology representations of finite groups
E.C. DADE	Algebraically stable modules
K. ERDMANN	On blocks whose defect groups are elementary abelian of order p^2
H.K. FARAHAT	Young and James ideals in a group algebra

E.L. GREEN	Remarks on projective resolutions
M. HAZEWINKEL	On the representations of the wild quiver $\cdot \rightarrow \cdot \circlearrowleft$
J.E. HUMPHREYS	Cartan invariants and decomposition numbers for Chevalley groups
Y. IWANAGA	Trivial extension of some artin algebras
H. JACOBINSKI	Hereditary covers and blocks
J.C. JANTZEN	Representations of semisimple groups and their Frobenius kernels
V.G. KAC	Infinite root systems, representations of quivers and invariant theory
H. KUPISCH	Symmetric algebras of finite representation type
P. LANDROCK	Some remarks on Loewy lengths of projective modules of a symmetric algebra
L. LEVY	Mixed modules over $\mathbb{Z}G$, G cyclic of prime order, and over related Dedekind pullbacks
R. MARTINEZ-VILLA	Algebras stably equivalent to ℓ-hereditary
G.O. MICHLER	On blocks of finite groups with abelian defect groups
R.V. MOODY	Hyperbolic Lie algebras and singularities
F. OKOH	A hereditary finite dimensional algebra that is not pure-hereditary
W. PLESKEN	Projective lattices over group orders as amalgamations of irreducible lattices
I. REINER	Solomon's conjecture and the functional equation for zeta function of orders
Ch. RIEDTMANN	Selfinjective algebras of class A_n and D_n

C.M. RINGEL	Indecomposable representations of wild quivers
K.W. ROGGENKAMP	Blocks of cyclic defect
A.V. ROJTER	Gelfand-Ponomarev algebra of a quiver
F.J. SERVEDIO	GL(V)-submodules of $V^{\otimes m}$ fixed by subgroups of S_m; forms of degree t in forms of degree r on V
D. SIMSON	Species and hereditary rings of finite representation type
H. TACHIKAWA	Representations of algebras of trivial extension
G. TODOROV**	Almost split sequences for TrD-periodic modules M, with no projectives in the class [M]
J. WASCHBÜSCH	Quasi-Frobenius algebras of finite representation type
P.J. WEBB	Distinguishing non-isomorphic relation modules
K. YAMAGATA	On artinian rings of local representation type

* denotes the speaker

** lecture delivered by I.M. Platzeck

LIST OF REGISTERED PARTICIPANTS

ABRAMS Gene, University of Oregon, Eugene, Oregon, U.S.A.

ALPERIN Jonathan L., University of Chicago, Chicago, Illinois, U.S.A.

AMDAL Ivar, University of Trondheim, Norway.

ASSEM Ibrahim, Carleton University, Ottawa, Ontario, Canada.

AUSLANDER Maurice, Brandeis University, Waltham, Mass., U.S.A.

BAUTISTA Raymundo, Universidad Nac. Autónoma de México, Mexico.

BECKER Helmut, Hochschule der Bundeswehr, München, West Germany.

BINGEN Franz, Vrije Universiteit Brussel, Belgium.

BLOCK Richard E., University of California, Riverside, Calif., U.S.A.

BONGARTZ Klaus, Universität Zürich, Switzerland.

BRENNER Sheila, University of Liverpool, United Kingdom.

BRITTEN Dan, University of Windsor, Ontario, Canada.

BURRY David, Yale University, New Haven, Connecticut, U.S.A.

BUTLER M.C.R., University of Liverpool, United Kingdom.

BUTSAN George, Math. Institute, AN USSR, Kiev, U.S.S.R.

CARLSON J.F., University of Georgia, Athens, U.S.A.

CIBILS Claude, Universidad Autónoma Metropolitana, Mexico City, Mexico.

CLIFF Gerald, University of Alberta, Edmonton, Alberta, Canada.

CURTIS Charles W., University of Oregon, Eugene, Oregon, U.S.A.

DADE Everett C., University of Illinois, Urbana, Illinois, U.S.A.

DAVIS Richard A., Columbia University, New York, U.S.A.

DIETERICH Ernst, Universität Bielefeld, West Germany.

DIPPER Richard, Universität Essen, West Germany.

DIXON John, Carleton University, Ottawa, Ontario, Canada.

DLAB Vlastimil, Carleton University, Ottawa, Ontario, Canada.

ERDMANN Karin, Universität Essen, West Germany.

FARAHAT H.K., University of Calgary, Calgary, Alberta, Canada.

FORD Charles, St. Louis University, Missouri, U.S.A.

FULLER Kent, R., University of Hawaii, Manoa, Hawaii, U.S.A.

GABRIEL Peter, Universität Zürich, Switzerland.

GODFREY Colin M., University of Massachusetts, Boston, Mass., U.S.A.

GREEN Edward L., Virginia Polytechnic Institute, Blacksburg, U.S.A.

GRIESS Robert L. Jr., University of Michigan, Ann Arbor, U.S.A./
Institute for Advanced Studies, Princeton, N.J., U.S.A.

GUSTAFSON William H., Texas Tech. University, Lubbock, Texas, U.S.A.

HAPPEL Dieter, Universität Bielefeld, West Germany.

HAZEWINKEL Michiel, Erasmus Univ. Rotterdam, Holland.

HUGHES David, University of Liverpool, United Kingdom.

HUMPHREYS J.E., University of Massachusetts, Amherst, Mass., U.S.A.

IWANAGA Yasuo, University of Tsukuba, Ibaraki, Japan/Carleton University,
Ottawa, Ontario, Canada.

JACOBINSKI H., Chalmers University of Technology, Göteborg, Sweden.

JANSEN Willem G., McMaster University, Hamilton, Ontario, Canada.

JANTZEN Jens C., Universität Bonn, West Germany

JANUSZ Gerald J., University of Illinois, Urbana, Ill., U.S.A.

JONES Alfredo, Universidade de São Paulo, Brasil.

KAC Victor, M.I.T., Cambridge, Mass., U.S.A.

KANEDA Masaharu, University of Oregon, Eugene, Oregon, U.S.A.

KLEINER Mark, New York, U.S.A.

KLEISLI Heinrich, Université de Fribourg, Switzerland.

KNODLE Stephen, State University of New York, Potsdam, N.Y., U.S.A.

KNÖRR Reinhard, University of Illinois, Urbana, Ill., U.S.A.

KOVÁCS Lászlo G., Australian National University, Canberra, Australia.

KRAEMER Julius, Universität München, West Germany.

KUBO Fujio, Southern Illinois University, Edwardsville, Ill., U.S.A.

KUPISCH H., Freie Universität, West Berlin.

LADY Lee E., University of Hawaii, Honolulu, Hawaii, U.S.A.

LANDROCK Peter, Aarhus University, Denmark.

LARRION Francisco, Universidad Nac. Autónoma de México, Mexico.

LEON Alejandra, Carleton University, Ottawa, Ontario, Canada.

LEUNG Tat-Wing, Queen's University, Kingston, Ontario, Canada.

LEVY Lawrence S., University of Wisconsin, Madison, Wisconsin, U.S.A.

MARMARIDIS Nikolaos, University of Creta, Greece.

MARTINEZ-VILA Roberto, Universidad Nac. Autónoma de México, Mexico.

MERKLEN Hector A., Universidade de São Paulo, Brasil.

MICHLER Gerhard O., Universitat Essen, West Germany.

MOLLIN Richard A., McMaster University, Hamilton, Ontario, Canada.

MOODY Robert V., University of Saskatchewan, Saskatoon, Canada.

MORTIMER Brian, Carleton University, Ottawa, Ontario, Canada.

OKOH Frank, York University, Downsview, Ontario, Canada.

OLSSON Jørn B., Universität Dortmund, West Germany.

O'NEILL John D., University of Detroit, Michigan, U.S.A.

OTAL Javier, Universidad de Zaragoza, Spain.

PLATZECK Maria Ines, Universidad Nacional del Sur, Bahia Blanca, Argentina.

PLESKEN Wilhelm, RWTH Aachen, West Germany.

PUTTASWAMAIAH B., Carleton University, Ottawa, Ontario, Canada.

REINER Irving, University of Illinois, Urbana, Ill., U.S.A.

REITEN Idun, University of Trondheim, Norway.

RIEDTMANN Christine, Universität Basel, Switzerland.

RIBES Luis, Carleton University, Ottawa, Ontario, Canada.

RINGEL Claus Michael, Universität Bielefeld, West Germany.

ROGGENKAMP Klaus W., Universität Stuttgart, West Germany.

ROJTER A.V., Math. Inst., AN USSR Kiev, U.S.S.R.

ROLDAN Oscar E., Carleton University, Ottawa, Ontario, Canada.

RUMP Wolfgang, Ohio State University, Columbus, Ohio, U.S.A.

SALMERON Leonardo, Universidad Nac. Autónoma de México, Mexico.

SANTHAROUBANE Louis-Joseph, Université de Paris, France.

SCHNEIDER Gerhard, Universität of Essen, West Germany.

SERVEDIO Frank J., William Paterson College, Wayne, New Jersey, U.S.A.

SHIAO Long-Shung, Carleton University, Ottawa, Ontario, Canada.

SIBLEY David, Pennsylvania State University, University Park, U.S.A.

SIMSON Daniel, University of Torun, Poland.

SMALØ Sverre O., University of Trondheim, Norway.

STAMBACH Urs, Eidg.Technische Hochschule, Zürich, Switzerland.

STOLTZFUS Neal W., Université de Genève, Switzerland/Louisiana State University, Baton Rouge, Louisiana, U.S.A.

TACHIKAWA H., University of Tsukuba, Ibaraki, Japan.

TINBERG Nalsey B., Southern Illinois University, Edwardsville, U.S.A.

WASCHBUSCH Josef, Freie Universität, West Berlin.

WEBB P.J., Queen Mary College, London, United Kingdom.

WIEDEMANN Alfred, Universität Stuttgart, West Germany.

YAMAGATA Kunio, University of Tsukuba, Ibaraki, Japan.

YOKONUMA Takeo, University of Saskatchewan, Saskatoon, Sask., Canada/
Sophia University, Tokyo, Japan.

UNISERIAL FUNCTORS

Maurice Auslander*) and Idun Reiten

Introduction

Throughout this paper modΛ denotes the category of finitely generated Λ-modules over an artin algebra Λ. We say that a covariant or contravariant functor F from modΛ to abelian groups is uniserial if the subfunctors of F are totally ordered by inclusion, i.e. if F_1 and F_2 are subfunctors of F, then either $F_1 \subset F_2$ or $F_2 \subset F_1$. If F is of finite length, then, as in the case of modules, F is uniserial if and only if the only subfunctors of F are the elements of the radical series of F, or equivalently, the elements of the socle series for F. The main purpose of this paper is to initiate a study of the connections between the existence and structure of various uniserial functors and the representation theory of Λ.

The first uniserial functors to be studied in a systematic way were the simple functors. This study [4] led to right and left almost split morphisms as well as almost split sequences, notions which are playing an increasingly important role in the representation theory of artin algebras. We recall that associated with each nonprojective indecomposable module C is a unique nonsplit exact sequence $0 \to A \xrightarrow{g} B \xrightarrow{f} C \to 0$ called an almost split sequence which is an invariant of the module C [5]. In particular, if $B = \coprod_{i=1}^{n} B_i$ is a decomposition of B into indecomposable modules, then $\alpha(C) = n$ and $\beta(C) = n$ minus

---

*) Written while a Guggenheim Fellow with the partial support of NSF MCS 77 04 951.

the number of projective B_i, are invariants of C which seem to have something to do with the complexity of the morphisms to C. The connection between these invariants and uniserial functors is illustrated by the following result (see Theorem 3.1): Suppose C is an indecomposable module over an artin algebra of finite representation type. If $(\ ,\underline{C})$, the representable functor $(\ ,C)$ modulo projectives, is nonzero uniserial, then $\beta(C) \leq 1$ and $\beta(C') \leq 2$ for all indecomposable C' such that $(\underline{C}',\underline{C}) \neq 0$.

While the interplay between the invariants $\alpha(C)$ and $\beta(C)$ for indecomposable nonprojective C and the uniseriality of certain types of functors is interesting in its own right, it is hoped that these results will help in describing and perhaps classifying certain types of artin algebras, especially those of finite representation type. For example Tachikawa in [17,18] studied artin algebras with the property that each indecomposable module has a simple socle. Here we show that such algebras can be characterized by a) $(\ ,S)$ is a uniserial functor for all simple modules S or b) $\alpha(C) \leq 2$ for all indecomposable modules C, $\underline{r}$ P is indecomposable for all indecomposable projectives P where $\underline{r}$ is the radical of P, and Λ is of finite type.

We give a brief outline of the contents of each section. Section 1 is devoted to a preliminary investigation of the connection between uniseriality of modules and functors. In section 2 we study the following general problem. Given a minimal projective presentation $(\ ,B) \to (\ ,C) \to G \to 0$ of a finitely presented functor G and a subfunctor F with projective cover $(\ ,E) \to F$, we construct projective presentations for F and for G/F and give conditions under which they are minimal. Specialized to semisimple subfactors of G, we get in certain cases a construction of sums of almost split sequences. These results are applied in section 3 to get criteria for uniseriality of functors in terms of the invariants α and β.

$\alpha(\Lambda)$ is defined to be the supremum of the $\alpha(C)$ for C indecomposable nonprojective. $\beta_L(\Lambda)(\beta_R(\Lambda))$ is defined to be the supremum of the $\beta(C)$ for C indecomposable nonprojective in modΛ (modΛ^{op}), and $\beta(\Lambda) = \sup(\beta_L(\Lambda), \beta_R(\Lambda))$. In section 4 our results are applied to get necessary conditions for $\alpha(\Lambda) \leq 2$ and for $\beta(\Lambda) \leq 2$ when Λ is of finite representation type.

The interest in the study of β is closely connected with problems about stable equivalence. In section 5 we apply our results to prove that if an artin algebra Λ is stably equivalent to a Nakayama algebra, then each indecomposable projective Λ-module or Λ^{op}-module is injective or uniserial.

In section 6 we prove that for an artin algebra Λ (,S) is uniserial for all simple Λ-modules S if and only if each indecomposable Λ-module has a simple socle, along with other characterizations of this class of algebras.

We would like to thank Dr. S.O. Smalø for his helpful suggestions.

1. <u>Uniserial modules and uniserial functors.</u>

Let Λ be an arbitrary artin algebra. Our purpose in this section is to investigate what it means about an indecomposable Λ-module C that some of the functors (,C), (,$\underline{C}$), ($\underline{C}$,) etc are uniserial. Here we recall that (,$\underline{C}$) denotes the functor from modΛ to abelian groups, given on objects by (,$\underline{C}$)(X) = ($\underline{X}$,$\underline{C}$) = $\mathrm{Hom}_\Lambda(X,C)/P(X,C)$, where P(X,C) is the subgroup of the maps factoring through projective Λ-modules. ($\underline{C}$,) is defined analogously.

While it is easy to see that (,C) being uniserial implies that C is uniserial, not all uniserial C have the property that (,C) is uniserial. For it is not hard to see that even if Λ is

Nakayama, (,C) is not in general uniserial. In fact, for Nakayama algebras (,C) is uniserial for an indecomposable Λ-module C if and only if C is simple, or C is projective with all proper submodules projective.

Since (,$\underline{C}$) is a factor of (,C), namely we have an exact sequence (,P) → (,C) → (,$\underline{C}$) → 0, where P is a projective cover for C, it is natural to ask what (,$\underline{C}$) being uniserial means about C. The case of a Nakayama algebra Λ again gives examples of uniserial C with (,$\underline{C}$) not uniserial. For it is not hard to see that if Λ is Nakayama and C is indecomposable nonprojective, then (,$\underline{C}$) is uniserial if and only if C is either P/$\underline{r}$P or P/socP, where P is an indecomposable projective and $\underline{r}$ is the radical of Λ and socP is the socle of P. Also, unlike the situation for (,C), the functor (,$\underline{C}$) being uniserial does not necessarily imply that C is uniserial. For example, suppose $\Lambda = k[x_1,\cdots,x_n]/(x_1,\cdots,x_n)^2$ where k is a field and $x_1,\cdots,x_n$ are indeterminates. Let S be the unique simple Λ-module. Then there is an almost split sequence $0 \to S \to \Lambda^n \to C \to 0$. This implies that (,$\underline{C}$) is simple and hence uniserial. Since $n = \ell(C/\underline{r}\,C)$ and $n^2-1 = \ell(\text{soc } C)$, where ℓ denotes length, this shows that C can be very far from being uniserial even though (,$\underline{C}$) is uniserial.

While this discussion shows that the connections between the uniseriality of C and of (,C) or (,$\underline{C}$) is somewhat tenuous in general, there are nonetheless interesting results along these lines in some special cases, as we will see after the following preliminary result. We recall that (,$\overline{C}$) denotes the factor of (,C) where we divide out by the maps factoring through injectives.

PROPOSITION 1.1. Let C be an indecomposable module over Λ, an arbitrary artin algebra.

(a) If $(\underline{C},\ )$ is nonzero and uniserial, then C is uniserial.

(b) If $(\ ,\overline{C})$ is nonzero and uniserial, then C is uniserial.

Proof: (a) Let C be an indecomposable non-projective Λ-module such that $(\underline{C},\ )$ is uniserial. Assume that there is some i such that $\underline{r}^iC/\underline{r}^{i+1}C$ is not simple. Then there are submodules K and L of $\underline{r}^iC$, which contain $\underline{r}^{i+1}C$, such that $K \not\subset L$ and $L \not\subset K$. Consider the diagram

$$\begin{array}{ccc} & \overset{p}{\nearrow} & C/K \\ C & & \\ & \underset{q}{\searrow} & C/L, \end{array}$$

where p and q are epimorphisms. This induces the diagram

$$\begin{array}{ccc} (\underline{C/K},\ ) & \overset{(\underline{p},\ )}{\searrow} & \\ & & (\underline{C},\ ) \\ (\underline{C/L},\ ) & \underset{}{\overset{(\underline{q},\ )}{\nearrow}} & \end{array}$$

Here the morphism $\underline{g}\colon \underline{X} \to \underline{Y}$ is the one associated with $g\colon X \to Y$ in a natural way.
Since $(\underline{C},\ )$ is uniserial, there is a map $t\colon C/L \to C/K$ (or $t\colon C/K \to C/L$) such that $(\underline{q},\ )(\underline{t},\ ) = (\underline{p},\ )$. Let $g = p - tq\colon C \to C/K$. We then have $\underline{g} = \underline{0}$, and hence a commutative diagram

$$\begin{array}{ccccc} & & Q & & \\ & \overset{u}{\nearrow} & & \overset{v}{\searrow} & \\ C & & \xrightarrow{\ g\ } & & C/K, \end{array}$$

where Q is projective.

Choose now $x \in L \setminus K$. Then $g(x) = p(x) - tq(x) = p(x) \neq C$. Since $\underline{r}^{i+1}C \subset K$, we have that $\underline{r}^{i+1}(C/K) = 0$. Assume now that $g(C) \subset \underline{r}(C/K)$. Then $g(\underline{r}^i C) = \underline{r}^i g(C) \subset \underline{r}^{i+1}(C/K) = 0$. This is a contradiction since $x \in L \subset \underline{r}^i U$ and $g(x) \neq 0$. We can then conclude that $g(C) \not\subset \underline{r}(C/K)$, so that $u(C) \not\subset \underline{r}Q$. But then C must be projective. This contradiction shows that C is uniserial.

b) Dual of (a).

As an immediate consequence of this proposition we have

<u>COROLLARY 1.2</u>. If Λ is selfinjective, then an indecomposable nonprojective module C is uniserial provided $(\, ,\underline{C})$ is uniserial.

Returning to the functors $(\, ,\underline{C})$ and $(\overline{C}, \,)$ we have the following. Here D denotes the ordinary duality for an artin algebra, Tr the transpose and Ω^1 the first syzygy.

<u>PROPOSITION 1.3.</u> Suppose C is an indecomposable Λ-module with Λ an arbitrary artin algebra.

a) Suppose $(\, ,\underline{C})$ is a nonzero uniserial functor. Then we have the following.

(i) $(\, ,\overline{DTrC})$ and DTrC are uniserial.

(ii) $\Omega^1 C/\underline{r}\,\Omega^1 C$ is simple and so $\Omega^1 C$ is indecomposable.

(iii) If $\Omega^1 C$ is not projective, then $(\underline{\Omega^1 C}, \,)$ and $\Omega^1 C$ are uniserial.

(iv) Let $0 \to \Omega^1 C \to P \to C \to 0$ be exact with $P \to C$ a projective cover of C. If $\operatorname{soc} P = \operatorname{soc} \Omega^1 C$, then C is uniserial.

b) Suppose $(\bar{C}, \ )$ is a nonzero uniserial functor.

Then we have the following.

(i) $(\underline{\mathrm{TrDC}}, \ )$ and TrDC are uniserial.

(ii) $\operatorname{soc}(\Omega^{-1} C)$ is simple.

(iii) If $\Omega^{-1} C$ is not injective, then $(\ , \overline{\Omega^{-1} C})$ and $\Omega^{-1} C$ are uniserial.

(iv) Let $0 \to C \to I \to \Omega^{-1} C \to 0$ be exact, where I is injective envelope for C. If $I/\underline{r}I \approx \Omega^{-1} C/\underline{r}\, \Omega^{-1} C$, then C is uniserial.

Proof: a)

i) The equivalence of categories $\mathrm{DTr}\colon \underline{\mathrm{mod}\Lambda} \to \overline{\mathrm{mod}\Lambda}$, shows that $(\ , \underline{C})$ is a nonzero uniserial functor if and only if $(\ , \overline{\mathrm{DTrC}})$ is a nonzero uniserial functor. By Proposition 1.1 we know that the non-injective indecomposable module DTrC is uniserial if $(\ , \overline{\mathrm{DTrC}})$ is uniserial.

ii) From [5] we know that $\Omega^1 C/\underline{r}\Omega^1 C$ is isomorphic to soc (DTrC). Therefore DTrC being uniserial implies $\Omega^1 C/\underline{r}\Omega^1 C$ is simple, which also shows that $\Omega^1 C$ is indecomposable.

iii) We know that there is a duality between the finitely presented contravariant functors vanishing on projectives and the finitely presented covariant functors vanishing on injectives (see [1,p.33]). Under this duality the functor $(\ ,\underline{C})$ corresponds to the functor $\mathrm{Ext}^1(C, \ )$. Therefore $(\ ,\underline{C})$ being uniserial is equivalent to $\mathrm{Ext}^1(C, \ )$ being uniserial. Let $0 \to \Omega^1 C \to P \to C \to 0$ be exact with P projective. From the exact sequence $0 \to (C, \ ) \to (P, \ ) \to (\Omega^1 C, \ ) \to \mathrm{Ext}^1(C, \ ) \to 0$ it follows that there is an epimorphism $\mathrm{Ext}^1(C, \ ) \to (\underline{\Omega^1 C}, \ )$. Hence $\mathrm{Ext}^1(C, \ )$ being uniserial implies that $(\underline{\Omega^1 C}, \ )$ is uniserial. Since by ii) $\Omega^1 C$ is indecomposable, the fact that $\Omega^1 C$ is not projective and $(\underline{\Omega^1 C}, \ )$is uniserial implies by Proposition 1.1 that $\Omega^1 C$ is uniserial.

iv) Instead of proving a) iv) we observe that its dual b) iv) can be proved in the same way as Proposition 1.1. We here use that if $i:C \to I$ is an injective envelope, then by assumption $i(C) \subset \underline{r}\, I$.

b) Dual of a).

As an immediate consequence of part a) ii) above and [2,Corollary 1.7] we have the following.

<u>COROLLARY 1.4</u>. For an artin algebra Λ there is a bound on the length of the indecomposable nonprojective C such that $(\ ,\underline{C})$ is uniserial.

In Proposition 1.3 we showed that if C is an indecomposable nonprojective module with $\Omega^1 C$ not projective, then $\Omega^1 C$ is uniserial if $(\ ,\underline{C})$ is uniserial. It is therefore natural to wonder if the hypothesis that $\Omega^1 C$ is not projective is necessary. While we do not know the complete answer to this question, we do have the following special case.

PROPOSITION 1.5. Let S be a simple Λ-module.

a) If $(\ ,\underline{S})$ is uniserial, then P, the projective cover of S, is uniserial.

b) If $(\overline{S}, \ )$ is uniserial, then I, the injective envelope of S, is uniserial.

Proof: a) Assume that $P \to S$ is a projective cover for S and that $(\ ,\underline{S})$ is uniserial. Then $\operatorname{soc}(\ ,S) = \operatorname{Im}((\ ,P) \to (\ ,S))$ which is simple, and hence $(\ ,S)$ is also uniserial. Let K_1 and K_2 be proper submodules of P. From the fact that $(\ ,S)$ is uniserial, it follows that if $\ell(K_1) = \ell(K_2)$, then there is an automorphism $\sigma: P \to P$ such that $\sigma(K_1) = K_2$. From this it follows that if P' is a factor of P, then socP' has the form nT for some simple module T. In particular each $\underline{r}^iP/\underline{r}^{i+1}P$ is of the form $n\,T_i$ for some simple T_i. We now use this fact to show that it suffies to prove a) under the additional hypothesis that Λ has at most two nonisomorphic simple modules.

For suppose P is not uniserial. Then there is an integer i such that $\underline{r}^iP/\underline{r}^{i+1}P \approx nT$ with T simple and $n \geq 2$, or what is the same thing, there are morphisms $f_1: Q \to P$ and $f_2: Q \to P$ such that $f_1 \neq f_2g$ and $f_2 \neq f_1g$ for any $g \in \operatorname{End}(Q)$, where Q is a projective cover for T. If we let $\Gamma = \operatorname{End}(P \amalg Q)^{op}$, then $(P \amalg Q, P)$, the projective cover over Γ for the simple Γ-module $(P \amalg Q, S)$, is not uniserial even though $(\ ,(P \amalg Q, S))$ is uniserial. Hence it suffices to prove a) under the additional hypothesis that Λ has at most two nonisomorphic simple modules S and T.

We now assume that Λ has at most two nonisomorphic simple modules. By Proposition 1.3 $\underline{r}P/\underline{r}^2P$ is simple, and we can assume that $Q = \underline{r}P$ is projective.

Further we can clearly assume that $\underline{r}^2P \neq 0$, since if $\underline{r}^2P = 0$, P is uniserial. We now show that $\underline{r}^2P$ is indecomposable.

Since $\underline{r}P$ is indecomposable and the inclusion map $i: rP \to P$ is the only irreducible morphism from $\underline{r}P$ to a projective, we have that $0 \to \underline{r}P \to P \amalg TrD(\underline{r}^2P) \to TrD(\underline{r}P) \to 0$ is an almost split sequence (see [7,Prop.2.4]). Thus we obtain the exact commutative diagram

$$\begin{array}{ccccccccc}
 & & & & 0 & & 0 & & \\
 & & & & \downarrow & & \downarrow & & \\
 & & & & TrD(\underline{r}^2P) & = & TrD(\underline{r}^2P) & & \\
 & & & & \downarrow & & \downarrow & & \\
0 & \to & \underline{r}P & \to & P \amalg TrD(\underline{r}^2P) & \to & TrD(\underline{r}P) & \to & 0 \\
 & & \| & & \downarrow & & \downarrow g & & \\
0 & \to & \underline{r}P & \longrightarrow & P & \longrightarrow & S & \longrightarrow & 0 \\
 & & & & \downarrow & & \downarrow & & \\
 & & & & 0 & & 0 & &
\end{array}$$

$* = (\ ,(TrD)^2(\underline{r}^2P))$

Since $TrD(\underline{r}P)$ is indecomposable, g is right minimal, that is, $(\ ,TrD(\underline{r}P)) \to Im(\ ,g)$ is a projective cover. Hence $(TrD)^2(\underline{r}^2P)$ is indecomposable since $*$ is a projective cover for soc (Coker(,g)) which is simple since (,S) is uniserial (see Proposition 2.6). Hence $\underline{r}^2P$ is indecomposable since neither $\underline{r}^2P$ or $TrD(\underline{r}^2P)$ have non-trivial injective summands.

Now $P/\underline{r}^3P$ is a projective $\Lambda/\underline{r}^3$-module which is a projective cover for S with the property that (,S) is uniserial viewed as a functor on $\text{mod}\Lambda/\underline{r}^3$. Therefore we have that $\underline{r}^2P/\underline{r}^3P$ is simple since it is semisimple and indecomposable by our previous argument. Therefore we have shown that $P/\underline{r}P$, $\underline{r}P/\underline{r}^2P$ and $\underline{r}^2P/\underline{r}^3P$ are simple, $\underline{r}P$ is an indecomposable projective and $P/\underline{r}P$ and $\underline{r}P/\underline{r}^2P$ are the only simple Λ-modules. From this it easily follows that P is uniserial. This finishes the proof of a).

b) Dual of a).

2. Projective resolutions of functors.

Let H be a finitely presented contravariant functor from modΛ to abelian groups, where Λ is an artin algebra, and consider a projective resolution $0 \to (\ ,A) \to (\ ,B) \to (\ ,C) \to H \to 0$. We know that each simple functor H is finitely presented [4, Prop.3.2], and if $H(\Lambda) = 0$, the associated sequence $0 \to A \to B \to C \to 0$ is an almost split sequence. Clearly if we have a minimal projective resolution of a semisimple functor H, with $H(\Lambda) = 0$, we have a corresponding exact sequence in modΛ, which is a sum of almost split sequences.

Let $0 \to (\ ,A) \to (\ ,B) \to (\ ,C) \to G \to 0$ be a projective resolution of a finitely presented functor G. If an epimorphism $(\ ,E) \to F$ is given, where F is a subfunctor of G, we shall in this section show how to construct projective resolutions for F and G/F. We shall also investigate when our new resolutions are minimal. We can use this method to construct projective and minimal projective resolutions for semisimple subfactors of G, and hence get information about almost split sequences. In particular, we get information about the invariants α and β given by almost split sequences. Here we recall that if C is indecomposable nonprojective in modΛ and $0 \to A \to B \to C \to C$ an almost split sequence, then $\alpha(C) = \sigma(B)$, the number of indecomposable summands in a direct sum decomposition of B, and $\beta(C)$ the number of nonprojective summands.

Let now G be a finitely presented contravariant functor and $0 \to (\ ,A) \to (\ ,B) \xrightarrow{(\ ,f)} (\ ,C) \to G \to 0$ a projective resolution for G.

If

$$\begin{array}{ccc} U & \xrightarrow{g_1} & E \\ {\scriptstyle g_2}\downarrow & & \downarrow{\scriptstyle h} \\ B & \xrightarrow{f} & C \end{array}$$

is a pullback diagram, it is not hard to see that we have an exact commutative diagram

$$\begin{array}{ccccccc}
 & & & & 0 & & \\
 & & & & \downarrow & & \\
(\ ,U) & \longrightarrow & (\ ,E) & \longrightarrow & \operatorname{Coker}(\ ,g_1) & \longrightarrow & 0 \\
\downarrow & & \downarrow & & \downarrow & & \\
(\ ,B) & \longrightarrow & (\ ,C) & \longrightarrow & G & \longrightarrow & 0,
\end{array}$$

that is, Coker $(\ ,g_1)$ is a subfunctor of G.

It is on the other hand not hard to see, using the defining properties of pullback, that any finitely generated subfunctor of G has a projective presentation given by a pullback.

PROPOSITION 2.1. Let Λ be an artin algebra and $(\ ,B) \xrightarrow{(\ ,f)} (\ ,C) \xrightarrow{v} G \to 0$ an exact sequence of functors, with B and C in $\operatorname{mod}\Lambda$. Let $h: E \to C$ be a map in $\operatorname{mod}\Lambda$, and denote by F the image of $v(\ ,h)$ in G. Then the pullback diagram

$$\begin{array}{ccc}
B \times_C E & \longrightarrow & E \\
\downarrow & & \downarrow h \\
B & \xrightarrow{f} & C
\end{array}$$

induces an exact commutative diagram

$$\begin{array}{ccccccccccc}
 & & & & & & & & 0 & & \\
 & & & & & & & & \downarrow & & \\
0 & \longrightarrow & (\ ,A) & \longrightarrow & (\ ,B \times_C E) & \longrightarrow & (\ ,E) & \xrightarrow{u} & F & \longrightarrow & 0 \\
 & & \downarrow \| & & \downarrow & & \downarrow (\ ,h) & & \downarrow & & \\
0 & \longrightarrow & (\ ,A) & \longrightarrow & (\ ,B) & \longrightarrow & (\ ,C) & \xrightarrow{v} & G & \longrightarrow & 0.
\end{array}$$

We shall later get more information about the above projective resolution for F, and get criteria for when it is minimal. But first we construct a projective presentation for G/F and investigate when it is minimal. We here recall that the radical of G, $\underline{r}G$, is the intersection of the maximal subfunctors of G. (See for example [4] for further properties of the radical.)

PROPOSITION 2.2. Let Λ be an artin algebra and $0 \to (\ ,A) \to (\ ,B) \xrightarrow{(\ ,f)} (\ ,C) \xrightarrow{v} G \to 0$ an exact sequence of functors where A, B and C are in modΛ. Let further $h: E \to C$ be a map in modΛ and $F = \operatorname{Im} v(\ ,h)$. $g_1 : B \times_C E \to B$ denotes the map given by the pullback.

(a) We have an exact sequence $0 \to (\ , L) \to (\ ,B' \amalg E) \to (\ ,C) \to G/F \to 0$, where B' is a summand of B such that $(\ ,B') \to \operatorname{Coker}(\ ,g_1)$ is a projective cover.

(b) $B \times_C E \approx L \amalg B''$, where $B = B' \amalg B''$.

(c) If $(\ ,E) \to F$ is a projective cover, then $(\ ,B' \amalg E) \to (\ ,C)$ is a projective cover for the image.

(d) If $(\ ,E) \to F$ and $(\ ,C) \to G$ are projective covers and $F \subset \underline{r}G$, then the projective resolution of G/F given in (a) is minimal.

Proof. Consider the exact commutative diagram

$$
\begin{array}{ccccccccc}
 & & 0 & & & & & & \\
 & & \downarrow & & & & & & \\
 & & (\ ,A) & & & & 0 & & \\
 & & \downarrow & & & & \downarrow & & \\
 & & (\ ,B\times_C E) & \xrightarrow{(\ ,g_1)} & (\ ,B) & \longrightarrow & \operatorname{Coker}(\ ,g_1) & \longrightarrow & 0 \\
 & & \downarrow {\scriptstyle (\ ,g_2)} & & \downarrow {\scriptstyle (\ ,f)} & & \downarrow & & \\
 & & (\ ,E) & \xrightarrow{(\ ,h)} & (\ ,C) & \longrightarrow & \operatorname{Coker}(\ ,h) & \longrightarrow & 0 \\
 & & \downarrow & & \downarrow & & \downarrow & & \\
0 & \longrightarrow & F & \longrightarrow & G & \longrightarrow & G/F & \longrightarrow & 0 \\
 & & \downarrow & & \downarrow & & \downarrow & & \\
 & & 0 & & 0 & & 0 & &
\end{array}
$$

We decompose $B = B' \amalg B''$ such that the induced map $(,B') \to \mathrm{Coker}(,g_1)$ is a projective cover, and let $p':B \to B'$ be the projection map. By diagram chasing we get an exact sequence
$0 \to (,L) \to (,B' \amalg E) \to (,C) \to G/F \to 0$, where L is the image of $(-g_2, p'g_1)$: $Bx_C E \to E \amalg B'$. Since we also have an exact sequence
$0 \to (,Bx_C E) \to (,B \amalg E) \to (,C) \to G/F \to 0$, we get by Schanuels lemma an isomorphism
$L \amalg B \amalg E \approx (Bx_C E) \amalg B' \amalg E$, and hence $Bx_C E \approx L \amalg B''$.
If $(,E) \to F$ is a projective cover and since $(,B') \to \mathrm{Coker}(,g_1)$ is a projective cover, it is not hard to see that $(,L) \subset \underline{r}(,B' \amalg E)$, so that $(,B' \amalg E) \to (,C)$ is a projective cover for the image. If $F \subset \underline{r}G$ and $(,C) \to G$ is a projective cover, then clearly $(,C) \to G/F$ is a projective cover. This finishes the proof.

To decide when the projective resolution we have constructed for a subfunctor F is minimal, we shall need some more information about minimal projective resolutions for socG.

<u>LEMMA 2.3.</u> Let $0 \to (,A) \to (,B) \to (,C) \to G \to 0$ be a projective resolution, where $0 \to A \xrightarrow{g} B \xrightarrow{f} C \to 0$ is exact. Then $f:B \to C$ is right minimal (i.e. $(,B) \to \mathrm{Im}(,f)$ is a projective cover) if and only if $\ell(\mathrm{soc}G) = \sigma(A)$.

<u>Proof</u>: There is a duality between the finitely presented contravariant functors vanishing on projectives and the finitely presented covariant functors vanishing on injectives. Since f is onto, $G(\Lambda) = 0$, and the correspondent of G under this duality is H, where H is given by the exact sequence $0 \to (C,) \to (B,) \to (A,) \to H \to 0$ [1, p. 153].

Hence we have $\ell(\mathrm{soc}G) = \ell(H/\underline{r}H)$. Clearly $\ell(H/\underline{r}H) = \sigma(A)$ if and only if $(A, \ ) \to H$ is a projective cover. Since it is not hard to see that $(A, \ ) \to H$ is a projective cover if and only if $f:B \to C$ is minimal, we have our desired result.

<u>PROPOSITION 2.4</u>. Let $0 \to A \xrightarrow{g} B \xrightarrow{f} C \to 0$ be an exact sequence with f right minimal and $A \neq 0$. Let $A = \coprod_{i=1}^{n} A_i$, where the A_i are indecomposable, and let $G = \mathrm{Coker}\,(\ ,f)$.

(a) No A_i is injective.

(b) Let $0 \to A_i \to E_i \to \mathrm{TrD}A_i \to 0$ be almost split sequences. Then there is some commutative exact diagram

$$\begin{array}{ccccccccc}
0 & \to & \coprod_{i=1}^{n} A_i & \to & \coprod_{i=1}^{n} E_i & \to & \coprod_{i=1}^{n} \mathrm{TrD}A_i & \to & 0 \\
 & & \| & & \downarrow & & \downarrow h & & \\
0 & \to & A & \xrightarrow{f} & B & \xrightarrow{f} & C & \to & 0
\end{array}$$

(c) Any above diagram has the property that the composition $(\ , \coprod_{i=1}^{n} \mathrm{TrD}A_i) \xrightarrow{(\ ,h)} (\ ,C) \to G$ has socG as its image, and the induced morphism $(\ , \coprod_{i=1}^{n} \mathrm{TrD}A_i) \to \mathrm{soc}G$ is a projective cover.

<u>Proof</u>: We know that the extension $x:0 \to A \xrightarrow{g} B \xrightarrow{f} C \to 0$ has the property that f is right minimal if and only if $\mathrm{Ext}^1(C,h)(x) \neq 0$ for each nonzero split epimorphism $h:A \to A'$ [1]. (a) is a direct consequence of this. Further it shows that if $f:B \to C$ is minimal and $A = \coprod A_i$ is some decomposition with the A_i indecomposable, then no induced map $A_i \to B$ is a split monomorphism. (b) now follows using elementary properties of almost split sequences [5].

Since the diagram in (b) is a pullback diagram, it

follows from earlier remarks that we have the exact commutative diagram

$$\begin{array}{ccccccc}
 & & & & 0 & & \\
 & & & & \downarrow & & \\
(\ ,\coprod_{i=1}^{n} E_i) & \xrightarrow{(\ ,\coprod t_i)} & (\ ,\coprod_{i=1}^{n} \mathrm{TrDA}_i) & \longrightarrow & \coprod_{i=1}^{n} \mathrm{Coker}(\ ,t_i) & \longrightarrow & 0 \\
\downarrow & & \downarrow & & \downarrow & & \\
(\ ,B) & \xrightarrow{(\ ,f)} & (\ ,C) & \longrightarrow & \mathrm{Coker}(\ ,f) & \longrightarrow & 0
\end{array}$$

Since each Coker $(\ ,t_i)$ is simple,

$\mathrm{Im}(\coprod_{i=1}^{n}$ Coker $(\ ,t_i) \to$ Coker $(\ ,f))$ is semisimple and

therefore contained in soc Coker $(\ ,f)$.
Also, by Lemma 2.3 we have that $\ell(\mathrm{soc}(\mathrm{Coker}(\ ,f)) =$

$\sigma(A) = \ell(\coprod_{i=1}^{n}$ Coker $(\ ,t_i))$. Therefore the induced

morphism $\coprod_{i=1}^{n}$ Coker $(\ ,t_i) \to \mathrm{soc}(\mathrm{Coker}(\ ,f))$ is an

isomorphism. Part (c) now follows from the fact that each $(\ ,\mathrm{TrDA}_i) \to$ Coker $(\ ,t_i)$ is a projective cover.

We remark that it is possible to extend the last two results to the case of an exact sequence $0 \to A \xrightarrow{g} B \xrightarrow{f} C$, where f is not necessarily an epimorphism. Here G must be replaced by G_o, the unique maximal finitely generated subfunctor of G vanishing on projectives. For we have an exact sequence $0 \to (\ ,A) \to (\ ,B) \to (\ ,\mathrm{Imf}) \to G_o \to 0$, and if $f:B \to C$ is minimal, then $f:B \to \mathrm{Imf}$ is minimal.

On the basis of the above we can now give criteria for the projective resolution we have constructed of a subfunctor to be minimal. For simplicity we shall here assume that $G(A) = 0$, so that the associated sequence $0 \to A \to B \to C \to 0$ is exact.

PROPOSITION 2.5. Let
$0 \to (,A) \xrightarrow{(,g)} (,B) \xrightarrow{(,f)} (,C) \xrightarrow{u} G \to 0$ be a projective presentation for a finitely presented functor G, and $h:E \to C$ such that $v = u(,h):(,E) \to G$ is a projective cover for $F = \operatorname{Im} v$.

(a) $Bx_C E \approx L \amalg B''$, where $(,TrDL)$ is a projective cover for soc G/F and B'' is a summand of B, as defined in Proposition 2.2.

(b) The projective resolution
$0 \to (,A) \to (,Bx_C E) \to (,E) \to F \to 0$ is a minimal projective resolution for F if and only if $F \supset \operatorname{soc} G$.

We have now seen that if we are given a minimal projective presentation for a finitely presented functor G, and we have a subfunctor F with a projective cover $(,E) \to F$, we can construct projective presentations for F and G/F, which under certain conditions are minimal. Clearly we can repeat the process and get projective resolutions for subfactors of G. Especially interesting from our point of view are the semisimple subfactors, because of the close connection of their minimal projective resolution with almost split sequences. Because of complication in notation, we shall not state these results in full generality, but shall as an illustration, to be applied in the next section, give the following.

PROPOSITION 2.6. Let
$0 \to (,A) \to (,B) \xrightarrow{(,f)} (,C) \to G \to 0$ be a minimal projective presentation for a finitely presented functor G. Assume that $f:B \to C$ is an epimorphism and that C is indecomposable. Let $u:(,E) \to F$ be a projective cover, where F is a proper subfunctor of G. Let further F_1 be a semisimple subfunctor of G/F, $(,C_1) \to F_1$ a projective cover, and L such that $(,L)$ is a projective cover for $\operatorname{soc}((G/F)/F_1)$.

(a) We then have a projective resolution $0 \to (\ ,DTrC_1) \to (\ ,DTrL \amalg E_1 \amalg B_1) \to (\ ,C_1) \to F_1 \to 0$, which is minimal if $F_1 = \operatorname{soc} G/F$. Here E_1 is a summand of E, B_1 a summand of B' in our previous notation, hence also of B.

(b) If the image of a lifting of $(\ ,C_1) \to F_1$ to a map $(\ ,C_1) \to G$ contains F, then $E_1 = E$.

(c) $B_1 = 0$ if the image of a lifting of u to a map $(\ ,E) \to (\ ,C)$ contains $\operatorname{Im}((\ ,B) \to (\ ,C))$.

(d) $B_1 = B$ if we have a minimal projective resolution $(\ ,B) \amalg (\ ,E) \to (\ ,C) \to G/F \to 0$ and $\operatorname{Im}((\ ,C_1) \amalg (\ ,E) \to (\ ,C))$ contains $\operatorname{Im}((\ ,B) \to (\ ,C))$, where $(\ ,C_1) \to (\ ,C)$ is a lifting of $(\ ,C_1) \to F_1$.

Proof: Since f is onto, $G(\Lambda) = 0$, and since C is indecomposable, $G/\underline{r}G$ is simple, so that any proper subfunctor of G is contained in $\underline{r}G$. Hence the presentation $(\ ,B') \amalg (\ ,E) \to (\ ,C) \to G/F \to 0$ is minimal. (a) follows from Proposition 2.5. Under the assumption of (b), the image of $(\ ,C_1) \amalg (\ ,B) \to (\ ,C)$ is the preimage of F_1 in $(\ ,C)$, so that we have a projective presentation $(\ ,C_1) \amalg (\ ,B) \to (\ ,C) \to (G/F)/F_1 \to 0$. Hence no part of $(\ ,E) \to (\ ,C)$ occurs in a minimal projective presentation, so that $E_1 = E$.

If $\operatorname{Im}((\ ,E) \to (\ ,C))$ contains $\operatorname{Im}((\ ,B) \to (\ ,C))$, then $B' = 0$, and hence $B_1 = 0$. If $\operatorname{Im}((\ ,C_1) \amalg (\ ,E) \to (\ ,C))$ contains $\operatorname{Im}((\ ,B) \to (\ ,C))$, no summand of $(\ ,B) \to (\ ,C)$ is needed to cover the preimage of F_1 in $(\ ,C)$. This proves (d).

3. Uniserial functors and the invariants α and β.

In this section we shall study the connection between uniseriality of a finitely presented functor H and values of $\alpha(C)$, $\beta(C)$ when C is indecomposable with H(C) nonzero. To give some insight into how these results fit in, we shall deduce our results from the general considerations in the previous section, even though we could have proved some of them more directly.

We denote by socH the socle of H and define $\mathrm{soc}^i H$ in the usual way. It is not hard to see that $\mathrm{soc}^\infty H = \bigcup_{i=1}^{\infty} \mathrm{soc}^i H$ is uniserial if and only if each $\mathrm{soc}^i H/\mathrm{soc}^{i-1} H$ is simple or zero, and that $H/\underline{r}^\infty H$ is uniserial if and only if each $\underline{r}^i H/\underline{r}^{i+1} H$ is simple or zero. Here $\underline{r}^\infty H = \bigcap_{i=1}^{\infty} \underline{r}^i H$.

We shall concentrate our study on three types of finitely presented functors, and we start out with the functors of type $(\ ,\underline{C})$, where C is indecomposable and not projective.

THEOREM 3.1. Let C be an indecomposable non-projective module over an artin algebra Λ.

(a) Assume that $\mathrm{soc}^\infty(\ ,\underline{C})$ is uniserial, and let C_k be such that $(\ ,C_k)$ is a projective cover for $\mathrm{soc}^k(\ ,\underline{C})/\mathrm{soc}^{k-1}(\ ,\underline{C})$. Assume that $C_k \neq 0$.

(i) We have an almost split sequence $0 \to \mathrm{DTr}C_k \to C_{k-1} \amalg \mathrm{DTr}C_{k+1} \amalg P_k \to C_k \to 0$, where P_k is projective.

(ii) $\beta(C_k) \leq 2$, $\beta(C_1) \leq 1$, and if $(\ ,\underline{C})$ has finite length, then $\beta(C) \leq 1$.

(b) Assume that $(\,,\underline{C})/\underline{r}^{\infty}(\,,\underline{C})$ is uniserial, and let A_k be such that $(\,,A_k)$ is a projective cover for $\underline{r}^k(\,,\underline{C})/\underline{r}^{k+1}(\,,\underline{C})$ and $A_k \neq 0$.

(i) We have an almost split sequence

$$0 \to DTrA_k \to A_{k+1} \amalg DTrA_{k-1} \amalg P_k \to A_k \to 0,$$

where P_k is projective.

(ii) $\beta(A_k) \leq 2$ and $\beta(C) \leq 1$.

Proof: (a) We have the minimal projective presentation $(\,,P) \to (\,,C) \to (\,,\underline{C}) \to 0$, where P is a projective cover of C. Let $F = soc^{k-1}(\,,\underline{C})$ where k is such that $F \neq (\,,\underline{C})$. Then the projective cover $(\,,C_k)$ for $soc\,(\,,\underline{C})/F$ is also a projective cover for $soc^k(\,,\underline{C})$. Our result now follows from Proposition 2.6. P_k is here a summand of P, and hence projective.

(b) Consider Proposition 2.6 with $G = (\,,\underline{C})$, $F = \underline{r}^{k+1}(\,,\underline{C})$, and assume that $F_1 = \underline{r}^k(\,,\underline{C})/\underline{r}^{k+1}(\,,\underline{C})$ is nonzero. Since $(\,,\underline{C})/\underline{r}^{\infty}(\,,\underline{C})$ is uniserial, we have that $F_1 = soc\, G/F$. Since clearly $(\,,A_k)$ is a projective cover for $\underline{r}^k(\,,\underline{C})$, we have a minimal projective resolution

$$0\to(\,,DTrA_k)\to(\,,DTrA_{k-1}\amalg A_{k+1}\amalg P_k)\to(\,,A_k)\to\underline{r}^k(\,,\underline{C})/\underline{r}^{k+1}(\,,\underline{C})\to 0,$$

where P_k is projective, hence an almost split sequence

$$0\to DTrA_k\to DTrA_{k-1}\amalg A_{k+1}\amalg P_k\to A_k\to 0.$$

(ii) is now a direct consequence.

On the basis of the above result we get the following characterization of when $(\,,\underline{C})/\underline{r}^{\infty}(\,,\underline{C})$ is uniserial.

PROPOSITION 3.2. Let C be an indecomposable non-projective module over an artin algebra Λ. Then $(\ ,\underline{C})/\underline{r}^{\infty}(\ ,\underline{C})$ is uniserial if and only if the following condition holds.

(*) There is a sequence of Λ-modules $C_{-1} = 0$, $C_o = C$, C_1, C_2,$\cdots$ which are zero or indecomposable, such that if C_i is not zero, there is an almost split sequence
$0 \to DTrC_i \to C_{i+1} \amalg DTrC_{i-1} \amalg P_i \to C_i \to 0$ with P_i projective.

Proof: If $(\ ,\underline{C})/\underline{r}^{\infty}(\ ,\underline{C})$ is uniserial, C has condition (*) by Theorem 3.1(b).

Assume conversely that C has condition (*). We then have an almost split sequence
$0 \to DTrC \to C_1 \amalg P_1 \to C \to 0$, where P_1 is projective and C_1 is zero or indecomposable nonprojective. If C_1 is zero, we are done, since $(\ ,\underline{C})$ is then simple. If C_1 is indecomposable, $\underline{r}(\ ,\underline{C})/\underline{r}^2(\ ,\underline{C})$ has $(\ ,C_1)$ as projective cover and is hence simple [7,Prop.1.3]. Consider then the almost split sequence
$0 \to DTrC_1 \to DTrC \amalg C_2 \amalg P_2 \to C_1 \to 0$, where P_2 is projective, and C_2 by assumption is indecomposable or zero. Since the composite map $DTrC \to C_1 \to C$ factors through a projective module, $(\ ,C_2)$ is a projective cover of $\underline{r}^2,(\ ,\underline{C})/\underline{r}^3(\ ,\underline{C})$, which hence has length at most one. Continuing this way, we get that each $\underline{r}^k(\ ,\underline{C})/\underline{r}^{k+1}(\ ,\underline{C})$ has length at most one, so that $(\ ,\underline{C})/\underline{r}^{\infty}(\ ,\underline{C})$ is uniserial.

To an indecomposable nonprojective Λ-module C we can associate a (left) diagram in the following way (see [20]). For each indecomposable nonprojective Λ-module X having an irreducible map to C we draw an arrow $X \to C$. For each indecomposable nonprojective Λ-module $Y \not\approx DTrC$ having an irreducible map to X we draw an arrow $Y \to X$, and for each indecomposable nonprojective Λ-module $Z \not\approx DTrX$ we draw an arrow $Z \to Y$, etc.

Under some assumptions which by the work of Bautista are known to hold for algebras of finite type over an algebraically closed field, we get as a direct consequence of Proposition 3.2 the following diagrammatic description of uniseriality for $(\ ,C)/\underline{r}^{\infty}(\ ,C)$. We here recall [6] that a map $h:X \to Y$ is irreducible if it is neither a split monomorphism nor a split epimorphism, and given any commutative diagram

$$\begin{array}{ccc} & Z & \\ {}^{s}\nearrow & & \searrow^{t} \\ X & \xrightarrow{\ h\ } & Y, \end{array}$$

then either s is a split monomorphism or t is a split epimorphism. We denote by [C] the irreducible component of C (see [2]).

<u>PROPOSITION 3.3</u>. Let C be an indecomposable nonprojective module over an artin algebra Λ. Assume that if Z is in [C] and $0 \to X \to Y \to Z \to 0$ is an almost split sequence, then Y has no repeated indecomposable nonprojective summands.

Then $(\ ,\underline{C})/\underline{r}^{\infty}(\ ,\underline{C})$ is uniserial if and only if the associated (left) diagram for C is of type A_n or A_∞.

In general we are not able to characterize uniseriality of $(\ ,\underline{C})/\underline{r}^{\infty}(\ ,\underline{C})$ by $\beta(C) \leq 1$ and $\beta(C') \leq 2$ for C' in Supp $(\ ,\underline{C})/\underline{r}^{\infty}(\ ,\underline{C})$. Here X is said to be in Supp G for a functor G if X is indecomposable and G(X) is not zero. The next lemma, which is the basis for such a description for selfinjective algebras, is formulated somewhat more generally to be applicable also in the next section.

LEMMA 3.4. Let C be an indecomposable nonprojective module over an artin algebra Λ, with $\beta(C) \leq 1$. For C' in Supp($,\underline{C})/\underline{r}^{\infty}(,\underline{C})$ we assume that $\beta(C') \leq 2$ and that if $\alpha(C') > 2$, the middle term in an almost split sequence with right hand term C' has a nonzero projective injective summand. Then $(,\underline{C})/\underline{r}^{\infty}(,\underline{C})$ is uniserial.

Proof: Since $\beta(C) \leq 1$, we have an almost split sequence $0 \to DTrC \to C_1 \amalg P_1 \to C \to 0$, where P_1 is projective and C_1 is indecomposable nonprojective or zero. We can clearly assume that C_1 is not zero. We then have an almost split sequence
$0 \to DTrC_1 \to DTrC \amalg C_2 \amalg P_2 \to C_1 \to 0$, where P_2 is projective and C_2 has no nonzero projective summand. If $\alpha(C_1) \leq 2$, C_2 must be zero or indecomposable. If $\alpha(C_1) > 2$, there is by assumption a nonzero projective injective summand of the middle term of the almost split sequence with right hand term C_1. Clearly DTrC is not projective injective, and C_2 is by assumption not projective. P_2 must then have a nonzero projective injective summand, and we know from [6, Prop.4.11] that none of the other summands of the middle term are projective. Hence DTrC is not projective, and since $\beta(C_1) \leq 2$, C_2 must then be zero or indecomposable. Continuing this way, we can use Proposition 3.2 to conclude that $(,\underline{C})/\underline{r}^{\infty}(,\underline{C})$ is uniserial.

Since for a selfinjective algebra each projective module is injective, we get as an immediate consequence of Theorem 3.1 and Lemma 3.4.

THEOREM 3.5. Let C be an indecomposable nonprojective module over a selfinjective artin algebra Λ. Then $(,\underline{C})/\underline{r}^{\infty}(,\underline{C})$ is uniserial if and only if $\beta(C) \leq 1$ and $\beta(C') \leq 2$ for C' in Supp $(,\underline{C})/\underline{r}^{\infty}(,C)$.

Let now $f:B \to C$ be an epimorphism, where B and C are indecomposable. We shall study the functor $G = \text{Coker}(,f)$ in two special cases, but first we give the following result.

PROPOSITION 3.5. Let $f:B \to C$ be an epimorphism which is not an isomorphism. Assume that B and C are indecomposable and let $G = \text{Coker}(,f)$.

(a) If $\text{soc}^{\infty}G$ is uniserial and C_k such that $(,C_k)$ is a projective cover for $\text{soc}^k G/\text{soc}^{k-1}G$, then $\alpha(C_1) \leq 2$, $\alpha(C_k) \leq 3$ if $C_k \neq 0$, and if $\text{soc}^{k-1}G \neq G$, $\text{soc}^k G = G$, then $\alpha(C_k) \leq 2$.

(b) If $G/\underline{r}^{\infty}G$ is uniserial and A_k is such that $(,A_k)$ is a projective cover for $\underline{r}^k G/\underline{r}^{k+1}G$, then $\alpha(A_k) \leq 3$ if A_k is not zero.

Proof: This follows from Proposition 2.6, using that $DTrC_{k+1}$, C_{k-1} and B are indecomposable if not zero, and the same for A_{k+1}, $DTrA_{k-1}$ and B.

In two important special cases we get better results.

<u>THEOREM 3.7.</u> Let $f:B \to C$ be an epimorphism with B and C indecomposable, and assume that $g:\mathrm{Ker}\, f \to B$ is irreducible. Let $G = \mathrm{Coker}(\ ,f)$.

(a) $\mathrm{soc}^{\infty}G$ is uniserial if and only if socG is simple and $\alpha(C') \leq 2$ for C' in Supp $\mathrm{soc}^{\infty}G$.

(b) If G is of finite length, then G is uniserial if and only if socG is simple and $\alpha(C') \leq 2$ for C' in Supp G.

(c) If G is uniserial of finite length, then $\alpha(C) \leq 1$.

<u>Proof</u>: Assume first that $\mathrm{soc}^{\infty}G$ is uniserial, and let C_i be such that $C_i \neq 0$ and $(\ ,C_i)$ is a projective cover for $\mathrm{soc}^iG/\mathrm{soc}^{i-1}G$. Then $(\ ,C_i)$ is a projective cover for soc^iG. Let $h:X \to C$ with X indecomposable and assume that $\mathrm{Im}(\ ,h) \not\subset \mathrm{Im}(\ ,f)$. Since $g:\mathrm{Ker}f \to B$ is irreducible, we then have that $\mathrm{Im}(\ ,h) \supset \mathrm{Im}(\ ,f)$[6,Prop.2.7]. This shows that $(\ ,C_i)$ is a projective cover for the preimage of soc^iG in $(\ ,C)$, so that we have a minimal projective presentation $(\ ,C_i) \to (\ ,C) \to G/\mathrm{soc}^i G \to 0$. It is then easy to see by Proposition 2.6 that we have minimal projective presentations
$(\ ,\mathrm{DTr}C_2 \amalg B) \to (\ ,C_1) \to \mathrm{soc}G \to 0$ and
$(\ ,\mathrm{DTr}C_{i+1} \amalg C_{i-1}) \to (\ ,C_i) \to \mathrm{soc}^iG/\mathrm{soc}^{i-1}G \to 0$.

From this we see that $\alpha(C') \leq 2$ for C' in Supp $\mathrm{soc}^{\infty}G$ and that $\alpha(C_i) \leq 1$ if $\mathrm{soc}^iG = G$ and $C_i \neq 0$. We here point out that since $f:B \to C$ is onto, no C' in Supp G is projective.

Assume conversely that $\alpha(C') \leq 2$ for C' in Supp $\mathrm{soc}^{\infty}G$. We still have the above minimal projective presentation for socG. We have assumed that socG is simple, so that C_1 is indecomposable. If G = socG, we are done. Otherwise, C_2 is not zero, and since $\alpha(C_1) \leq 2$, we conclude that C_2 is indecomposable. Since socG is uniserial, socG = $\underline{r}(\mathrm{soc}^2G)$, so that ($,C_2$) is clearly a projective cover for soc^2G. Hence we have a minimal projective presentation

$$(\ ,DTrC_3 \amalg C_1) \to (\ ,C_2) \to \mathrm{soc}^2G/\mathrm{soc}G \to 0.$$

Since $\alpha(C_2) \leq 2$, C_3 is zero or indecomposable. Continuing this way, we get that all nonzero C_i are indecomposable, so that $\mathrm{soc}^{\infty}G$ is uniserial.

We have the following immediate consequence, which also contains more information on the uniseriality of certain types of ($,\underline{C}$).

<u>COROLLARY 3.8</u>. Let $g:U \to P$ be an irreducible map where U and P are indecomposable and P is projective. Let C = Coker g, and assume that ($,\underline{C}$) has finite length. Then ($,\underline{C}$) is uniserial if and only if $\alpha(C') \leq 2$ for C' in Supp ($,\underline{C}$). And if ($,\underline{C}$) is uniserial of finite length, then $\alpha(C) \leq 1$.

<u>THEOREM 3.9</u>. Let $f:B \to C$ be an irreducible epimorphism, with B and C indecomposable, and let G = Coker(,f).

Let C_1 be such that ($,C_1$) covers socG. Then

(a) $\mathrm{soc}^{\infty}G$ is uniserial if and only if socG is simple, $\alpha(C_1) \leq 1$ and $\alpha(C') \leq 2$ for C' in Supp $\mathrm{soc}^{\infty}G$.

(b) If G is of finite length, then G is uniserial if and only if socG is simple, $\alpha(C_1) \leq 1$ and $\alpha(C') \leq 2$ for C' in Supp G.

<u>Proof</u>: (a) Let $h:X \to C$ with X indecomposable and h not an isomorphism, such that Im(,h) $\not\subset$ Im(,f). Since f is irreducible, we must then have that Im(,f) $\not\subset$ Im(,h).

Using this observation and Proposition 2.6 the theorem can be proved in a way similar to Theorem 3.6.

4. The invariants α and β.

Let Λ be an artin algebra and C an indecomposable nonprojective Λ-module. We defined in the previous section the invariants $\alpha(C)$ and $\beta(C)$, and we define globally $\alpha(\Lambda)$ to be the supremum of the $\alpha(C)$, $\beta(\Lambda) = \sup(\beta_L(\Lambda), \beta_R(\Lambda))$, where $\beta_L(\Lambda)$ ($\beta_R(\Lambda)$) is defined to be the supremum of the $\beta(C)$ for C in modΛ (modΛ^{op}). These invariants give a way of describing the complication of the structure of maps between modules, and it would be interesting to have a classification of artin algebras from this point of view. As we shall see in the next section, the study of β is especially interesting in connection with questions about stable equivalence. For algebras of finite representation type $\alpha(\Lambda)$ and $\beta(\Lambda)$ should not be to big, and we have conjectured that $\alpha(\Lambda) \leq 4$ and $\beta(\Lambda) \leq 3$ for such algebras. It is known to be true for hereditary algebras of finite type that $\beta(\Lambda) \leq \alpha(\Lambda) \leq 3$, and for selfinjective algebras of finite type that $\alpha(\Lambda) \leq 4$, $\beta(\Lambda) \leq 3$. The last result follows from the work in [16] for selfinjective algebras over an algebraically closed field, and from [20] for arbitrary selfinjective algebras.

In the previous section we have seen a close connection between uniseriality of certain functors and the existence of several C with $\alpha(C) \leq 2$ or $\beta(C) \leq 2$. We shall here use this connection to get necessary conditions on the structure of Λ when Λ is of finite type and $\alpha(\Lambda) \leq 2$ or $\beta(\Lambda) \leq 2$. For the algebras of finite type with $\beta(\Lambda) \leq 2$ we shall also use our previous results to give a complete description of all uniserial functors of type (,$\underline{C}$). For completeness, we start with characterizing the algebras Λ with $\alpha(\Lambda) \leq 1$ and $\beta(\Lambda) \leq 1$.

We say that $\alpha(\Lambda) = 0$ if Λ is semisimple, and if Λ is not semisimple, it is clear that $\alpha(\Lambda) > 0$.

PROPOSITION 4.1. Let Λ be an artin algebra. Then $\alpha(\Lambda) \leq 1$ if and only if Λ is Nakayama of Loewy length at most 2.

Proof: Assume first that $\alpha(\Lambda) \leq 1$. If X is indecomposable nonprojective, there can be no irreducible monomorphism $Y \to X$. Let now A be an indecomposable noninjective Λ-module, and consider the almost split sequence $0 \to A \to B \to C \to 0$. Since B is indecomposable and $A \to B$ is an irreducible mono, B must be projective. It then follows that A is simple [5, Theorem 5.5]. This shows that each indecomposable noninjective Λ-module is simple, and dually, each indecomposable nonprojective Λ-module is simple. If P is an indecomposable nonsimple projective (injective) Λ-module, it must be injective (projective) and of length 2. Hence Λ is Nakayama of Loewy length at most 2.

Assume conversely that Λ is Nakayama of Loewy length at most 2. Then each indecomposable nonprojective module S is simple. Clearly $\alpha(S) = 1$, hence $\alpha(\Lambda) \leq 1$.

Before we go on to describe when $\beta(\Lambda) \leq 1$, we give the following result about the connection between α and β for a given algebra Λ.

LEMMA 4.2. Let Λ be an artin algebra, and let $\Lambda = P \amalg Q$ with P projective injective and Q having no nonzero injective summands. Then $I = \operatorname{soc} P$ is a twosided ideal in Λ, and if $\beta(\Lambda) \leq n$, then $\alpha(\Lambda/I) \leq n$.

Proof: We first prove that I is an ideal in Λ. Let then $g:\Lambda \to \Lambda$ be a Λ-homomorphism, and we want to show that $g(I) \subset I$. Write $\Lambda = P \amalg Q_1 \amalg \cdots \amalg Q_s$, where the Q_i are indecomposable projective, and $P = P_1 \amalg \cdots \amalg P_t$, with the P_j indecomposable. Assume that for some i, $g(\mathrm{soc}P_i) \not\subset P$. Then some projection map $g(\mathrm{soc}P_i) \to Q_j$ is not zero, and hence some projection map $g(P_i) \to Q_j$ is not zero. We must then have a monomorphism $P_i \to Q_j$, and this is a contradiction since P_i is injective. This shows that $g(I) \subset I$, and hence that I is an ideal in Λ.

Now let C be an indecomposable nonprojective Λ/I-module, and let $0 \to A \to B \to C \to 0$ be almost split in modΛ. Since $C \not\approx P_i/\mathrm{soc}P_i$ for P_i projective injective, we know that no P_i is a summand of B [6,Prop.4.11]. Since it is easy to see that $P_1,\cdots,P_t$ are the only indecomposable Λ-modules which are not Λ/I-modules, the above sequence is in mod(Λ/I), and is consequently almost split in mod(Λ/I).Assume that $\alpha(C) > n$. Since $\beta(C) \leq n$, B must have an indecomposable projective summand L, which is not injective by our previous discussion. Write $B = L \amalg K$. Since $A \to L$ is a monomorphism, $K \to C$ is mono, so that K has no nonzero injective summands. Using that $\beta(DA) \leq n$, we can then conclude that $\alpha(C) \leq n$ and the proof is done.

We can now get the following description.

PROPOSITION 4.3. Let Λ be an artin algebra.

(a) $\beta(\Lambda) = 0 \Leftrightarrow \Lambda$ is Nakayama of Loewy length at most 2.

(b) $\beta(\Lambda) = 1 \Leftrightarrow \Lambda$ is Nakayama of Loewy length 3.

Proof: (a) If $\beta(\Lambda) = 0$, then $\alpha(\Lambda/I) = 0$, where the ideal I is as above. Λ/I is then semisimple, and Λ is hence Nakayama of Loewy length at most 2.

If Λ is Nakayama of Loewy length at most 2, the only nonprojective indecomposable Λ-modules are simple, and the middle term in an almost split sequence must be projective injective.

(b) If $\beta(\Lambda) \leq 1$, we have $\alpha(\Lambda/I) \leq 1$, so that Λ/I is Nakayama of Loewy length at most 2. It then follows that Λ is Nakayama of Loewy length at most 3.

The converse also follows easily.

We now go on to apply our study of uniserial functors to get necessary conditions for $\alpha(\Lambda) \leq 2$ and for $\beta(\Lambda) \leq 2$. We shall need the following preliminary results.

LEMMA 4.4. Let Λ be an artin algebra, and assume that $\alpha(\text{TrD}P) \leq 2$ for an indecomposable projective noninjective Λ-module P. Then $\underline{r}P$ is indecomposable or a direct sum of two indecomposable modules.

Proof: We have an almost split sequence $0 \to P \to Q \amalg \text{TrD}(\underline{r}P) \to \text{TrD}P \to 0$, where Q is projective [7,Prop.2.4]. Since $\alpha(\text{TrD}P) \leq 2$, $\text{TrD}(\underline{r}P)$, and hence $\underline{r}P$, is either indecomposable or a direct sum of two indecomposable modules.

LEMMA 4.5. Let Λ be an artin algebra of finite type, and assume that the following conditions hold.

(i) If C is an indecomposable Λ-module such that $(\underline{C},\underline{P/U}) \neq 0$ for some indecomposable projective Λ-module P and an indecomposable summand U of $\underline{r}P$, then $\alpha(C) \leq 2$.

(ii) $\alpha(\mathrm{TrDP}) \leq 2$ for each indecomposable projective noninjective Λ-module P.

Then for each indecomposable projective nonuniserial Λ-module P, $\underline{r}P$ is a direct sum of two uniserial modules.

Proof: By Corollary 3.8, (i) implies that $(\ ,\underline{P/U})$ is a uniserial functor when U is an indecomposable summand of $\underline{r}P$. If P is indecomposable projective, we know by Lemma 4.4 that $\underline{r}P$ is indecomposable or a direct sum of two indecomposable modules. If U is not projective it follows from Proposition 1.3 that U is uniserial. If U is projective, we have an almost split sequence $0 \to U \to P \amalg \mathrm{TrD}(\underline{r}U) \to \mathrm{TrDU} \to 0$ [7,Prop.2.4], where $\underline{r}U$ must be indecomposable by (ii). If $\underline{r}U$ is not projective, the above argument gives that $\underline{r}U$, and hence U, is uniserial. If $\underline{r}U$ is projective, we get that $\underline{r}^2U$ is indecomposable. Continuing this way, we can conclude that U is uniserial.

As an immediate consequence of the above, using that $\alpha(\Lambda^{op}) = \alpha(\Lambda)$ we get the following main result which gives necessary conditions for $\alpha(\Lambda) \leq 2$.

THEOREM 4.6. Let Λ be an artin algebra of finite type and assume that $\alpha(\Lambda) \leq 2$. Then for each indecomposable nonuniserial projective Λ-module or Λ^{op}-module P, $\underline{r}P$ is the direct sum of two uniserial modules.

It would be interesting to know some "minimal" class of C such that if $\alpha(C) \leq 2$ for C in this class, then $\alpha(\Lambda) \leq 2$. If for an algebra of finite type the converse of Theorem 4.6 is true, which it probably is, then the C described in Lemma 4.5 would give such a testing class.

We mention that there are examples of algebras Λ of infinite type such that $\alpha(\Lambda) = 2$ and the indecomposable projective Λ-modules and Λ^{op}-modules have the above structure. For $\Lambda = k[x,y]/(x,y)^2$, where k is a field and x and y indeterminates, we have $\alpha(\Lambda) \leq 2$. If C is indecomposable of type P/U, where P is indecomposable projective and U is a summand of $\underline{r}P$, we even have that $(\ ,\underline{C})$ is uniserial. At the same time this gives examples of uniserial functors of infinite length. Also $\alpha(\Lambda /\mathrm{soc}\Lambda) \leq 2$ for the group algebras Λ studied in [10] are such examples. We do not know, however, if the assumption of finite type can be left out of Theorem 4.6. In general we do not even know if $\alpha(P/U) \leq 1$ for Λ with $\alpha(\Lambda) \leq 2$, where P is indecomposable projective and U a summand of $\underline{r}P$. In the case of hereditary algebras, however, we can leave out the assumption of finite type, and then also the converse of Theorem 4.6 holds [19].

The corresponding necessary conditions for an artin algebra Λ of finite type to satisfy $\beta(\Lambda) \leq 2$ are now a direct consequence of Theorem 4.6 and Lemma 4.2.

<u>THEOREM 4.7</u>. Let Λ be an artin algebra of finite type satisfying $\beta(\Lambda) \leq 2$. Let P be an indecomposable projective Λ-module or Λ^{op}-module.

(i) If P is injective nonuniserial, $\underline{r}P/\mathrm{soc}P$ is a direct sum of two uniserial modules.

(ii) If P is noninjective nonuniserial, $\underline{r}P$ is a direct sum of two uniserial modules.

Proof: Let I be the ideal of Λ generated by the socles of the projective injective Λ-modules. Since by Lemma 4.2 we have $\alpha(\Lambda/I) \leq 2$, we can apply Theorem 4.6.

We shall now give a description of all uniserial functors of type $(\ ,\underline{C})$, for an artin algebra Λ of finite type with $\beta(\Lambda) \leq 2$. For this the following preliminary result about algebras of finite type with $\alpha(\Lambda) \leq 2$ will be useful.

LEMMA 4.8. Assume that Λ is of finite type and that $\alpha(\Lambda) \leq 2$.

(a) Let $0 \to A \overset{g}{\to} B \overset{f}{\to} C \to 0$ be an exact sequence where A and B are indecomposable and $g:A \to B$ is irreducible. Then there is an exact sequence $0 \to U \overset{h}{\to} P \to C \to 0$, where U and P are indecomposable, P is projective and $h:U \to P$ is irreducible.

(b) If C is an indecomposable nonprojective Λ-module, then $\alpha(C) \leq 1$ if and only if $C = P/U$, where P is indecomposable projective and U an indecomposable summand of $\underline{r}P$.

Proof: (a) Let $0 \to A \overset{g}{\to} B \overset{f}{\to} C \to 0$ have the assumed properties. Let C be fixed, and assume that $F = \text{Im}(\ ,f) \subset (\ ,C)$ is of minimal length. This is possible since $(\ ,C)$ is of finite length. Assume that B is not projective. Then there is a map $g':A_1 \to B$ such that we have an almost split sequence $0 \to \text{DTr}B \to A \amalg A_1 \overset{(g,g')}{\to} B \to 0$. Since the composition $A \to B \to C$ is zero, the composite map $g_1:A_1 \to B \to C$ must be onto, and since $\text{Im}(\ ,(g,g')) = \underline{r}(\ ,B)$[8,Prop.1.3], we get $\text{Im}(\ ,g_1) = \underline{r}F$. Since $g:A \to B$ is irreducible and B is indecomposable, F is a waist in $(\ ,C)$, that is, any subfunctor of $(\ ,C)$ either contains or is contained in F

[6,Prop.2.8]. Since $F/\underline{r}F$ is simple, $\underline{r}F$ must also be a waist in $(\ ,C)$, and it follows from [6,Prop.2.8] that $h:\mathrm{Ker}g_1 \to A_1$ is irreducible. Since A_1 is indecomposable, the map $g_1:A_1 \to C$ is minimal. Hence by Proposition 2.4 $(\ ,\mathrm{TrD}(\mathrm{Ker}g_1))$ is a projective cover for $\mathrm{soc}((\ ,C)/\underline{r}F)$, and since F is a waist in $(\ ,C)$, it is easy to see that $\mathrm{soc}((\ ,C)/\underline{r}F) = F/\underline{r}F$ is simple. This shows that $\mathrm{Ker}g_1$ is indecomposable. We now have a contradiction to the minimal choice of F.

(b) One implication follows from (a) and the other from Corollary 3.9.

We can now give our desired description.

THEOREM 4.9. Let Λ be an artin algebra of finite type with $\beta(\Lambda) \leq 2$, and C an indecomposable non-projective Λ-module. Then $(\ ,\underline{C})$ is uniserial if and only if one of the following conditions holds.

(i) $C = P/U$, where P is indecomposable projective noninjective and U an indecomposable summand of $\underline{r}P$.

(ii) $C = P/U$, where P is indecomposable projective injective with uniserial submodules U and V such that $\underline{r}P = U+V$, $U \cap V = \mathrm{soc}P$.

(iii) $C = P/\mathrm{soc}P$ for an indecomposable projective injective uniserial Λ-module P.

(iv) $C = \mathrm{TrD}U$, where U is a summand of $\underline{r}P$ for an indecomposable projective noninjective Λ-module P.

<u>Proof</u>: If I is the ideal of Lemma 4.2, we know that $\alpha(\Lambda/I) \leq 2$, and the almost split sequences in mod(Λ/I) stay almost split in modΛ. If $0 \to A \to B \to C \to 0$ is almost split in modΛ and $\alpha(C) > 2$, we then know that B must have a nonzero projective injective summand. By Lemma 3.4 we then know that $(\ ,\underline{C})$ is uniserial if and only if $\beta(C) = 1$. The indecomposable nonprojective Λ/I-modules C with $\alpha(C) = 1$ are exactly those described in (i) and (ii), by Lemma 4.8. C is an indecomposable nonprojective Λ-module which is a projective Λ/I-module if and only if $C = P/\text{soc}P$ for an indecomposable projective injective Λ-module P. Since we have an almost split sequence $0 \to \underline{r}P \to P \amalg \underline{r}P/\text{soc}P \to P/\text{soc}P \to 0$ [6,Prop.4.11], $\beta(P/\text{soc}P) \leq 1$ if and only if P is uniserial.

The proof is finished by considering the indecomposable nonprojective Λ/I-modules C with $\alpha(C) = 2$, $\beta(C) \leq 1$.

For indecomposable selfinjective algebras of finite type we shall show that the existence of a nonzero uniserial functor $(\ ,\underline{C})$ implies that $\beta(\Lambda) \leq 2$. In view of Theorem 4.9 we then get a description of all uniserial functors of the type $(\ ,\underline{C})$ for selfinjective algebras of finite type. This is based upon the following interesting result.

<u>PROPOSITION 4.10</u>. Let Λ be an indecomposable selfinjective algebra of finite type, and C an indecomposable nonprojective Λ-module. If X is an indecomposable nonprojective Λ-module, then DTr^iX is in Supp$(\ ,\underline{C})$ for some integer i.

<u>Proof</u>: Consider the set of indecomposable Λ-modules X which are such that DTr^iX is in Supp$(\ ,\underline{C})$ for some i, or X is projective and there is an irreducible map $X \to X'$ where DTr^iX' is in Supp$(\ ,\underline{C})$.

We want to show that this set is closed under irreducible maps. Let first X be a nonprojective module in the set, so that $DTr^{i}X \in$ Supp(,$\underline{C}$). From our method for computing almost split sequences in section 2, we have an almost split sequence
$0 \to DTr^{i+1}X \to TrDY_2 \amalg Y_1 \amalg P \to DTr^{i}X \to 0$, where P is projective and each indecomposable summand of Y_1 and Y_2 is in Supp(,$\underline{C}$). We then have almost split sequences
$0 \to DTrX \to DTr^{-i-1}Y_2 \amalg DTr^{-i}Y_1 \amalg P' \to X \to 0$ and
$0 \to X \to DTr^{-i-2}Y_2 \amalg DTr^{-i-1}Y_1 \amalg P'' \to TrDX \to 0$, where P' and P'' are projective. Each indecomposable summand of the middle terms is then again in our set. If X is an indecomposable projective module in the set, we have an irreducible map $X \to X'$ with X' in the set. Here $X' = X/\mathrm{soc}X$ and if $X \to Y$ is irreducible then $Y \approx X'$. And if $Y \to X$ is irreducible, then $Y \approx DTrX'$, so that Y is in the set. We have now shown that our set is an irreducible component in modΛ. Since Λ is of finite type, our set must consist of all indecomposables in modΛ (see [8,Corollary 1.8]. This finishes our proof.

As an immediate consequence we have the following surprising result.

THEOREM 4.11. Let Λ be an indecomposable selfinjective algebra of finite type. If there is a nonzero uniserial functor (,$\underline{C}$) in $((\mathrm{mod}\Lambda)^{op},\mathrm{Ab})$, then $\beta(\Lambda) \leq 2$.

Proof: Let X be an indecomposable nonprojective Λ-module. By Proposition 4.10 $DTr^{i}X$ is in Supp(,$\underline{C}$) for some i. $\beta(DTr^{i}X) \leq 2$ by Theorem 3.1, and since $\beta(X) = \beta(DTr^{i}X) \leq 2$, we are done. (See Lemma 5.2 and [7]).

5. Algebras stably equivalent to Nakayama algebras.

Two artin algebras Λ and Λ' are said to be stably equivalent if the module categories modulo projectives $\underline{\mathrm{mod}}\Lambda$ and $\underline{\mathrm{mod}}\Lambda'$ are equivalent. It is an interesting question to describe what it means that two artin algebras are stably equivalent, and to describe the class of algebras stably equivalent to a given class of algebras. For hereditary algebras there is a good description of the algebras stably equivalent to them [3,13], and also more generally for algebras such that every map between two indecomposable projectives is zero or a monomorphism [12].

We have earlier used uniserial functors to get information about the algebras Λ with $\beta(\Lambda) \leq 2$. Here we shall apply our previous results to get necessary conditions for an artin algebra to be stably equivalent to a Nakayama algebra. In this connection we point out that in the case of selfinjective Nakayama algebras over an algebraically closed field, necessary and sufficient conditions are known [11]. One part of our proof will be a direct consequence of our earlier results and the fact that β behaves nicely under stable equivalence. Namely we show that if C and C'are indecomposable nonprojective modules corresponding to each other under a stable equivalence, then $\beta(C) = \beta(C')$. This behavior accounts for much of the interest in the study of the invariant β.

The second part of the proof will be based on the following. If Λ is a Nakayama algebra there is a close relationship between the number of C with $\beta(C) = 0$, $\beta(C) = 1$ and the number of nonprojective simple modules. Even though it is an important open problem in general whether stably equivalent algebras have the same number of nonprojective simple modules, it is known to be true for algebras stably equivalent to Nakayama algebras [14]. We will hence get a similar relationship between the number of C with $\beta(C) = 0$, $\beta(C) = 1$ and the number of nonprojective simple modules, also for algebras stably equivalent to Nakayama algebras. We can then use counting arguments to improve our necessary conditions.

We shall precede our main theorem with some preliminary lemmas, some of which formalize the ideas we have already discussed.

LEMMA 5.1. Let $\gamma:\underline{\mathrm{mod}}\Lambda \to \underline{\mathrm{mod}}\Lambda'$ be a stable equivalence between artin algebras, and denote also by γ the induced correspondence between the indecomposable nonprojective modules. Then for an indecomposable nonprojective Λ-module C, $\beta(C) = \beta(\gamma C)$. In particular, $\beta(\Lambda) = \beta(\Lambda')$.

Proof: Let $0 \to A \to B \amalg P \to C \to 0$ be an almost split sequence in modΛ, where P is projective and B has no nonzero projective summand. Then we have an almost split sequence $0 \to A' \to \gamma B \amalg Q \to \gamma C \to 0$ in modΛ', where Q is projective [7,section 2], and this shows that $\beta(C) = \beta(\gamma C)$. Since we also have an equivalence of categories $\underline{\mathrm{mod}\Lambda}^{op} \to \underline{\mathrm{mod}\Lambda'}^{op}$, we can conclude that $\beta(\Lambda)=\beta(\Lambda')$[3].

LEMMA 5.2. Let Λ be a Nakayama algebra, n_1 the number of indecomposable projective Λ-modules of length 2, and n_2 the number of length at least 3. Then there are n_1 indecomposable nonprojective C with $\beta(C)=0$ and $2n_2$ indecomposable nonprojective C with $\beta(C) = 1$.

Proof: It is not hard to see that $\beta(C) = 0$ if and only if $C = P/\underline{r}P$ with P indecomposable projective of length 2 and that $\beta(C) = 1$ if and only if $C = P/\underline{r}P$ or $C = P/\mathrm{soc}P$, with P indecomposable projective of length at least 3.

<u>LEMMA 5.3</u>. If Λ is stably equivalent to a Nakayama algebra Γ, and P is an indecomposable projective Λ-module, then $\underline{r}P/\underline{r}^2P$ does not contain two copies of any simple Λ-module S.

<u>Proof</u>: Assume that S_1 and S_2 are different isomorphic simple submodules of $P/\underline{r}^2P$, and let $i:S_1 \to P/\underline{r}^2P$ and $j:S_2 \to P/\underline{r}^2P$ be the corresponding inclusions. Since Λ and Γ are stably equivalent, we know that there is an equivalence $\gamma:\overline{\text{mod}\Lambda} \to \overline{\text{mod}\Gamma}$, between the module categories modulo injectives [3]. Consider the diagram

$$\begin{array}{ccc} \overline{S}_1 & \xrightarrow{\overline{i}} & \overline{P/\underline{r}^2P} \\ & & \\ \overline{S}_2 & \nearrow_{\overline{j}} & \end{array}$$

Since a monomorphism between indecomposable noninjective modules can not factor through an injective module, $\overline{j}\overline{s} \neq \overline{i}$ for any $s: S_1 \to S_2$ and $\overline{i}\,\overline{t} \neq \overline{j}$ for any $t:S_2 \to S_1$. Here $\overline{f}: \overline{A} \to \overline{B}$ denotes the morphism in $\overline{\text{mod}\Lambda}$ corresponding to $f:A \to B$ in modΛ. On the other hand, consider the diagram

$$\begin{array}{ccc} \overline{\gamma S_1} & \xrightarrow{\gamma(\overline{i})} & \overline{\gamma(P/\underline{r}^2P)} \\ & & \\ \overline{\gamma S_2} & \nearrow_{\gamma(\overline{j})} & \end{array}$$

Since Γ is Nakayama and $\gamma S_1 \approx \gamma S_2$, it is not hard to see that there must be a morphism $s:\gamma S_1 \to \gamma S_2$ such that $\gamma(\overline{j})\overline{s} = \gamma(\overline{i})$ or a morphism $t:\gamma S_2 \to \gamma S_1$ such that $\gamma(\overline{i})\overline{t} = \gamma(\overline{j})$. This contradiction finishes the proof of the lemma.

We shall now prove the main result of this section.

THEOREM 5.4. Let Γ be an artin algebra stably equivalent to a Nakayama algebra Λ, and let P be an indecomposable projective Γ-module or Γ^{op}-module. If P is not uniserial, then P is injective and $\underline{r}P/socP$ is a direct sum of two uniserial modules.

Proof: Since $\beta(\Lambda) \leq 2$ and Γ is stably equivalent to Λ, we have that $\beta(\Gamma) \leq 2$. Let P be an indecomposable projective Γ-module. Since Γ is clearly of finite type, we know from Theorem 4.7 that if P is nonuniserial injective, then $\underline{r}P/socP$ is a direct sum of two uniserial modules, and if P is nonuniserial noninjective, then $\underline{r}P$ is a direct sum of two uniserial modules. We want to show that this last possibility does not occur and we shall outline the main steps in the proof of this.

Since Γ is stably equivalent to a Nakayama algebra and Λ and Γ hence have the same number of simple nonprojective modules[14], we know using Lemma 5.1 and Lemma 5.2 that there are numbers n_1 and n_2 such that $n_1 + n_2$ is the number of nonprojective simple Γ-modules, there are n_1 indecomposable C with $\beta(C) = 0$ and $2n_2$ indecomposable C with $\beta(C) = 1$.

Let C be an indecomposable nonprojective Γ-module with $\beta(C) \leq 1$. Since $\beta(\Gamma) = \beta(\Lambda) \leq 2$, we know (see Lemma 3.3 and Theorem 4.9) that either C is a factor of an indecomposable projective P or there is some irreducible map $P \to C$ where P is projective. In either case we associate with C this indecomposable projective. We note that if $C/\underline{r}C$ is simple and there is an irreducible map $P \to C$ with P projective, then C is a factor of P [6,Lemma 4.3]. It is not hard to see that if C is associated with more than one indecomposable projective, then $\beta(C) = 0$, and C is associated with two indecomposable projectives.

We now consider the various types of indecomposable projectives and investigate which indecomposable C with $\beta(C) = 0$ and with $\beta(C) = 1$ are associated with each of them.

(1) Let P be injective nonuniserial. Associated with P are then P/U and P/V, where U and V are uniserial, $\underline{r}P = U + V$, $U \cap V = \mathrm{soc}P$. Here we have $\beta(P/U) = 1 = \beta(P/V)$, hence exactly 2 indecomposables C with $\beta(C) = 1$.

(2) Let P be injective uniserial. If P has length at least 3, then $\beta(P/\underline{r}P) = 1$ and $\beta(P/\mathrm{soc}P) = 1$. If P has length 2, then $\beta(P/\underline{r}P) = 0$. In the first case there are associated exactly 2 indeomposables C with $\beta(C) \neq 1$. In the second case there is 1 indecomposable C with $\beta(C)=0$, and it is not hard to see that there is no other indecomposable projective associated with C.

(3) Let P be uniserial noninjective. Then P is associated with $P/\underline{r}P$ and TrDP. It is not hard to show that if $P/\underline{r}P \approx \mathrm{TrD}P$, then $\beta(P/\underline{r}P) = 0$, hence there is 1 indecomposable C associated with P with $\beta(C)=0$. Here no other projective will be associated with $P/\underline{r}P$. If $P/\underline{r}P \not\approx \mathrm{TrD}P$, $\beta(P/\underline{r}P) = 1$. Then either $\beta(\mathrm{TrD}(\underline{r}P))$ is 1, or $\beta(\mathrm{TrD}(\underline{r}P))= 0$. In the second case $\mathrm{TrD}(\underline{r}P)$ may also be associated with another indecomposable projective. So we get either 2 indecomposables C with $\beta(C)=1$ or 1 with $\beta(C)=1$ and 1 with $\beta(C)=0$, where the last one may be counted twice.

(4) Let P be nonuniserial noninjective. We have $\underline{r}P = U \amalg V$, where $U \not\approx V$ by Lemma 5.3. It is not hard to see by our previous results that the indecomposable C with $\beta(C) \leq 1$ associated with P are P/U, P/V, TrD(U), TrD(V), where two of them may be isomorphic, but $P/U \not\approx P/V$, $\mathrm{TrD}(U) \not\approx \mathrm{TrD}(V)$. Studying the various possibilities in more detail, we will be able to exclude this case, since there will be too many indecomposables C with $\beta(C)=0$ or $\beta(C)=1$. For in cases (1) and (2) there is the correct relationship between the number of indecomposable projectives (or simples) and the indecomposable C with $\beta(C)=0$ or $\beta(C)=1$. And in case (3) the possibility of using double counting will also turn out not to allow (4) to accur. This finishes our sketch of the proof of Theorem 5.4.

6. Algebras where (,S) is uniserial for S simple.

One of the hopes in the study of uniserial functors is to classify classes of algebras in terms of certai types of functors being uniserial. In this section we shall illustrate this type of result by showing how a class of algebras studied by Tachikawa [17,18] can be described in terms of (,$\underline{S}$) being uniserial for all simple Λ-modules S, or equivalently, (,S) uniserial for all simple Λ-modules S. At the same time we get a new description of these algebras in terms of irreducible maps, and get some more insight into one of the cases of algebras with $\alpha(\Lambda) \leq 2$. We recall that in section 1 we have shown that for (,$\underline{S}$) (or (,S)) to be uniserial, it is necessary that the projective cover P of S is uniserial. We have the following main result.

THEOREM 6.1. For an artin algebra Λ the following are equivalent.

(a) Each indecomposable Λ-module has simple socle.

(b) Λ is of finite type with $\alpha(\Lambda) \leq 2$, and $\underline{r}P$ is indecomposable for each indecomposable projective Λ-module P.

(c) (,S) is uniserial for all simple Λ-modules S.

(d) If X is an indecomposable nonsimple Λ-module, there is some irreducible monomorphism $Y \to X$.

<u>Proof</u>: (a) ⇒ (b). Assume that each indecomposable Λ-module has simple socle. Let P be an indecomposable projective Λ-module. Then $P/\underline{r}^2P$ is indecomposable, so it has simple socle. Hence $\underline{r}P/\underline{r}^2P$ is simple, so that $\underline{r}P$ is indecomposable.

Let 0 → A → B → C → 0 be an almost split sequence in modΛ. Since A and C are indecomposable, socA and socC are simple, so that socB has length at most 2, and hence $\alpha(C) \leq 2$. It then follows that $\alpha(\Lambda) \leq 2$. Clearly Λ is of finite type.

(b) ⇒ (c). Let S be a simple nonprojective Λ-module. Since $\underline{r}P = \Omega^1 S$ is indecomposable, where P is a projective cover for S, and $(\ ,\mathrm{TrD}\Omega^1 S)$ is a projective cover for $\mathrm{soc}(\ ,\underline{S})$ (Proposition 2.4), $\mathrm{soc}(\ ,\underline{S})$ is simple. Since we have the exact sequence $0 \to \underline{r}P \xrightarrow{i} P \to S \to 0$, where i is irreducible and $\underline{r}P$ is indecomposable, we use that $\alpha(\Lambda) \leq 2$ and $(\ ,\underline{S})$ has finite length to conclude that $(\ ,\underline{S})$ is uniserial (Theorem 3.7), or equivalently (,S) is uniserial.

(c) ⇒ (d). Assume that (,S) is uniserial, where S is a simple Λ-module, and let X be in Supp(,S) and X not isomorphic to S. We then have an epimorphism f:X → S. Since(,S) is uniserial, Im(,f) is a waist in (,S), that is, every subfunctor H of (,S) either contains or is contained in Im(,f). Hence the morphism Kerf → X is irreducible [6,Prop.2.8]. Since every indecomposable Λ-module is in Supp(,S) for some simple Λ-module S, we are done.

(d) ⇒ (a). Assume to the contrary that there is some indecomposable Λ-module C such that socC has length n > 1, and choose C minimal with this property. Since C is not simple, there is by assumption an irreducible monomorphism L → C, with L indecomposable. Consider the exact sequence 0 → L → C → M → 0. Since we know [6,Prop.2.6] that M is indecomposable and each irreducible map t:Y → M is an epimorphism, M must be simple. But then socL = socC, and since L is indecomposable, this is a contradiction. This finishes the proof of Theorem 6.1.

Let Λ be an artin algebra, and C a simple Λ-module. Then each irreducible map X → C is an epimorphism. And as the above result shows there are artin algebras Λ such that this is not true if C is not simple, and it even describes the algebras where this is not true for C non-simple. Similarly, it is clear that if C is an indecomposable projective Λ-module, then all irreducible maps X → C with X indecomposable are monomorphisms. And there are algebras where the projectives are the only modules which have this property. The following result characterizes such algebras. [9, Proposition 2.4.]

PROPOSITION 6.2. Let Λ be an artin algebra. Then every indecomposable Λ-module C has simple top, i.e., $C/\underline{r}C$ is simple, if and only for every indecomposable non-projective Λ-module C there is some irreducible epimorphism X → C, with X indecomposable.

While the equivalence of (a) in (d) in Theorem 6.1 is somewhat related to Proposition 6.2, the proofs are completely different. The proof of Proposition 6.2 when Λ is of finite type is quite elementary, but in the general case the theory of preprojective modules is used in the proof given in [9].

Since it is clear that Λ is a Nakayama algebra if and only if every indecomposable Λ-module has simple socle and simple top, we end the section with the following interesting description of Nakayama algebras, which follows directly from our previous results.

THEOREM 6.3. For an artin algebra Λ the following are equivalent.

(a) Λ is Nakayama.

(b) For every indecomposable nonsimple Λ-module C there is some irreducible monomorphism $X \to C$, and for every indecomposable nonprojective Λ-module C there is some irreducible epimorphism $X \to C$, with X indecomposable.

(c) $\alpha(\Lambda) \leq 2$, and if C is an indecomposable non-projective nonsimple Λ-module, then $\alpha(C) = 2$. If $0 \to DTrC \to B_1 \amalg B_2 \to C \to 0$ is almost split, one of the maps $B_i \to C$ is mono and the other one epi.

REFERENCES

1. Auslander, M.: Functors and morphisms determined by objects. Proc.Conf.Temple Univ.1976. Lecture Notes in Pure and Applied Math. Vol.37, M.Dekker, N.Y.1978.

2. Auslander, M.: Applications of morphisms determined by modules. Proc.Conf. Temple Univ.1976. Lecture Notes in Pure and Applied Math. Vol. 37, M.Dekker, N.Y. 1978.

3. Auslander, M., Reiten I.: Stable equivalence of artin algrbras, Conf.on orders, group rings and related topics. Springer-Verlag 353, 8-71 (1973).

4. Auslander, M., Reiten I.: Stable equivalence of dualizing R-varieties. Adv.in Math. Vol.12, No.3, 306-366 (1974).

5. Auslander,M., Reiten,I.: Representation theory of artin algebras III: Almost split sequences. Comm.in Algebra, Vol.3, No.3, 239-294 (1975)

6. Auslander,M., Reiten,I.: Representation theory of artin algebras IV: Invariants given by almost split sequences. Comm.in Algebra 5, 443-518 (1977)

7. Auslander,M., Reiten,I.: Representation theory of artin algebras V: Methods for computing almost split sequences and irreducible morphisms. Comm.in Algebra 5, 519-554 (1977)

8. Auslander,M., Reiten,I.: Representation theory of artin algebras VI: A functorial approach to almost split sequences. Comm.in Algebra 11,279-291 (1977)

9. Auslander, M., Smalø, S.: Preprojective modules: An introduction and some applications. These Proceedings.

10. Butler, M., Shahzamanian: The construction of almost split sequences III: Modules over two classes of tame local algebras.

11. Gabriel,P., Riedtmann,C.: Group representations without groups. Comm.Math.Helvetici 54, 240-287 (1979)

12. Martinez, R.: Algebras stably equivalent to l-hereditary algebras. These Proceedings.

13. Platzeck,M.I.: On algebras stably equivalent to an hereditary algebra, Can.J.Math. 30, No.4, 817-829 (1978)

14. Reiten, I.: A note on stable equivalence and Nakayama algebras. Proc.Amer.Math.Soc. 71,2, 157-163 (1978)

15. Reiten, I.: Almost split sequences, Proc. Antwerp Ring Theory Conference 1978.

16. Riedtmann, C.: Algebren, Darstellungsköcher, Überlagerungen und zurück.

17. Tachikawa, H.: On rings for which every indecomposable right module has a unique maximal submodule. Math.Z.71, 200-222 (1959)

18. Tachikawa,H.: Balancedness and left serial algebras of finite type. Proc.ICRA 1974, Springer Lecture Notes 488, 351-378 (1975)

19. Todorov, G.: Almost split sequences in the representation theory of certain classes of artin algebras, Ph.D.Thesis, Brandeis Univ.(1978)

20. Todorov, G.: Almost split sequences for TrD-periodic modules. These Proceedings.

Brandeis University
Department of Mathematics
Waltham, MA 02154, U.S.A.

Universitetet i Trondheim, NLHT
Matematisk institutt
7055 Dragvoll, Norway

PREPROJECTIVE MODULES: AN INTRODUCTION AND SOME APPLICATIONS

M. AUSLANDER[1] AND S. O. SMALØ[2]

Introduction

In this paper we want to give a brief and informal introduction of the notion of a preprojective partition of ind Λ, the set of isomorphism classes of finitely generated indecomposable modules over an artin algebra Λ. This leads to the theory of preprojective modules. One of the main points of interest in these modules is that they are "close" to being projective and the preprojective partition gives a measure of this closeness. Also, given any artin algebra Λ, one should be able to classify the preprojective modules.

By duality one obtains a dual partition called the preinjective partition and in terms of this partition the preinjective modules are defined.

By specializing our notion of preprojective modules to hereditary tensor algebras it will coincide with Dlab-Ringel's use of the same term [5]. For artin algebras stably equivalent to an hereditary artin algebra M. I. Platzeck [7] studied modules which were clearly analogous to these preprojective modules introduced by Dlab-Ringel. In that case these modules also coincide with our use of the term preprojective modules. Finally, the results one obtains by using this notion extend results

---

1) Written while a Guggenheim Fellow with the partial support of NSF MCS 77 04 951.

2) Supported by the Norwegian Research Council.

found by Gabriel - Roiter [6,8] for rings of finite representation type.

Section one is devoted to giving the necessary definitions involved. The existence and uniqueness theorems for the preprojective and the preinjective partition are given without proofs and some examples to illustrate the concepts. In addition to this, section one will contain some results which describe how to find the preprojective and the preinjective partition.

In section two we will deal with the problem of determining which modules A in ind Λ can have the property that if $f: B \to A$ is irreducible with B indecomposable, then f is an injective morphism. We prove that if it is only the indecomposable projective modules which satisfy this property, then all the indecomposable Λ-modules have simple top.

In section three some ideals associated with the preprojective and the preinjective partition of ind Λ will be introduced. These ideals will then be used to give a description of when Λ is of finite representation type in terms of the preprojective and the preinjective modules.

Section four is devoted to give a lower bound on the number of nonempty classes in the preprojective partition of ind Λ when Λ is an artin algebra of finite representation type.

1. The preprojective and the preinjective partition.

Let Λ be an artin algebra and let ind Λ denote the set of isomorphism classes of finitely generated indecomposable Λ-modules. To motivate the introduction of the preprojective partition, let us look at ind Λ and the subset $\underline{\underline{P}}_0$ of ind Λ consisting of the indecomposable projective modules. Then each of the following three conditions characterizes $\underline{\underline{P}}_0$ completely.

PROPOSITION 1.1.

a) $P \in \underline{\underline{P}}_0$ if and only if given any surjective morphism $f : A \to B$ between finitely generated Λ-modules and a morphism $g : P \to B$ there exists an $h : P \to A$ such that the diagram

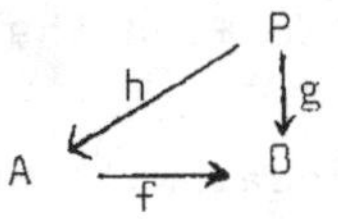

commutes.

b) $P \in \underline{\underline{P}}_0$ if and only if every surjective morphism $f : M \to P$ for a finitely generated Λ-module M is a splittable surjection.

c) $\underline{\underline{P}}_0$ is the unique minimal subset of ind Λ with the property that for all modules A in ind Λ, there exists a P which is a direct sum of copies of modules from $\underline{\underline{P}}_0$ and a surjective morphism $P \to A$.

Of these characterizations of $\underline{\underline{P}}_0$ a) is the one which is most often used. However, we will concentrate on the two other descriptions and in particular property c) which will be the most important one from our point of view. For instance, if we look at the subset ind Λ - $\underline{\underline{P}}_0$ of ind Λ does it contain a minimal finite subset $\underline{\underline{A}}$ such that for all B in ind Λ - $\underline{\underline{P}}_0$ there exists a module A which is a direct sum of copies of modules from $\underline{\underline{A}}$ and a surjective morphism $f : A \to B$. This turns out to be true and also the modules in this minimal finite subset $\underline{\underline{A}}$ of ind Λ - $\underline{\underline{P}}_0$ are characterized by the following which is analogous to b) in Proposition 1.1.: A in ind Λ - $\underline{\underline{P}}_0$ is in $\underline{\underline{A}}$ if and only if a surjective morphism $B \to A$ is a splittable surjective morphism if B is a module which is a direct sum of copies of modules from $\operatorname{ind}\Lambda - \underline{\underline{P}}_0$.

We may formalize this as follows.

<u>DEFINITION</u>: Let $\underline{\underline{A}}$ be a subset of $\operatorname{ind}\Lambda$. A subset $\underline{\underline{B}}$ of $\underline{\underline{A}}$ is called a cover for $\underline{\underline{A}}$ if for all modules A in $\underline{\underline{A}}$ there exists a module B , which is a finite direct sum of copies of modules from $\underline{\underline{B}}$, and a surjective morphism $B \to A$.

We saw that $\underline{\underline{P}}_0$ is a cover for $\operatorname{ind}\Lambda$, but $\underline{\underline{P}}_0$ is also minimal with this property. This indicates the following definition.

<u>DEFINITION</u>: Let $\underline{\underline{A}}$ be a subset of $\operatorname{ind}\Lambda$. A cover $\underline{\underline{B}}$ of $\underline{\underline{A}}$ is minimal if no proper subset of $\underline{\underline{B}}$ is a cover for $\underline{\underline{A}}$.

Let $\underline{\underline{A}}$ be a subset of $\operatorname{ind}\Lambda$, then $\underline{\underline{A}}$ itself is a cover for $\underline{\underline{A}}$. This shows that covers exist for all subsets $\underline{\underline{A}}$ of $\operatorname{ind}\Lambda$, but minimal covers do not always exist. However, if there exists a minimal cover for a subset $\underline{\underline{A}}$ of $\operatorname{ind}\Lambda$ we have the following uniqueness theorem.

<u>THEOREM 1.2.</u> Let $\underline{\underline{A}}$ be a subset of $\operatorname{ind}\Lambda$ and assume that $\underline{\underline{B}}$ and $\underline{\underline{B}}'$ are minimal covers. Then $\underline{\underline{B}} = \underline{\underline{B}}'$. Morover, $\underline{\underline{B}}$ consists of the modules B in $\underline{\underline{A}}$ such that if there is a surjective morphism $f : A \to B$ with A a direct sum of copies from $\underline{\underline{A}}$, then f is a splittable surjection.

<u>Proof</u>: (See [4] Theorem 2.3.)

By dualizing the notion of cover and minimal cover we obtain the following.

<u>DEFINITION</u>: Let $\underline{\underline{A}}$ be a subset of ind Λ. A subset $\underline{\underline{B}}$ of $\underline{\underline{A}}$ is called a cocover for $\underline{\underline{A}}$ if for all modules A in $\underline{\underline{A}}$ there exists a module B, which is a finite direct sum of copies of modules from $\underline{\underline{B}}$, and an injective morphism $A \to B$.

<u>DEFINITION</u>: Let $\underline{\underline{A}}$ be a subset of ind Λ. A cocover $\underline{\underline{B}}$ of $\underline{\underline{A}}$ is minimal if no proper subset of $\underline{\underline{B}}$ is a cocover for $\underline{\underline{A}}$.

As for minimal covers we have the following uniqueness theorem for minimal cocovers when they exist.

<u>THEOREM 1.3.</u> Let $\underline{\underline{A}}$ be a subset of ind Λ and assume that $\underline{\underline{B}}$ and $\underline{\underline{B}}'$ are minimal cocovers. Then $\underline{\underline{B}} = \underline{\underline{B}}'$. Moreover, $\underline{\underline{B}}$ consists of the modules B in $\underline{\underline{A}}$ such that if there is an injective morphism $f : B \to A$ with A a finite direct sum of copies from $\underline{\underline{A}}$, then f is a splittable injection.

<u>Proof</u>: (See [4] Theorem 2.3.)

We now want to define what we mean by a preprojective and a preinjective partition of ind Λ.

<u>DEFINITION</u>: A partition $\underline{\underline{P}}_o, \underline{\underline{P}}_1, \ldots, \underline{\underline{P}}_n, \ldots, \underline{\underline{P}}_\infty$ of ind Λ is called a preprojective partition of ind Λ if

i) $\bigcup_{i=o}^{\infty} \underline{\underline{P}}_i = \text{ind } \Lambda$

ii) $\underline{\underline{P}}_i \cap \underline{\underline{P}}_j = \emptyset$ if $i \neq j$

iii) For all $n < \infty$ $\underline{\underline{P}}_n$ is a finite minimal cover for ind $\Lambda - \bigcup_{j<n} \underline{\underline{P}}_j$.

A partition $\underline{\underline{I}}_o, \underline{\underline{I}}_1, \ldots, \underline{\underline{I}}_n, \ldots, \underline{\underline{I}}_\infty$ of ind Λ is called a preinjective partition of ind Λ if

i) $\bigcup_{i=o}^{\infty} \underline{\underline{I}}_i =$ ind Λ

ii) $\underline{\underline{I}}_i \cap \underline{\underline{I}}_j = \emptyset$ if $i \neq j$

iii) For all $n < \infty$ $\underline{\underline{I}}_n$ is a finite minimal cocover for ind $\Lambda - \bigcup_{j<n} \underline{\underline{I}}_j$.

We have the following existence and uniqueness theorem for a preprojective and a preinjective partition of ind Λ.

THEOREM 1.4. Let Λ be any artin algebra. Then there exists a unique preprojective partition $\underline{\underline{P}}_o, \underline{\underline{P}}_1, \ldots, \underline{\underline{P}}_n, \ldots, \underline{\underline{P}}_\infty$ and a unique preinjective partition $\underline{\underline{I}}_o, \underline{\underline{I}}_1, \ldots, \underline{\underline{I}}_n, \ldots, \underline{\underline{I}}_\infty$ of ind Λ.

It is in terms of these partitions of ind Λ that the preprojective and the preinjective modules are defined.

DEFINITION: The modules in $\bigcup_{i<\infty} \underline{\underline{P}}_i$ are called the preprojective modules and the modules in $\bigcup_{i<\infty} \underline{\underline{I}}_i$ are called the preinjective modules.

If we look at property iii) in the definition of a preprojective partition we see that the preprojective modules in $\underline{\underline{P}}_n$, $n < \infty$, act as projective modules in $\bigcup_{i=n}^{\infty} \underline{\underline{P}}_i$ according to b) and c) in Proposition 1.1 describing $\underline{\underline{P}}_o$, the set of projective modules in ind Λ. Further, these subsets $\underline{\underline{P}}_n$ of ind Λ are obtained inductively by first throwing away the projectives $\underline{\underline{P}}_o$, then throwing away the finite minimal cover $\underline{\underline{P}}_1$ of ind $\Lambda - \underline{\underline{P}}_o$ and so on. By dualizing one obtains the preinjective partition.

We now want to look at some examples which will illustrate the concepts we are dealing with.

<u>EXAMPLE 1.</u> $\Lambda = Z/p^nZ$ where Z denotes the integers, p is a prime number and $n \geq 1$.

The preprojective partition of ind Λ is then

$\underline{\underline{P}}_o = \{\Lambda\} = \{Z/p^nZ\}$, $\underline{\underline{P}}_1 = \{Z/p^{n-1}Z\}$, $\underline{\underline{P}}_2 = \{Z/p^{n-2}Z\}$

... $\underline{\underline{P}}_m = \{Z/p^{n-m}Z\}$,..., $\underline{\underline{P}}_{n-1} = \{Z/pZ\} = \{\Lambda/\text{rad }\Lambda\}$,

which is also the preinjective partition of ind Λ in this case.

Intuitively, this is the most natural way of ordering the modules in ind Λ according to how far they are from being projective. Since Λ in this case is of finite representation type, the partition ends at $\underline{\underline{P}}_k$ for some $k < \infty$.

<u>EXAMPLE 2.</u> $\Lambda = \begin{pmatrix} k & 0 & 0 \\ k & k & 0 \\ k & 0 & k \end{pmatrix}$ where k is a field.

$$\underline{\underline{P}}_o = \{P_1 , P_2 , P_3\} = \left\{ \begin{pmatrix} k \\ k \\ k \end{pmatrix} , \begin{pmatrix} 0 \\ k \\ 0 \end{pmatrix} , \begin{pmatrix} 0 \\ 0 \\ k \end{pmatrix} \right\}$$

$$\underline{\underline{P}}_1 = \{P_1/P_2 , P_1/P_3\} = \left\{ \begin{pmatrix} k \\ 0 \\ k \end{pmatrix} , \begin{pmatrix} k \\ k \\ 0 \end{pmatrix} \right\} = \{\mathrm{TrD}\ P_3 , \mathrm{TrD}\ P_2\}$$

where Tr denotes the transpose and D the ordinary duality.

$$\underline{\underline{P}}_2 = \{P_1/\underline{r}P_1\} = \left\{ \begin{pmatrix} k \\ 0 \\ 0 \end{pmatrix} \right\} = \{\mathrm{TrD}\ P_1\} .$$

$$\underline{\underline{I}}_o = \{I_1 , I_2 , I_3\} = \left\{ \begin{pmatrix} k \\ 0 \\ 0 \end{pmatrix} \begin{pmatrix} k \\ k \\ 0 \end{pmatrix} \begin{pmatrix} k \\ 0 \\ k \end{pmatrix} \right\} ,$$

$$\underline{\underline{I}}_1 = \left\{ \begin{pmatrix} k \\ k \\ k \end{pmatrix} \right\} = \{\mathrm{DTr}\ I_1\} , \quad \underline{\underline{I}}_2 = \left\{ \begin{pmatrix} 0 \\ k \\ 0 \end{pmatrix} \begin{pmatrix} 0 \\ 0 \\ k \end{pmatrix} \right\}$$

For hereditary artin algebras it is known that if A is a preprojective module there exists a projective module P and an n such that $A = (\mathrm{TrD})^n P$. Λ in Example 2 is hereditary. This example then shows thát even though the class of all preprojective modules are $\{(\mathrm{TrD})^n P\}$ where $n \geq 0$ and P is an indecomposable projective Λ-module, the preprojective partition does not coincide with the partition

$$\underline{\underline{P}}_o , (\mathrm{TrD}\ \underline{\underline{P}}_o) , (\mathrm{TrD}^2\ \underline{\underline{P}}_o), \ldots, \quad \mathrm{ind}\ \Lambda - \bigcup_{i<\infty} (\mathrm{TrD}^i\ \underline{\underline{P}}_o)$$

of ind Λ when Λ is hereditary.

<u>EXAMPLE 3.</u> $\Lambda = k[X_1,\dots,X_n]/(X_1,\dots,X_n)^2$, where k is a field.

$\underline{\underline{P}}_0 = \{\Lambda\}$, $\underline{\underline{P}}_1 = \{\mathrm{TrD}\,(\Lambda/\underline{r})\}$, $\underline{\underline{P}}_2 = \{\mathrm{TrD}\,\Lambda\}$ if $n \geq 2$

($\underline{\underline{P}}_2 = \emptyset$ if $n = 1$) . $\underline{\underline{P}}_3 = \{\mathrm{TrD}^2\,(\Lambda/\underline{r})\}$,..., where $\underline{r}$

is the radical of Λ .

In Example 3 Λ is not hereditary. Further, if $n \geq 2$, then $\Lambda/\underline{r}$ is not isomorphic to any $\mathrm{TrD}^n P$ for P a projective Λ-module and n a natural number. Then $\mathrm{TrD}(\Lambda/\underline{r})$ which is preprojective is not of the form $(\mathrm{TrD})^n P$ for some projective module P and natural number n. It is also easy to see that in this case $\Lambda/\underline{r} \in \underline{\underline{P}}_\infty$ and $\Lambda/\underline{r} \in \underline{\underline{I}}_\infty$. Another property the preprojective Λ-modules satisfy for this special algebra is the following: If A is a preprojective Λ-module and $x \in A$, $x \notin \underline{r}\,A$, then (x), the submodule of A generated by x is projective.

We now want to indicate how to find the preprojective partition and the preprojective modules. The proof of the existence and the way one finds thies partition are closely related to the theory of almost split sequences and irreducible morphisms found and studied by Auslander - Reiten, [2,3].

<u>DEFINITION</u>: Let A and B be Λ-modules. A morphism $f : A \to B$ is called an irreducible morphism if f is neither a splittable monomorphism nor a splittable epimorphism and whenever the diagram

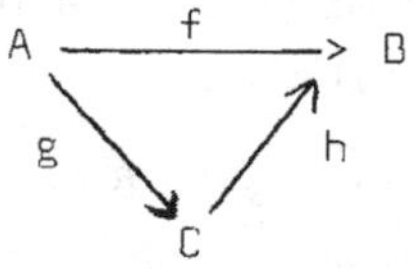

commutes either g is a splittable monomorphism or h is a splittable epimorphism.

For basic properties of irreducible morphisms, see [3].

We now have the following useful result.

PROPOSITION 1.5. Let Λ be any artin algebra and let $\underline{\underline{P}}_0$, $\underline{\underline{P}}_1$,..., $\underline{\underline{P}}_n$,..., $\underline{\underline{P}}_\infty$ denote the preprojective partition of ind Λ. Then

i) $A \in \underline{\underline{P}}_0$ if and only if A is an indecomposable projective module.

ii) $A \in \underline{\underline{P}}_1$ if and only if $A = \mathrm{TrD}B$ for some summand B of the radical of Λ as left module.

iii) If $A \in \underline{\underline{P}}_n$, then there exists a B in $\underline{\underline{P}}_j$ for some $j < n$ and an irreducible morphism $B \to A$.

The converse of iii) in Proportition 1.5 is not valid in general, i.e. there exist examples where A is an indecomposable module not in $\underline{\underline{P}}_n$ and an irreducible morphism $f : B \to A$ with B in $\bigcup_{j<n} \underline{\underline{P}}_j$. A natural question to ask now is the following: Assume B is a preprojective module and $f : B \to A$ an irreducible morphism with A an indecomposable module. Is then A necessary preprojective ? J. Alperin has given examples which show that the answer is no [1]. A general open problem is then to give a "nice" characterization of those artin algebras Λ for which the answer to the above question is affirmative.

Proposition 1.5 gives a complete description of the modules in $\underline{\underline{P}}_0$ and $\underline{\underline{P}}_1$. We now want to indicate how we can find $\underline{\underline{P}}_n$, $n < \infty$, when all the $\underline{\underline{P}}_j$, $j < n$ are given. First look at the set $\underline{\underline{A}}_n$ of all indecomposable modules A such that there exists an irreducible morphism $f : B \to A$ for some B in $\bigcup_{i<n} \underline{\underline{P}}_i$. $\bigcup_{i<n} \underline{\underline{P}}_i$ is finite so the set $\underline{\underline{A}}_n$ will also be finite. (See [3]). Further it is possible to prove that $\underline{\underline{A}}_n$ is a cover for ind $\Lambda - \bigcup_{i<n} \underline{\underline{P}}_i$ (see [4]), and therefore it contains a cover with a minimal number of elements which then is the unique minimal cover $\underline{\underline{P}}_n$ of ind $\Lambda - \bigcup_{i<n} \underline{\underline{P}}_i$. To determine this minimal cover one can start with any A in $\underline{\underline{A}}_n$ and see if $\underline{\underline{A}}_n - \{A\}$ is a cover for $\underline{\underline{A}}_n$. A is then in the minimal cover $\underline{\underline{P}}_n$ if and only if $\underline{\underline{A}}_n - \{A\}$ is not a cover for $\underline{\underline{A}}_n$. Hence, by going through all modules in $\underline{\underline{A}}_n$ in an arbitrary order one can find the members of $\underline{\underline{P}}_n$.

The next result give some information about the nonpreprojective modules in ind Λ when Λ is of infinite representation type.

<u>PROPOSITION 1.6</u>. If Λ is an artin algebra of infinite representation type and if $\underline{\underline{P}}_0, \underline{\underline{P}}_1, \ldots, \underline{\underline{P}}_\infty$ denotes the preprojective partition of ind Λ then $\underline{\underline{P}}_\infty$ does not contain any minimal cover and is therefore infinite.

<u>Proof</u>: See [4] Proposition 7.1.

2. Relationship between the preprojective modules and injective irreducible morphisms.

In this section we want to use the existence of a unique preprojective partition of ind Λ for an artin algebra Λ to study the problem of existence of modules A in ind Λ such that all irreducible morphisms $f : B \to A$ with B in ind Λ are injective morphisms. As a starting point for this discussion we have the following result taken from [3].

PROPOSITION 2.1. Let Λ be an artin algebra and let mod Λ denote the category of finitely generated Λ-modules. Then A in ind Λ is projective if and only if whenever there exists an irreducible morphism $f : B \to A$ in mod Λ, f is an injective morphism.

A natural question to ask now is the following: If in the above proposition we restrict B to be indecomposable, will then the indecomposable projective modules be characterized in the same way ? The next result gives a partial answer to this.

PROPOSITION 2.2. Let Λ be any artin algebra and assume there exists a preprojective module A in ind Λ such that $\ell(A/\underline{r}A) = n \geq 2$ where $\underline{r}$ is the radical of Λ and ℓ denotes the length as Λ-module. Then there exists a preprojective module A' with $\ell(A'/\underline{r}A') = n' \geq n$ such that $f : B \to A'$ is injective whenever f is an irreducible morphism and B is an indecomposable Λ-module.

Proof: Since A is preprojective, A is in $\underline{\underline{P}}_m$ for some $m < \infty$ where $\underline{\underline{P}}_0, \underline{\underline{P}}_1, \ldots, \underline{\underline{P}}_\infty$ denotes the preprojective partition of ind Λ. Now choose m' minimal with the property that there exists an A' in $\underline{\underline{P}}_{m'}$, with $\ell(A'/\underline{r}A') \geq n \geq 2$. Let $f : B \to A'$ be an irreducible morphism with B an indecomposable module in mod Λ. Assume that f is a surjective morphism. Then B is in $\bigcup_{i<m'} \underline{\underline{P}}_i$ since any surjective morphism $X \to A$ with X a direct sum of copies from $\bigcup_{i=m'}^{\infty} \underline{\underline{P}}_i$ would be a splittable surjection. Therefore by minimal choice of m', $\ell(B/\underline{r}B) < n$. From [3] we have that any irreducible morphism is either injective or surjective, so f has to be injective, which completes the proof of the proposition.

We will need the following result. (See [4] Proposition 5.11.)

LEMMA 2.3. Let Λ be an artin algebra of infinite representation type. Then there is no bound on the length of the preprojective modules in ind Λ and hence no bound on the length of $A/\underline{r}A$ for preprojective modules A in ind Λ.

As a direct consequence of Proposition 2.2 and Lemma 2.3 we get a result for artin algebras of infinite representation type which we leave to the reader to prove.

COROLLARY 2.4. Let Λ be an artin algebra of infinite representation type. Then there is no bound on the length of the preprojective modules A such that $f : B \to A$ is an injective morphism whenever f is irreducible and B is an indecomposable Λ-module.

We will now end this discussion about when irreducible morphisms are injective by a result characterizing those artin algebras where the question raised before Proportition 2.2 is answered in the affirmative. The proof of this is easy using Proposition 2.2 and Lemma 2.3 and is left to the reader.

PROPOSITION 2.5. For an artin algebra Λ, the following are equivalent.

i) All modules in ind Λ have simple top i.e. M in ind Λ implies $M/\underline{r}M$ is simple.

ii) If M is a nonprojective module in ind Λ then there exists an indecomposable module A and a surjective irreducible morphism $f : A \to M$.

By dualizing using preinjective modules one obtains dual results.

3. Characterization of artin algebras of finite representation type.

In this section we are going to study how the interaction between the preprojective and the preinjective modules in ind Λ, for an artin algebra Λ, determines the representation type of Λ. Our aim is to give some conditions on the intersection between the set of preprojective and the set of preinjective modules which are equivalent to the algebra being of finite representation type. The idea is that one in the algebra Λ can produce a chain of ideals $0 \subseteq \underline{\underline{a}}_0 \subseteq \underline{\underline{a}}_1 \subseteq \underline{\underline{a}}_2 \subseteq \cdots\cdots$ associated with the preinjective partition such that all but possibly a finite number of indecomposable $\Lambda/\underline{\underline{a}}_i$ - modules are annihilated by the ideal $\underline{\underline{a}}_{i+1}/\underline{\underline{a}}_i$. Then, under certain assumptions, the chain of ideals exhaust the whole algebra, in which case the algebra has to be of finite representation type. In general, the ascending chain of ideals becomes stable say at $\underline{\underline{a}}_i$ when ind $\Lambda/\underline{\underline{a}}_i$ does not contain any projective modules which are preinjective.

We start out by a theorem giving different characterizations of the preprojective modules some of which are independent of the preprojective partition.

<u>THEOREM 3.1</u>. Let Λ be an artin algebra. Then the following are equivalent for a module A in ind Λ.

i) A is preprojective.

ii) There exists an n such that whenever there is a surjective morphism $f : B \to A$ for some module B, B contains a summand from $\bigcup_{i<n} \underline{\underline{P}}_i$.

iii) There exists a finite set $\underline{\underline{A}}$ of ind Λ such that if there is a surjective morphism $f : B \to A$ for some module B then B contains a summand from $\underline{\underline{A}}$.

iv) There exists a maximal submodule M of A such that $\tau_B(A) \subset M$ for all but a finite number of modules B in ind Λ where $\tau_B(A)$, the trace of B in A, is the submodule of A generated by $\{\operatorname{Im} f \mid f \in \operatorname{Hom}(B,A)\}$.

v) There exists a simple module S and a morphism $f : A \to S$ such that $\operatorname{Im}(B,f) = 0$ for all but a finite number of modules B in ind Λ.

vi) $\bigcap_{i<\infty} \tau_{\underline{\underline{P}}_i}(A) \neq A$ where

$$\tau_{\underline{\underline{P}}_i}(A) = \sum_{B \in \underline{\underline{P}}_i} \tau_B(A) .$$

vii) $\tau_{\underline{\underline{P}}_\infty}(A) \neq A$.

Proof: See [4] Theorem 5.1.

The main theorem we want to prove in this section is the following, which is analogous to Theorem 6.1 in [4] .

THEOREM 3.2. For an artin algebra Λ the following statements are equivalent

i) Λ is of finite representation type

ii) All indecomposable modules are preinjective

ii') All indecomposable modules are preprojective

iii) All preprojective modules are preinjective

iii') All preinjective modules are preprojective

iv) All preprojective modules with simple top (i.e. all preprojective modules A with $A/\underline{r}A$ simple, $\underline{r}$ = Rad Λ) are preinjective iv') All preinjective modules with simple socle are preprojective

Proof: The proof of the theorem will go as follows: i) → ii) → iii) → iv) → i). The proof of the rest is then just the dual cycle. The only hard part here is iv) → i), which will be done after we have proven some intermediate results.

i) → ii) is trivial since ind Λ is finite

ii) → iii) is also trivial since if every indecomposable module is preinjective, then certainly every preprojective module is preinjective

iii) → iv) is also a triviality

Remembering the preinjective partition $\underline{\underline{I}}_0$, $\underline{\underline{I}}_1$, $\underline{\underline{I}}_2$, ..., of ind Λ we want to associate some ideals in Λ to this partition. For a module M we have that the annihilator of M, $\text{ann}M = \bigcap_{f:\Lambda\to M} \text{Ker}f$. In a similar way we may define the annihilator of $\underline{\underline{I}}_i$, $\underline{\underline{a}}_i = \text{ann}\underline{\underline{I}}_i = \bigcap_{M\in\underline{\underline{I}}_i} \text{ann}M$ or more generally we can define the annihilator of any subcategory $\underline{\underline{C}}$ of ind Λ by $\text{ann}\underline{\underline{C}} = \bigcap_{M\in\underline{\underline{C}}} \text{ann}M$.

Since Λ is artin and therefore satisfies the descending chain condition, we know that it is enough to look at finite intersections. Let $\underline{\underline{C}}$ be a full subcategory of ind Λ. $\underline{\underline{c}} = \text{ann}\underline{\underline{C}} \subseteq \Lambda$ is of course the minimal ideal such that $\Lambda/\underline{\underline{c}}$ can be embedded in a direct sum of copies of modules from $\underline{\underline{C}}$.

Let $\underline{\underline{a}}_i = \text{ann}\underline{\underline{I}}_i$. Since $\underline{\underline{a}}_o = \text{ann}\underline{\underline{I}}_o$ and $\underline{\underline{I}}_o$ consists of the injective modules in ind Λ we know that $\underline{\underline{a}}_o = 0$. The ideals $\underline{\underline{a}}_i$ are related by inclusion.

PROPOSITION 3.3. Let $\underline{\underline{a}}_i$ be defined as above. Then $\underline{\underline{a}}_i \subseteq \underline{\underline{a}}_{i+1}$.

Proof: We only have to show that $\Lambda/\underline{\underline{a}}_{i+1}$ can be embedded in a sum of copies of modules from $\underline{\underline{I}}_i$. By the definition of $\underline{\underline{a}}_{i+1}$ we know that $\Lambda/\underline{\underline{a}}_{i+1}$ can be embedded in a direct sum of copies of modules from $\underline{\underline{I}}_{i+1}$, but every module in $\underline{\underline{I}}_{i+1}$ can be embedded in a direct sum of copies of modules from $\underline{\underline{I}}_i$, hence also $\Lambda/\underline{\underline{a}}_{i+1}$; i.e. $\underline{\underline{a}}_{i+1} \supseteq \underline{\underline{a}}_i$.

So we have gotten the following chain of ideals in Λ

$$0 = \underline{\underline{a}}_o \subseteq \underline{\underline{a}}_1 \subseteq \cdots \subseteq \underline{\underline{a}}_n \subseteq \cdots$$

Now the question is, when do we get a proper inclusion ? Proposition 3.4 gives an answer to this.

PROPOSITION 3.4. Let the notation be as above. Then $\underline{\underline{a}}_n \neq \underline{\underline{a}}_{n+1}$ if and only if $\underline{\underline{I}}_n$ contains a summand of $\Lambda/\underline{\underline{a}}_n$.

Proof: Assume first that $\underline{\underline{a}}_n \neq \underline{\underline{a}}_{n+1}$. Then $\Lambda/\underline{\underline{a}}_n$ can be embedded in a direct sum of copies of modules from $\underline{\underline{I}}_n$, but not in a direct sum of copies of modules from $\underline{\underline{I}}_{n+1}$. The first of these observations shows that $\Lambda/\underline{\underline{a}}_n$ has no indecomposable direct summand in $\bigcup_{i<n} \underline{\underline{I}}_i$. The

second one implies that $\Lambda/\underline{\underline{a}}_n$ has a direct summand from $\bigcup_{i\leq n} \underline{\underline{I}}_i$ i.e. $\Lambda/\underline{\underline{a}}_n$ has a direct summand from $\underline{\underline{I}}_n$.

The other implication is as easy as the first one and is left to the reader.

As a corollary of this we get:

COROLLARY 3.5. $\underline{\underline{a}}_i = 0$ for all i if and only if no indecomposable projective Λ-module is preinjective.

Since $\underline{\underline{I}}_n$ is a cocover for ind $\Lambda - \bigcup_{i<n} \underline{\underline{I}}_i$ we also get that $\text{ann}\underline{\underline{I}}_n = \text{ann} \bigcup_{i=n}^{\infty} \underline{\underline{I}}_i$. So in fact all but a finite number of the indecomposable modules in mod Λ are annihilated by $\underline{\underline{a}}_i$; i.e. by passing from Λ to $\Lambda/\underline{\underline{a}}_i$ we only lose a finite set of indecomposable modules which all are preinjective over Λ.

We are now in a positon where we can complete the proof of Theorem 3.2 i.e. prove that iv) → i). As follows from the discussion above, we only lose a finite number of nonisomorphic indecomposable modules by passing from Λ to $\Lambda/\underline{\underline{a}}_i$; i.e. we need to prove that if iv) is satisfied, the ascending chain $\underline{\underline{a}}_0 \subseteq \underline{\underline{a}}_1 \subseteq \dots \subseteq \underline{\underline{a}}_n \subseteq$ of ideals does not become stable before we have reached the whole ring.

To prove this, assume that $\underline{\underline{a}}_n = \underline{\underline{a}}_i$ for all $n > i$ but $\underline{\underline{a}}_i \neq \Lambda$. Look at $\Lambda/\underline{\underline{a}}_i$ and decompose it into indecomposable summands A_i . The finite set $\underline{\underline{A}}$ consisting of nonisomorphic indecomposable summands of $\Lambda/\underline{\underline{a}}_i$ is clearly a minimal cover for ind $\Lambda - \bigcup_{i<i} \underline{\underline{I}}_j$. Since $\bigcup_{j=i+1}^{\infty} \underline{\underline{I}}_j$ and $\bigcup_{j=0}^{\infty} \underline{\underline{I}}_j$ only differ in a finite set of isomorphism classes, every module which is preprojective over $\Lambda/\underline{\underline{a}}_i$ is preprojective over Λ, hence $\Lambda/\underline{\underline{a}}_i$ is itself a sum of preprojective modules over Λ which in addition have simple top. But then by assumption $\Lambda/\underline{\underline{a}}_i$ contains a preinjective module as a summand. Also, since $\Lambda/\underline{\underline{a}}_i$ can be embedded in a direct sum of copies of modules from $\underline{\underline{I}}_i$, the preinjective summands of $\Lambda/\underline{\underline{a}}_i$ have to appear in some $\underline{\underline{I}}_n$ $n \geq i$ that is $\underline{\underline{a}}_n \neq \underline{\underline{a}}_{n+1}$ for an $n \geq i$. This is a contradition. This completes the proof that iv) → i) as well as the proof of the theorem.

An open question in connection with the characterization of algebras of finite representation type in Theorem 3.2 is the following: Let Λ be an artin algebra and assume that all indecomposable Λ-modules are either preprojective or preinjective. Is then Λ necessary of finite representation type ? This problem was first raised by C. M. Ringel.

4. Further properties of preprojective partitions.

Let Λ be an artin algebra of finite representation type and let $\underline{\underline{P}}_o$, $\underline{\underline{P}}_1$,..., $\underline{\underline{P}}_n$ be the preprojective partition of ind Λ such that $\underline{\underline{P}}_n \neq \emptyset$ and $\underline{\underline{P}}_{n+1} = \emptyset$. In Proposition 1.5 a description of $\underline{\underline{P}}_o$ and $\underline{\underline{P}}_1$ was given. We will now give a partial description of $\underline{\underline{P}}_n$ and $\underline{\underline{P}}_{n-1}$. In addition to this, a lower bound on n is given in terms of the length of the projective modules. Finally, we want to discuss certain types of decompositions of any finitely generated module over any artin algebra Λ .

PROPOSITION 4.1. Let Λ be an artin algebra of finite representation type and $\underline{\underline{P}}_0, \underline{\underline{P}}_1, \ldots, \underline{\underline{P}}_n$ the preprojective partition of ind Λ where $\underline{\underline{P}}_n \neq \emptyset$ and $\underline{\underline{P}}_{n+1} = \emptyset$. Then

i) all modules in $\underline{\underline{P}}_n$ are simple.

ii) if M is a module in $\underline{\underline{P}}_{n-1}$ then either M is simple or M is of Loewy length two with $\underline{r}M = \text{Soc } M$ a simple module. Here $\underline{r}$ is the radical of Λ and Soc M denotes the socle of M.

Proof:

i) This is a triviality and left to the reader.

ii) Since $\underline{\underline{P}}_{n-1}$ is a minimal cover for $\underline{\underline{P}}_{n-1} \cup \underline{\underline{P}}_n$ we have that all proper quotients of a module in $\underline{\underline{P}}_{n-1}$ are either a direct sum of modules in $\underline{\underline{P}}_n$ or zero. Hence all proper quotients of a module in $\underline{\underline{P}}_{n-1}$ are either semisimple or zero. That means that if a module M in ind Λ is in $\underline{\underline{P}}_{n-1}$, then either M is simple or M is of Loewy length two with all proper quotients semisimple. Is is easy to see that the latter is equivalent to M being of Loewy length two with $\underline{r}M = \text{Soc } M$, a simple Λ-module.

We now want to give a lower bound on the number of classes in a preprojective partition of ind Λ

when Λ is an artin algebra of finite representation type.

PROPOSITION 4.2. Let Λ be an artin algebra of finite representation type and let $\underline{\underline{P}}_0, \underline{\underline{P}}_1, \underline{\underline{P}}_2, \ldots, \underline{\underline{P}}_n$ be the preprojective partition of ind Λ. Then $\underline{\underline{P}}_i \neq \emptyset$ if $i < \max \{\ell(Q) \mid Q$ a projective module in ind $\Lambda\}$. Here ℓ denotes the length of a module.

Proof: Assume $i < \max \{\ell(Q) \mid Q$ a projective module in ind $\Lambda\}$. Let Q be a projective module such that $\ell(Q) > i$. Then there exists a chain of modules $\{M_i\}_{i=0,\ldots,\ell(Q)-1}$ in ind Λ and proper surjective morphisms $f_i : M_i \to M_{i+1}$. It follows directly by definition of minimal cover that $M_k \in \bigcup_{j=k}^{\infty} \underline{\underline{P}}_j$ and hence $\underline{\underline{P}}_i \neq \emptyset$ when $i \leq \ell(Q) - 1$.

One may dualize the two last results formally and obtain similar results for the preinjective partition of ind Λ when Λ is a ring of finite representation type.

Some interesting questions in connection with the last proposition are the following:

QUESTION 1. Is there an upper bound on the number of classes in a preprojective and a preinjective partition of ind Λ depending only of the length of Λ and Λ^{op} for an artin algebra of finite representation type ?

QUESTION 2. What is the connection between $\underline{\underline{P}}_n$ and $\underline{\underline{I}}_n$ for an artin algebra Λ where $\underline{\underline{I}}_0, \underline{\underline{I}}_1, \ldots$ is the preinjective partition of ind Λ. For instance, if Λ is of finite representation type in then $\underline{\underline{P}}_n = \emptyset$ if and only of $\underline{\underline{I}}_n = \emptyset$?

In the end of the paper we now want to look at decompositons of any finitely generated module over an artin algebra Λ. As a starting point for this assume Λ is an artin algebra of finite representation type. Let $\underline{\underline{P}}_0, \underline{\underline{P}}_1, \ldots, \underline{\underline{P}}_n$ be the preprojective partition of ind Λ. If we let M be the sum of one copy of each of the indecomposable Λ-modules we may write $M = \amalg P_i$ where each P_i is a direct sum of one copy of each of the modules in $\underline{\underline{P}}_i$. Since the preprojective partition of ind Λ is unique, this decomposition of M as a direct sum of modules is also unique up to an isomorphism. Further, each of the modules P_i has the following property:

$$\tau_{P_i'}(P_i) \neq P_i \quad \text{for all proper summands } P_i' \text{ of } P_i .$$

DEFINITION: For an arbitrary module M we will say that M is a covering indecomposable module if $\tau_{M'}(M) \neq M$ for all proper summands M' of M.

Observe that a covering indecomposable module does not contain two copies of the same indecomposable module in a direct sum decomposition into indecomposable modules.

DEFINITION: Let M be any module. A decomposition $M = M_1 \amalg M_2 \amalg \ldots$ of M where each M_i is covering indecomposable and $\tau_{M_i}(M_{i+1}) = M_{i+1}$ is called a cover decomposition of M.

We then have the following theorem:

THEOREM 4.3. Let Λ be any artin algebra and let M be a finitely generated Λ-module. Then M has a unique up to isomorphism cover decomposition, i.e. if $M = M_1 \amalg \ldots \amalg M_n$ and $M = M_1' \amalg \ldots \amalg M_m'$ are cover decompositions of M, then $n = m$ and $M_i \simeq M'_i$.

Proof: See [4] Proposition 2.6.

This is really an analogue of the invariant factor decomposition of a finite abelian group. To see this let M be a finite abelian group, let $|M|$ denote the order of M and let $M = M_1 \amalg M_2 \amalg \ldots \amalg M_n$ be the invariant factor decomposition. Then $|M_i| \,\big|\, |M_{i-1}|$ and each M_i does not contain two different indecomposable modules whose order is divisible by the same prime number p. But that means simply that each M_i is covering indecomposable and that $\tau_{M_i}(M_{i+1}) = M_{i+1}$.

We may dualize the concept of covering decomposition in the following way.

DEFINITION: For an arbitrary module M we will say that M is cocover indecomposable if $\cap \{\mathrm{Ker}\, f \mid f \in \mathrm{Hom}(M,M')\} \neq 0$ for all proper summands M' of M.

Observe that a cocover indecomposable module M does not contain two copies of the same module in a direct sum decomposition.

DEFINITION: Let M be a module. A decomposition $M = M_1 \amalg M_2 \amalg \dots$ of M is called a cocover decomposition of M if each M_i is cocover indecomposable and $\cap \{\mathrm{Ker}\, f \mid f \in \mathrm{Hom}(M_{i+1}, M_i)\} = 0$.

We now have the following existence and uniqueness theorem.

THEOREM 4.4. Let Λ be any artin algebra and let M be a finitely generated Λ-module. Then M has a unique up to isomorphism cocover decomposition.

Proof: See [4] Proposition 2.6.

This is also an analogue of the invariant factor decomposition. For abelian groups the cover and cocover decomposition coincide but that does not happen in general as the following example shows. Let Λ be a basic nonsemisimple hereditary artin algebra. Then as a Λ-module, Λ is cover indecomposable but not cocover indecomposable.

REFERENCES

[1] ALPERIN, J.L.: A preprojective module which is not strongly preprojective. To appear.

[2] AUSLANDER, M., REITEN, I.: Representation theory of algebras III. Almost split sequences. Comm. in Algebra. Vol. 3 (1975), 239-294.

[3] AUSLANDER, M., REITEN, I.: Representation theory of artin algebras IV. Invariants given by almost split sequences. Comm. in Algebra. Vol. 5 (1977), 441-518.

[4] AUSLANDER, M., SMALØ, S.O.: Preprojective modules over artin algebras. To appear in Journal of Algebra.

[5] DLAB, V., RINGEL, C.M.: The representation theory of tame hereditary artin algebras. Representation theory of algebras. (Proc. Conf. Temple University, Philadelphia, PA, 1976). Lecture Notes in Pure and Applied Math. Vol. 37, M. Dekker, N.Y., 1978.

[6] GABRIEL, P.: Indecomposable Representations II. Symposia Mathematica, Vol. XI. Academic Press, London (1973), 81-104.

[7] PLATZECK, M.I.: Representation theory of algebras stably equivalent to hereditary artin algebras. Transactions of the A.M.S. 238 (1978), 89-128.

[8] ROITER, A.V.: Unboundness of the dimension of the indecomposable representations of an algebra which has infinitely many indecomposable representations. Izv. Akad. SSSR. Ser. Mat. 32 (1968), 1275-1282.

Brandeis University
Department of Mathematics
Waltham, MA 02154
U.S.A.

Universitetet i Trondheim, NLHT
Matematisk institut
7055 Dragvoll
Norway

SECTIONS IN AUSLANDER-REITEN QUIVERS

R. Bautista

If Λ is an artin algebra, by the Auslander-Reiten quiver Γ_Λ of $\mathrm{mod}(\Lambda)$ we understand the following:

Γ_Λ is an oriented graph (quiver), the points of Γ_Λ are the isomorphism classes $[M]$ for M indecomposable Λ-module, we put an arrow from $[M]$ to $[N]$ if there exists an irreducible map $f: M \to N$. If Γ is any quiver we can introduce the concept of modulation of Γ, in such case we can construct a category $A(\Gamma)$ associated to Γ and the modulation. In the case of artin algebras Λ we can introduce a modulation for Γ_Λ. Thus we have a category $A(\Gamma_\Lambda)$. In some cases (for instance if Λ is an artin algebra on an algebraically closed field) we have a full and dense functor
$G: \mathrm{mod}(\Lambda) \to \mathrm{mod}(\Lambda)/\mathrm{rad}^\infty \mathrm{mod}(\Lambda)$.

Would be good to have a complete description of Ker G, we do not have such description, however we will give some information that is very useful. This information is in terms of certain paths in the Auslander-Reiten graph that we call sectional paths.

In this paper we will study connected subgraphs of Γ_Λ, such that they do not have any non sectional path, we call this subsets subsections.

In general would be good to have a good formula for the composition factors of trDM in terms of those of M. It seems that this is not always possible, however if S is a finite subsection without oriented cycles we will give an aproximation formula for the composition factors of trDM ($M \in S$) in terms of the composition factors of all the M_i in S. Moreover we will have good information of the "error" of the formula. As application we will give another proof of results of Ch. Riedtmann and G. Todorov on Dtr-periodic modules.

In some cases there exists a good subsection called C-section. In such case the above formula is exact. This fact will be basic for to prove that if Λ is ℓ-hereditary (see [3] for definition) of finite representation type, then the indecomposable finitely generated Λ-modules are completely determined by its composition factors. This will be proved in other paper.

1. Modulations.

We will follow [7] in our definitions. Let Γ be a quiver (directed graph) and k a commutative field. By a k-modulation of Γ we understand the following:

For each x point of Γ a division ring K_x with $k \subset$ center K_x and $[K_x : k] < \infty$.

For each arrow γ with start $\alpha(\gamma)$ and end $\beta(\gamma)$ a $K_{\beta(\gamma)} - K_{\alpha(\gamma)}$ bimodule M with k acting centrally on M_γ.

By a path δ on Γ we mean, either a point x of Γ or a finite collection of arrows $\gamma_t, \dots, \gamma_2\ \gamma_1$ with $\beta(\gamma_i) = \alpha(\gamma_{i+1})$ $i=1,\dots,t-1$. We put $\alpha(\gamma) = \alpha(\gamma_1)$, $\beta(\gamma) = (\gamma_t)$. We will put $\delta = \gamma_t, \dots, \gamma_2\gamma_1$. If δ_1 and δ_2 are paths on Γ and $\beta(\delta_1) = \alpha(\delta_2)$ we define in the obvious way $\delta_2\delta_1$.

Now, if $\delta = \gamma_t, \dots, \gamma_2\gamma_1$ we define M_δ by:

$$M_\delta = M_{\gamma_t} \otimes_{K_{\alpha(\gamma_t)}}, \dots, \otimes M_{\gamma_2} \otimes_{K'_{\rho(\gamma_1)}} M_{\gamma_1}$$

Obviously M_δ is a $K_{\beta(\delta)} - K_{\alpha(\delta)}$ bimodule. Now we introduce a category $I(\Gamma)$ associated to the k-modulation.

Objects of $I(\Gamma)$ = points of Γ.

$$\mathrm{Hom}(x,y) = \bigoplus_{\substack{\delta \text{ path on } \Gamma \\ \alpha(\delta) = x \\ \beta(\delta) = y}} M$$

If $\delta = \delta_2\delta_1$ with δ_1 and δ_2 paths in Γ we have a map $M_{\delta_2} \times M_{\delta_1} \to M_{\delta_2\delta_1}$, and these maps induce a map

$$\mathrm{Hom}(y,z) \times \mathrm{Hom}(x,y) \to \mathrm{Hom}(x,z)$$

Now is not difficult to see that $I(\Gamma)$ is preadditive category. By $I(\Gamma)$ we denote the additive category generated by $I(\Gamma)$. See [1].

Notation. If $u \in \mathrm{Hom}(x,y)$, as we know

$$u \in \bigoplus_{\substack{\delta \\ \alpha(\delta)=x \\ \beta(\delta)=y}} M_\delta .$$ Then u is written in unique way as

$$u = \sum_\gamma a_\gamma \qquad a_\gamma \in M_\gamma .$$

We will put $\mathrm{Supp}(u) = \{\gamma \mid \alpha(\delta) = x, \beta(\delta)=y \text{ and } a_\gamma \neq 0\}$.

We will put $\mathrm{gr}\ a_\gamma = \text{length}\ \gamma$, $\mathrm{gr}_m\ u = \min\{\mathrm{gr}\ a_\gamma \mid a_\gamma \neq 0\}$ $\mathrm{gr}_M\ u = \max\{\mathrm{gr}\ a_\gamma \mid a_\delta \neq 0\}$.

Let Λ be an indecomposable basic artin algebra.

Let R be the center of Λ, R is artinian and Λ indecomposable, so R is local and $R/\mathrm{rad}\ R = k$ is a field.

Let Γ_Λ be the Auslander-Reiten quiver of Λ. We have a k-modulation for Γ_Λ given in the following way:

If $x = [M]$ is a point of Γ_Λ we associate to x the division ring $K_x = \mathrm{End}(M)/\mathrm{rad}\ \mathrm{End}(M)$ (we choose some $M \in [M]$). If γ is an arrow from $x = [M]$ to $y = [N]$ we put $M_\gamma = \mathrm{rad}/\mathrm{rad}^2(M,N)$. Obviously M_γ is a $K_y - K_x$ bimodule, and k acts centrally on M_γ.

Consider the following property of artin algebras Λ.

A) There exists a full functor:

$$G\colon I(\Gamma_\Lambda) \quad \mathrm{mod}(\Lambda)/\mathrm{rad}^\infty \mathrm{mod}(\Lambda)$$

such that if N is indecomposable in $\mathrm{mod}(\Lambda)$, there exists $x \in \mathrm{Obj}(I(\Gamma_\Lambda))$ with $G(x) \simeq N$, moreover if γ is an arrow $G|M_\gamma\colon M_\gamma \to \mathrm{Hom}(X,Y)$ is mono, and induces an iso $M_\gamma \to \mathrm{Hom}(X,Y) \to \mathrm{Hom}(X,Y)/\mathrm{rad}^2(X,Y)$ as $K_Y - K_X$ bimodules.

Remark. $G\colon \mathcal{I}(\Gamma_\Lambda) \to \mathrm{mod}(\Lambda)/\mathrm{rad}^\infty$ induces a functor $G\colon \mathcal{A}(\Gamma_\Lambda) \to \mathrm{mod}(\Lambda)/\mathrm{rad}^\infty$ and G is full and dense.

We recall that if $G\colon\ C_1 \to C_2$ is an additive functor, then $\mathrm{Ker}\, G(C_1,C_2) = \{f\colon C_1 \to C_2 \mid G(f) = 0\}$; $\mathrm{Ker}\, G$ is an ideal of C_1.

PROPOSITION 1.1. Assume that in $\mathrm{mod}(\Lambda)$ there is no diagrams of the form:

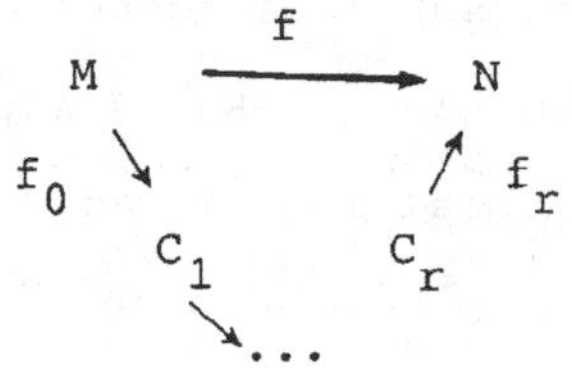

with M, N, C_i , indecomposables and f, f_i irreducible maps, then Λ has A) condition.

Proof. As in [7], (see [4]).

PROPOSITION 1.2. Let Λ be any artin algebra over an algebraically closed field k. Then Λ has A) condition.

Proof. See [4].

In all that follows of this paper we will assume that Λ has A) condition. Therefore we have a functor

$$G\colon \mathcal{A}(\Gamma_\Lambda) \to \mathrm{mod}(\Lambda)/\mathrm{rad}^\infty.$$

Our aim is give information about $\mathrm{Ker}\, G$.

DEFINITION 1.3. Let γ be a path from $[X]$ to $[Y]$ in

Γ_Λ of lenght ≥ 2, γ is called an a-path if there exists an arrow from [X] to [Y].

PROPOSITION 1.4. Let Λ be any artin algebra with conditon A), so we have a full and dense functor:

$$G: A(\Gamma_\Lambda) \to \mathrm{mod}(\Lambda)/\mathrm{rad}^\infty$$

Let f: [A] → [B] be a morphism in $A(\Gamma_\Lambda)$ such that G(f) = 0. Then if γ Supp(f) and γ does not contain any a-path, γ can be factorized through some path of the form:

$$[C] \to [L] \to [\mathrm{trDC}]$$

Proof. We will use induction on length γ. Here $f \in \mathrm{Ker}\, G$, length $\gamma \geq 2$. Then $f = \Phi g$.

With

$$0 \to \mathrm{DtrB} \xrightarrow{\Delta'} B_0 \xrightarrow{G(\Phi)=\Phi'} B \to 0$$

the almost split sequence for B.

Here $\Phi: [B_0] = \oplus n_i[B_i] \to [B]$ $\Phi = (\underline{t}_1, \underline{t}_2, \ldots, \underline{t}_s)$
$\underline{t}_j: n_j[B_j] \to [B]$ given by $\underline{t}_j = (t_{j,1}, \ldots, t_{j,n_j})$
$\mathrm{Supp}(t_{ji})$ = one arrow.

We put $G(\Phi) = (G(\underline{t}_1), \ldots, G(\underline{t}_s))$ $G(\underline{t}_j) =$
$= (G(t_{j,1}), \ldots, G(t_{j,n_j}))$.

We will put $\Phi' = G(\Phi)$ $\underline{t}_i' = G(\underline{t}_i)$ $t'_{j,i} = G(t_{j,i})$
$g' = G(g)$. We have that $\Phi' g' \in \mathrm{rad}^\infty$.

$$\therefore \quad \Phi' g' = u \in \mathrm{rad}^\infty.$$

From here using properties of almost split sequences we get:

$$\Phi' g' = \text{'u'} \qquad u' \in \text{rad}^{\infty}$$

$$\therefore \quad \Phi'(g'-u') = 0$$

Here $g' \equiv g' - u' \bmod \text{rad}^{\infty}$. We can assume $\Phi' g' = 0$.

Let Δ be such that $G(\Delta) = \Delta'$. Then using the fact that $\Phi' g' = 0$ we have that there exists $\sigma': A \to DtrB$ doing commutative the diagram:

$$\begin{array}{cccccccccc} & & & & & A & & & & \\ & & & \sigma' \swarrow & & & \searrow g' & & & \\ \text{(I)} & 0 & \to & DtrB & & \xrightarrow[\Phi']{} & B_0 & \xrightarrow[\Delta']{} & B & \to 0 \end{array}$$

Assume now that length $\gamma = 2$

$$g: \; A \to \bigoplus_{i=1}^{t} X_i = [B_0] \qquad g = \begin{pmatrix} g_1 \\ \vdots \\ g_t \end{pmatrix}$$

Here $\text{Supp}\, \Phi g \subsetneq \text{Supp}\, \Phi \cup \text{Supp}\, g = \{\gamma_1\gamma_2 \mid \gamma_1 \in \text{Supp}\, \Phi, \gamma_2 \in \text{Supp}\, g)$ and we know that length $\gamma = 2$.

There exists $\gamma_i \in \text{Supp}\, g$ and $\gamma_i' \in \text{Supp}\, \Phi$ with $\gamma = \gamma_i'\gamma_i$. But lenght $\gamma = 2$ $\therefore$ length $\gamma_i = 1$. Therefore there exists $g_i: [A] \to [X_i]$ with $\text{Supp}\, g_i = \gamma_i$ one arrow $\therefore$ g_i is irreducible, hence g_i' is irreducible.

Putting $\Delta' = \begin{pmatrix} d_1' \\ \vdots \\ d_t' \end{pmatrix}$ from $\Delta'\sigma' = g'$ we get

$g'_j = d'_j\, \sigma'$ in particular $g'_i = d'_i \sigma'$, therefore σ' is iso $\therefore$ $A \approx DtrB$ and we get our proposition.

Assume now the proposition proved for length $\gamma < n$, we will prove it for length $\gamma = n$.

Let σ be such that $G(\sigma) = \sigma'$. We have $g = \Delta\sigma \in \operatorname{Ker} G$.

Put $B_0 = \bigoplus_{i=1}^{s} n_i B_i = \bigoplus_{j=1}^{t} X_j$ B_i indecomposable $B_i \neq B_j$ if $i \neq j$, X_j indecomposable for $j = 1,\dots,t$.

Assume $f = \Sigma a_\gamma$ $a_\gamma \in M_\gamma$ $a_\gamma \neq 0$.

Let γ_0 be a path in $\operatorname{Supp}(f)$, γ_0 non containing any a-path and length $\gamma_0 = n$.

$$a_{j_0} = \sum_{u=1}^{n_i} C_{\rho_i,u} \cdot a'_{\gamma_i,u}$$

with $a'_{\rho_i,u} \neq 0$, γ_i a path from $[A]$ to $[B_i]$ and $a'_{\gamma_i,u} \in M_{\gamma_i}$ $u = 1,\dots,n_i$ ρ_i is a path from $[B_i]$ to $[B]$ of length one, $C_{\rho_i,u} \in M_{\rho_i}$ $u = 1,\dots,n_i$; $\gamma_0 = \gamma_i \rho_i$.

Let Δ be such that $G(\Delta) = \Delta'$. Then taking the decomposition $[B_0] = \bigoplus_{j=1}^{t} [X_j]$ Δ can be written in matrix form as follows:

$$\text{(II)} \qquad \Delta = \begin{pmatrix} d_1+r_1 \\ \vdots \\ d_t+r_t \end{pmatrix} = \begin{pmatrix} d_1 \\ \vdots \\ d_t \end{pmatrix} + \begin{pmatrix} r_1 \\ \vdots \\ r_t \end{pmatrix} = \Delta_0 + \underline{r}$$

with $\operatorname{Supp}(d_i)$ = one arrow, $gr_m\, r_i > 1$, clearly if $\tau \in \operatorname{Supp} \gamma_i$ τ is an a-path.

As before, putting $g' = G(g)$ we have the existence of $\sigma': A \to DtrB$ doing commutative the diagram (I).

Let $\sigma: [A] \to [DtrB]$ be such that $G(\sigma) = \sigma'$ $\therefore$ $(g - \Delta\sigma) \in \text{Ker } G$

$$g - \Delta\sigma: [A] \to \bigoplus_{i=1}^{s} n_i\,[B_i]$$

Put $(g - \Delta\sigma)_i: [A] \to n_i[B_i]$, $g - \Delta\sigma$ followed by the i-th projection on $n_i[B_i]$ and $(g - \Delta\sigma)_{i,v} = p_{iv}(g - \Delta\sigma)_i$, $p_{i,v}$ the v-th projection of $n_i[B_i]$ on $[B_i]$ $v = 1,\ldots,n_i$.

We have that $(g - \Delta\sigma)_{i,v} \in \text{Ker } G$
$(g - \Delta\sigma)_{i,v}: [A] \to [B_i]$. We know that $\gamma_0 = \gamma_i\rho_i$ with $\gamma_i \in \text{Supp}(g)$; ρ_i a path from $[B_i]$ to $[B]$ of length one. We have length $\gamma_i = n-1$.

Here γ_0 does not contain any a-path, so the same is true for γ_i.

If $\gamma_i \in \text{Supp}(g - \Delta\sigma)_{i,u}$, then by induction hypothesis γ_i can be factorized as in the conclusion, thus γ can be factorized as we want.

Assume now that $\gamma_i \notin \text{Supp}(g - \Delta\sigma)_{i,u}$.

Using (II) we have $\Delta = \Delta_0 + \underline{r}$

$(\Delta\sigma)_{i,u} = (\Delta_0\sigma)_{i,u} + (\underline{r}\sigma)_{i,u}$

$\sigma: [A] \to [DtrB]$

$\sigma = \sum_j h_{\rho'_j}$ $h_{\rho'_j} \in M_{\rho'_j}$ ρ'_j path from $[A]$ to $[DtrB]$

$(\Delta\sigma)_{i,u} = \sum_j d_{iu}\, h_{\rho'_j} + \sum_j r_{i,u}\, h_{\rho'_j}$

with $\text{Supp}(d_{i,u})$ = one arrow $gr_m\, r_{i,u} > 1$.

$(g - \Delta\sigma)_{i,u} = g_{i,u} - (\Delta\sigma)_{i,u}$

$g_{i,u} = \sum_{j'} a'_{j',u} \; \gamma'$ path from $[A]$ to $[B_i]$

$a'_{j',u} \in M_{j'}$.

Here $\gamma_i \notin \mathrm{Supp}\,(g - \Delta\sigma)_{i,u}$ but $\gamma_i \in \mathrm{Supp}\, g$, so $\gamma_i \in \mathrm{Sup}(\Delta\sigma)_{i,u}$. We know that

$$\mathrm{Supp}(\Delta\sigma)_{i,u} \subset \{\rho'_j\delta_i \;,\; \rho'_j\tau_e \;,\; \tau_e \in \mathrm{Supp}(\gamma_i)$$

$\mathrm{length}(\delta_i) = 1$, as we noted before the τ_e are a-paths, therefore $\gamma_i = \rho'_j\delta_i$, consequently $\gamma_0 = \rho'_j\delta_i\rho_i$ factorizes through

$$[DtrB] \xrightarrow{\delta_i} [B_i] \xrightarrow{\rho_i} [B]$$

From here our proposition follows.

2. Sectional paths.

Consider as before Λ an artin algebra with A) condition. Let Γ_Λ be the Auslander quiver associated to Λ and A_Λ the respective tensor category associated to Γ_Λ.

DEFINITION 2.1. $[M_1] \to [M_2] \to \dots \to [M_s]$ a path in Γ_Λ is called a sectional path if for any $i + 2 \leq s$ $M_{i+2} \not\cong \mathrm{trD}M_i$.

PROPOSITION 2.2. Assume Λ has A) condition. Then if $L \xrightarrow{f} M \xrightarrow{g} L$ are irreducible maps with $\ell(M) > \ell(L)$ then $L \cong \mathrm{trD}L$.

Proof. The path $[L] \to [M] \to [L]$ is not an a-path then if $L \not\cong trDL$ $[L] \to [M] \to [L]$ is a sectional path, so $gf \neq 0$. But $\operatorname{Im} gf \subsetneq L$

$$\operatorname{Im} gf = \bigoplus_{i=1}^{t} X_i \qquad X_i \text{ indecomposables}$$

and $gf = \sum_{i=1}^{t} h_i' h_i$ with $h_i : L \to X_i$, $h_i' : X_i \to L$ h_i and h_i' are not isomorphisms, so they are in rad. By A) condition there exist h_i^o and $h_i'^o$ such that $G(h_i^o) = h_i$ $G(h_i'^o) = h_i'$ and g^o, f^o with $G(g^o) = g$ $G(f^o) = f$. Therefore

$$g^o f^o - \Sigma\, h_i'^o\, h_i^o \in \operatorname{Ker} G$$

We can assume $\operatorname{Supp}(h_i'^o\, h_i^o) = \gamma_i$ a path, moreover it is clear that γ_i factorizes trough some $[X_j]$. Here $\ell(M) > \ell(L) > \ell(X_i)$, so $X_i \not\cong M$, consequently each γ_i is different of the path $[L] \to [M] \to [L]$, therefore $[L] \to [M] \to [L]$ can not be sectional path by proposition 2.2. But this is a contradiction, thus $L \cong trDL$.

DEFINITION 2.3. Let $S \subset \Gamma_\Lambda$ be a connected subgraph of Γ_Λ (no necessarily full subgraph). S is called subsection if any path in S is sectional.

Let $k = R/\operatorname{rad} R$, R = center of Λ. Construct a k-modulation for S. To each $[X]$ corresponds $K_X = \operatorname{End}(X)/\operatorname{rad} \operatorname{End}(X)$.

If $[X]$ and $[Y]$ are in S and $[X] \to [Y]$ is an arrow in S, we put $M_{[X],[X]} = \operatorname{rad}(X,Y)/\operatorname{rad}^2(X,Y)$.

$M_{[X],[Y]}$ is a $K_X^{op} - K_Y^{op}$-bimodule

(K_X = End(X)/rad End(X)).

Moreover k acts centrally on $M_{[X],[Y]}$,

$k \subset$ center of K_X^{op} and K_Y^{op}.

Assume now that S does not contain oriented cycles. Associated to the modulation of S we have a valuation for S:

If $[X] \to [Y]$ is an arrow, we will put

$$d_{X,Y} = \dim_{KX} \text{rad}/\text{rad}^2(X,Y),\ d_{Y,X} = \dim_{KY} \text{rad}/\text{rad}^2(Y,X)$$

Now we will use a slight different notation to that employed in paragraph 1. γ a path in S is i) a vertix [X] of S or a collection or arrows $\rho_1\rho_2,\dots,\rho_t$ with $\alpha(\rho_{i+1}) = \beta(\rho_i)$. We will put $\gamma = \rho_1\rho_2,\dots,\rho_t$. As in paragraph 1:

$$M_\gamma = M_{\rho_1} \otimes_{K_{\alpha(\rho_2)}} M_{\rho_2} \otimes \dots \otimes_{K_{\alpha(\rho_i)}} M_{\rho_t}$$

We define $A(S) = \bigoplus_{\gamma \text{ path in } S} M_\gamma$.

If $a_1 \in M_{\gamma_1}$ and $a_2 \in M_{\gamma_2}$ we will put $a_1a_2 = 0$ if $\beta(\gamma_1) \neq \alpha(\gamma_2)$ and $a_1 \otimes a_2 \in M_{\gamma_1} \otimes M_{\gamma_2} = M_{\gamma_1\gamma_2}$ if $\beta(\gamma_1) = \alpha(\gamma_2)$. Therefore we have that $A(S)$ is a k-hereditary artin algebra.

Assumme $\{S_1,\dots,S_t\}$ are all the vertices of S.

With each S_i we have an $A(S)$-right projective module P_i and a right inyective I_i such that

$P_i/\text{rad } P_i \approx \text{Soc } I_i = S_i$. Where S_i is the simple associated to S_i.

Consider now M any $A(S)$-module, $M_i = (P_i, M)$, M_i is a K_i-module, in particular a k-module.

We will put $\underline{\dim}\ M = (\dim_k M_1, \dim_k M_2, \ldots, \dim_k M_t)$. Observe that if we put $f_i = \dim_k S_i = \dim_k K_i$, then $f_i | \dim_k M_i$.

$\underline{\dim}\ M \in \mathbb{Z}^t$, but moreover as we note before $f_i | \dim_k M_i$. Any vector $\underline{X} \in \mathbb{Z}^t$ with $f_i | x_i$ will be called a m-vector.

It is easy to see that $\underline{X}$ is a m-vector if and only if $\underline{X} = \underline{\dim}\ M$ for some $A(S)$-module.

PROPOSITION 2.4. Assume we have a quiver S with k-modulation S and valuation (d_{ij}). Let $A(S)$ be the corresponding tensor algebra. Then if $S_1, \ldots, S_t$ are all the $A(S)$-simples, we have that there exist a linear transformation

$$\overline{C}: \mathbb{Z}^t \to \mathbb{Z}^t$$

such that $\underline{\dim}\ \text{trDM} = \overline{C}\ \underline{\dim}\ M$.

Moreover we have

$$* \qquad m_i + m_i' = \sum_{\substack{(i,j) \text{ arrow} \\ \text{in } S}} d_{ji}\, m_j + \sum_{\substack{(j',i) \text{ arrow} \\ \text{in } S}} d_{j'i}\, m_j'$$

$$\text{with} \quad \begin{pmatrix} m_1' \\ \vdots \\ m_t' \end{pmatrix} = \bar{C} \begin{pmatrix} m_1 \\ \vdots \\ m_t \end{pmatrix}$$

Proof. By [2] we know that $\bar{C}$ is defined by

$$\bar{C}\,[I_j] = -\,[P_j]$$

with I_j indecomposable injective and P_j indecomposable projective such that $P_j/\mathrm{rad}\,P_j \approx \mathrm{Soc}\,I_j$. Now is not difficult to see that (*) is true for $[I_j]$, then by linearlity the proposition follows. (Here the $[I_j]$ are a basic for $\mathbb{Z}^t \otimes_{\mathbb{Z}} \mathbb{Q}$).

If S_i is the simple corresponding to the point s_i in S, we have that the $A(S)$-right module given by

$$\bigoplus_{\substack{\gamma \text{ path} \\ \text{starting in } i}} M_j \subset A(S) \quad \text{is the projective } P_i.$$

We recall that if $\gamma = \gamma_1\gamma_2,\ldots,\gamma_r$, γ_i arrow, $M_\gamma = M_{\gamma 1} \otimes_{K_{\beta(\gamma_1)}} \cdots \otimes_{K_{\alpha(\gamma_1)}} M_{\gamma_r}$

$$P_{i,u} = (P_u, P_i) = \bigoplus_{\substack{\gamma \\ \alpha(\gamma)=i \\ \beta(\gamma)=u}} M_\gamma$$

We denote by d'_γ the k-dimension of M_γ and by d_{γ_i} (γ_i an arrow) the $K_{\alpha(\gamma_i)}$-dimension of M_{γ_i}.

If γ is the trivial path (i) we have $d'_\gamma = f_i$.

LEMMA 2.5. If $\gamma = \gamma_1\gamma_2,\dots,\gamma_r$ $\alpha(\gamma) = i$, we have

$$d'_\gamma = f_i d_{\gamma_1} d_{\gamma_2},\dots,d_{\gamma_r}$$

Proof. If $r=1$, the lemma is clear. Assume the result proved for $r-1$, then $\dim_k M_{\gamma_1,\dots,\gamma_{r-1}} = f_i d_{\gamma_1},\dots,d_{\gamma_{r-1}}$

$$\dim_k M_\gamma = \dim_k (M_{\gamma_1,\dots,\gamma_{r-1}} \otimes_{K_{\beta(\gamma_{r-1})}} M_{\gamma_r}) =$$

$$= \dim_k (M_{\gamma_1,\dots,\gamma_{r-1}}) \dim_{K_{\beta(\gamma_{r-1})}} M_{\gamma_r}$$

$$= f_i d_{\gamma_1},\dots,d_{\gamma_{r-1}} d_{\gamma_r}.$$

LEMMA 2.6. If P_i is the projective corresponding to the vertex S_i in S, then $\underline{\dim}\, P_i = (x_1,x_2,\dots,x_t)$ with

$$x_u = \sum_{\substack{\gamma \\ \alpha(\gamma)=i \\ \beta(\gamma)=u}} d'_\gamma$$

Proof. It follows from Lemma 2.5 and the remark stated before it.

Consider S a subsection of Γ_Λ. Assume $\{S_1,S_2,\dots,S_t\}$ are the vertices of S. To each point $S_i \in S$ corresponds a module M_i, we have then the collection of modules $M_1,\ldots,M_t$ corresponding to S.

On the other hand, let $S_1,\ldots,S_m$ be all the Λ-simple modules. Assume Λ is an artin algebra over k a field. We will define functions $\ell_i\colon \operatorname{mod}(\Lambda) \to \mathbb{Z}^+$, $i = 1,\ldots,m$, as follows:

$\ell_i(M) = \dim_k(P_i,M)$ where P_i is the projective cover of S_i. We define $\underline{\dim}\, M$ as $(\ell_1(M),\ldots,\ell_m(M))$. Would be good in general to have a formula for $\underline{\dim}$ trDM in terms of $\underline{\dim}$ M. We do not have such forumla, however we will found an approximation formula for all the $\underline{\dim}\,(\mathrm{trDM}_1),\ldots,\underline{\dim}(\mathrm{trDM}_t)$ in terms of all the $\underline{\dim}(M_i)$.

An additive function $\ell\colon \operatorname{mod}(\Lambda) \to \mathbb{Z}^+$ is a function such that $\ell(N_1) + \ell(N_2) = \ell(M)$ if $0 \to N_1 \to M \to N_2 \to 0$ is an exact sequence in mod Λ. An additive function ℓ is called m-function if $\dim_k K_M | \ell(M)$. recall that $k = R/\mathrm{rad}\, R$, R = center of Λ, $K_M = \mathrm{End}(M)/\mathrm{rad}\,\mathrm{End}(M)$. In particular the $\ell_i(M)$ defined before are additive functions. If $\ell(M)$ = R-length of M, ℓ is a m-function.

PROPOSITION 2.7. Assume S is an subsection of Γ_Λ without oriented cycles. Assume moreover that if $M_1,\ldots,M_t$ are the modules in S, all the M_i are not injectives. Let ℓ be any m-additive function $\ell\colon \operatorname{mod}(\Lambda) \to \mathbb{Z}^+$.

Put

$$\underline{m} = \begin{pmatrix} \ell(M_1) \\ \ell(M_2) \\ \vdots \\ \ell(M_t) \end{pmatrix} \qquad \underline{m}' = \begin{pmatrix} \ell(trDM_1) \\ \ell(trDM_2) \\ \vdots \\ \ell(trDM_t) \end{pmatrix}$$

Then, if $\bar{C}: \mathbb{Z}^t \to \mathbb{Z}^t$ is the Coxeter linear transformation associated to A_S as in 2. we have:

$$\underline{m}' = \bar{C}\underline{m} + \underline{q}$$

and there exists a $A(S)$-right projective Q such that $\underline{q} = \underline{\dim}\ Q$.

Proof. Recall that if $i_1 \to i_2$ is an arrow in S $d_{i_1,i_2} = \dim_{K_{i_1}} M_{(i_1,i_2)}$ and $f_i = \dim K_i$. If

$i_1 \to i_2 \to \dots \to i_{s-1} \to i_s$ is a path γ in S, to this path we will associate the number:

$$d_\gamma = d_{(i_1,i_2)},\dots,d_{(i_{s-1},i_s)} \quad \text{and} \quad \gamma \text{ is trivial}$$

$d_\gamma = 1$.

$$d'_\gamma = f_i d_\gamma$$

We will put

$$\begin{pmatrix} m'_{1,c} \\ m'_{2,c} \\ \vdots \\ m'_{t,c} \end{pmatrix} = \bar{C} \begin{pmatrix} m_1 \\ m_2 \\ \vdots \\ m_t \end{pmatrix}$$

We will prove

$$(**)\quad m_i' = m_{i,c}' + \sum_u \sum_{\substack{\gamma \text{ path} \\ \alpha(\gamma)=u \\ \beta(\gamma)-i}} C_{(u)}\, d_\gamma'$$

$C_{(u)}$ non negative integers depending only of u.

By assumption S does not contain oriented cycles; so we can order the points of S putting $i < j$ if there exists a path from i to j. We will prove (**) for some minimal i:

Consider the almost split sequence for M_i:

$$0 \to M_i \to \oplus\, d_{ji}\, X_j \oplus T \to \mathrm{trDm}_i \to 0$$

with X_j such that $M_i \to X_j$ is in S. So we have

$$\ell(M_i) + \ell(\mathrm{trDM}_i) = \sum_{\substack{(i,j) \\ \text{arrow in } S}} d_{ji}\ell(X_j) + \ell(T)$$

Here ℓ is a m-additive function, $f_j | \ell(X_j)$, therefore $f_j d_{ji} | d_{ji}\ \ell(X_j)$, but $f_j d_{ji} = f_i d_{ij}$, consequently $f_i | d_{ji}\ \ell(X_j)$, here $f_i | \ell(M_i)$ and $f_i | \ell(\mathrm{trDM}_i)$. Thus $f_i | \ell(T)$, consequently $\ell(T) = C_{(i)}\ f_i = C_{(i)} d'_{\gamma_0}$ γ_0 the trivial path on i.

$$\therefore\quad m_i + m_i' = \sum_{\substack{(i,j) \\ \text{arrow in } S}} d_{ji} m_j + C_{(i)}\ d_{\gamma_0}$$

Now assume the proposition proved for $i' < i$ and we will prove it for i. Consider as before the almost split sequence associated to M_i:

$$0 \to M_i \to \oplus d_{ji} X_j \oplus T \to trDM_i \to 0$$

with X_j such that $[M_i] \to [X_j]$ is in S. Now, if $[Y] \to [M_i]$ is in S, $[M_i] \to [trDY]$ is not in S, therefore we have:

$$0 \to M_i \to \oplus d_{ji} X_j \oplus \oplus d_{ui} \; trDY_u \oplus T' \to trDM_i \to 0$$

with $[Y_u] \to [M_i]$ in S. We get:

$$\ell(M_i + \ell(trDM_i) = \sum_{\substack{(i,j)\\ \text{arrow in } S}} d_{ji} \ell(X_j) +$$

$$\sum_{\substack{(i,j)\\ \text{arrow in } S}} d_{ui} \; \ell(trDY_u) + \ell(T')$$

$\ell(trDY_u) = m'_u$; $u < i$, by induction:

$$m'_u = m'_{u,c} + \sum_{v} \sum_{\substack{\gamma\\ \alpha(\gamma)=v\\ \beta(\gamma)=u}} C_{(v)} \; d'_\gamma \; C_{(v)} \quad \text{non negative}$$

integer.

$$\therefore \quad m_i + m_i' = \sum_{\substack{(i,j) \\ \text{arrow in} \\ S}} d_{ji}\, m_j + \sum_{\substack{(u,i) \\ \text{arrow in} \\ S}} d_{ui}\, m_{uc}' +$$

$$+ \sum_{v} \sum_{\substack{\gamma \\ \alpha(\gamma)=v \\ \alpha(\gamma)=u}} c_{(v)}\, d_{u,i}\, d_\gamma' + \ell(T')$$

But if $\gamma_1 = u \to i$ and $\gamma_0 = \gamma\gamma_1$ $d'_{\gamma_0} = d_{u,i}\, d'$.

As before $\ell(T') = c_{(i)}\, f_i = c_i\, d_{\rho_0}$ ρ_0 the trivial path on i.

Here,

$$m + m_{i,c}' = \sum_{\substack{(i,j)\ \text{arrow} \\ \text{in}\ S}} d_{ji}\, m_j + \sum_{\substack{(u,i)\ \text{arrow} \\ \text{in}\ S}} d_{u,i}\, m_{u,c}'$$

$$\therefore \quad m_i' = m_{i,c}' + \sum_{v} \sum_{\substack{\gamma \\ \alpha(\gamma)=v \\ \beta(\gamma)=i}} c_{(v)}\, d_\gamma' \ .$$

As an application of proposition 2. we will see a new proof of the following proposition, proved for Ch. Riedtmann in [8], in the case of algebras over algebraically closed fields of finite representation type, and proved by G. Todorov in the general case [9]. The proof that we give here is inspirated in the Todorov's proof.

PROPOSITION 2.8. Assume S is a finite subsection consisting of periodic modules without oriented cycles. Then if $S \subsetneq S_0$, S_0 a subsection, S with his valuation

(d_{ij}) is a Dynkin diagram.

Proof. Assume S is not Dynkin, then S contains an extended Dynkin diagram. We can assume S is extended Dynkin diagram.

Consider $A(S)$ the tensor algebra associated to S. Let $M_1,\dots,M_t$, be the modules forming S.

Put $m_i^{(k)} = \ell(\mathrm{tr}D^k M_i)$ $\quad \ell(M) = R$ - length of M

and $\underline{m}^{(k)} = \begin{pmatrix} m_i^{(k)} \\ \vdots \\ m_t^{(k)} \end{pmatrix}$

For $k = 1$ we have:

$$\underline{m}^{(1)} = \bar{C}\underline{m} + \underline{q}_1 \qquad \underline{q}_1 = \underline{\dim}\, Q \quad Q \text{ an } A_S\text{-projec-}$$

tive.

By induction, we have for any ℓ:

$$\underline{m}^{(\ell)} = \bar{C}^{\ell}\, \underline{m} + \bar{C}^{\ell-1}\, \underline{q}_1 + \bar{C}^{\ell-1}\, \underline{q}_2 + \dots + \underline{q}_\ell$$

with $\underline{q}_1,\dots,\underline{q}$ the dimensions of A-projectives.

Here S is an extended Dynkin diagram, we know by [7] that there exists N such that for any $\underline{x} \in \mathbb{Z}^t$ $\bar{C}^N \underline{x} = \underline{x} + \underline{n}$ $\delta \in \mathbb{Z}$, $\underline{n}$ a fixed m-vector such that $\bar{C}\underline{n} = n$.

On the other hand, modules in S are periodic, we can find L such that

$$\underline{m}^{(L)} = \underline{m}$$

Putting $N_0 = NL$ we have

$$\underline{m}^{(N_0)} = \underline{m} \qquad \bar{C}^{N_0}\underline{m} = \underline{m} + \delta\underline{n}$$

Here $S \subsetneq S_0$, then $\underline{q}_1 \neq 0$. We have

$$\underline{m} = \underline{m}^{(N_0)} = \bar{C}^{N_0}\underline{m} + \bar{C}^{N_0-1}\underline{q}_1 + \bar{C}^{N_0-2}\underline{q}_2 + \ldots + \underline{q}_{N_0}$$

$$= \underline{m} + \underline{n} + \bar{C}^{N_0+1}\underline{q}_1 + \ldots + \underline{q}_{N_0}$$

Consequently $-\delta\underline{n} = \bar{C}^{N_0-1}\underline{q}_1 + \ldots + \underline{q}_{N_0} = \underline{r}$. Here $\underline{r}$ corresponds to the dimension of a sum of preprojectives, so $\bar{C}\underline{r} \neq \underline{r}$. Moreover $\underline{q}_1 \neq 0$ implies $\underline{r} \neq 0$. But $\bar{C}\underline{n} = \underline{n}$, this is a contradiction. This proves proposition 2.8.

REFERENCES

[1] Auslander, M.: Representation theory of Artin algebras I, Comm. in Algebra, Vol. I No. 3, 177-268 (1974).

[2] Auslander, M., Platzeck, M.I.: Representation theory of hereditary Artin algebras, Proc. Conf. on Representation theory (Philadephia 1976), Marcel Dekker, 389-424 (1978).

[3] Bautista, R.: Algebras close to hereditary algebras, Oberwolfach Conf. report, 17-21 (1977), preprint.

[4] Bautista, R.: Irreducible maps and the radical of a category, preprint.

[5] Bernstein, I.N., Gelfand, I.M., Ponomarev, V.A.: Coxeter functors and Gabriel's theorem, Uspechi Mat. Nauk 28 (1973) translated in Russian Math. Surveys, 17-23 (1973).

[6] Dlab, V., Ringel, C.M.: On algebras of finite representation type, J. Algebra 33, 306-394 (1975).

[7] Dlab, V., Ringel, C.M.: Indecomposable representations of graphs and algebras, Memoirs, Amer. Math. Soc. 173, Providence (1976).

[8] Riedtmann Ch.: Algebren, Darstellungsköcher, Ueberlagerungen und Zurück, Thesis 1979 (Zürich).

[9] Todorov, G.: Almost split sequences for trD-periodic modules M, with no projectives in the class [M], preprint (1979).

INSTITUTO DE MATEMATICAS
U. N. A. M.
México 20, D.F.
MEXICO

Zykellose Algebren sind nicht zügellos

Klaus Bongartz, University of Zurich, Switzerland

We prove, that for each natural number N there exists only a finite number of isomorphism classes of N-dimensional algebras of finite representation type without oriented cycles in its quiver.

Throughout the paper k denotes an algebraically closed commutative field and A a finite-dimensional basic k-algebra.

Lemma 1 a) Let $e^2 = e \in A$ be an idempotent. If A is of finite representation type, so is eAe.
b) Let A be an algebra of finite representation type without oriented cycles in its quiver K_A. If $f_1, f_2 \in A$ are two primitive orthogonal idempotents, the k-dimension of $f_1 A f_2$ is smaller or equal to 1.

Proof : The functor $R: \text{mod } A \longrightarrow \text{mod } eAe$

$$M \longrightarrow eM = \text{Hom}_A(Ae, M)$$

from the category of finite dimensional left A-modules to the one of the left eAe-modules has as left-adjoint

$$L: \text{mod } eAe \longrightarrow \text{mod } A$$

$$N \longmapsto Ae \otimes_{eAe} N$$

The isomorphism $1 \xrightarrow{\sim} RL$ implies, that eAe is of finite representation type.

To prove b) put $e = f_1 + f_2$. Then we get
$eAe = f_1Af_1 \oplus f_2Af_2 \oplus f_1Af_2 \oplus f_2Af_1$. Because K_A contains no cycles, one of the last two terms - say f_2Af_1 - has to be zero, and f_iAf_i is equal to kf_i for $i = 1,2$. Therefore $f_1Af_2 = \text{rad } eAe$ and $(\text{rad } eAe)^2 = 0$. As the quiver $1 \rightrightarrows 2$ is not of finite representation type, part b) follows by part a).

From now on we assume, that the algebra A is of finite representation type without oriented cycles in its quiver K_A.

We denote the point-set of a quiver K by K_0, the arrow-set by K_1.

The natural basis of the quiver - algebra $k[K_A]$ consists of the paths in K_A, and the multiplication of two paths is defined in the obvious way.

For the convenience of the reader we repeat the construction of a surjective algebra homomorphism from $k[K_A]$ to A in our special case : (See [1])

Decompose $1 \in A$ in primitive orthogonal idempotents $1 = e_1 + e_2 + \ldots + e_n$. The simple modules $Ae_i/\mathrm{rad}\, Ae_i$, $1 \leq i \leq n$, represent the isomorphism classes of simple A-modules. Therefore we can identify $(K_A)_0$ with the set $\{1,2,\ldots,n\}$. By assumption K_A contains no double - arrow $\cdot \rightrightarrows \cdot$. For each arrow $i \xrightarrow{\alpha} j$ we choose a non-zero element $\bar{\alpha} \in e_j A e_i$. By Lemma 1 and the definition of K_A we have $k\bar{\alpha} = e_j A e_i = e_j \,\mathrm{rad}\, A e_i$ and $e_j \,\mathrm{rad}^2 A e_i = 0$. To each $\lambda = (\lambda_\alpha) \in (k^*)^{(K_A)_1}$ there is a corresponding surjective algebra homomorphism $\Phi = \Phi_\lambda : k[K_A] \longrightarrow A$ given by $\Phi((i\|i)) = e_i$, $i \in (K_A)_0$ and $\Phi((j|\alpha|i)) = \lambda_\alpha \bar{\alpha}$, $\alpha \in (K_A)_1$. Here, k^* means the set of invertible elements of k, $(i\|i)$ the "lazy" path at the point i, and $(j|\alpha|i)$ the path from i to j, which is defined by the arrow α.

Let ${}_jW_i$ be the set of paths in K_A from i to j. Given $w, w' \in {}_jW_i$, such that $\Phi(w) \neq 0 \neq \Phi(w')$, we get by Lemma 1 $\Phi(w) = \mu(w,w')\,\Phi(w')$, where $\mu(w,w') \in k^*$. Using the same notations as before we state the following theorem.

<u>Theorem</u> : There is a choice of μ, such that $\mu(w,w') = 1$ in all occurring cases.

<u>Proof</u> : By induction on $|(K_A)_0|$. By assumption on K_A there is a point - say n - , where no arrow starts. Let $n-i \xrightarrow{\alpha_i} n$, $1 \leq i \leq m$ be the arrows ending at n.

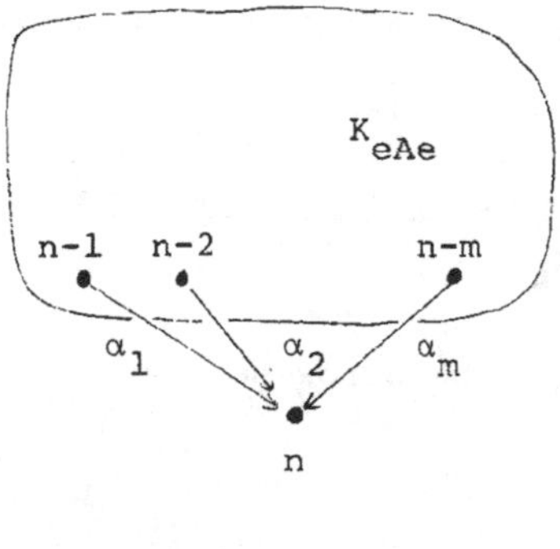

Put $e = e_1 + e_2 + \ldots + e_{n-1}$. We can identify K_{eAe} with the quiver obtained from K_A by deleting n and the arrows α_i, $1 \leq i \leq m$. Consider the following commutative diagram.

$$\begin{array}{ccc} k[K_A] = (1-(n||n))\, k[K_A] \oplus (n||n)\, k[K_A] & \xrightarrow{\Phi = \begin{bmatrix} \Phi_1 & 0 \\ 0 & \Phi_2 \end{bmatrix}} & eA \oplus e_n A = A \\ \uparrow \begin{bmatrix} 1\!\!1 \\ 0 \end{bmatrix} & & \uparrow \begin{bmatrix} 1\!\!1 \\ 0 \end{bmatrix} \\ k[K_{eAe}] = (1-(n||n))\, k[K_A] & \xrightarrow{\Phi_1} & eA = eAe . \end{array}$$

By induction, we can choose Φ_1 such that $\mu(w,w') = 1$ holds already, if the paths w, w' do not end at n. Changing $\lambda_{\alpha_1}, \lambda_{\alpha_2}, \ldots, \lambda_{\alpha_m}$ does not influence Φ_1.

We will choose appropriate $\lambda_{\alpha_1}, \ldots, \lambda_{\alpha_m}$ to satisfy the conditions of the theorem. But first we need two lemmas.

Lemma 2 Let $f = e_{i_1} + e_{i_2} + \ldots + e_{i_r}$ be an idempotent in A with $\{e_{i_1}, e_{i_2}, \ldots, e_{i_r}\} \subset \{e_1, \ldots, e_n\}$. Then we can identify $(K_{fAf})_0$ with $\{i_1, \ldots, i_r\}$. There is an arrow from i_s to i_t if and only if there is a $w \in {}_{i_t}W_{i_s}$ with $\Phi(w) \neq 0$ and if for all $i_k \in \{i_1, \ldots, i_r\} \setminus \{i_s, i_t\}$ and all paths $w_1 \in {}_{i_k}W_{i_s}$, $w_2 \in {}_{i_t}W_{i_k}$ the equality $\Phi(w_2 w_1) = 0$ holds.

Proof : The assertion concerning $(K_{fAf})_0$ is true, because $fAf = fAe_{i_1} \oplus \ldots \oplus fAe_{i_r}$ is a decomposition of fAf in a direct sum of pairwise non-isomorphic projective indecomposable modules.

By definition and Lemma 1, there is an arrow from i_s to i_t if and only if $e_{i_t} A e_{i_s}$ is not contained in $\mathrm{rad}^2 fAf$. The equality $\mathrm{rad}\, fAf = \bigoplus_{1 \le p,q \le r,\, p \neq q} e_{i_q} A e_{i_p}$ implies the lemma.

Fix two arrows α_p, α_q ending at n, and put

$$R(p,q) = \{ (k, w_p, w_q) \mid k \in (K_A)_0,\ w_p \in {}_{n-p}W_k,\ w_q \in {}_{n-q}W_k \text{ such that } \Phi(\alpha_p w_p) \neq 0 \neq \Phi(\alpha_q w_q) \} .$$

Because $\mathrm{Ker}\,\Phi \subset \mathrm{rad}^2 k[K_A]$ we have for each $(k, w_p, w_q) \in R(p,q)$, that $n-p \neq k \neq n-q$. Now we define a pre-order on the finite set $R(p,q)$ by the following :

$(k,w_p,w_q) \leq (k',w'_p,w'_q)$ $:\Leftrightarrow$ $\exists$ $\xi \in {}_k W_{k'}$ such that $\Phi(w_p\,\xi) = \Phi(w'_p)$ and $\Phi(w_q\,\xi) = \Phi(w'_q)$.

<u>Lemma 3</u> Let (k,w_p,w_q) and (k',w'_p,w'_q) be minimal elements in $R(p,q)$. Then $k = k'$.

<u>Proof</u> : By contradiction. Suppose $k \neq k'$. Put $f = e_k + e'_k + e_{n-p} + e_{n-q}$.

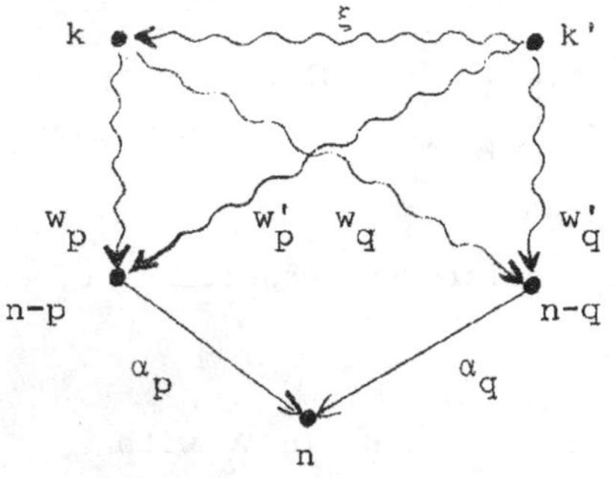

We claim that K_{fAf} contains the the quiver

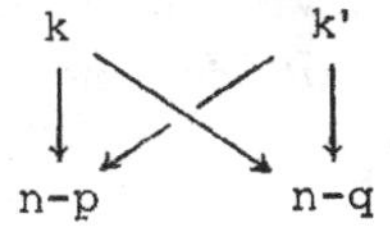

, which is not of finite representation type.

By symmetry, it is sufficient to show, that there is an arrow from k' to $n-p$. First of all, $w'_p \in {}_{n-p}W_{k'}$ and $\Phi(w'_p) \neq 0$. Suppose, there exist $v_p \in {}_{n-p}W_k$, $\xi \in {}_k W_{k'}$ such that $\Phi(v_p\,\xi) \neq 0$. By induction, we get $\Phi(v_p) = \Phi(w_p)$ and $\Phi(w_p\,\xi) = \Phi(w'_p)$. Now $\Phi(\alpha_p w'_p) \neq 0$, hence $\Phi(\alpha_p w_p\,\xi) \neq 0$, hence $\Phi(\alpha_p w_q\,\xi) \neq 0$ and $\Phi(w_q\,\xi) \neq 0$. Again by induction, we have $\Phi(w_q\,\xi) = w'_q$, i.e. $(k,w_p,w_q) \leq (k',w'_p,w'_q)$.

In the other case, suppose there exist $v_q \in {}_{n-q}W_{k'}$, $\xi \in {}_{n-p}W_{n-q}$ such that $\Phi(\xi\,v_q) \neq 0$. By induction we have $\Phi(\xi\,v_q) = \Phi(w'_p)$ and therefore $\Phi(\alpha_p\,\xi\,v_q) \neq 0$. But $\Phi(\alpha_q) = \mu(\alpha_q,\alpha_p\,\xi)\,\Phi(\alpha_p\,\xi)$ in contradiction to $\operatorname{Ker}\Phi \subset \operatorname{rad}^2 k[K_A]$.

Now, we finish the proof of the theorem. As A is of finite representation type, there are at most three arrows ending at n , i.e. $m \leq 3$.

<u>1st case (m = 1)</u> : The theorem holds for each choice of λ_{α_1} .

<u>2nd case (m = 2)</u> : If $R(1,2)$ is empty, the theorem holds for each choice of λ_{α_1} and λ_{α_2} . In the other case, let $(k,w_1,w_2) \in R(1,2)$ be a minimal element and choose λ_{α_1} and λ_{α_2} such that $\Phi(\alpha_1 w_1) = \Phi(\alpha_2 w_2)$

holds (for the "new" algebra homomorphism $\Phi = \Phi_\lambda$!)

If (k', w_1', w_2') is another minimal element in $R(1,2)$, then $k = k'$ by Lemma 3 and $\Phi(w_j') = \Phi(w_j)$ by induction, i.e. $\Phi(\alpha_1 w_1') = \Phi(\alpha_2 w_2')$. The theorem follows by the definition of "$\leq$" in $R(1,2)$.

<u>3rd case $(m=3)$</u> : Consider $R(1,2)$, $R(2,3)$ and $R(3,1)$. If one of these sets is empty, say $R(3,1)$, one can choose first λ_{α_2} and then λ_{α_3} such that for two minimal elements $(k, w_1, w_2) \in R(1,2)$ and $(k', w_2', w_3) \in R(2,3)$ the equalities $\mu(\alpha_1 w_1, \alpha_2 w_2) = \mu(\alpha_2 w_2', \alpha_3 w_3) = 1$ hold. Then the theorem follows like in the previous case.

In the other case choose minimal elements in each of the three sets : $(k, w_1, w_2) \in R(1,2)$, $(k', w_2', w_3') \in R(2,3)$, and $(k'', w_3'', w_1'') \in R(3,1)$.

We have $\{k, k', k''\} \cap \{n-1, n-2, n-3\} = \emptyset$, because $\mathrm{Ker}\,\Phi \subset \mathrm{rad}^2\, k[K_A]$. Suppose first, that k, k', k'' are pairwise different and put

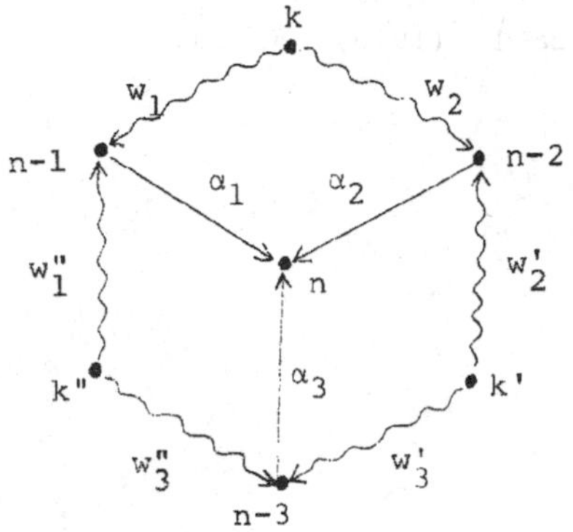

$$f = e_k + e_{k'} + e_{k''} + e_{n-1} + e_{n-2} + e_{n-3}.$$

The quiver K_{fAf} does not contain the quiver

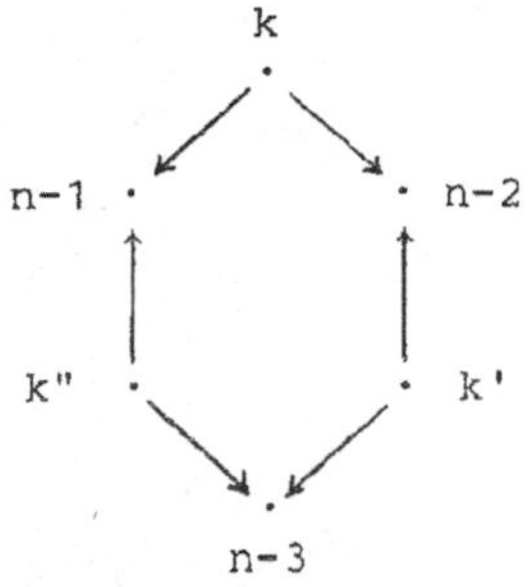

, which is not of finite representation type.

Assume, for instance, that there is no arrow from k' to $n-2$. Hence we must have a path w in K_A from k' to $n-2$ with $\Phi(w) \neq 0$, which passes through k, $n-1$, $n-3$ or k''. The possibilities $n-1$ and $n-3$ are excluded by the condition $\mathrm{Ker}\,\Phi \subset \mathrm{rad}^2\, k[K_A]$. For instance we can assume, that there is a $\xi \in {}_k W_{k'}$ such that $\Phi(w_2 \xi) \neq 0$, i.e. $(k', w_3', w_1 \xi) \in R(3,1)$. Choose λ_{α_2} and λ_{α_3} such that $\mu(\alpha_1 w_1, \alpha_2 w_2) = 1$ and $\mu(\alpha_1 w_1'', \alpha_3 w_3'') = 1$ hold. As in the proof of the second case we can conclude, that $\mu(\alpha_1 v_1, \alpha_3 v_3) = 1$ for all $(i, v_3, v_1) \in R(3,1)$. In particular, we have $\Phi(\alpha_3 w_3') = \Phi(\alpha_1 w_1 \xi) = \Phi(\alpha_2 w_2 \xi) = \Phi(\alpha_2 w_2')$. The theorem follows easily.

At the end, we show by an example that we really need the assumption "A is of finite representation type". Let K be the following quiver :

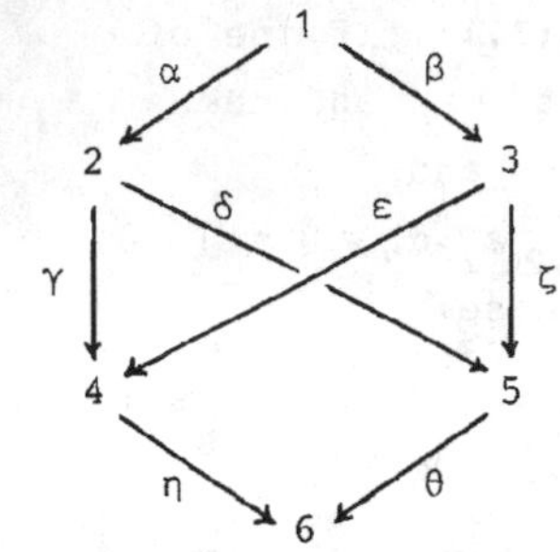

For $x \in k^*$ we define

$A_x := k[K]/I_x + \mathrm{rad}^3\, k[K]$,

where I_x is the ideal generated by

$\delta\alpha - \zeta\beta$, $\gamma\alpha - \varepsilon\beta$, $\theta\delta - \eta\gamma$, $\eta\varepsilon - x\theta\zeta$.

Then $A_x \xrightarrow{\sim} A_y$ holds if and only if $x = y$ or $x = y^{-1}$.

[1] Gabriel: Indecomposable representations II, Istituto Nazionale di alta Matematica, Symposia Mathematica 11 (1973), 81-104

GENERALIZATIONS OF THE BERNSTEIN-GELFAND-PONOMAREV REFLECTION FUNCTORS

Sheila Brenner and M.C.R. Butler

Introduction

Reflection functors were introduced into the representation theory of quivers by Bernstein, Gelfand and Ponomarev in their work on the 4-subspace problem and on Gabriel's Theorem and there have been several generalisations, see [13], [6], [10] and [2]. The aim of this paper is to present a further extension of the concept and to give some applications to quivers with relations (QWR's). A special case of this theory has been developed by Marmaridis [19] and applied to certain QWR's; indeed some of the methods used in his Thesis [18] may also be regarded as applications of these functors, though they are not presented in that way.

Associated with any representation of a quiver is a dimension vector, and the dimension vectors of indecomposable modules are the positive roots of the quadratic form associated to the quiver (see e.g. [6], [10], [15]). Similar results seem to hold for certain QWR's. Some applications of reflection functors involve the study of the transformations of dimension vectors they induce. It turns out that there are applications of our functors which make use of the analogous transformations which we like to think of as a change of basis for a fixed root-system - a tilting of the axes relative to the roots which results in a different subset of roots lying in the positive cone. (An example is considered in some detail in Chapter 4, §2). For this reason, and because the word 'tilt' inflects easily, we call our functors <u>tilting functors</u> or simply <u>tilts</u>.

Chapter 1 contains the general theory. Tilting functors are defined in §2; their main properties are stated in §3 and proved in subsequent sections. They relate module categories over artin algebras Λ and Λ' and occur in adjoint pairs mod $\Lambda \underset{G}{\overset{F}{\rightleftarrows}}$ mod Λ' where $F = \mathrm{Hom}_\Lambda(X,-)$ and $G = X \otimes_{\Lambda'} -$ are determined by a bimodule ${}_\Lambda X_{\Lambda'}$. Definition 1 requires ${}_\Lambda X_{\Lambda'}$ to satisfy a set of conditions T_* and also its dual T'_*. The main purpose of this requirement is to ensure that the full subcategories ImG of mod Λ and ImF of mod Λ' are 'large' and isomorphic.

In the applications that we have considered one of the algebras - Λ say - is given and a construction is required for Λ' and ${}_\Lambda X_{\Lambda'}$. In fact it is sufficient to construct ${}_\Lambda X$, to define $\Lambda' = (\mathrm{End}_\Lambda X)^{op}$ (as in [2]) and to verify T_* since Theorem I states that each of T_* and T'_* implies the other.

The construction of certain suitable bimodules, and the study of special properties of the associated adjoint tilting functors is the subject of Chapter 2. Our constuction generalises that of Auslander, Platzeck and Reiten, [2]. Starting with a suitable non-injective simple Λ-module S, with projective cover P(S), we take ${}_\Lambda X$ to be the sum of TrDS and the indecomposable projectives other than P(S) (and there is a dual construction starting from a suitable simple Λ'-module). Marmaridis' functors are constructed similarly, though instead of considering adjoint pairs he obtains results for the case when either Λ or Λ' is l-hereditary. Theorem IX contains the definitions and basic properties, and Theorem X gives further conditions, appropriate to many applications to QWR's, for desireable properties of Λ to give rise to desireable properties of Λ' or vice-versa.

In Chapter 3 we define triangulable algebras and, as a special case, quivers with relations. For such algebras we obtain a convenient formula for the bilinear form b_Λ on the Grothendieck space Q Gr Λ defined by the alternating sum of dimensions of Ext groups. We draw attention to certain classes of QWR's for which finite representation type is equivalent to the quadratic form q_Λ associated to b_Λ being positive definite on the positive cone in QGrΛ. Another important property of b_Λ is that the linear transformation induced by a tilting functor induces an isometry of the associated quadratic form. We end this Chapter by giving an expression for the dimension type of the dual of transpose of a module involving the Coxeter matrix; this generalises the well-known formula for the case of a hereditary algebra.

Chapter 4 is devoted to examples and applications. The tilting functors used are a special case of those considered in Chapter 2 where the simple Λ-module S is projective or, dually, where the simple Λ'-module S' is injective (but not usually both) and we call these APR tilts. (Marmaridis [18] applies also non-APR tilts). Theorem XI in §3 shows, using Kac's results [16] and the Nazarova-Roiter Theorem [21], that if Λ is a QWR with q_Λ positive definite on positive vectors, and if also Λ tilts to a quiver, then it is of finite representation type. In §4 we discuss the problem of deciding which QWR's tilt to quivers using APR tilts. This problem seems to require more knowledge of the effect of APR tilts on the components of the Auslander-Reiten graph. The remaining sections are concerned with tame quivers with one relation (and related QWR's) and connections with Ringel's work [23] are discussed. In particular §6 discusses some properties of certain QWR's whose quadratic form has co-rank 2.

Notation and terminology

The notation mentioned above will remain fixed throughout this paper, so that Λ, Λ' are artin algebras over a common central artinian subring k, modΛ, modΛ' are their respective categories of finitely generated left modules, $X = {}_\Lambda X_{\Lambda'}$ is a bimodule, and $F = \mathrm{Hom}_\Lambda(X,-) : \mathrm{mod}\Lambda \to \mathrm{mod}\Lambda'$ and $G = X \otimes_{\Lambda'} - : \mathrm{mod}\Lambda' \to \mathrm{mod}\Lambda$ are the associated adjoint functor pair.

We need much of the machinery introduced by Auslander and Reiten in part III of their seminal series of papers [3]. For any algebra A, the opposite algebra is denoted A^{op} (so that mod A^{op} is the category of right A-modules), and the radical of A is denoted by rad A. The duality on mod k is denoted by D, and the same symbol will be used for its restrictions to various subcategories of modk; for example, for the induced contravariant equivalences of mod Λ with mod Λ^{op}. For Λ-modules M and N, we write M* for the Λ^{op}-module $\mathrm{Hom}_\Lambda(M,\Lambda)$, ev_N^M for the k-morphism $M^* \otimes_\Lambda N \to \mathrm{Hom}_\Lambda(M,N)$, $\xi \otimes y \mapsto (x \mapsto \xi(x)y)$, and $\underline{\mathrm{Hom}}_\Lambda(M,N) = \mathrm{Coker}(ev_N^M)$. Then $\underline{\mathrm{mod}}$ Λ is the 'stable' category with objects the Λ-modules and morphisms the elements of $\underline{\mathrm{Hom}}_\Lambda(M,N)$. For any Λ-module M, the Λ^{op}-module $\mathrm{Tr}_\Lambda M$, the transpose of M, is defined by selecting two terms $P_1 \to P_0 \to M \to 0$ of a minimal Λ-projective resolution of M and setting $\mathrm{Tr}_\Lambda M = \mathrm{Coker}(P_0^* \to P_1^*)$.

Some special notations will be used in connection with full subcategories of modules. If M is isomorphic to a direct summand of N, we write $M|N$. For any Λ-module N, add N denotes the full subcategory of mod Λ with objects M such that $M|N^n$ for some n, and in case $N = {}_\Lambda N_{\Lambda'}$ is a bimodule we make the convention that add N = $\mathrm{add}\,{}_\Lambda N \leq$ mod Λ and add N^{op} = add $N_{\Lambda'} \leq$ mod Λ'^{op}. Thus add Λ, add Λ^{op}, add $D\Lambda$, add $D\Lambda^{op}$ are, respectively, the full subcategories of projectives in mod Λ, projectives in mod Λ^{op}, injectives in mod Λ, injectives in mod Λ^{op}. Note that since $G\Lambda' = X \otimes_{\Lambda'} \Lambda' \simeq {}_\Lambda X$ and $FD\Lambda = \mathrm{Hom}_\Lambda(X, D\Lambda) \simeq D(\Lambda \otimes_\Lambda X) \simeq D(X_{\Lambda'}) = {}_{\Lambda'}DX$, we obtain by naturality categorical formulae such as G add Λ' = add X and F add $D\Lambda$ = add DX.

Finally if C is a subcategory of mod Λ, $\underline{Q}C$ and $\underline{S}C$ denote the full subcategories with objects the quotients of modules in C and the submodules of modules in C, respectively.

CHAPTER 1 - GENERAL THEORY

1. Generalities on adjoint functors

Let $X = {}_{\Lambda}X_{\Lambda'}$ be a bimodule and $F = \mathrm{Hom}_{\Lambda}(X,-)$ and $G = X \otimes_{\Lambda'} -$ the adjoint functor pair determined by X. For each pair of modules $M \in \mathrm{mod}\,\Lambda$, $M' \in \mathrm{mod}\,\Lambda'$, the associativity isomorphism

$$\mathrm{Hom}_{\Lambda}(GM',M) \simeq \mathrm{Hom}_{\Lambda'}(M',FM),$$

will be denoted by $\omega = \omega_M^{M'}$, and the adjunctions by $\epsilon_M = \omega^{-1}(1_{FM}) : GFM \to M$ and $\mu_{M'} = \omega(1_{GM'}) : M' \to FGM'$. Note that $F(\epsilon_M)\mu_{FM} = 1_{FM}$ and $\epsilon_{GM'}G(\mu_{M'}) = 1_{GM'}$. Write $F_{M,N}$ and $G_{M',N'}$ for the maps of Hom-groups induced by F and G respectively.

PROPOSITION 1. For each Λ-module M the following statements are equivalent:

(a) $M \in \underline{Q}\mathrm{Im}G$; (b) $M \in \underline{Q}G\,\mathrm{add}\,\Lambda' = \underline{Q}\,\mathrm{add}\,X$;

(c) there is a short exact sequence $0 \to M_1 \to X_0 \to M \to 0$ with $X_0 \in \mathrm{add}\,X$ on which F is exact;

(d) $\forall N \in \mathrm{mod}\,\Lambda$, $F_{M,N}$ is injective;

(e) ϵ_M is surjective.

Proof Obviously (b), (c), (e) $\Rightarrow$ (a). (a) $\Rightarrow$ (b) since each M' is quotient of a Λ'-projective $P' \in \mathrm{add}\,\Lambda'$ and G is right exact. The formula $F_{M,N} = \omega_N^{FM} \circ \mathrm{Hom}_{\Lambda}(\epsilon_M,N)$ gives (d) $\Leftrightarrow$ (e), so it remains to prove that (a) $\Rightarrow$ (e) and (b) $\Rightarrow$ (c). First we show that (a) $\Rightarrow$ (e). Let $\phi : GM' \to M$ be surjective. By naturality $\epsilon_M GF(\phi) = \phi\epsilon_{GM'}$ and $\epsilon_{GM'}$ is a split epimorphism, so ϵ_M is surjective, as (e) asserts. We use a well-known argument to show that (b) $\Rightarrow$ (c). Let $M \in \underline{Q}\,\mathrm{add}\,X$, choose an epimorphism $\eta : Y \to M$ with $Y \in \mathrm{add}\,X$, then choose a Λ'-projective module P' and a morphism $\pi' : P' \to FM$ such that $P' \overset{\pi'}{\to} FM \to \mathrm{Coker}\,F(\eta)$ is surjective. Then $(\pi',F(\eta)) : P' \amalg FY \to FM$ is surjective. Now $GP' \in G\,\mathrm{add}\,\Lambda' = \mathrm{add}\,X$, so $X_0 = GP' \amalg FY \in \mathrm{add}\,X$ and $\xi = (\epsilon_M G(\pi'),\eta) : X_0 \to M$ is surjective. Further $F(\xi)(\mu_{P'},1_{FY}) = (\pi',F(\eta))$ is surjective, from which we see that $F(\xi)$ is surjective. Now the exact sequence $0 \to \mathrm{Ker}\xi \to X_0 \to M \to 0$ has the properties required to satisfy (c).

In dualising Proposition 1 we replace G add Λ' in (b) by F add $D\Lambda$ = add DX, so obtaining

PROPOSITION 1'. *For each Λ'-module M', the following statements are equivalent*:

(a) $M' \in \underline{S}\mathrm{Im}F$; (b) $M' \in \underline{S}F$ add $D\Lambda = \underline{S}$ add DX;

(c) *there is a short exact sequence* $0 \to M' \to X'_0 \to M'_1 \to 0$ *with* $X'_0 \in$ add DX *on which* G *is exact*;

(d) $\forall N' \in \mathrm{mod}\,\Lambda'$, $G_{N',M'}$ *is injective*;

(e) $\mu_{M'}$ *is injective*.

The next two propositions are concerned with the analysis of the adjunction morphisms ϵ_M and $\mu_{M'}$.

PROPOSITION 2. *Let* M *be a* Λ-*module*. *Then*

(a) $\mathrm{Im}\,\epsilon_M \in \underline{Q}\mathrm{Im}G$;

(b) *the map* $F\mathrm{Im}\,\epsilon_M \to FM$ *induced by the inclusion of* $\mathrm{Im}\,\epsilon_M$ *into* M *is an isomorphism*;

(c) *the sequence*

$$0 \to F\mathrm{Ker}\epsilon_M \to FGFM \xrightarrow{F(\epsilon_M)} FM \to 0$$

is split exact;

(d) *there are exact connected sequences*

$$0 \to \mathrm{Ext}^1_\Lambda(X,\mathrm{Ker}\epsilon_M) \to \mathrm{Ext}^1_\Lambda(X,GFM) \to \mathrm{Ext}^1_\Lambda(X,\mathrm{Im}\epsilon_M) \to \mathrm{Ext}^2_\Lambda(X,\mathrm{Ker}\epsilon_M) \to \dots ,$$

$$0 \to F\mathrm{Coker}\epsilon_M \to \mathrm{Ext}^1_\Lambda(X,\mathrm{Im}\epsilon_M) \to \mathrm{Ext}^1_\Lambda(X,M) \to \mathrm{Ext}^1_\Lambda(X,\mathrm{Coker}\epsilon_M) \to \dots .$$

Proof $\mathrm{Im}\epsilon_M$ is a quotient of $GFM \in \mathrm{Im}G$ so (a) holds. (b) and (c) depend on the fact that $F(\epsilon_M)$ is a split epimorphism which factors through the inclusion $F\mathrm{Im}\epsilon_M \to FM$. Then (b) and (c) imply (d) on using them to simplify the exact connected sequences for $F = \mathrm{Hom}_\Lambda(X,-)$ over the short exact sequences $0 \to \mathrm{Ker}\epsilon_M \to GFM \to \mathrm{Im}\epsilon_M \to 0$ and $0 \to \mathrm{Im}\epsilon_M \to M \to \mathrm{Coker}\epsilon_M \to 0$.

The dual Proposition 2' for a Λ'-module M' of Proposition 2 may be written down by interchanging F and G, replacing ϵ_M by $\mu_{M'}$, and Ext^n_Λ by $\mathrm{Tor}^{\Lambda'}_n$, and reversing all arrows.

COROLLARY $\mathrm{Im}F = F(\underline{Q}\mathrm{Im}G)$ *and* $\mathrm{Im}G = G(\underline{S}\mathrm{Im}F)$.

Proof The first formula is a consequence of Proposition 2, (a) and (b), and the second is obtained similarly from Proposition 2′.

2. The tilting conditions T_* and T'_*

This section lists the conditions T_* and T'_* to be satisfied by the bimodule ${}_{\Lambda}X_{\Lambda'}$, in order that the adjoint functor pair F,G be a pair of tilting functors. We also mention some useful consequent and equivalent conditions, and note the examples discussed in [2].

The algebra maps referred to in T_0 and T'_0 are, respectively, the maps

$$\Lambda' \to (\mathrm{End}_{\Lambda}X)^{op}, \qquad \lambda' \mapsto (x \mapsto x\lambda'),$$

$$\Lambda \to \mathrm{End}_{\Lambda'^{op}}X, \qquad \lambda \mapsto (x \mapsto \lambda x).$$

DEFINITION 1. We say that the bimodule $X = {}_{\Lambda}X_{\Lambda'}$ satisfies T_* if the following five conditions are satisfied:

T_0 : $\Lambda' \to (\mathrm{End}_{\Lambda}X)^{op}$ is an algebra isomorphism.

T_1 : $\mathrm{pd}_{\Lambda}X = 1$.

T_2 : $\mathrm{Ext}^1_{\Lambda}(X,X) = 0$.

T_3 : ${}_{\Lambda}X$ has a projective cover in add X.

T_4 : $\forall M \in \mathrm{mod}\ \Lambda$, $FM = 0$ and $\mathrm{Ext}^1_{\Lambda}(X,M) = 0 \Rightarrow M = 0$.

The conditions T'_* on X are obtained by interchanging Λ and Λ' and F and G, and we refer to them individually as the 'adjoint duals' of the T_n's. They are

T'_0 : $\Lambda \to \mathrm{End}_{\Lambda'^{op}}X$ is an algebra isomorphism.

T'_1 : $\mathrm{pd}_{\Lambda'^{op}}X = 1$.

T'_2 : $\mathrm{Tor}^{\Lambda'}_1(X,DX) = 0$.

T'_3 : $X_{\Lambda'}$ has a projective cover in add X^{op}.

T'_4 : $\forall M' \in \mathrm{mod}\ \Lambda'$, $GM' = 0$ and $\mathrm{Tor}^{\Lambda'}(X,M') = 0 \Rightarrow M' = 0$.

Note that the second variable, ${}_{\Lambda}X \cong G\Lambda'$, in T_2 has been replaced in T_2' by its adjoint dual, $FD\Lambda \cong DX$. Notice also that, in applying T_2' and T_4', it is sometimes useful to replace 'projective' conditions on $X_{\Lambda'}$ by 'injective' conditions on ${}_{\Lambda'}DX$ by making use of the duality isomorphism of $\mathrm{Tor}^{\Lambda'}(X,-)$ with $\mathrm{Ext}_{\Lambda'}(DX,-)$.

Now we give a number of conditions consequent upon or equivalent to the T_n's and T_n''s. These will be used later and are also useful in recognising tilting functors.

PROPOSITION 3. (a) T_0 *holds if and only if, for each* Λ'-*projective module* P', *the adjunction* $\mu_{P'}$ *is an isomorphism.* (b) *Suppose* T_0 *holds. Then* (i) $F\ \mathrm{add}\ X = \mathrm{add}\ \Lambda'$ (*the category of* Λ'-*projectives*); (ii) $\mathrm{Hom}_{\Lambda}(\mathrm{add}\ X, X) = \mathrm{add}\ \Lambda'^{op}$ (*the category of* Λ'^{op}-*projectives*); (iii) $\forall X_0 \in \mathrm{add}\ X$ *and* $\forall N \in \mathrm{mod}\ \Lambda$, $F_{X_0,N} : \mathrm{Hom}_{\Lambda}(X_0,N) \to \mathrm{Hom}_{\Lambda'}(FX_0,FN)$ *is an isomorphism.*

Proof (a) This follows by naturality from the observation that $\mu_{\Lambda'} : \Lambda' \to FG\Lambda'$ coincides with the algebra morphism $\Lambda' \to (\mathrm{End}_{\Lambda}X)^{op}$. (b) Note first that $X \cong G\Lambda'$ and $\mathrm{add}\ X = \mathrm{add}\ G\Lambda'$ in general. Suppose T_0 holds, then from (a) $\mathrm{add}\,\Lambda' = FG\ \mathrm{add}\ \Lambda' = F\ \mathrm{add}\ X$, which is (b)(i). Also,

$$\begin{aligned}\mathrm{add}\ \Lambda'^{op} &= \mathrm{Hom}_{\Lambda'}(\mathrm{add}\Lambda',\Lambda') = \mathrm{Hom}_{\Lambda'}(F\ \mathrm{add}\ X,\Lambda')\ \text{(using (b)(i))}\\ &= \mathrm{Hom}_{\Lambda}(\mathrm{add}\ X, G\Lambda') \qquad \text{(since } F, G \text{ are adjoint)}\\ &= \mathrm{Hom}_{\Lambda}(\mathrm{add}\ X, X).\end{aligned}$$

This is (b)(ii). Finally T_0 implies that $FX \cong {}_{\Lambda'}\Lambda'$, so (b)(iii) is obtained by naturality.

DEFINITION 2. *Throughout the rest of this paper*

$$0 \to R \overset{\rho}{\to} P \overset{\pi}{\to} X \to 0$$

is an exact sequence in which π *is a* Λ-*projective cover of* X; *it will be called the standard projective resolution of* X.

PROPOSITION 4. (a) $T_1 \Longleftrightarrow R$ *is projective*;
(b) $T_3 \Longleftrightarrow P \in \mathrm{add}\ X$;
(c) *for each simple* Λ-*module* S, *the maps*

$$\mathrm{Hom}_{\Lambda}(X,S) \to \mathrm{Hom}_{\Lambda}(P,S) \quad \text{and} \quad \mathrm{Hom}_{\Lambda}(R,S) \to \mathrm{Ext}^1_{\Lambda}(X,S)$$

induced from the resolution of X are isomorphisms;
(d) $\{T_2$ and $T_3\} \Rightarrow \mathrm{Ext}^1_\Lambda(X,P) = 0$.

Proof (a) and (b) are obvious, (c) is an immediate consequence of the fact that π is a projective cover, and (d) follows from (b).

The next result is useful for verifying T_4.

PROPOSITION 5. Assume T_1, T_2, T_3 hold. Then T_4 holds if and only if each Λ-projective indecomposable module occurs as a direct summand of P or of R but not of both.

Proof For each simple Λ-module S, choose a Λ-projective module P_S with S as top. Then $P_S|P \Leftrightarrow \mathrm{Hom}_\Lambda(P,S) \neq 0$, and similarly for R since, by T_1, it too is projective. First we show that $P_S|P \Rightarrow P_S \nmid R$. Let $P_S|P$. Then $\mathrm{Ext}^1_\Lambda(X,P_S) = 0$ by Proposition 4(d), which implies $\mathrm{Ext}^1_\Lambda(X,S) = 0$ (since, by T_1, $\mathrm{Ext}^1_\Lambda(X,-)$ is right exact), so that we conclude from Proposition 4(c) that $\mathrm{Hom}_\Lambda(R,S) = 0$. Thus $P_S \nmid R$. Now suppose that T_4 holds and that $P_S \nmid P$; then $\mathrm{Hom}_\Lambda(P,S) = 0 \Rightarrow$ $\mathrm{Hom}_\Lambda(X,S) = 0$ (by Proposition 4(c)) $\Rightarrow \mathrm{Ext}^1_\Lambda(X,S) \neq 0$ (by T_4) $\Rightarrow \mathrm{Hom}_\Lambda(R,S) \neq 0$ (by Proposition 4(c)) $\Rightarrow P_S|R$. This completes the 'only if' part of the proposition. For the 'if' part, suppose the conditions on projective indecomposables are satisfied and let M be a Λ-module with $FM = \mathrm{Hom}_\Lambda(X,M) = 0$ and $\mathrm{Ext}^1_\Lambda(X,M) = 0$. By T_3, $\mathrm{Hom}_\Lambda(P,M) = 0$, which by Definition 2 implies that the connecting homomorphism $\mathrm{Hom}_\Lambda(R,M) \to \mathrm{Ext}^1_\Lambda(X,M)$ is an isomorphism, thus that $\mathrm{Hom}_\Lambda(R,M) = 0$. Since ${}_\Lambda\Lambda|P \amalg R$, it follows that $M \cong \mathrm{Hom}_\Lambda(\Lambda,M) = 0$. So T_4 holds.

EXAMPLE We conclude this section by showing that the method of Auslander, Platzeck and Reiten gives a pair of tilting functors. As in [2], suppose Λ is a basic artin algebra with a simple module, S, which is Λ-projective but not Λ-injective, and let

$$0 \to S \xrightarrow{\rho_0} P_0 \xrightarrow{\pi_0} \mathrm{Tr}_{\Lambda^{op}}DS \to 0$$

be the almost split sequence with S as kernel. Then it is known that P_0 is also projective, π_0 is a projective cover map, and $S \nmid P_0$ and $\mathrm{Tr}_{\Lambda^{op}}DS \ncong S$ (indeed $\mathrm{Tr}_{\Lambda^{op}}DS$ cannot be projective). Let ${}_\Lambda\Lambda \cong S \amalg P_1$, so that $S \nmid P_1$. Set $P = P_0 \amalg P_1$, $X = \mathrm{Tr}_{\Lambda^{op}} DS \amalg P_1$, $\pi = \pi_0 \amalg \mathrm{id}$, $\Lambda' = (\mathrm{End}_\Lambda X)^{op}$, $F = \mathrm{Hom}_\Lambda(X,-)$, $G = X \otimes_{\Lambda'} -$. Then, we claim that ${}_\Lambda X_{\Lambda'}$ satisfies T_*, so that F and G are a pair of adjoint tilting functors. Indeed, T_0 and T_1 are obvious. For T_2, we have

$$\mathrm{Ext}^1_\Lambda(X,X) = \mathrm{Ext}^1_\Lambda(\mathrm{Tr}_{\Lambda^{op}} DS, X)$$

$$= \mathrm{Coker}(\mathrm{Hom}_\Lambda(P_0,X) \overset{\rho_0}{\to} \mathrm{Hom}_\Lambda(S,X)),$$

and ρ_0 is surjective since $S \nmid X$ and ρ_0 is the injection in an almost split sequence. Finally, Proposition 5 gives T_4, for $S \amalg P \cong S \amalg P_0 \amalg P_1$ has ${}_\Lambda\Lambda$ as a summand, but S is indecomposable but not a summand of P_0 or of P_1.

3. Statement of main results

Our main results are set out in this section in the form of eight theorems I - VIII. The notations used earlier are used again without further explanation.

THEOREM I *The bimodule X satisfies T_* if and only if it satisfies T'_*.*

THEOREM II *Suppose X satisfies T_*. Then:*

(1) *For each Λ-module M, the following six statements are equivalent:*

   (a) $M \in \mathrm{Im}\, G$; (b) $\mathrm{Ext}^1_\Lambda(X,M) = 0$;

   (c) M *has an* F-*exact* X-*projective resolution (that is, there is an exact-sequence* $\cdots \to X_n \to \cdots \to X_1 \to X_0 \to M \to 0$ *with each* $X_n \in \mathrm{add}\, X$ *such that* $\cdots \to FX_n \to \cdots \to FX_1 \to FX_0 \to FM \to 0$ *is exact in* mod Λ');

   (d) $\forall N \in \mathrm{mod}\,\Lambda$, $F_{M,N} : \mathrm{Hom}_\Lambda(M,N) \to \mathrm{Hom}_{\Lambda'}(FM,FN)$ *is an isomorphism*;

   (e) $\epsilon_M : GFM \to M$ *is an isomorphism*;

   (f) $\mathrm{Hom}_\Lambda(M, D\mathrm{Tr}_\Lambda X) = 0$.

(2) *The full subcategories of* mod Λ,

$\mathrm{Im}\, G$, $G\, \mathrm{Im}\, F$, $\mathrm{Im}\, GF$, $\underline{Q}\, \mathrm{Im}\, G$, $\underline{Q}\, \mathrm{add}\, X$, $\underline{Q}G\, \mathrm{add}\, \Lambda'$, *coincide and contain* $\mathrm{add}\, X = G\, \mathrm{add}\, \Lambda'$ *and* $\mathrm{add}\, D\Lambda = G\, \mathrm{add}\, DX$

(the latter being the full subcategory of injective Λ-modules).

THEOREM III Suppose X satisfies T_*. Then
(a) the functors F, G induce inverse equivalences of the full subcategories Im G and Im F of mod Λ and mod Λ', respectively;
(b) there are natural isomorphisms

$$\mathrm{Ext}^n_\Lambda(M,N) \rightleftarrows \mathrm{Ext}^n_{\Lambda'}(FM,FN)$$

for all M, N $\in$ Im G and all $n \geq 0$.

THEOREM IV Suppose X satisfies T_*. For each Λ-module M, the adjunction ϵ_M : GFM $\to$ M is injective, $F(\epsilon_M)$ is an isomorphism, F Coker $\epsilon_M = 0$, and the map $\mathrm{Ext}^1_\Lambda(X,M) \to \mathrm{Ext}^1(X, \mathrm{Coker}\ \epsilon_M)$ is an isomorphism.

THEOREM V Suppose T_* holds and M is a Λ-module. Then
(a) $\mathrm{Ext}^1_\Lambda(X, D\mathrm{Tr}_\Lambda M) \cong D\ \mathrm{Hom}_\Lambda(M,X)$.
(b) $D\mathrm{Tr}_\Lambda M \in$ Im G iff $\mathrm{Hom}_\Lambda(M,X) = 0$.
(c) Assume that each indecomposable summand A of $D\mathrm{Tr}_\Lambda X$ has an almost split sequence $0 \to A \to B \to \mathrm{Tr}_\Lambda DA \to 0$ in which B is projective (equivalently, each indecomposable summand of $D\mathrm{Tr}_\Lambda X$ is a simple module of the form $\Lambda e/re$ where e is a primitive idempotent of Λ and $rer = 0$). Then $D\mathrm{Tr}_\Lambda M \in$ Im G if M has no indecomposable summands in add($\Lambda \amalg X$).

THEOREM VI Suppose T_* holds, M $\in$ Im G, and $\mathrm{Tor}^\Lambda_p(\mathrm{Tr}_\Lambda X,M) = 0$ for p = 1, 2 ..., k (the last conditions are, by duality, equivalent to $\mathrm{Ext}^p_\Lambda(M, D\mathrm{Tr}_\Lambda X) = 0$ for p = 1, 2, ..., k). Then F induces isomorphisms

$$\mathrm{Ext}^p_\Lambda(M,N) \rightrightarrows \mathrm{Ext}^p_{\Lambda'}(FM,FN)$$

for p = 0, 1, ..., k-1 and for every Λ-module N.

THEOREM VII Suppose T_* holds, M $\in$ Im G, and $\mathrm{Tor}^\Lambda_p(\mathrm{Tr}_\Lambda X,M) = 0$ for p = 1, 2 (the last conditions are equivalent to $X^* \otimes_\Lambda M \to FM$ being an isomorphism). Then (a) $D\mathrm{Tr}_{\Lambda'}FM \cong FD\mathrm{Tr}_\Lambda M$; (b) if further, $D\mathrm{Tr}_\Lambda M \in$ Im G (equivalently, $\mathrm{Hom}_\Lambda(M,X) = 0$), then the isomorphism

$$\mathrm{Ext}^1_\Lambda(M,\mathrm{DTr}_\Lambda M) \rightleftarrows \mathrm{Ext}^1_{\Lambda'}(FM,F\mathrm{DTr}_\Lambda M)$$

maps almost split sequences to almost split sequences.

The last theorem gives better results than Theorem VII on almost split sequences in the case of the special tilting functors to be studied in Chapters 2, 3, 4. Recall that, for any Λ-module M with minimal projective resolution $\dots \to P_1 \to P_0 \to M \to 0$, the second syzygy module Ker $(P_1 \to P_0)$ is denoted by $\Omega^2 M$.

THEOREM VIII Suppose T_* holds and that the kernel R of the standard projective resolution of ${}_\Lambda X$ is indecomposable. Suppose that the indecomposable Λ-module M has a projective cover in Im G and that $\Omega^2 M \in \mathrm{Im}\, G$. Then (a) $\mathrm{DTr}_{\Lambda'}FM \cong F\mathrm{DTr}_\Lambda M$, and (b) if also $\mathrm{DTr}_\Lambda M \in \mathrm{Im}\, G$ (equivalently, $\mathrm{Hom}_\Lambda(M,X) = 0$), then the isomorphism

$$\mathrm{Ext}^1_\Lambda(M,\mathrm{DTr}_\Lambda M) \rightleftarrows \mathrm{Ext}^1_{\Lambda'}(FM,F\mathrm{DTr}_\Lambda M)$$

maps almost split sequences to almost split sequences.

4. Proofs, the first stage

We begin by proving all of Theorem II, except (1)(f) which will be considered in §6, Theorem III(a), and some corollaries needed later. It is assumed throughout that X satisfies T_*.

Proof of Theorem II(1)(a),...,(e). Let M be a Λ-module. It suffices to verify the following implications:

$$\begin{array}{ccccc} (b) \Leftrightarrow & M \in \underline{Q}\,\mathrm{Im}\, G & \Rightarrow & (c) \\ & \Uparrow & & \Downarrow \\ & (a) \Leftarrow (e) & \Leftarrow & (d) \end{array}$$

$M \in \underline{Q}\,\mathrm{Im}\, G \Rightarrow$ (b). Using Proposition 1, we select an F-exact exact sequence $0 \to M_1 \to X_0 \to M \to 0$ with $X_0 \in \mathrm{add}\, X$. By T_1 and T_2,

$$0 = \mathrm{Ext}^1_\Lambda(X,X_0) \to \mathrm{Ext}^1_\Lambda(X,M) \to 0$$

is exact, so (b) holds.

(b) $\Rightarrow M \in \underline{Q}\,\mathrm{Im}\, G$. We consider the analysis of ϵ_M : $GFM \to M$ given in Proposition 2. By part (a) of that proposition, and the implication just established, we have $\mathrm{Ext}^1_\Lambda(X,\mathrm{Im}\,\epsilon_M) = 0$, and then T_1 and Proposition 2(d) show that F Coker $\epsilon_M = 0$ and $\mathrm{Ext}^1_\Lambda(X, \mathrm{Coker}\,\epsilon_M) = 0$. Now T_4 implies that Coker $\epsilon_M = 0$, so ϵ_M is surjective and $M \in \underline{Q}\mathrm{Im}\, G$.

$M \in \underline{Q}\, \mathrm{Im}\, G \Rightarrow$ (c). Again use Proposition 1, and select an F-exact exact sequence $0 \to M_1 \to X_0 \to M \to 0$ with $X_0 \in \mathrm{add}\, X$. This gives an exact sequence

$$0 \to FM_1 \to FX_0 \to FM \overset{0}{\to} \mathrm{Ext}^1_\Lambda(X,M_1) \to \mathrm{Ext}^1_\Lambda(X,M),$$

in which the connecting map is 0. Since $M \in \underline{Q}\,\mathrm{Im}\,G$, (b) $\Rightarrow M_1 \in \underline{Q}\,\mathrm{Im}\,G$. Proceeding iteratively, we can find F-exact short exact sequences $0 \to M_{n+1} \to X_n \to M_n \to 0$ with $X_0, X_1, \ldots$ all in add X and $M = M_0, M_1, \ldots$ all in $\underline{Q}$ Im G, and these may be spliced to obtain the required F-exact X-resolution for M. This proves (c).

(c) $\Rightarrow$ (d). Let $X_1 \to X_0 \to M \to 0$ be F-exact with X_1, $X_0 \in \mathrm{add}\, X$. We obtain a commutative diagram (one for each Λ-module N)

$$\begin{array}{ccccccc}
0 \to & \mathrm{Hom}_\Lambda(M,N) & \to & \mathrm{Hom}_\Lambda(X_0,N) & \to & \mathrm{Hom}_\Lambda(X_1,N) \\
 & \downarrow F_{M,N} & & \downarrow F_{X_0,N} & & \downarrow F_{X_1,N} \\
0 \to & \mathrm{Hom}_{\Lambda'}(FM,FN) & \to & \mathrm{Hom}_{\Lambda'}(FX_0,FN) & \to & \mathrm{Hom}_{\Lambda'}(FX_1,FN) ,
\end{array}$$

with exact rows, in which (by Proposition 3) $F_{X_0,N}$ and $F_{X_1,N}$ are isomorphisms. Hence $F_{M,N}$ is an isomorphism.

(d) $\Rightarrow$ (e). $F_{M,N} = \omega^M_N \circ \mathrm{Hom}_\Lambda(\epsilon_M,N)$ holds for any two Λ-modules M,N, and ω^M_N is an isomorphism. So (d) $\Rightarrow$ (e) is obvious.

(e) $\Rightarrow$ (a) $\Rightarrow M \in \underline{Q}$ Im G. These implications are obvious.

<u>Proof of Theorem II(2)</u>. The above proof shows that Im G = $\underline{Q}$ Im G, and (a) $\Leftrightarrow$ (e) shows that Im G = G Im F = Im GF. Proposition 1 shows that Im G = $\underline{Q}$ Im G = $\underline{Q}$ G(add Λ') = $\underline{Q}$ addX, so that add X $\subset$ Im G. Finally, (a) $\Leftrightarrow$ (b) shows that Im G contains all the injective Λ-modules.

This completes the proof of Theorem II, with (1)(f) deleted.

As immediate corollaries of Theorem II(1),(a) $\Leftrightarrow$ (d) and (a) $\Leftrightarrow$ (b), respectively, we note:

<u>COROLLARY 1</u> Theorem III(a) holds.

<u>COROLLARY 2</u> F <u>leaves exact exact sequences</u> $0 \to M \to M_1 \to M_2 \to 0$ with $M \in \operatorname{Im} G$.

There are three less trivial corollaries.

<u>COROLLARY 3</u> T_0' <u>holds</u>.

<u>Proof</u> The map $\Lambda \to \operatorname{End}_{\Lambda'^{op}} X$ in T_0' will be denoted by h. Now, for any Λ^{op}-module M, there is a sequence of maps.

$$GFDM = G\operatorname{Hom}_\Lambda(X,DM) \cong GD(M \otimes_\Lambda X) \cong D^2(X \otimes_{\Lambda'} D(M \otimes_\Lambda X))$$

$$\cong D\operatorname{Hom}_{\Lambda'^{op}}(X, D^2(M \otimes_\Lambda X) \cong D\operatorname{Hom}_{\Lambda'^{op}}(X, M \otimes_\Lambda X),$$

of which the first and third are associativity isomorphisms and the second and fourth are duality isomorphisms, $A \cong D^2A$, for suitable A. Let η_M be the composite. Set $M = \Lambda$. Then it is easily verified that the diagram

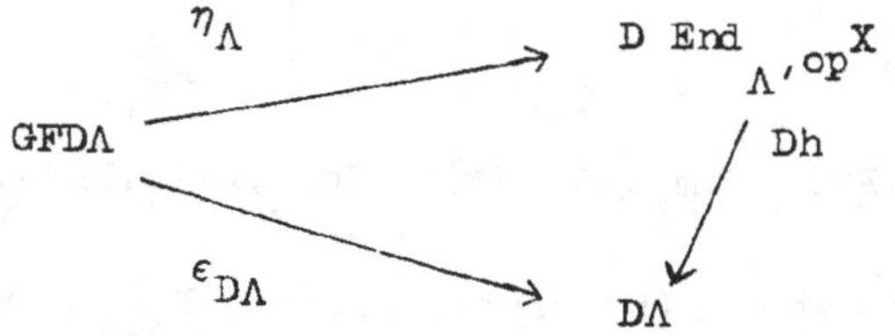

commutes and, since $D\Lambda \in \operatorname{Im} G$ by Theorem II, $\epsilon_{D\Lambda}$ is an isomorphism, again by Theorem II. Hence Dh and h are isomorphisms.

<u>COROLLARY 4</u> <u>If</u> $\operatorname{Tor}_1^{\Lambda'}(X,M') = 0$, <u>then</u> $M' \in \operatorname{Im} F$ <u>and</u> $\mu_{M'}$ <u>is an isomorphism</u>.

<u>Proof</u> Choose an exact sequence $P_2' \to P_1' \to P_0' \to M' \to 0$ with $P_i' \in \operatorname{add} \Lambda'(i = 0, 1, 2)$. Since $\operatorname{Tor}_1^{\Lambda'}(X,M') = 0$ it follows that $GP_2' \to GP_1' \to GP_0' \to GM' \to 0$ is exact, with terms in $\operatorname{Im} G = \underline{Q} \operatorname{Im} G$. Hence $FGP_1' \to FGP_0' \to FGM' \to 0$ is exact, using Corollary 2. Now Proposition 3(a) implies that $\mu_{P'}$: $P' \to FGP'$ is an isomorphism for $P' \in \operatorname{add} \Lambda'$ and so $\mu_{M'} : M' \to FGM'$ is also an isomorphism.

<u>COROLLARY 5</u> T_4' <u>holds</u>.

Proof Suppose $GM' = 0 = \mathrm{Tor}_1^{\Lambda'}(X,M')$. By Corollary 4, $M' \cong FM$ with $M \cong GM' \in \mathrm{Im}\, G$. Hence $GFM = 0$ and $M \in \mathrm{Im}\, G$, so Theorem II, (1), (a) <=> (e) shows that $M = 0$. Hence $M' \cong FM = 0$, and T_4' follows.

5. Proofs of Theorems I and IV

Throughout this section we suppose T_* holds, and shall first prove Theorem I, that is, that T_*' holds; note that we have already obtained T_0', T_4' as Corollaries 3, 5 in §4. The proof will be obtained by studying exact sequences induced from the standard projective resolution of X in Definition 2. We shall use repeatedly the bimodule isomorphism ${}_{\Lambda'}\Lambda'_{\Lambda'} \cong {}_{\Lambda'}FX_{\Lambda'}$ which is a consequence of T_0.

PROPOSITION 6 (a) There is an exact sequence

$$0 \to \Lambda'_{\Lambda'} \overset{\pi'}{\to} P' \overset{\rho'}{\to} T' \to 0$$

in mod Λ'^{op} in which $P' = \mathrm{Hom}_\Lambda(P,X)$ is both Λ'^{op} projective and in add X^{op}, $T' = \mathrm{Hom}_\Lambda(R,X)$, $\rho' = \mathrm{Hom}_\Lambda(\rho,X)$, and π' is composite of $\mathrm{Hom}_\Lambda(\pi,X)$ and the isomorphism $\Lambda' \cong (\mathrm{End}_\Lambda X)^{op}$.

(b) There is an exact sequence

$$0 \to FR \overset{F(\rho)}{\to} FP \overset{F(\pi)}{\to} FX \to \mathrm{Ext}_\Lambda^1(X,R) \to 0$$

in which FX and FP are Λ'-projective, and $F(\pi)$: $FP \to \mathrm{Im}\, F(\pi)$ is an essential epimorphism.

Proof The exactness of the sequence in (a) follows from T_2, P' is projective by Proposition 3, and in add X^{op} since it is a Λ'^{op}-summand of $\mathrm{Hom}_\Lambda(\Lambda^n,X) \cong X^n_{\Lambda'}$ for some natural number n. The exactness of the sequence in (b) is a consequence of Proposition 4(d), the projectivity of FX and FP depends on T_0 and Propositions 3 and 4, and the last part is obtained by applying the following lemma to the map $\pi : P \to X$:

LEMMA 1 Let $M \in \mathrm{add}\, X$ and σ : $M \to N$ be an essential epimorphism. Then $F(\sigma)$: $FM \to \mathrm{Im}\, F(\sigma)$ is an essential epimorphism.

Proof Let $r = \mathrm{rad}\, \Lambda$; then $r' = \mathrm{rad}\, \Lambda' \supset \mathrm{Im}(F(rX) \to FX)$

since r is nilpotent. Now $K = \mathrm{Ker}\ \sigma \subset rM$ so the left exactness of F implies that $\mathrm{Ker}\ F(\sigma) = FK \subset F(rM)$. Since $M \in \mathrm{add}\ X$ and $\mathrm{Im}(F(rX) \to FX) \subset r'$, we find that $\mathrm{Im}(F(rM) \to FM) \subset r'FM$, and so $\mathrm{Ker}\ F(\sigma) \subset r'FM$. This gives the conclusion of the lemma.

In the next set of formulae, the natural transformations

$$\eta^Y : \mathrm{Hom}_\Lambda(Y,X) \otimes_{\Lambda'} M' \to \mathrm{Hom}_{\Lambda'}(FY,M')$$

are given by

$$\eta^Y(\chi \otimes_{\Lambda'} m')(\phi) = (\chi o \phi)'m',$$

where $(\chi o \phi)'$ is the unique element of Λ' mapping (because of T_0) to $\chi o \phi \in FX$. Note that

<u>LEMMA 2</u> η^Y <u>is an isomorphism for</u> $Y \in \mathrm{add}\ X$.
This is clear from T_0 for $Y = X$, and follows by naturality for arbitrary $Y \in \mathrm{add}\ X$.

<u>PROPOSITION 7</u> <u>For each</u> Λ'-<u>module</u> M', <u>there exists</u>
(a) <u>a commutative diagram</u>,

$$\begin{array}{ccccccc} \Lambda' \otimes_{\Lambda'} M' & \xrightarrow{\pi'} & P' \otimes_{\Lambda'} M' & \xrightarrow{\rho'} & T' \otimes_{\Lambda'} M' & \to & 0 \\ \downarrow \eta^X & & \downarrow \eta^P & & \downarrow \eta^R & & \\ \mathrm{Hom}_{\Lambda'}(FX,M') & \xrightarrow{F(\pi)} & \mathrm{Hom}_{\Lambda'}(FP,M') & \xrightarrow{F(\rho)} & \mathrm{Hom}_{\Lambda'}(FR,M') & & \end{array},$$

<u>in which the top row is exact, and</u> η^X <u>and</u> η^P <u>are isomorphisms which induce, respectively</u>,
(b) <u>an isomorphism</u> $\mathrm{Tor}_1^{\Lambda'}(T',M') \simeq \mathrm{Hom}_{\Lambda'}(\mathrm{Ext}_\Lambda^1(X,R),M')$,
(c) <u>and an exact sequence</u>

$$0 \to \mathrm{Ext}_\Lambda^1(\mathrm{Ext}_\Lambda^1(X,R),M') \to T' \otimes_{\Lambda'} M \xrightarrow{\eta^R} \mathrm{Hom}_{\Lambda'}(FR,M')$$

$$\to \mathrm{Ext}_{\Lambda'}^2(\mathrm{Ext}_\Lambda^1(X,R),M') \to 0 .$$

<u>Proof</u> The existence of the diagram in (a) and the exactness of the top row are formal consequences of Proposition 6. Since $X, P \in \mathrm{add}\ X$, Lemma 2 shows that η^X and η^P are

isomorphisms. It follows that η^X induces an isomorphism $\text{Ker}\ \pi' \xrightarrow{\sim} \text{Ker}\ F(\pi)$, which reduces to (b) using Proposition 6 to identify these kernels. For (c), observe that, from (a), $\text{Ker}\ \eta^R \cong \text{Ker}\ F(\rho)/\text{Im}\ F(\pi)$ and $\text{Coker}\ \eta^R \cong \text{Coker}\ F(\rho)$, so the terms in the exact sequence are obtained from (a).

The following proposition records a number of consequences of these formulae. Recall that the notation $A \sim B$ means that each indecomposable summand of A is a summand of B, and vice versa.

<u>PROPOSITION 8</u> (a) $\text{Hom}_{\Lambda'}(\text{Ext}^1_\Lambda(X,R),\Lambda') = 0$,

(b) $\text{Ext}^1_\Lambda(X,R) \neq 0$ <u>and has no</u> <u>Λ'-projective summands</u>,

(c) $G\ \text{Ext}^1_\Lambda(X,R) = 0$, (d) $\text{Ext}^1_\Lambda(X,R) \cong \text{Tr}_{\Lambda'^{op}} T' \sim \text{Tr}_{\Lambda'^{op}} X$,

(e) $T' \cong \text{Tr}_{\Lambda'}\text{Ext}^1_\Lambda(X,R) \neq 0$ <u>and has no projective summands</u>,

(f) $\text{pd}_{\Lambda'^{op}} T' = 1$, (g) $X_{\Lambda'} \sim P' \amalg T'$, (h) $\text{pd}_{\Lambda'^{op}} X = 1$.

(i) $\text{Hom}_{\Lambda'}(FP,\text{Ext}^1_\Lambda(X,R)) = 0$.

<u>Proof</u> (a) is obtained by setting $M' = \Lambda'$ in Proposition 7 (b), and then (a) => (b) once we prove that $\text{Ext}^1_\Lambda(X,R) \neq 0$. To do this, choose a non-zero map $R \to S$ for some suitably chosen simple module S. Then Proposition 4 shows that $\text{Ext}^1_\Lambda(X,S) \neq 0$, so that the non-vanishing of $\text{Ext}^1_\Lambda(X,R)$ follows from the right exactness, (T_1), of $\text{Ext}^1_\Lambda(X,-)$. For (c), since G is right exact, we obtain from Proposition 6 the formula $G\ \text{Ext}^1_\Lambda(X,R) \cong \text{Coker}\ GF(\pi)$; since $GF(\pi)\epsilon_P = \epsilon_X\pi$ and $P,X \in \underline{Q}\ \text{Im}\ G$, Theorem II(1) shows that ϵ_P, ϵ_X are isomorphisms. Now π is surjective, and so $GF(\pi)$ is also surjective, which fact gives (c). Next we prove (e). From (b), $\text{Tr}_{\Lambda'}\text{Ext}^1_\Lambda(X,R)$ is non-zero and has no projective summands and it follows from Proposition 6 that

$$\text{Tr}_{\Lambda'}\text{Ext}^1_\Lambda(X,R) \cong \text{Coker}\ \text{Hom}_{\Lambda'}(F(\pi),\Lambda')$$
$$\cong \text{Coker}\ (\eta^P\pi'\ (\eta^X)^{-1}) \quad \text{with } M' = \Lambda' \text{ in Proposition 7(a)}$$
$$\cong T'.$$

as required for (e). Since Proposition 6(a) gives a projective resolution of T′, and T′ is not projective by (e), we deduce (f). For (g), one simply notes that $X_{\Lambda'} \cong \mathrm{Hom}_{\Lambda}(\Lambda,X) \sim \mathrm{Hom}_{\Lambda}(P \amalg R,X)$ (by Proposition 5) $\cong P' \amalg T'$. Obviously (f) and (g) imply (h). Also (b) and (e) give the first isomorphism in (d), and (g) gives $\mathrm{Tr}_{\Lambda'^{op}}T' \sim \mathrm{Tr}_{\Lambda'^{op}}X$ on noting that Tr vanishes on projectives. Only (i) remains to be proved.

By Proposition 6(b), (i) is equivalent to the statement that $\mathrm{Hom}_{\Lambda'}(FP,F(\pi))$ is surjective. Now $P \in \mathrm{add}\, X$, so Theorem II(1) ensures that $F_{P,-}: \mathrm{Hom}_{\Lambda}(P,-) \to \mathrm{Hom}_{\Lambda'}(FP,F-)$ is an equivalence of functors. Thus

$$\mathrm{Hom}_{\Lambda'}(FP,F(\pi)) = F_{P,X} \circ \mathrm{Hom}_{\Lambda}(P,\pi) \circ F_{P,P}^{-1},$$

so its surjectivity is a trivial consequence of the surjectivity of π and projectivity of P. This completes the proof of Proposition 8.

<u>COROLLARY 1</u> <u>Theorem</u> I <u>holds</u>.

<u>Proof</u> We must prove T_1', T_2', T_3'. Of course T_1' is just Proposition 8(h). To obtain T_2', we have

$$\mathrm{Tor}_1^{\Lambda'}(X,DX) \sim \mathrm{Hom}_{\Lambda'}(\mathrm{Ext}_{\Lambda}^1(X,R),DX)$$ by Propositions 8(g) and 7(b)

$$\cong DG\,\mathrm{Ext}_{\Lambda}^1(X,R)$$

$$= 0$$ by Proposition 8(c).

Finally, for T_3', Propositions 8(g) and 6(a) show that $X_{\Lambda'}$ has a projective cover in $\mathrm{add}(P' \amalg P') = \mathrm{add}\, P' \subset \mathrm{add}\, X^{op}$.

<u>COROLLARY 2</u> <u>Theorem</u> IV <u>holds</u>.

<u>Proof</u> We use the results in Proposition 2 on the analysis of $\epsilon_M : GFM \to M$. Since T_* holds then, by Theorem I, so does T_*' and, hence, the adjoint dual of Theorem II holds. Since $FM \in \mathrm{Im}\, F$, the dual of Theorem II(1) shows that μ_{FM} is is an isomorphism. Since $F(\epsilon_M)\mu_{FM} = 1_{FM}$, $F(\epsilon_M)$ is an isomorphism and so Proposition 2(c) gives $F\mathrm{Ker}\epsilon_M = 0$. Also, by Theorem II, Proposition 2(a), and T_1, the exact sequences of Proposition 2(d) degenerate to $\mathrm{Ext}_{\Lambda}^1(X,\mathrm{Ker}\,\epsilon_M) = 0$, $F\,\mathrm{Coker}\,\epsilon_M = 0$, and exactness of $0 \to \mathrm{Ext}_{\Lambda}^1(X,M) \to \mathrm{Ext}_{\Lambda}^1(X,\mathrm{Coker}\,\epsilon_M) \to 0$.

The statements of Theorem IV follow now on using T_4 to conclude that Ker $\epsilon_M = 0$.

6. Proofs of Theorems II and V

For Theorem II, we have yet to prove the equivalence of (1)(f) with each of (1)(a)-(e). This, and the proof of Theorem V, will be achieved by studying the transformation $ev_N^M : M^* \otimes_\Lambda N \to \mathrm{Hom}_\Lambda(M,N)$ defined in the Introduction for suitable pairs M,N. We assume throughout that T_* and T'_* hold.

PROPOSITION 9 *For each Λ-module M, there exist*

(a) *an isomorphism* $\mathrm{Tr}_\Lambda X \otimes_\Lambda M \cong \mathrm{Ext}^1_\Lambda(X,M)$,

(b) *an exact sequence*

$$0 \to \mathrm{Tor}_2^\Lambda(\mathrm{Tr}_\Lambda X, M) \to X^* \otimes_\Lambda M \overset{ev_M^X}{\to} FM \to \mathrm{Tor}_1^\Lambda(\mathrm{Tr}_\Lambda X, M) \to 0,$$

(c) *an isomorphism* $\mathrm{Hom}_\Lambda(X,M) \cong \mathrm{Tor}_1^\Lambda(\mathrm{Tr}_\Lambda X, M)$.

Proof The standard projective resolution of X gives an exact sequence

$$0 \to X^* \overset{\pi^*}{\to} P^* \overset{\rho^*}{\to} R^* \to \mathrm{Tr}_\Lambda X \to 0$$

defining $\mathrm{Tr}_\Lambda X$. On tensoring with M, we obtain a commutative diagram

$$\begin{array}{ccccccc} X^* \otimes_\Lambda M & \to & P^* \otimes_\Lambda M & \to & R^* \otimes_\Lambda M & \to \mathrm{Tr}_\Lambda X \otimes_\Lambda M \to 0 \\ \downarrow ev_M^X & & \downarrow ev_M^P & & \downarrow ev_M^R & \\ 0 \to (FM=)\mathrm{Hom}_\Lambda(X,M) & \to & \mathrm{Hom}_\Lambda(P,M) & \to & \mathrm{Hom}_\Lambda(R,M) & \end{array}$$

in which the top row is exact except, possibly, at $P^* \otimes_\Lambda M$, the bottom row is exact, and ev_M^P and ev_M^R are isomorphisms (P and R being projective). Thus ev_M^R induces an isomorphism of $\mathrm{Tr}_\Lambda X \otimes_\Lambda M$ onto Coker $(\mathrm{Hom}_\Lambda(P,M) \to \mathrm{Hom}_\Lambda(R,M)) \cong \mathrm{Ext}^1_\Lambda(X,M)$. This proves (a), and (b) follows by diagram chasing since P^* and R^* are Λ^{op}-projectives. Finally (b) implies (c), for

$\underline{\mathrm{Hom}}_\Lambda(X,M)$ is just the cokernel of ev_M^X.

The completion of the proof of Theorem II is part of the following corollary.

COROLLARY. For each Λ-module M, the following statements are equivalent: (a) $M \in \mathrm{Im}\, G$, (b) $\mathrm{Tr}_\Lambda X \otimes_\Lambda M = 0$, (c) $\mathrm{Hom}_\Lambda(M, D\mathrm{Tr}_\Lambda X) = 0$.

Proof From Theorem II, (a) $\Leftrightarrow \mathrm{Ext}^1_\Lambda(X,M) = 0$, so Proposition 9 gives the equivalence of (a) and (b). (b) $\Leftrightarrow$ (c) since $\mathrm{Hom}_\Lambda(M, D\mathrm{Tr}_\Lambda X) \cong D(\mathrm{Tr}_\Lambda X \otimes_\Lambda M)$.

We state formally the statements adjoint dual to these; they may be proved using the natural transformations

$$N' \otimes_{\Lambda'} M' \to \mathrm{Hom}_{\Lambda'}(N'^*, M'), \qquad n' \otimes_{\Lambda'} m' \mapsto (\phi' \mapsto \phi'(n')m'),$$

in which $N' = N'_{\Lambda'}$, $M' = {}_{\Lambda'}M'$ and $N'^* = \mathrm{Hom}_{\Lambda'^{op}}(N', \Lambda')$. These transformations are isomorphisms for injective M' (to see this, take $M' = D\Lambda'$) and the kernels are $D\,\overline{\mathrm{Hom}}_{\Lambda'}(M', DN')$. We find

PROPOSITION 9' For each Λ'-module M', there exist:

(a) an isomorphism $\mathrm{Tor}_1^{\Lambda'}(X,M') \cong \mathrm{Hom}_{\Lambda'}(\mathrm{Tr}_{\Lambda'^{op}} X, M')$,

(b) an exact sequence, with $X^{*\prime} = \mathrm{Hom}_{\Lambda'^{op}}(X, \Lambda')$,

$$0 \to \mathrm{Ext}^1_{\Lambda'}(\mathrm{Tr}_{\Lambda'^{op}} X, M') \to GM' \to \mathrm{Hom}_{\Lambda'}(X^{*\prime}, M') \to \mathrm{Ext}^2_{\Lambda'}(\mathrm{Tr}_{\Lambda'^{op}} X, M') \to 0.$$

(c) An isomorphism $D\overline{\mathrm{Hom}}_{\Lambda'}(M', DX) \cong \mathrm{Ext}^1_{\Lambda'}(\mathrm{Tr}_{\Lambda'^{op}} X, M')$.

COROLLARY For each Λ'-module M', the following statements are equivalent: (a) $M' \in \mathrm{Im}\, F$, (b) $\mathrm{Hom}_{\Lambda'}(\mathrm{Tr}_{\Lambda'^{op}} X, M') = 0$, (c) $DM' \otimes_{\Lambda'} \mathrm{Tr}_{\Lambda'^{op}} X = 0$.

The proof of Theorem V needs an analysis of the maps ev_X^M, and also the use of two terms of a minimal projective resolution of the arbitrary Λ-module M.

PROPOSITION 10 *Let* M *be a* Λ-*module and*

$P_1 \xrightarrow{\mu_1} P_0 \xrightarrow{\mu_0} M \to 0$ *two terms of a minimal projective resolution of* M. *There exist*

(a) *an exact sequence*

$$0 \to M^* \otimes_\Lambda X \xrightarrow{ev_X^M} \mathrm{Hom}_\Lambda(M,X) \to \mathrm{Ext}^1_\Lambda(M,R) \to \mathrm{Ext}^1_\Lambda(M,P) \quad \dots ,$$

(b) *isomorphisms*

$$\mathrm{Ext}^1_\Lambda(X,\mathrm{DTr}_\Lambda M) \cong \mathrm{D}\underline{\mathrm{Hom}}_\Lambda(M,X) \cong \mathrm{DCoker}\ ev_X^M ,$$

(c) *and an exact sequence*

$$0 \to \mathrm{FDTr}_\Lambda M \to \mathrm{DHom}_\Lambda(P_1,X) \to \mathrm{DHom}_\Lambda(P_0,X) \to \mathrm{DHom}_\Lambda(M,X) \to 0.$$

Proof The standard projective resolution of X determines an exact and commutative diagram

$$\begin{array}{ccccccccc}
 & & M^*\otimes_\Lambda R & \to & M^*\otimes_\Lambda P & \to & M^*\otimes_\Lambda X & \to & 0 \\
 & & \downarrow ev_R^M & & \downarrow ev_P^M & & \downarrow ev_X^M & & \\
0 & \to & \mathrm{Hom}_\Lambda(M,R) & \to & \mathrm{Hom}_\Lambda(M,P) & \to & \mathrm{Hom}_\Lambda(M,X) & \to & \mathrm{Ext}^1_\Lambda(M,R) \to \mathrm{Ext}^1_\Lambda(M,P) \to \dots
\end{array}$$

in which the first two vertical maps are isomorphisms. This gives (a) at once. For any Λ-module N, define ϵ_N^X to be the composite

$$\mathrm{DHom}_\Lambda(N,X) \xrightarrow{D(ev_X^N)} D(N^*\otimes_\Lambda X) \cong \mathrm{Hom}_\Lambda(X,DN^*) = \mathrm{FDN}^*.$$

From (a) we deduce that ϵ_N^X is surjective with Ker ϵ_N^X equal to DCoker $ev_X^N \cong \mathrm{D}\underline{\mathrm{Hom}}_\Lambda(M,X)$. On applying these transformations to the two given terms of the minimal projective resolution of M, we obtain the commutative diagram

$$\begin{array}{ccccc}
\mathrm{DHom}_\Lambda(P_1,X) & \to & \mathrm{DHom}_\Lambda(P_0,X) & \to & \mathrm{DHom}_\Lambda(M,X) \to 0 \\
\downarrow \epsilon^X_{P_1} & & \downarrow \epsilon^X_{P_0} & & \downarrow \epsilon^X_M \\
\mathrm{FDP}^*_1 & \overset{\mathrm{FD}\mu^*_1}{\to} & \mathrm{FDP}^*_0 & \overset{\mathrm{FD}\mu^*_0}{\to} & \mathrm{FDM}^*
\end{array} ,$$

in which the top row is exact; also $\epsilon^X_{P_1}$ and $\epsilon^X_{P_0}$ are isomorphisms since P_1 and P_0 are projective. Obviously the bottom row is obtained by applying the left exact functor $F = \mathrm{Hom}_\Lambda(X,-)$ to part of the exact sequence

$$0 \to \mathrm{DTr}_\Lambda M \to \mathrm{DP}^*_1 \to \mathrm{DP}^*_0 \to \mathrm{DM}^* \to 0 ,$$

in which DP^*_1 and DP^*_0 are injective Λ-modules. Hence

$$\mathrm{Ker}\ \mathrm{FD}\mu^*_1 \cong \mathrm{FDTr}_\Lambda M \quad \text{and} \quad \mathrm{Ker}\ \mathrm{FD}\mu^*_0/\mathrm{Im}\ \mathrm{FD}\mu^*_1 \cong \mathrm{Ext}^1_\Lambda(X,\mathrm{DTr}\,M) ,$$

and then (b) and (c) follow from the properties of this diagram and of ϵ^X_M.

<u>COROLLARY</u> <u>Theorem</u> V <u>holds</u>.

<u>Proof</u> The first part (a), is just Proposition 10(b), and part (b) follows from (a) and the equivalence (a) <=> (b) in Theorem II(1). For (c), we can assume M is indecomposable, and $M \notin \mathrm{add}\,(\Lambda \amalg X)$. Let $X_0 = \mathrm{Tr}_{\Lambda^{op}}\mathrm{Tr}_\Lambda X$, so that $X \cong X_0 \amalg P(X)$ with $P(X)$ projective. Then the hypothesis of (c) means that the almost split sequence of $\mathrm{DTr}_\Lambda X$ has the form

$$0 \to \mathrm{DTr}_\Lambda X \to B \to X_0 \to 0$$

with B projective, and the hypothesis on M implies that every map $M \to X_0$ factorises through B. Thus $\underline{\mathrm{Hom}}_\Lambda(M,X) = 0$, so part (b) shows that $\mathrm{DTr}_\Lambda M \in \mathrm{Im}\ G$.

7. <u>Proofs of Theorems III(b), VI and VII</u>

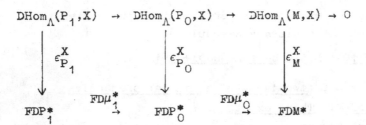

The proofs to be given in this section require the choices of projective resolutions to compute derived functors. We assume throughout that T_* holds and that M is an arbitrary Λ-module. Let k be an integer greater than 1, or $k = \infty$, and let

$$P(k) : P_{k-1} \overset{\mu_{k-1}}{\to} \cdots \to P_1 \overset{\mu_1}{\to} P_0 \overset{\mu_0}{\to} M \to 0$$

denote k terms of a minimal projective resolution of M. Further, we may and do select k terms of an F-exact X-projective resolution

$$X(k) : X_{k-1} \overset{\lambda_{k-1}}{\to} \cdots \to X_1 \overset{\lambda_1}{\to} X_0 \overset{\lambda_0}{\to} GFM \to 0$$

of GFM such that

$$FX(k) : FX_{k-1} \overset{F\lambda_{k-1}}{\to} \cdots \to FX_1 \overset{F\lambda_1}{\to} FX_0 \overset{F\lambda_0}{\to} FGFM \to 0$$

are the first k terms of a minimal Λ'-projective resolution of $FGFM \cong FM$. (This is possible: first select any minimal Λ'-projective resolution of FM. Since T_* and T_*' hold, all its terms are in Im F and the projective terms are in F add X and then Theorem III(a) and its adjoint dual ensure that the image under G of this resolution has the properties required of X(k)).

PROPOSITION 11 (a) Suppose $M \in \operatorname{Im} G$. There is a complex morphism $\theta(k) : P(k) \to X(k)$ extending the isomorphism (Theorem II) $\epsilon_M^{-1} : M \to GFM$.

(b) Suppose $\operatorname{Tor}_p^\Lambda(\operatorname{Tr}_\Lambda X, M) = 0$ for $p = 1, 2, \ldots, k$ (equivalently, that $\operatorname{Ext}_\Lambda^p(M, D\operatorname{Tr}_\Lambda X) = 0$ for $p = 1, 2, \ldots, k$). There is a complex morphism $\phi(k) : X(k) \to P(k)$ covering the morphism $\epsilon_M : GFM \to M$.

Proof (a) P(k) is a projective resolution of M.

(b) Assume the hypothesis of the lemma, and let $M_p = \operatorname{Im} \mu_p$ in P(k), so that $M_0 = M$. By dimension shifting, we have

$$\operatorname{Tor}_1^\Lambda(\operatorname{Tr}_\Lambda X, M_p) = 0 = \operatorname{Tor}_2^\Lambda(\operatorname{Tr}_\Lambda X, M_p) \text{ for } p = 0, 1, \ldots, k-2,$$

and then Proposition 9(b) shows that the maps

$$ev_{M_p}^X : X^* \otimes_\Lambda M_p \to \operatorname{Hom}_\Lambda(X, M_p) = FM_p$$

are isomorphisms for $p = 0, 2, \ldots, k - 2$. For each such p, the exact sequence $P_{p+1} \to P_p \to M_p \to 0$, the right exactness of $X^* \otimes_\Lambda -$, and the fact that ev_N^X is an isomorphism for projective N, therefore ensure that $FP_{p+1} \to FP_p \to FM_p \to 0$ is also an exact sequence. Thus the sequence $FP(k)$ is exact. Since $FX(k)$ is part of a projective resolution of $FGFM$, the morphism $F\epsilon_M : FGFM \to FM$ lifts to a complex morphism $\phi'(k) : FX(k) \to FP(k)$. Finally, every term in $X(k)$ is in Im G so Theorem II(1) can be used to show that there is a complex morphism $\phi(k): X(k) \to P(k)$ such that $F\phi(k) = \phi'(k)$, and also that $\phi(k)$ lifts ϵ_M as required.

<u>COROLLARY 1</u> <u>Suppose</u> M <u>satisfies the hypotheses of Proposition</u> 11, (a) <u>and</u> (b). <u>The complex morphisms</u> $\theta(k)$ <u>and</u> $\phi(k)$ <u>are complex isomorphisms</u>.

<u>Proof</u> Under the combined hypotheses, $\phi(k)\ \theta(k)$ is a complex endomorphism of $P(k)$ covering $\epsilon_M \epsilon_M^{-1} = id_M$. Since $P(k)$ is a minimal projective resolution, it follows that $\phi(k)\ \theta(k)$ is an automorphism. Similarly, $F\theta(k)\ F\phi(k)$ covers an automorphism of $FGFM$, so is itself an automorphism, so that Theorem II(1) shows that $\theta(k)\ \phi(k)$ is an automorphism of $X(k)$. Thus $\theta(k)$, $\phi(k)$ are both isomorphisms.

<u>COROLLARY 2</u> <u>Theorem</u> VI <u>holds</u>.

<u>Proof</u> Let N be any Λ-module and let M satisfy the hypotheses of Theorem VI. Then Corollary 1 yields a complex isomorphism

$$\mathrm{Hom}_\Lambda(P(k),N) \xrightarrow{\sim} \mathrm{Hom}_\Lambda(X(k),N),$$

and Theorem II(1) a complex isomorphism

$$\mathrm{Hom}_\Lambda(X(k),N) \xrightarrow{\sim} \mathrm{Hom}_{\Lambda'}(FX(k),FN).$$

Theorem VI simply records the induced isomorphisms of homology of these complexes.

<u>COROLLARY 3</u> <u>Theorem</u> VII <u>holds</u>.

<u>Proof</u> Since $FX(2)$ provides two terms of a minimal Λ'-projective resolution of $FGFM \cong FM$, we obtain an exact sequence

$$0 \to DTr_{\Lambda'}FM \to DHom_{\Lambda'}(FX_1,\Lambda') \to DHom_{\Lambda'}(FX_0,\Lambda')$$

$$\to DHom_{\Lambda'}(FGFM,\Lambda') \to 0$$

the maps being induced by the λ_i's. Since $\Lambda' = FX$ and X_1, X_0, GFM ϵ Im G, Theorem II(1)(d) shows that it is isomorphic to the exact sequence

$$0 \to DTr_{\Lambda'}FM \to DHom_{\Lambda}(X_1,X) \to DHom_{\Lambda}(X_0,X)$$

$$\to DHom_{\Lambda}(GFM,X) \to 0.$$

The hypotheses of Theorem VII ensure that Corollary 1 above holds with k = 2, so we have complex isomorphisms $P(2) \underset{\phi(2)}{\overset{\theta(2)}{\rightleftarrows}} X(2)$; therefore comparison of the last exact sequence with that in Proposition 10(c) shows that $\theta(2)$ and $\phi(2)$ induce isomorphisms $FDTr_{\Lambda}M \underset{\phi}{\overset{\theta}{\rightleftarrows}} DTr_{\Lambda'}FM$. This is Theorem VII(a). To obtain (b) <u>we shall suppose that Theorem III(b)</u> has already been proved. Then the hypotheses of Theorem VII(b) ensure that

$$Ext^1(M,DTr_{\Lambda}M) \overset{F}{\to} Ext^1(FM,FDTr_{\Lambda}M)$$

is an isomorphism, and since we have just shown that $FDTr_{\Lambda}M \cong DTr_{\Lambda'}FM$, we have an isomorphism

$$Ext^1_{\Lambda}(M,DTr_{\Lambda}M) \overset{\sim}{\to} Ext^1_{\Lambda'}(FM,DTr_{\Lambda'}FM)$$

which may easily be verified to commute with the actions of $(End_{\Lambda}M)^{op}$ on the left and of the isomorphic algebra $(End_{\Lambda'}FM)^{op}$ on the right. Thus the isomorphism maps the $(End_{\Lambda}M)^{op}$-socle of the left hand side to the $(End_{\Lambda'}FM)^{op}$-socle of the right hand side, so it maps almost split sequences to almost split sequences.

The proof of Theorem III(b) makes use of the theory of spectral sequences of double complexes ([9], Chapter XV).

Let M, N be Λ-modules, choose $X_{\bullet} = X(\infty)$ to be any F-exact X-projective resolution of GFM, and $Q^{\bullet}$ to be an injective resolution of N. We consider the double complex

$$D^{\bullet\bullet} = \mathrm{Hom}_{\Lambda}(X_{\bullet},Q^{\bullet})$$

and observe immediately that, since each $X_p \in \mathrm{add}\, X$, Theorem II gives an isomorphism of double complexes,

$$D^{\bullet\bullet} = \mathrm{Hom}_{\Lambda}(X_{\bullet},Q^{\bullet}) \cong \mathrm{Hom}_{\Lambda'}(FX_{\bullet},FQ^{\bullet}).$$

This is a first quadrant double complex and determines a pair of spectral sequences both converging to the total homology $H^{\bullet}$ of the associated single complex. By H_1 we denote the homology with respect to the first variable, with the second variable fixed, and by H_2 homology computed with roles of the variables reversed. Then clearly we have

$$H_1^p D^{\bullet\bullet} = \mathrm{Hom}_{\Lambda}(H_1^p X_{\bullet},Q^{\bullet}) \quad \text{since } Q^{\bullet} \text{ is an injective resolution,}$$

$$= \begin{cases} \mathrm{Hom}_{\Lambda}(GFM,Q^{\bullet}) & \text{for } p = 0 \\ 0 & \text{for } p > 0, \end{cases}$$

so that

$$(H_2H_1)^{pq}D^{\bullet\bullet} = \begin{cases} \mathrm{Ext}_{\Lambda}^{q}(GFM,N) & \text{for } p = 0 \\ 0 & \text{for } p > 0 \end{cases}$$

So the spectral sequence with this "E_2" term collapses and the total homology of $D^{\bullet\bullet}$ is given by

$$H^{\bullet} = \mathrm{Ext}_{\Lambda}^{\bullet}(GFM,N).$$

Next we calculate the terms $E_2^{pq} = (H_1H_2)^{pq}D^{\bullet\bullet}$ of the second spectral sequence using the formula $\mathrm{Hom}_{\Lambda'}(FX_{\bullet},FQ^{\bullet}) \cong D^{\bullet\bullet}$. Since each term of $FX_{\bullet}$ is Λ'-projective,
$H_2^q D^{\bullet\bullet} \cong \mathrm{Hom}_{\Lambda'}(FX_{\bullet},H^qFQ^{\bullet})$. Also $Q^{\bullet}$ is an injective resolution of N and $F = \mathrm{Hom}_{\Lambda}(X,-)$, so since $\mathrm{pd}_{\Lambda}X = 1$ we find that

$$H^0FQ^{\bullet} = FN, \quad H^1FQ^{\bullet} = \mathrm{Ext}_{\Lambda}^1(X,N), \text{ and } \quad H^qFQ^{\bullet} = 0 \text{ for } q > 1.$$

Therefore, since FX. is a Λ'-projective resolution of FGFM, we obtain

$$E_2^{po} = (H_1H_2)^{po}D^{\bullet\bullet} = \mathrm{Ext}^p_{\Lambda'}(FGFM,FN) ;$$

$$E_2^{p1} = (H_1H_2)^{p1}D^{\bullet\bullet} = \mathrm{Ext}^p_{\Lambda'}(FGFM,\mathrm{Ext}^1_\Lambda(X,N));$$

$$E_2^{pq} = 0 \quad \text{for } q > 1.$$

For a spectral sequence $E_2^{\bullet\bullet} \Rightarrow H^\bullet$ in which $E_2^{pq} = 0$ except for q = 0, 1 there is a long exact sequence ([9], p.329)

$$0 \to E_2^{10} \to H^1 \to E_2^{01} \to E_2^{20} \to H^2 \to E_2^{11} \to \cdots$$

$$\cdots \to E^{n0} \to H^n \to E^{n-1,1} \to E_2^{n+1,0} \to \cdots ,$$

and an isomorphism $\quad E_2^{00} \xrightarrow{\sim} H^0$,

so in this case, we arrive at the statement:

PROPOSITION 12 *For any two* Λ-*modules* M *and* N, *there is a long exact sequence of natural transformations*

$$0 \to \mathrm{Ext}^1_{\Lambda'}(FGFM,FN) \overset{\alpha_1}{\to} \mathrm{Ext}^1_\Lambda(GFM,N) \overset{\beta_1}{\to} \mathrm{Hom}_{\Lambda'}(FGFM,\mathrm{Ext}^1_\Lambda(\mathrm{Ext}^1_\Lambda(X,N))$$

$$\overset{\delta_2}{\to} \mathrm{Ext}^2_{\Lambda'}(FGFM,FN) \overset{\alpha_2}{\to} \cdots$$

$$\cdots \overset{\delta_n}{\to} \mathrm{Ext}^n_{\Lambda'}(FGFM,FN) \overset{\alpha_n}{\to} \mathrm{Ext}^n_\Lambda(GFM,N) \overset{\beta_n}{\to} \mathrm{Ext}^{n-1}_{\Lambda'}(FGFM,\mathrm{Ext}^1_\Lambda(X,N))$$

$$\overset{\delta_{n+1}}{\to} \cdots$$

and an isomorphism

$$\mathrm{Hom}_{\Lambda'}(FGFM,FN) \overset{\alpha_0}{\cong} \mathrm{Hom}_\Lambda(GFM,N).$$

It is easy to verify that α_0 is just the inverse of the isomorphism $F_{GFM,N}$ whose existence is proved in Theorem II,(1)(d), and that α_n is induced by the inverse of this isomorphism.

COROLLARY *Theorem* III(b) *holds*.

Proof Let $M,N \in \operatorname{Im} G$, then by Theorem II, $GFM \cong M$, $FGFM \cong FM$ and $\operatorname{Ext}^1_\Lambda(X,N) = 0$. Thus each α_n reduces to an isomorphism, $\operatorname{Ext}^n_{\Lambda'}(FM,FN) \cong \operatorname{Ext}^n_\Lambda(M,N)$, as Theorem III(b) asserts.

Proposition 12 can be used to obtain various relations between dimensions in mod Λ and mod Λ'. For example, on recalling that $\operatorname{Ext}^1_\Lambda(X,X) = 0$, one sees that the α_n's are all isomorphisms whenever $N \in \operatorname{add} X$, so that if

$$\operatorname{pd}_\Lambda GFM = p ,$$

then $\operatorname{Ext}^n_{\Lambda'}(FM,\ ) = 0$ on Λ'-projectives for $n > p$. Thus *if Λ'-modules have finite projective dimension it follows that*

$$\operatorname{pd}_\Lambda GFM = p \Rightarrow \operatorname{pd}_{\Lambda'} FM \leqslant p.$$

8. The proof of Theorem VIII

We assume throughout the section that T_* holds, and that the kernel R of the standard projective resolution of X is indecomposable. Further, we take a Λ-module M and two terms

$$P_1 \xrightarrow{\mu_1} P_0 \xrightarrow{\mu_0} M \to 0$$

of a minimal projective resolution of M. Theorem VIII will be deduced from the following lemma.

LEMMA 3 *Suppose* P_0 *and* $\Omega^2 M$ *belong to* Im G. *There exists an exact and commutative diagram*

$$\begin{array}{ccccccccc}
0 \to \Omega^2 M & \to & P_1 & \xrightarrow{\mu_1} & P_0 & \xrightarrow{\mu_0} & M & \to & 0 \\
\| & & \downarrow{\scriptstyle \theta_1} & & \downarrow{\scriptstyle \theta_0} & & \| & & \\
0 \to \Omega^2 M & \to & X_1 & \xrightarrow{\lambda_1} & X_0 & \xrightarrow{\lambda_0} & M & \to & 0
\end{array}$$

in which θ_1 *is injective,* θ_0 *is a split monomorphism,* X_0, X_1 *and* Coker $\theta_0 \cong$ Coker θ_1 *all belong to* add X, *and the map* $FX_1 \to \operatorname{Im} F(\lambda_1)$ *induced by* λ_1 *is a projective cover in* mod Λ'.

Proof Since P_0 is a projective in Im G, then $P_0 \in \operatorname{add} X$. Since R is indecomposable and P_1 projective, P_1 and μ_1 may be expressed in the forms

$$P_1 = Q \amalg R^n \quad \text{and} \quad \mu_1 = (\mu_Q, \mu_R)$$

where $R \nmid Q$, so that $Q \in \text{add } X$. We define

$$X_1 = Q \amalg P^n \quad \text{and} \quad \theta_1 = \begin{pmatrix} 1 & 0 \\ 0 & \rho^n \end{pmatrix},$$

and construct the diagram above by performing a pushout construction along $\theta_1 : P_1 \to X_1$. Since θ_1 is, clearly, injective with cokernel $X_1 = Q \amalg P^n \xrightarrow{(0,\pi^n)} X^n$, also θ_0 is injective and has cokernel $\pi_0 : X_0 \to X^n$ satisfying $\pi_0 \lambda_1 = (0, \pi^n)$. By construction $X_1 \in \text{add } X$, and since add X is closed under extensions, so also is $X_0 \in \text{add } X$. By T_2, short exact sequences split in add X, so θ_0 is a split morphism. Now it remains only to prove the last statement.

Since $\Omega^2 M \in \text{Im } G$ and X_0, $X_1 \in \text{add } X$, the bottom row is mapped by F to an exact sequence

$$0 \to F\Omega^2 M \to FX_1 \overset{F\lambda_1}{\to} FX_0 \overset{F\lambda_0}{\to} FM \to 0$$

and FX_1 and FX_0 are Λ'-projectives. To prove the last part of the lemma, we verify that if $\xi' : P' \to FX_1$ is a map from a Λ'-projective module P' and if

$$P' \overset{(F\lambda_1)\xi'}{\to} FX_0 \to FM \to 0$$

is exact, then ξ' is a split epimorphism. Now $P' \in F \text{ add } X$, so by Theorem II(1), $\xi' = F\xi$, where $\xi : Y \to X_1$ and $Y \in \text{add } X$. On applying G to the exact sequence

$FY \overset{F(\lambda_1\xi)}{\to} FX_0 \overset{F\lambda_0}{\to} FM \to 0$ and using Theorem II(1)(a) <=> (e),

we find that the sequence $Y \overset{\lambda_1\xi}{\to} X_0 \overset{\lambda_0}{\to} M \to 0$ is also exact. Now consider the map $(0 1)\xi : Y \to P^n$ (the second factor of X_1). It has composite

$$\pi^n (0 1)\xi = (0\ \pi^n)\xi = \pi_0 \lambda_1 \xi$$

with $\pi^n : P^n \to X^n$, and this is surjective since

$$\text{Im } \pi_0\lambda_1\xi = \pi_0 \text{ Im } \lambda_1\xi = \pi_0 \text{ Im } \lambda_1 = X^n .$$

However π^n is a projective cover, so $(0\ 1)\xi$ is a split epimorphism. In particular, Y and ξ may be taken in the form

$$Y = Z \amalg P^n \quad \text{and} \quad \xi = \begin{pmatrix} \xi_Q & \xi_P \\ 0 & 1 \end{pmatrix},$$

with $Z \in \operatorname{add} X$. Since $(0\ \pi^n)\xi = (0\ \pi^n) : Z \amalg P^n \to X^n$, ξ induces a map

$$\eta = \begin{pmatrix} \xi_Q & \xi_R \\ 0 & 1 \end{pmatrix} : Z \amalg R^n \to Q \amalg R^n = P_1$$

such that $\theta_1\eta = \xi\begin{pmatrix} 1 & 0 \\ 0 & \rho^n \end{pmatrix}$, which entails $\xi_R = \xi_P\rho^n$. Using again the exactness of $Y \xrightarrow{\lambda_1\xi} X_0 \xrightarrow{\lambda_0} M \to 0$, one see that $Z \amalg R^n \xrightarrow{\mu_1\eta} P_0 \xrightarrow{\mu_0} M \to 0$ is also exact, and then the fact that $\mu_1 : P_1 \to \operatorname{Im} \mu_1$ is a projective cover implies that η is a split epimorphism. Hence ξ_Q is a split epimorphism, and if ζ is one of its right inverses, we find that η and ξ have right inverses $\begin{pmatrix} \zeta & -\zeta\xi_R \\ 0 & 1 \end{pmatrix}$ and $\begin{pmatrix} \zeta & -\zeta\xi_P \\ 0 & 1 \end{pmatrix}$, respectively. Now $\xi' = F\xi$ also has a right inverse, so is a split epimorphism. This completes the proof of the lemma.

<u>COROLLARY</u> <u>Theorem</u> VIII <u>holds</u>

<u>Proof</u> Apply $D\operatorname{Hom}_\Lambda(-,X)$ to the diagram constructed in Lemma 3. Since, by T_2, $\operatorname{Ext}^1_\Lambda(-,X)$ vanishes on add X, in particular on Coker θ_0 and Coker θ_1, there results an exact and commutative diagram

$$\begin{array}{ccccc}
0 & & 0 & & \\
\downarrow & & \downarrow & & \\
D\operatorname{Hom}_\Lambda(P_1,X) & \xrightarrow{\mu_{1\#}} & D\operatorname{Hom}_\Lambda(P_0,X) & \xrightarrow{\mu_{0\#}} & D\operatorname{Hom}_\Lambda(M,X) \to 0 \\
\downarrow \theta_{1\#} & & \downarrow \theta_{0\#} & & \| \\
D\operatorname{Hom}_\Lambda(X_1, X) & \xrightarrow{\lambda_{1\#}} & D\operatorname{Hom}_\Lambda(X_0,X) & \xrightarrow{\lambda_{0\#}} & D\operatorname{Hom}_\Lambda(M,X) \to 0 \\
\downarrow & & \downarrow & & \\
D\operatorname{Hom}_\Lambda(\operatorname{Coker}\theta_1,X) & \xrightarrow{\sim} & D\operatorname{Hom}_\Lambda(\operatorname{Coker}\theta_0,X) & & \\
\downarrow & & \downarrow & & \\
0 & & 0 & &
\end{array}$$

Hence there is an isomorphism θ : Ker $\mu_{1\#} \xrightarrow{\sim}$ Ker $\lambda_{1\#}$. By Proposition 10, Ker $\mu_{1\#}$ = $\mathrm{FDTr}_\Lambda M$. Since X_0, X_1, X ϵ add X, $F_{X_1,X}$ and $F_{X_0,X}$ are isomorphisms. Hence Ker $\lambda_{1\#}$ is isomorphic to Ker $D\mathrm{Hom}_{\Lambda'}(F\lambda_1, FX) \cong$ Ker $D(F\lambda_1)^*$, and the last statement of Lemma 3 shows that this last module is isomorphic to $\mathrm{DTr}_{\Lambda'}\mathrm{FM}$. Thus $\mathrm{DTr}_{\Lambda'}\mathrm{FM} \cong \mathrm{FDTr}_\Lambda M$, which is part (a) of Theorem VIII. Part (b) follows from (a) just as in the proof of the same point in Theorem VII.

CHAPTER 2 - TILTING FUNCTORS DEFINED AT SIMPLE MODULES

1. Statement of results

In this chapter we study a construction for a pair of tilting functors associated with a suitable simple module, this construction generalising directly the one due to Auslander, Platzeck, and Reiten described at the end of §2 in Chapter 1. Throughout the chapter, Λ is a basic artin algebra with radical r, e is a primitive idempotent and S the simple module with projective cover Λe. The construction, and its properties, are given in

THEOREM IX Let S, e satisfy the following three conditions: (i) S is not injective, (ii) $(DS)^* = 0$, (iii) $e\Lambda$ is not a summand of the projective cover of er in mod Λ^{op}. Define the algebra Λ' and bimodule $X = {}_\Lambda X_{\Lambda'}$ by the formulae

$$ {}_\Lambda X = Tr_{\Lambda^{op}} DS \amalg \Lambda(1-e) \quad \text{and} \quad \Lambda' = (End_\Lambda X)^{op}. $$

Let e′ be the idempotent in Λ' which has image $Tr_{\Lambda^{op}} DS$ and kernel $\Lambda(1-e)$, and S′ the simple Λ'-module with projective cover $\Lambda' e'$. Then:

(a) $F = Hom_\Lambda(X,-)$ and $G = X \otimes_{\Lambda'} -$ are an adjoint pair of tilting functors, and

$$ Im\ G = \{M \mid M \in mod\ \Lambda \quad \text{and} \quad Hom_\Lambda(M,S) = 0\}\ , $$

$$ Im\ F = \{M' \mid M' \in mod\ \Lambda' \quad \text{and} \quad Hom_{\Lambda'}(S',M') = 0\}\ ; $$

(b) S′,e′ satisfy the 'adjoint duals' of the conditions (i), (ii), (iii), namely,

(i)′ S′ is not projective, (ii)′ $S'^* = 0$, (iii)′ $\Lambda' e'$ is not a summand of the projective cover of r′e′ in mod Λ';

(c) $S' \cong Ext^1_\Lambda(X,S) \cong Tr_{\Lambda'^{op}} X$, $\quad S \cong Tor_1^{\Lambda'}(X,S')$,

and $X_{\Lambda'} \cong Tr_{\Lambda'} S' \amalg (1-e')\Lambda'$.

If S is a projective but non-injective simple module, then (i), (ii) and (iii) hold, so that Theorem IX does constitute a generalisation of parts of the Auslander-

Platzeck-Reiten theory. That theory also gives a condition, namely that $(\mathrm{Tr}_{\Lambda^{op}} DS)^* = 0$, for S′ to be injective, in which case F and G are called, in [2], partial Coxeter functors. This result can be generalised as follows.

<u>COROLLARY</u> <u>Assume the hypotheses and notations of Theorem IX. Then</u>

(a) <u>the following three statements are equivalent</u>:

(1) S <u>is projective</u>; (2) S′ <u>has projective dimension</u> 1;

(3) $(\mathrm{Tr}_{\Lambda'} S')^* = 0$;

(b) <u>the following three statements are equivalent</u>:

(1) S′ <u>is injective</u>; (2) S <u>has injective dimension</u> 1;

(3) $(\mathrm{Tr}_{\Lambda^{op}} DS)^* = 0$.

The second result, Theorem X, in this chapter is motivated by applications to quivers (with or without relations) with no oriented cycles. Let $\mathcal{E}$ denote a full set of orthogonal primitive idempotents for Λ, including e.

<u>DEFINITION 3</u> (a) <u>A subset</u> $\mathcal{G}$ <u>of</u> $\mathcal{E}$ <u>is said to be semisimple if the subalgebra</u> $\sum_{g_1, g_2 \in \mathcal{G}} g_1 \Lambda g_2$ <u>of</u> Λ <u>is semisimple</u>.

(b) <u>If</u> $\mathcal{G}$ <u>and</u> $\mathcal{H}$ <u>are subsets of</u> $\mathcal{E}$, <u>we say that</u> $\mathcal{H}$ <u>is triangulated at</u> $\mathcal{G}$ <u>if for each</u> $h \in \mathcal{H} \setminus \mathcal{G}$, <u>either</u> $g \Lambda h = 0$ <u>for each</u> $g \in \mathcal{G}$, <u>or</u> $h \Lambda g = 0$ <u>for each</u> $g \in \mathcal{G}$.

Each primitive idempotent, e, in the algebra of a quiver with no oriented cycles, or more generally, in the 'triangulated matrix algebras' occurring in Chapter 3 and 4, is semisimple, and also $\mathcal{E} \setminus \{e\}$ is triangulated at e. In an arbitrary basic artin algebra the primitive idempotent e has these properties if the simple module associated with it is either projective or injective. The purpose of Theorem X is to give, in the context of Theorem IX, some further conditions on S, e, and $\mathcal{E}$ which, firstly, ensure that e is semisimple and that $\mathcal{E} \setminus \{e\}$ is triangulated at e, and secondly, imply that Λ' inherits under tilting the 'adjoint duals' of these conditions relative to S′, e′ and a full set $\mathcal{E}'$ of orthogonal primitive idempotents which includes e′.

One obviously sufficient pair of conditions is that S be a projective simple of injective dimension 1 (so that F and G are partial Coxeter functors), for the Corollary to Theorem IX then shows that S′ is an injective simple of projective dimension 1. However, these conditions are too restrictive for quivers with non-trivial relations, for which a typical non-trivial tilting may well occur at a projective simple of injective dimension 2 or more. Two more definitions are required to formulate conditions to cover such cases.

DEFINITION 4 *Assume the hypotheses and notation of Theorem* IX.

(a) *Let* $\mathcal{F}$ *denote the set of* $f \in \mathcal{E}$ *such that* $f\Lambda$ *is a summand of the projective cover of* er.

(b) *For each* $g \in \mathcal{E} \setminus \{e\}$, *let* g' *denote the idempotent in* $\Lambda' = (\mathrm{End}_\Lambda X)^{op}$ *which has image the summand* Λg *of* X *and kernel the complementary summand* $\mathrm{Tr}_{\Lambda^{op}} DS \amalg \Lambda(1 - g - e)$. *For each subset* $\mathcal{G}$ *of* $\mathcal{E}$, *let* $\mathcal{G}' = \{g' \mid g \in \mathcal{G}\}$.

THEOREM X *Let* S, e *satisfy the hypotheses of Theorem* IX. *Then*:

(a) S *is projective and of injective dimension* 1 *if and only if* S′ *is injective and of projective dimension* 1, *and in these circumstances* $\mathcal{E} \setminus \{e\}$ *is triangulated at* e, $\mathcal{E}' \setminus \{e'\}$ *is triangulated at* e′, *and both* e *and* e′ *are semisimple*.

(b) $\mathcal{F}$ *is semisimple and* $\mathcal{E} \setminus \{e\}$ *is triangulated at* $\mathcal{F}$ *if and only if* $\mathcal{F}'$ *is semisimple and* $\mathcal{E}' \setminus \{e'\}$ *is triangulated at* $\mathcal{F}'$, *and in these circumstances*, $\mathcal{E} \setminus \{e\}$ *is triangulated at* e, $\mathcal{E}' \setminus \{e'\}$ *is triangulated at* e′, *and both* e *and* e′ *are semisimple*.

In the last section, we note some improvements to some of the Theorems of Chapter 1 on the effect of tilting on Ext-groups and almost split sequences obtainable for the special types of tilting functors considered in this chapter. These are most striking when either S is projective or S′ is injective.

2. *Proof of Theorem IX*

The proofs of Theorems IX and X will require detailed analysis of the structure of Λ', and we begin by introducing suitable notation.

Firstly, suppose that S is non-injective and, for brevity, write $T = Tr_{\Lambda^{op}} DS$. Then, using the notation introduced in §1, we have

$$_{\Lambda}X = T \amalg \Lambda(1-e) = T \amalg \coprod_{g \in \mathcal{E} \setminus \{e\}} \Lambda g ,$$

and $\mathcal{E}$ determines a full set $\mathcal{E}'$ of orthogonal primitive idempotents in $\Lambda' = (End_{\Lambda}X)^{op}$ as in Definition 4(b). The following result is obvious.

<u>LEMMA 4</u> <u>For</u> $g, h \in \mathcal{E}$, <u>the following formulae hold</u>:

$$g'\Lambda'h' \cong Hom_{\Lambda'}(\Lambda'g', \Lambda'h') \cong \begin{cases} (End_{\Lambda}T)^{op} & \underline{\text{for}}\ g=h=e \\ Hom_{\Lambda}(\Lambda g, T) & \underline{\text{for}}\ g \neq h = e \\ Hom_{\Lambda}(T, \Lambda h) & \underline{\text{for}}\ e=g \neq h \\ Hom_{\Lambda}(\Lambda g, \Lambda h) \cong g\Lambda h & \underline{\text{for}}\ g,h \neq e \end{cases}$$

Thus computation in Λ' can be reduced to computation of Λ-modules using an explicit representation of T. For this, we select once and for all two terms of a minimal Λ^{op}-projective resolution,

$$Q \xrightarrow{i} e\Lambda \xrightarrow{j} DS \rightarrow 0,$$

of the simple non-projective Λ^{op}-module DS, and define $T = Coker\ i^*$. Using the isomorphisms $(\Lambda e)^* \cong e\Lambda$, $(e\Lambda)^* \cong \Lambda e$, and $A \cong A^{**}$ for projective A, we obtain a pair of exact sequences

$$0 \rightarrow T^* \xrightarrow{\tau^*} Q \xrightarrow{i} e \xrightarrow{j} DS \rightarrow 0 ,$$

$$0 \rightarrow (DS)^* \xrightarrow{j^*} e \xrightarrow{i^*} Q^* \xrightarrow{\tau} T \rightarrow 0 ,$$

the latter providing two terms of the minimal projective resolution of T. We refer to them as the <u>standard projective resolutions of</u> DS <u>and of</u> T. As in Definition 4(a), let $\mathcal{F}$ be the set of idempotents f in $\mathcal{E}$ such that $f\Lambda$ is a summand of the projective cover of er. Since this projective cover is Q, and since Q^* is projective cover of T, we have an alternative characterisation of $\mathcal{F}$;

$\mathcal{F} = \{f \mid f \in \mathcal{E}$ and Λf is a summand of the projective cover of $T\}$.
Next suppose also that $(DS)^* = 0$. Then $\mathrm{pd}_\Lambda T = 1 = \mathrm{pd}_\Lambda X$ and we obtain the standard projective resolution

$$0 \to R \overset{\pi}{\to} P \overset{\rho}{\to} X \to 0$$

of $X = T \amalg \Lambda(1-e)$ (Chapter 1, Definition 2) in the form

$$R = \Lambda e, \quad P = Q^* \amalg \Lambda(1-e), \quad \pi = \begin{pmatrix} i^* \\ 0 \end{pmatrix}, \quad \rho = \begin{pmatrix} \tau & 0 \\ 0 & 1 \end{pmatrix}.$$

<u>Proof of Theorem IX</u> Assume that (S,e) satisfy (i), (ii), (iii) in Theorem IX. We first show that X satisfies T_*. T_0 holds by definition of Λ', and we noted above that (ii) implies $\mathrm{pd}_\Lambda X = 1$, so that T_1 holds. (iii) simply states that $e \notin \mathcal{F}$, so as noted above, $R = \Lambda e$ is not a summand of Q^*. Since Λ is basic, R is not a summand of P, so $P \in \mathrm{add}\,\Lambda(1-e) \subset \mathrm{add}\,X$. Hence T_3 holds, and since clearly each projective indecomposable is a summand of P or of R but not of both, Proposition 5 enables us to complete the proof of T_* by demonstrating T_2, that is, that $\mathrm{Ext}^1_\Lambda(X,X) = 0$. This is equivalent to surjectivity of the map $\mathrm{Hom}_\Lambda(Q^*,X) \to \mathrm{Hom}_\Lambda(\Lambda e,X)$ induced by i^*, and since $P \overset{\pi}{\to} X$ is surjective and Q^* and Λe are both projective, to surjectivity of the map $\mathrm{Hom}_\Lambda(Q^*,P) \to \mathrm{Hom}_\Lambda(\Lambda e,P)$ induced by i^*. On applying $*$, we see that the last map is isomorphic to $\mathrm{Hom}_\Lambda(P^*,Q) \overset{i}{\to} \mathrm{Hom}_\Lambda(P^*,e\Lambda)$, and this is surjective, for its cokernel $\mathrm{Hom}_\Lambda(P^*,DS)$ vanishes since $R = \Lambda e$, the projective cover of S, is not a summand of P. Thus T_* holds and F, G are an adjoint pair of tilting functors. Since $D\mathrm{Tr}_\Lambda X \cong S$, the characterisation of Im G in Theorem IX(a) is given by Theorem II(1)(f).

Next we prove the formula $S' \cong \mathrm{Ext}^1_\Lambda(X,S)$ of part (c) of Theorem IX, noting that the left action of Λ' on the Ext-group is induced by its right action on X. We have

$$e'\mathrm{Ext}^1_\Lambda(X,S) \cong \mathrm{Ext}^1_\Lambda(T,S) \; ,$$

$$(1-e')\,\mathrm{Ext}^1_\Lambda(X,S) \cong \mathrm{Ext}^1_\Lambda(\Lambda(1-e),S) = 0 \; ,$$

since $Xe' = T$ and $X(1-e') = \Lambda(1-e)$. Thus the inclusion of $\mathrm{Ext}^1_\Lambda(T,S)$ into $\mathrm{Ext}^1_\Lambda(X,S)$ is an isomorphism of groups respecting the action of the subalgebra $(\mathrm{End}_\Lambda T)^{op}$ of Λ' on these groups. Now $T = \mathrm{Tr}_{\Lambda^{op}} DS$, and S is simple, so ([3], Part III, Proposition 5.1) $\mathrm{Ext}^1_\Lambda(T,S)$ is simple as an $(\mathrm{End}_\Lambda T)^{op}$-module. Hence $\mathrm{Ext}^1_\Lambda(X,S)$ is a simple Λ'-module, and since it is annihilated by $1-e'$, it must be isomorphic to S'.

Next we show that $S' \cong \mathrm{Ext}^1_\Lambda(X,R)$. By (iii),

$$0 = \mathrm{Hom}_{\Lambda^{op}}(e\Lambda,\ er/er^2) \cong ere/er^2e \cong \mathrm{Hom}_\Lambda(\Lambda e,\ re/r^2e),$$

so that $\mathrm{Hom}_\Lambda(re,S) = 0$. Hence $re \in \mathrm{Im}\, G$, so $\mathrm{Ext}^1_\Lambda(X,re) = 0$ by Theorem II(1). Now $0 \to re \to R \to S \to 0$ is exact, and since $\mathrm{Ext}^1_\Lambda(X,\ )$ is right exact (by T_1), we have an isomorphism $\mathrm{Ext}^1_\Lambda(X,R) \cong \mathrm{Ext}^1_\Lambda(X,S) \cong S'$.

The formula $S' \cong \mathrm{Ext}^1_\Lambda(X,R)$ links with results of §5 in Chapter 1. Thus Proposition 8(a) and (b) immediately give parts (i)′ and (ii)′, respectively, of Theorem IX(b), and since Proposition 6(c) shows that the Λ'-projective cover of $r'e'$ is a summand of FP, Proposition 8(i) leads to part (iii)′ of Theorem IX(b). Proposition 8(e) now reads

$$\mathrm{Tr}_{\Lambda'} S' \cong T' = \mathrm{Hom}_\Lambda(R,X),$$

from which we deduce that

$$\begin{aligned} X_{\Lambda'} &\cong \mathrm{Hom}_\Lambda(\Lambda,X) \cong \mathrm{Hom}_\Lambda(R,X) \amalg \mathrm{Hom}_\Lambda(\Lambda(1-e),X) \\ &\cong T' \amalg (1-e')\Lambda' \\ &\cong \mathrm{Tr}_{\Lambda'} S' \amalg (1-e')\Lambda' . \end{aligned}$$

This formula then gives $S' \cong \mathrm{Tr}_{\Lambda'^{op}} X$. We have now proved all the assertions of Theorem IX except for the characterisation of $\mathrm{Im}\, F$ in (a) and the formula $S \cong \mathrm{Tor}_1^{\Lambda'}(X,S')$, and these of course are just the adjoint duals of the formulae for Im G and S′ so follow from part (b) of the theorem.

COROLLARY 1 *Using the notation and hypotheses of Theorem IX and its proof, the standard projective resolution of DS determines two terms of a minimal projective resolution of S′,*

$$0 \to FR \xrightarrow{F(i^*)} FQ^* \xrightarrow{F(\tau)} FT \to S' \to 0 .$$

Proof Proposition 6(b) and the formulae above for P, X, ρ, π show that the sequence is exact and that $F(\tau)$ is a projective cover of its image. Since $FT = \mathrm{Hom}_\Lambda(X,T) \cong \Lambda' e'$, the map $FT \to S'$ is also a projective cover.

COROLLARY 2 *The corollary of Theorem IX in §1 holds.*

Proof We just prove (a). For any module M, we have $\Omega^2 M = (\mathrm{Tr}M)^*$, so Corollary 1 above shows that $(\mathrm{Tr}_{\Lambda'}S')^* \cong FR \cong \Omega^2 S'$. Since S′ is not projective, the equivalence of the 3 statements {S′ has projective dimension 1}, $\{(\mathrm{Tr}_{\Lambda'}S')^* = 0\}$, and $FR = 0$ is evident. Finally, we show that $FR = 0$ is equivalent to projectivity of S, that is, to $re = 0$. Since $re \in \mathrm{Im}\, G$, Theorem II implies that $re = 0$ iff $F(re) = 0$ and also that $0 \to F(re) \to FR \to FS \to 0$ is exact. But also $FS = 0$, so $FR = 0$ iff $F(re) = 0$ iff $re = 0$, as asserted.

3. Proof of Theorem X

LEMMA 5 *Suppose that* $\mathcal{E}\setminus\{e\}$ *is triangulated at* e *and that* $e\Lambda$ *is not a summand of the projective cover of* er. *Then* e *is semisimple.*

Proof We need to show that $ere = \mathrm{rad}\, e\Lambda e = 0$. The condition on the projective cover of er, hence of er/er^2, is equivalent to $ere = er^2e$. Now

$$ere = er^2e = \sum_{g \in \mathcal{E}\setminus e} erg.gre \; + (ere)^2$$

and the right hand side reduces to $(ere)^2$ supposing $\mathcal{E}\setminus\{e\}$ to be triangulated at e. Since $ere = (ere)^2$ is nilpotent, we find $ere = 0$, as required.

<u>Proof of Theorem X(a)</u> The first part follows from the corollary to Theorem IX, and the triangulability of $\mathcal{E}\setminus\{e\}$ at e follows from the observation that, for $g \neq e$,

$$g\Lambda e \cong \mathrm{Hom}_\Lambda(\Lambda g, S) = 0.$$

Lemma 5 then gives the last statement.

<u>Proof of Theorem X(b)</u> Since $e \notin \mathcal{F}$, Lemma 4 shows that $\mathcal{F}$ is semisimple if and only if $\mathcal{F}'$ is semisimple, and that $\mathcal{E}\setminus\{e\}$ is triangulated at e if and only if $\mathcal{E}'\setminus\{e'\}$ is triangulated at e'. With Lemma 5 in mind, we can finish the proof by showing, for example, that $\mathcal{E}'\setminus\{e'\}$ is triangulated at e'. Let $g' \in \mathcal{E}'\setminus\{e'\}$. By Lemma 4,

$$g'\Lambda'e' \cong \mathrm{Hom}_\Lambda(\Lambda g, T) \quad \text{and} \quad e'\Lambda'g \cong \mathrm{Hom}_\Lambda(T, \Lambda g).$$

We have to show that one at least of these modules is 0, and for this purpose make use of the standard projective resolution

$$0 \to \Lambda e \to Q^* \xrightarrow{\tau} T \to 0$$

of T used in §2. Note that $\mathcal{F}$ is precisely the set of primitive idempotents, f, such that $\Lambda f | Q^*$. Let $\phi = \Sigma f$. Then $Q^* \sim \Lambda\phi$. Now $\mathcal{F}$ is semisimple and $\mathcal{E}\setminus\{e\}$ triangulated at $\mathcal{F}$. Hence:

either $g\Lambda\phi = 0$, or $g\Lambda\phi \neq 0$ and $\phi rg = 0$.

Suppose $g\Lambda\phi \cong \mathrm{Hom}_\Lambda(\Lambda g, \Lambda\phi) = 0$. Then $\mathrm{Hom}_\Lambda(\Lambda g, Q^*) = 0$, and since Λg is projective, we have $\mathrm{Hom}_\Lambda(\Lambda g, T) = g'\Lambda'e' = 0$. Suppose $g\Lambda\phi \neq 0$, so that $\phi rg = 0$, hence $\mathrm{Hom}_\Lambda(Q^*, rg) = 0$. Then $\mathrm{Hom}_\Lambda(T, rg) = 0$ and since rg is the unique maximal submodule of Λg and T has no projective summands, we find that

$$e'\Lambda'g' \cong \mathrm{Hom}_\Lambda(T, \Lambda g) \cong \mathrm{Hom}_\Lambda(T, rg) = 0.$$

This proves that $\mathcal{E}'\setminus\{e'\}$ is triangulated at e'. By duality, $\mathcal{E}\setminus\{e\}$ is triangulated at e, and Theorem X(b) follows.

4. Remarks on exact sequences

Suppose that F, G are an adjoint pair of tilting functors determined as in Theorem IX by simple modules S and S′. We describe briefly in this section some improvements to the results in Chapter 1, Theorems VI, VII and VIII, on the induced mappings of Ext-groups and of almost split sequences. The best results occur when F or G is an APR-tilting functor, that is, when either S is projective or S′ is injective.

Consider, first, Theorem VIII in the context of Theorem IX, so that T_* does hold, and $R = \Lambda e$ is indecomposable. The condition on M, that it has a projective cover P(M) in Im G, reduces to $\mathrm{Hom}_\Lambda(P(M),S) = 0$, that is, to $\mathrm{Hom}_\Lambda(M,S) = 0$, and therefore to $M \in \mathrm{Im}\, G$. Also the condition $\underline{\mathrm{Hom}}_\Lambda(M,X) = 0$ reduces to $\underline{\mathrm{Hom}}_\Lambda(M,T) = 0$, and in the special case that S is projective, this in turn reduces to $T \nmid M$ (see the proof of Theorem V(c)). Thus we obtain a re-formulation:

THEOREM VIII* *Suppose Theorem IX holds and the indecomposable* Λ-*module* M *satisfies the following conditions*:

$$\mathrm{Hom}_\Lambda(M,S) = 0 \quad \text{and} \quad \mathrm{Hom}_\Lambda(\Omega^2 M,S) = 0.$$

Then (a) $D\mathrm{Tr}_{\Lambda'}FM \cong FD\mathrm{Tr}_\Lambda M$, *and (b), if also* $\underline{\mathrm{Hom}}_\Lambda(M,T) = 0$, *then the isomorphism*

$$\mathrm{Ext}^1_\Lambda(M,D\mathrm{Tr}_\Lambda M) \cong \mathrm{Ext}^1_{\Lambda'}(FM,D\mathrm{Tr}_{\Lambda'}FM)$$

maps almost split sequences to almost split sequences.

In particular, if S *is projective, the conditions on* M *reduce to* $M \not\cong S$, $S \nmid \Omega^2 M$, *and (for (b))* $M \not\cong T$.

Next we consider the effect of these special tiltings on more general exact sequences, noting first that when Theorem IX holds, the only Λ-modules annihilated by F are those of the form S^n. (Since X has each projective indecomposable, except Λe, as a summand, then $FA = 0$ implies that the only composition factors of A are copies of S; but Theorem IX(iii) => $\mathrm{Hom}_\Lambda(re,S) = 0$ => $\mathrm{Ext}^1_\Lambda(S,S) = 0$, so A must be a direct summ of copies of S). In particular, it follows from Theorem IV that for any Λ-module, N, Coker $\epsilon_N \cong S^{n(N)}$ and hence $\mathrm{Ext}^1_\Lambda(X,N) \cong \mathrm{Ext}^1_\Lambda(X,S)^{n(N)}$. Now Theorem IX(c) shows that

$$\mathrm{Ext}^1_\Lambda(X,N) \cong S'^{n(N)} .$$

This formula can be used to simplify the long exact sequence of Proposition 12 for any pair of Λ-modules M and N. The simplification is most spectacular if S′ is injective, that is, if S has injective dimension 1, for then the terms $\mathrm{Ext}^p_{\Lambda'}(FGFM, \mathrm{Ext}^1_\Lambda(X,N)) = 0$ for $p \geq 1$, irrespective of whether or not $N \in \mathrm{Im}\, G$. We state the result formally as follows:

<u>PROPOSITION 12*</u> <u>Suppose Theorem</u> IX <u>holds and the simple module</u> S <u>has injective dimension</u> 1. <u>For each</u> $M \in \mathrm{Im}\, G$ <u>and each</u> Λ-<u>module</u> N <u>we have an exact sequence of natural transformations</u>

$$0 \to \mathrm{Ext}^1_{\Lambda'}(FM,FN) \to \mathrm{Ext}^1_\Lambda(M,N) \to \mathrm{Hom}_{\Lambda'}(FM,S')^{n(M)}$$

$$\to \mathrm{Ext}^2_{\Lambda'}(FM,FN) \to \mathrm{Ext}^2_\Lambda(M,N) \to 0$$

<u>and isomorphisms</u>

$$\mathrm{Ext}^p_{\Lambda'}(FM,FN) \to \mathrm{Ext}^p_\Lambda(M,N)$$

<u>for</u> $p = 0$ <u>and</u> $p \geq 3$.

CHAPTER 3 - TRIANGULATED ALGEBRAS AND QUADRATIC FORMS

1. Triangulable algebras and quivers with relations

We call a basic artin algebra Λ over a central subfield k a <u>triangulable algebra</u> if

(i) the endomorphism ring of each indecomposable projective is a skew field, and

(ii) there is an ordering $P_1, P_2, \dots, P_n$ of the distinct indecomposable projectives such that $\mathrm{Hom}_\Lambda(P_i, P_j) = 0$ whenever $i > j$. (See also Definition 3 in Chapter 2.)

Let Λ be a triangulable algebra. We order a complete set of orthogonal primitive idempotents $e_1, \dots, e_n$ so that the projectives $P_i \simeq \Lambda e_i$ have the ordering of (ii) above. Define

$$\Lambda_{ij} = e_i \Lambda e_j \quad \text{and} \quad \Lambda_i = \Lambda_{ii}.$$

Then $\Lambda_i \simeq \mathrm{End}_\Lambda P_i$ is a skewfield and $\Lambda_{ij} \simeq \mathrm{Hom}_\Lambda(P_i, P_j)$ is a $\Lambda_i - \Lambda_j{}^{op}$- bimodule. We shall view Λ as the <u>triangulated matrix algebra</u>

$$\Lambda = \begin{pmatrix} \Lambda_1 & \Lambda_{12} & \cdots & \Lambda_{1n} \\ & \Lambda_2 & \cdots & \Lambda_{2n} \\ & & \ddots & \vdots \\ \bigcirc & & & \Lambda_n \end{pmatrix}.$$

(Of course this description in terms of a matrix of skewfields and bimodules is not complete without the <u>multiplication maps</u> $\phi_{ijk} : \Lambda_{ij} \otimes_{\Lambda_j} \Lambda_{jk} \to \Lambda_{ik}$, which satisfy the obvious associativity conditions).

The radical r of Λ consists of the strictly upper triangular matrices and the semisimple quotient Λ/r can be identified with $\Lambda_1 \amalg \Lambda_2 \amalg \dots \amalg \Lambda_n$.

Define $f_i = \dim_k \Lambda_i$, $d_{ij} = \dim_{\Lambda_i} \Lambda_{ij}$, $d_{ji} = \dim_{\Lambda_j{}^{op}} \Lambda_{ij}$.

Then $f_i d_{ij} = \dim_k \Lambda_{ij} = f_j d_{ji}$. Thus Λ determines a valued graph G_Λ with vertices 1, 2, ..., n; the vertex i is assigned the weight f_i and to each pair $i < j$ with $d_{ij} \neq 0$ there is assigned the weighted arrow $i \xrightarrow{d_{ij}} j$. The graph determines, and is determined by the two matrices

$$D = \begin{pmatrix} 1 & d_{12} & \cdots & d_{1n} \\ & 1 & \cdots & d_{2n} \\ & & \ddots & \vdots \\ \bigcirc & & & 1 \end{pmatrix} \quad \text{and} \quad F = \begin{pmatrix} f_1 & & & \bigcirc \\ & f_2 & & \\ & & \ddots & \\ \bigcirc & & & f_n \end{pmatrix}$$

We shall be particularly interested in the case when $\Lambda_i = k$ for all i. In this case it is convenient to choose a basis E_{ik} for each of the spaces $\Lambda_{ik}/\sum_j \operatorname{im} \phi_{ijk}$ and to replace G_Λ by the graph D_Λ with the same vertices but with an arrow from i to j corresponding to each such basis element. Suppose $i = i_0, i_1, \ldots, i_r = j$ are successive points in an oriented path in D_Λ and that the basis element $e_{i_{s-1}i_s} \in E_{i_{s-1}i_s}$ corresponds to the arrow from i_s to i_{s-1}. Then we shall write $e_{i_0i_1} e_{i_1i_2} \cdots e_{i_{r-1}i_r}$ for the element of Λ defined inductively by

$$e_{i_0i_1} e_{i_1i_2} \cdots e_{i_{r-1}i_r}$$

$$= \phi_{i_0i_{r-1}i_r}\left(\left[e_{i_0i_1} e_{i_1i_2} \cdots e_{i_{r-2},i_{r-1}}\right] \otimes e_{i_{r-1},i_r}\right).$$

We shall use the same notation for an oriented path in D_Λ, and often refer to such an element of Λ as <u>an oriented path</u>. Clearly the oriented paths generate Λ over k. In case the set of oriented paths is a k-basis of Λ, either Λ or D_Λ is frequently called a quiver. In case the oriented paths are not linearly independent over k, we call a vanishing linear combination of them a <u>relation</u> on D_Λ, and Λ (or D_Λ together with a basis of the space of relations) a <u>quiver with relations</u> or more shortly, a QWR. (Note that since for us both quivers and QWR's are triangulable, they contain no oriented cycles.) If $d_{ij} = 1$ or 0 and ϕ_{ijk} is an isomorphism for all $i \leq j \leq k$, we shall call Λ (or D_Λ and the relations) a <u>fully commutative quiver</u>.

2. The quadratic form for a triangulable algebra

Let Λ be a triangulated algebra. We use the notation of the previous section.

A Λ-module M has a direct decomposition $M = \amalg M_i$ in mod k, where $M_i = e_i M$ is a Λ_i-space. Thus M determines, and is determined by, an assignment to each vertex i of the k-space M_i, and to each arrow $i < j$ of a 'structure map' $\Lambda_{ij} \otimes_{\Lambda_j} M_j \to M_i$ (satisfying, together with the multiplication maps, appropriate associativity conditions).

The dimension type $\underline{x} = \underline{\dim}\, M$ of the Λ-module M is defined to be the vector in $\mathbb{Q}^n$ with coordinates

$$x_i = \dim_{\Lambda_i} M_i \qquad (i = 1, 2, .., n).$$

Since $M_i \simeq \mathrm{Hom}_\Lambda(P_i, M)$, x_i is the multiplicity of $S_i = P_i/rP_i$ as a composition factor of M. Let $\mathbb{Q}Gr(\Lambda)$ be the Grothendieck space in which M determines the element [M]. The map $M \to \underline{\dim}\, M$ induces an isomorphism $\underline{\dim} : \mathbb{Q}Gr(\Lambda) \rightrightarrows \mathbb{Q}^n$ determined by the choice of $[S_1], .., [S_n]$ as a basis of $\mathbb{Q}Gr(\Lambda)$.

Clearly the global dimension of Λ is at most $n - 1$. Hence we have a map mod $\Lambda \times$ mod $\Lambda \to \mathbb{Q}$, defined by

$$(M,N) \mapsto \sum_{p=0}^{\infty} (-1)^p \dim_k \mathrm{Ext}^p_\Lambda(M,N)$$

This induces a bilinear form b_Λ on $\mathbb{Q}Gr(\Lambda)$. Let B denote the matrix of b_Λ referred to the simples $[S_1], ..., [S_n]$ as a basis of $\mathbb{Q}Gr(\Lambda)$ so that

$$b_\Lambda(\underline{x}, \underline{y}) = \underline{x}\, B \underline{y}^t .$$

It is easily verified, by considering $b_\Lambda(\underline{\dim}\, P_i, \underline{\dim}\, P_j)$ for all $i, j = 1, 2, .., n$, that

$$B = (D^t)^{-1} F .$$

Thus b_Λ depends only on the valued graph of Λ, and not on the multiplication maps.

This homological form seems to have been introduced first by Ringel [22], who showed that, for an hereditary algebra Λ its associated quadratic form q_Λ coincides with the Tits form of Λ [12]. By writing the matrix elements of $(D^t)^{-1} = \mathrm{adj}D^t$ as sums of monomials, it is not hard to verify that for a fully commutative quiver Λ, q_Λ coincides with the form obtained using a naive version of the Tits argument.

An indecomposable hereditary algebra is known to be of finite, tame or wild representation type according to whether q_Λ is positive definite, positive semi-definite or indefinite [10]; the corank of q_Λ, (n-rank q_Λ), can take only the values 0, 1, or 0, respectively. For Λ of finite or tame type the valued graphs are (essentially) the Dynkin or extended Dynkin graphs, for which there are root systems consisting of all integral vectors (that is $\underline{x} \in \mathbb{Q}^n$) such that $q_\Lambda(\underline{x}) = 0, f_1, f_2, \ldots, f_n$. The positive roots (that is $\underline{x} \neq \underline{0}$ with each $x_i \geq 0$) are known to be precisely the dimension types of the indecomposable modules and for Λ of finite type the map $M \mapsto \underline{\dim}\, M$ is a (1 - 1)-correspondence of indecomposables with positive roots.

Recently Kac [15] has shown how to define roots also in the case of wild quivers. He has verified the conjecture of Bernstein, Gelfand and Ponomarev that if M is an indecomposable representation of a quiver Λ, $q_\Lambda(\underline{\dim}\, M) \leq 1$ and that, up to isomorphism, there is exactly one indecomposable module M with $\underline{\dim}\, M = \underline{x}$ if $\underline{x}$ is a positive real root ($q_\Lambda(\underline{x}) = 1$), and that there are (over an algebraically closed field) infinitely many non-isomorphic indecomposables with $\underline{\dim}\, M = \underline{x}$ if $\underline{x}$ is a positive imaginary root ($q(\underline{x}) \leq 0$).

The fully commutative quivers of finite type have been found by Loupias [17] and by Zavadsky and Shkabara [26]. They are precisely those Λ for which $q_\Lambda(\underline{x}) > 0$ for all positive $\underline{x}$.

Bautista [4] has found the ℓ-hereditary algebras (algebras in which maps between projectives are injective or zero) of finite representation type. (They may be thought of as commutative species). They include the fully commutative quivers and again they are precisely those for which $q_\Lambda(\underline{x}) > 0$ for all positive $\underline{x}$.

We shall say more about this and about some fully commutative quivers (and other QWR's) of tame type in the next chapter.

It is not hard to find QWR's where representation type is not given by the definiteness of q_Λ (see e.g. [7]). Since q_Λ depends only on the dimensions of the Λ_{ij}, and not on the structure maps, this is not surprising. It seems that, if the relations are sufficiently 'natural' or 'symmetric' in some sense, the form does give representation type. We comment on this in relation to tilting in the next chapter.

3. The quadratic form and tilting

Let Λ be the triangulated algebra of section 1 and suppose that $e = e_p$ and $S = S_p$ satisfy the conditions of Theorem IX (since $e_p r e_p = 0$, condition (iii) holds for each p in a triangulated algebra). The standard projective resolution of $T = \mathrm{Tr}_{\Lambda^{op}} DS_p$ can be taken in the form

$$0 \to \Lambda e_p \to Q^* = \coprod_{q=p+1}^{q=n} \Lambda e_q \otimes_{\Lambda_q} U_{qp} \to T \to 0 ,$$

where it is not difficult to show, using the definition of T, that U_{qp} is the Λ_q-dual space of the $\Lambda_p - \Lambda_q^{op}$-bimodule

$$V_{pq} = e_p r e_q / e_p r^2 e_q .$$

Let $X = T \amalg \Lambda(1-e_p)$ and use the usual notation F and G for tilting functors. For $M \in \mathrm{mod}\,\Lambda$, we write $M = \coprod_{q=1}^{n} M_q$, with

$M_q = e_q M \cong \mathrm{Hom}_\Lambda(\Lambda e_q, M)$. Similarly, for the Λ'-module

$FM = \mathrm{Hom}_\Lambda(X,M)$, we have a decomposition determined by the Λ'-idempotents $e'_1, .., e'_n$ canonically associated (as in the proof of Theorem IX and Theorem X) with $e_1, .., e_n$. Thus, for $q \neq p$, $(FM)_q \cong M_q$, and for $q = p$, the above presentation of T gives an exact sequence

$$0 \to (FM)_p = \mathrm{Hom}_\Lambda(T,M) \to \coprod_{q=p+1}^{n} V_{pq} \otimes_{\Lambda_q} M_q \to M_p \to \mathrm{Ext}^1_\Lambda(T,M) \to 0$$

If $M \in \mathrm{Im}\, G$, then $\mathrm{Ext}^1_\Lambda(T,M) = 0$, by Theorem II, and then $\underline{x} = \underline{\dim}\, M$ and $\underline{x}' = \underline{\dim}\, FM$ are related by $x'_q = x_q$ for $q \neq p$, and

$$x'_p = \sum_{q=p+1}^{n} v_{pq} x_q - x_p ,$$

where $v_{pq} = \dim_{\Lambda_p} V_{pq}$.

According to Theorem II, Im G contains all the injective modules and so, since Λ has finite global dimension, the image of Im G in $QGr(\Lambda)$ contains a basis of $QGr(\Lambda)$. Similarly, Im F contains all the projective Λ'-modules so, assuming Λ' has finite global dimension (in particular, assuming it to be triangulable) the image of Im F spans $QGr(\Lambda')$. Thus, when both Λ and Λ' are triangulable, the tilting induces a linear isomorphism $\phi_p : QGr(\Lambda) \to QGr(\Lambda')$, and Theorem III implies that, for M, N $\in$ Im G,

$$b_\Lambda(\underline{\dim}\, M, \underline{\dim}\, N) = b_{\Lambda'}(\underline{\dim}\, FM, \underline{\dim}\, FN).$$

Hence $b_\Lambda(\underline{x},\underline{y}) = b_{\Lambda'}(\phi_p(\underline{x}),\phi_p(\underline{y}))$. Note that the formula above for ϕ_p shows that ϕ_p and ϕ_p^{-1} have the same reflection matrix representing them relative to the simple modules as bases of Grothendieck groups.

In the case when S_p is a simple projective module of injective dimension 1, F is a reflection functor (partial Coxeter functor) and ϕ_p is the usual reflection at S_p (c.f. Theorem X(a)) with respect to the symmetrised form

$$\bar{b}_\Lambda(\underline{x},\underline{y}) = \tfrac{1}{2}(b_\Lambda(\underline{x},\underline{y}) + b_\Lambda(\underline{y},\underline{x}))$$

In case Λ is hereditary and triangulated as above, the dimension type of DTrM for a module M with no projective summands is given by

$$\underline{\dim}\, \mathrm{DTr}M = c\ \underline{\dim}\, M$$

where $c = \phi_n \phi_{n-1} \cdots \phi_1$, ([10], [8]).

For any triangulated algebra Λ, hereditary or not, it is possible to define the reflections in the symmetrised bilinear form and the Coxeter element exactly as above. The formula

$$\underline{\dim}\ \mathrm{DTrM} = c\ \underline{\dim}\ M + \underline{\dim}(D\Lambda \otimes_\Lambda M) + \sum_{i=2}^{d} (-1)^i\ \underline{\dim}(D\Lambda \otimes_\Lambda P_i) ,$$

where $0 \to P_d \to \cdots \to P_1 \to P_0 \to M \to 0$

is a minimal projective resolution of $M \in \mathrm{mod}\ \Lambda$, is straightforward to prove. It is sometimes useful when $g\ell.\dim \Lambda = 2$ (then $D\Lambda \otimes_\Lambda M$ is injective or zero) especially when Λ is of finite representation type.

CHAPTER 4 - EXAMPLES AND APPLICATIONS

1. Notation

Throughout this section we shall be concerned with quivers with relations. For any QWR there is a subset $\mathcal{C}$ of pairs of vertices (i,j), $i < j$ such that every relation is a linear combination of oriented paths from j to i for some $(i,j) \in \mathcal{C}$.

We shall frequently draw diagrams to represent QWR's and indicate a basis of the space of relations by dotted lines. In cases where we draw sequences of diagrams we shall not always write down the relations where they have maximum symmetry. Thus the diagram

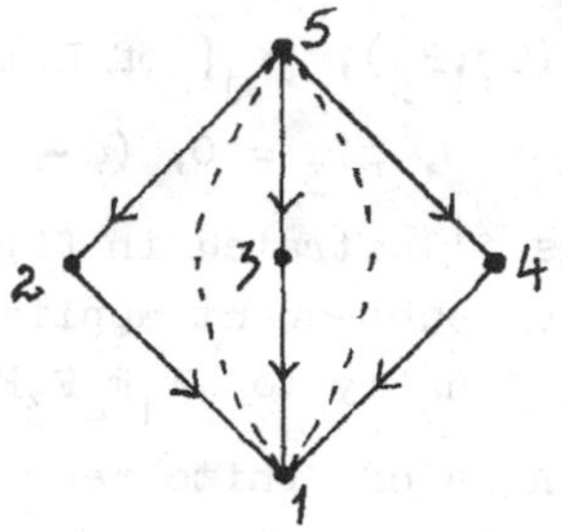

denotes the (12 dimensional) k-algebra Λ with basis e_{ii}, $1 \leq i \leq 5$, e_{1j} and e_{j5}, $2 \leq j \leq 4$, and e_{15}, satisfying $e_{ij}e_{k\ell} = \delta_{jk}e_{i\ell}$ (where δ is the Kronecker symbol). The two relations $e_{12}e_{25} = e_{13}e_{35}$ and $e_{13}e_{35} = e_{14}e_{45}$ are subsumed in the definition of e_{15}; Λ is a fully commutative quiver.

Most of the tilting functors we use in this section will satisfy the conditions of Theorem IX and the additional condition that Λe is simple projective, or the dual conditions with also $D(\Lambda)e$ simple injective. We shall call these <u>APR tilts</u> and, if $e = e_i$, we denote them, respectively, by F_i and G_i.

Note that, in general, the inverse of an APR tilt is not APR. If Λe_i is a simple projective then by Theorem X(a), $D(\Lambda')e_i'$ is simple injective if and only if i is not the end point of a relation, and the dual result holds if and only if i is not the initial point of a relation. When these conditions are satisfied the tilts are the reflection functors, called partial Coxeter functors in [3].

2. An example

Consider the fully commutative quiver illustrated above and let Λ be the corresponding algebra. Then $P_1 = \Lambda e_{11}$ is simple projective and $T_1 = \mathrm{TrD}P_1$ has dimension type $\begin{matrix} & 0 & \\ 1 & 1 & 1 \\ & 2 & \end{matrix}$. We find $\dim \mathrm{Hom}(T_1, P_i) = \begin{cases} 2 & \text{if } i = 5 \\ 0 & \text{if } 2 \leq i \leq 4 \end{cases}$

and $\dim \mathrm{Hom}(P_j, T_1) = \begin{cases} 1 & \text{if } 2 \leq j \leq 4 \\ 0 & \text{if } j = 5 \end{cases}$. Further we may choose bases $\{\phi, \psi\}$ of $\mathrm{Hom}(T_1, P_5)$, $\{\epsilon_j\}$ of $\mathrm{Hom}(P_j, T_1)$, $j = 2, 3, 4$, such that $\phi\epsilon_2 = 0$, $\psi\epsilon_3 = 0$, $(\phi - \psi)\epsilon_4 = 0$. The QWR corresponding to $F_1\Lambda$ is illustrated in figure 1, together with the results of subsequent application of F_2, F_3, F_4 and F_1, leading finally to $F_1F_4F_3F_2F_1\Lambda = \tilde{D}_4$.

The original algebra Λ is of finite representation type whereas $F_1\Lambda$ (and <u>a fortiori</u> each subsequent algebra) is not. However we do know all about the indecomposable representations of the final algebra $\tilde{D}_4$. It turns out that the indecomposable representations of Λ are those annihilated by the composite tilting functor $F = F_1F_4F_3F_2F_1$, namely P_1, P_2, P_3, P_4 and $\mathrm{TrD}P_1$, and 17 others whose dimension types are the 17 positive vectors of form $\phi^{-1}\underline{x}$, where ϕ is the linear transformation of dimension vectors induced by F and $\underline{x}$ is the dimension type of an indecomposable representation of $\tilde{D}_4$. If we allow $\underline{x}$ to run through all the roots, both positive and negative, of $\tilde{D}_4$ we find exactly 22 positive vectors of the form $\phi^{-1}\underline{x}$ and these give the dimension-types of the indecomposable Λ-modules.

We can look on ϕ as a change of axes and suppose the $\tilde{D}_4$ roots remain fixed while we 'tilt' the axes so that different subsets of the roots lie in the positive cone.

This aspect of tilting is well-known [1] for the application of a reflection functor to a (finite type) hereditary algebra. A left exact reflection functor annihilates a projective - and "creates" an injective. The corresponding linear transformation maps the dimension type $\underline{x}$ of the projective to minus that of the corresponding injective (in the new algebra) and therefore maps $-\underline{x}$ to the dimension of the 'created' injective.

3. Finite representation type

THEOREM XI Let Λ be the algebra of a quiver with relations which can be transformed into a quiver $\mathcal{K}$ by a finite sequence of APR tilts. If the associated quadratic form q_Λ is positive definite on positive vectors, then Λ is of finite representation type.

Proof Let F be the composite of the APR tilting functors taking Λ to the algebra of $\mathcal{K}$. Since each APR tilt annihilates exactly one indecomposable representation, F annihilates only a finite number of indecomposable Λ-modules.

Let M be an indecomposable Λ-module such that $FM \neq 0$. Then FM is indecomposable and $q_{\mathcal{K}}(\underline{\dim}\ FM) = q_\Lambda(\underline{\dim}\ M) > 0$.

Kac [15] has shown that if N is an indecomposable representation of a quiver $\mathcal{K}$, then $q_{\mathcal{K}}(\dim N) \leq 1$. Hence we have $q_\Lambda(\underline{\dim}\ M) = 1$. Since q_Λ is positive definite on positive vectors there is only a finite number of positive integral vectors satisfying $q_\Lambda(\underline{x}) = 1$. It follows from the Theorem of Nazarova and Roiter (proof of the Brauer-Thrall conjecture) [21] that Λ is of finite representation type.

REMARK We stated earlier that the fully commutative quivers of finite representation type are precisely those whose quadratic form is positive definite on positive vectors. We shall see in the next section that they do, indeed, all tilt to quivers. It seems that for these QWR's the dimension types of the indecomposables are precisely those positive vectors of form $\phi^{-1}\underline{x}$ where ϕ is the linear transformation induced by F and $\underline{x}$ is a root of $\mathcal{K}$ - any non-positive root satisfying this condition giving the dimension type of an indecomposable annihilated by F.

4. QWR's which tilt to quivers

A necessary condition that a QWR tilts to a quiver is that it has a 'suitable' quadratic form. In particular the form must be positive definite, positive semi-definite of co-rank 1, or indefinite of full rank. It would be interesting to know other invariants of the quadratic forms under tilting. However no criterion involving the quadratic form only can be sufficient to ensure that a QWR tilts to a quiver. Consider the diagram

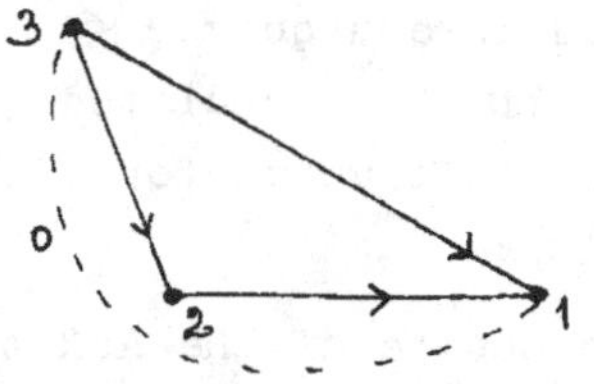

with $e_{12}e_{23} = 0$ which has the same quadratic form as A_3 but, as is easily checked, cannot be tilted to a quiver. We have no convenient criterion which enables us to be certain whether or not a QWR can be tilted to a quiver.

The results of sections 3 and 5 indicate that it would be particularly useful to have a criterion for deciding whether a QWR can be transformed to a quiver using APR tilts. Although this still eludes us we can formulate a condition which seems to guarantee that such a transformation can be effected using only left-exact or only right-exact APR tilts. First we need a definition.

Let G be the Auslander-Reiten graph of a QWR Λ with primitive idempotents $e_1, e_2, \ldots, e_n$. A <u>complete section</u> is a connected subgraph H of G with exactly n vertices; these are either (i) for each i, $1 \leqslant i \leqslant n$, a module of form $(\mathrm{TrD})^{r_i}(\Lambda e_i)$ with $r_i \geqslant 0$, or (ii) for each i, $1 \leqslant i \leqslant n$, a module of form $(\mathrm{DTr})^{r_i}(De_i\Lambda)$ with $r_i \geqslant 0$. The edges of H are all the irreducible maps between its vertices. (This is very similar to Bautista's exact Coxeter section, [5]).

It seems that, if a connected component of the Auslander-Reiten graph of a QWR Λ consists of a complete section H, together with its translates under DTr and TrD then, provided H is a quiver, there is a sequence of APR tilts, all left-exact or all right-exact according as H satisfies (i) or (ii) above, which transforms Λ to a quiver.

David Hughes [14] has pointed out that it follows from Bautista's results [4] that the Auslander-Reiten graph of a fully commutative quiver of finite representation type does satisfy the above condition (and so do the remaining ℓ-hereditary algebras of finite representation type; these therefore tilt to species).

Hughes has also pointed out that, provided we relax the requirement that the APR tilts are either all left-exact or all right-exact, the condition is not necessary. As an example he considers the fully commutative quiver K and sequence of APR tilts illustrated in figure 2. Since there is no connected component of the Auslander-Reiten graph of K which contains either all the projective or all the injectives, there is no complete section.

5. Tame quivers with one relation

The tame fully commutative quivers with one relation were listed (with three doubtful cases) by Marmaridis [18]; a complete list is given by Shkabara [24]. Zavadsky [25] describes the tame QWR's without cycles which have one zero-relation. Ringel [23] has recently given a very elegant treatment of all tame quivers with one relation.

We shall omit from our discussion here (tame) quivers with a cycle in which there is just one relation which is a zero relation. These are just the ones in which the relation can be opened up to give a tame quiver (e.g. those which are stably equivalent to hereditary algebras), together with three diagrams from Ringel's list (see figure 3). The two diagrams of figure 4 fit naturally with those of figure 3 and we shall say a little about them in section 7.

The remaining quivers with one relation can be divided into four types according to the nature of their quadratic form.

A. The quadratic form is semi-definite of corank 1. The QWR can be transformed to a tame quiver by a finite sequence of APR tilts. These (with one exception mentioned in section 7) are Ringel's concealed quivers (the quiver is concealed in the form of a complete section in one or both of the 'projective' or 'injective' components of the Auslander-Reiten graph). Ringel's transforming functors are composites of APR tilts.

B. The quadratic form is indefinite of maximal rank, but non-negative on positive vectors and vanishes on exactly one positive ray. These QWR's can be reduced to a wild quiver by a finite sequence of APR tilts. Kac's results [16] imply that if $\underline{x}$ is the dimension vector of an indecomposable, then $q_\Lambda(\underline{x}) = 0$ or 1. They also imply that there is at most one indecomposable with dimension vector $\underline{x}$ satisfying $q_\Lambda(\underline{x}) = 1$ and that for $\underline{x}$ satisfying $q_\Lambda(\underline{x}) = 0$ there is a '1-parameter family', but no 2-parameter family. These are Ringel's finite extensions.

C. The quadratic form is indefinite of maximal rank, but non-negative on positive vectors and vanishes on exactly two positive rays. These are Ringel's glueings. They cannot be transformed to a quiver using only left-exact or only right-exact APR tilts. It is not known whether there are mixed sequences of left-exact and right-exact APR tilts which reduce them to quivers. It seems that a more detailed knowledge of how left-exact APR tilts influence the DTr orbits of injectives (and the corresponding information for right-exact tilts) is needed to answer this question.

D. The quadratic form is semi-definite of co-rank 2. These are Ringel's non-domestic tame QWR's (apart from one each from figures 3 and 4). These certainly do not tilt to quivers since no quiver has a co-rank 2 quadratic form. We shall discuss some of these further in the next section.

6. Tame QWR's with co-rank 2 quadratic forms

Since we do not have any easy criterion for when the quadratic form gives good information about representation type and dimensions of indecomposable representations, let us start by considering the fully commutative quivers with one relation and semi-definite co-rank 2 quadratic form. These fall into three disjoint families: $\mathcal{D} = \bigcup_{n\geq 5} \mathcal{D}_n$ where each member of $\mathcal{D}_n$ is obtained by factorising one of the arrows in the central portion of a $\tilde{D}_n$ ($n \geq 5$); N, which contains three diagrams each with nine points; and T, which contains 14 diagrams each with ten points.

For each of these diagrams there are many ways of constructing sequences of APR tilts whose composites are endofunctors. Many other QWR's which are not fully commutative and/or which have more than one relation occur at intermediate stages. We shall say that two QWR's are equivalent if each can be transformed to the other by a sequence of APR tilts. For each of the sets $\mathfrak{D}_n$($n \geq 5$), N and T all members belong to the same equivalence class. It turns out that each of these classes contains an easily recognised member, the $\tilde{D}_{n-1}$-squid, the $\tilde{E}_7$-squid and the $\tilde{E}_8$-squid (see Figure 5) corresponding to $\mathfrak{D}_n$, N and T, respectively. There is no fully commutative quiver with one relation corresponding to the $\tilde{E}_6$-squid, though two of Ringel's non-domestic QWR's are equivalent to it. Indeed (with the two exceptions mentioned) each of Ringel's non-domestic QWR's is equivalent to one of these squids, which we call 3-squids, corresponding to the 3 zero-relations (which correspond to 3 symmetrically placed one dimensional subspaces of the 2-dimensional space of maps forming the 'body' of the squid). Note that we have $\tilde{D}_n$-squids only for $n \geq 5$; the $\tilde{D}_4$-squid discussed by Donovan and Freislich [11] is not a 3-squid and does not fit into our pattern. (It shares some features with the diagrams of figures 3 and 4. In particular it can be transformed to a non-triangulable algebra using APR tilts. Note too that it is not fully symmetric and that a fully symmetric squid with four limbs needs a body consisting of three arrows, and is therefore wild.)

The equivalence class of a 3-squid has the property that once the indecomposable representations are known for one member Σ of it, the indecomposable representations of any other member Σ' are easily determined by tilting. One finds a sequence of APR tilts $\tau_1, \tau_2, \ldots, \tau_r$ whose composite Φ transforms Σ' to Σ. The indecomposable representations of Σ' are the indecomposable summands of modules in the image of $\operatorname{ind}\Sigma$ under $\Psi = \tau_r^+ \tau_{r-1}^+ \cdots \tau_1^+$, where τ_i^+ denotes the adjoint of τ_i (and is not APR), together with the r (easily constructed) indecomposables annihilated by Φ. (Of course $\operatorname{ind}\Sigma$ may be replaced by $\operatorname{ind}\operatorname{im}\Phi$; this avoids the necessity to decompose the images under Ψ, see Theorem III. Theorem IX can be used to recognise $\operatorname{im}\Phi \subset \operatorname{mod}\Sigma$.)

Nazarova and Roiter [20] have shown how to describe the indecomposables in each $\tilde{D}_n$-squid class; Ringel's methods also give quite a lot of information for the class of each 3-squid.

Here we show how tilting functors can be applied to diagrams in the class of the $\tilde{E}_7$-squid to construct a 'one-parameter family' of representations corresponding to each positive vector in the (2-dimensional) null space of the appropriate quadratic form. (The term 'one-parameter family' is used rather loosely. Here we mean a family indexed by powers of irreducible polynomials. When we speak of the dimension-type of the family we mean the smallest dimension-type of a member of the family; the dimension-type of any other member is an integral multiple of this.)

We shall consider the diagram P =

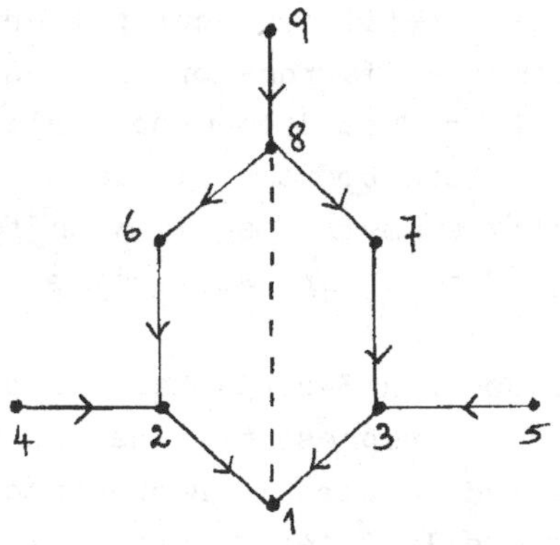

(where the dotted line indicates the commutation relation). This contains $\tilde{D}_6$ and $\tilde{E}_7$ as subdiagrams and the one-parameter families of indecomposables for these give one-parameter families for P with dimension types $n_1 = (2,2,2,1,1,1,1,0,0)$ and $n_2 = (0,2,2,1,1,3,3,4,2)$. We observe that $a = \frac{1}{2}(n_1 + n_2)$ is also an integral null vector and that it gives the $\tilde{D}_4$ one-parameter family obtained by identifying the spaces at 2, 6, 8, 7 and 3. The three vectors a, n_1, n_2 form a minimal generating set for the semi-lattice of positive integral null vectors. Each element of the semi-lattice may be represented either in the form $rn_1 + sa$ or in the form $rn_2 + sa$ where, in each case, r and s are positive integers.

We construct endofunctors θ, ϕ, ψ, $\chi^{(t)}$ $(t = 0, 1, 2,\ldots)$ from sequences of APR tilts and, for all but the first, the duality operator D,

$$\theta = F_7F_6F_5F_3F_4F_2F_1$$

$$\phi = D\, G_7G_9G_8F_4F_2F_1$$

$$\psi = D\, G_7G_6F_4F_5F_2F_3G_8F_1$$

$$\chi^{(t)} = D\, F_2F_1F_6F_5F_3F_9(\sigma G_7G_3G_5G_6)^t F_4F_2F_5F_3F_1G_9G_8$$

where σ is the permutation (13568742) which is applied to all vertices before proceeding to apply the remaining tilts so that, for example,

$$\chi^{(1)} = D\, F_1F_3F_8F_6F_5F_9G_7G_3G_5G_6F_4F_2F_5F_3F_1G_9G_8 \,.$$

These sequences of tilts are illustrated in figures 6 and 7. We use the same symbols to denote the linear transformations of dimension vectors induced by the functors. It is easy to verify that their effects on n_1 and n_2 are given by

$$\theta n_1 = n_1 \qquad \theta n_2 = n_1 + 2a$$

$$\phi n_1 = a \qquad \phi n_2 = 2n_1 + a$$

$$\psi n_1 = n_2 \qquad \psi n_2 = n_2 + 2a$$

$$\chi^{(t)} n_1 = n_2 + (t+1)a, \quad \chi^{(t)} n_2 = n_2 + (t+3)a \,.$$

Thus we have

$$rn_1 + sa = \begin{cases} \theta[(r-s)n_1 + sa] & \text{if } r \geq s \\ \phi[(s-r)n_1 + ra] & \text{if } r \leq s \end{cases}$$

$$rn_2 + sa = \begin{cases} \psi[(r-s)n_1 + sa] & \text{if } r \geq s \\ \chi^{(t)}[\{(t+2)r - s\}n_1 + \{s-(t+1)r\}a] & \\ & \text{if } (t+1)r \leq s \leq (t+2)r. \end{cases}$$

It follows that each positive null ray can be generated from n_1 using the linear transformations θ, ϕ, ψ, $\chi^{(t)}$. Similarly, starting with the one-parameter family of indecomposables for $\tilde{D}_6$, regarded as a sub-diagram of P, we can use the functors θ, ϕ, ψ, $\chi^{(t)}$ to generate a one-parameter family for each null vector. Of course there are often several ways of generating a one-parameter family corresponding to a given null ray; at present it is not known whether or not they are isomorphic.

A number of other points are apparent from figures 6 and 7:

(a) The sequences illustrated do not contain the $\tilde{E}_7$-squid. The reader is invited to remedy this by constructing for herself the sequence of diagrams corresponding to the composite endofunctor $(G_3G_7G_2G_6G_9G_8G_5G_4)^3$.

(b) The numbering used in the diagrams is the obvious one in which the 'new' point corresponding to the simple module Λe_i (or $D(\Lambda)e_i$) used in the tilting is labelled i and the remaining points are unchanged. Most of the endofunctors used induce a permutation of labels and we can, indeed, find an endofunctor corresponding to each element of the permutation group S_9. It is not known whether this has any significance - but it is sometimes a nuisance in recognising dimension vectors.

(c) For most of the diagrams there are two obvious positive null-vectors which form a basis of the null space. Sometimes these generate the positive semi-lattice and sometimes (as with our starting point) they do not. So far 5 is the largest number of elements we have found in a minimal generating set. This occurs e.g. for the diagram

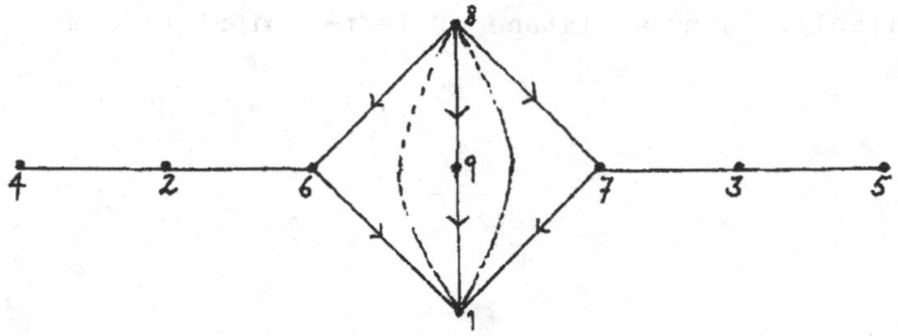

in which the dotted lines indicate that the three composite maps from 8 to 1 are equal (and which occurs in the sequence described in (a) above).

7 The diagrams of figures 3 and 4

These diagrams are closely related. Each may be transformed to one of type 3(c) by sequences of APR tilts, and each tilts to a non-triangulable algebra. It is easy to see that 4(b) is (at least) non-domestic since it is possible to embed the one-parameter family of $\tilde{A}_n$ into it as well as to wind round (once or more times) a suitable $\tilde{D}$. (In the latter case the endomorphism ring of the 'smallest' element is not a field.) Here the dimension vectors for the one-parameter families (i.e. for the 'smallest' members of each family) are all multiples of that for the $\tilde{A}_n$, corresponding to the fact that the quadratic form for such a diagram has co-rank 1. Since 4(b) can be transformed to 3(c) by APR tilts, it follows that 3(c) is also (at least) non-domestic. Ringel [23] has shown that these two are indeed non-domestic tame and the remaining three domestic.

The three domestic diagrams together with one other, are Ringel's regular extensions. This 'odd' one is a concealed quiver with respect to right-exact APR tilts (i.e. the component of its Auslander-Reiten graph which contains the hereditary injectives, contains a complete section), but not with respect to left exact ones (or vice-versa, according to orientation). We have implicitly included it in class A in section 5. It would be nice to be certain that the three diagrams here cannot be dealt with similarly, using a mixture of left- and right-exact APR tilts.

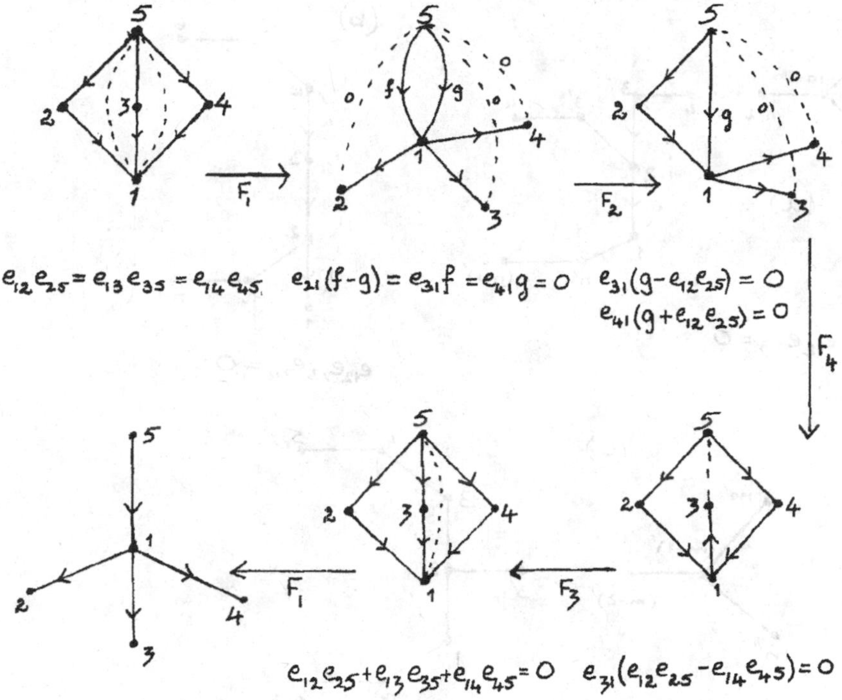

Figure 1

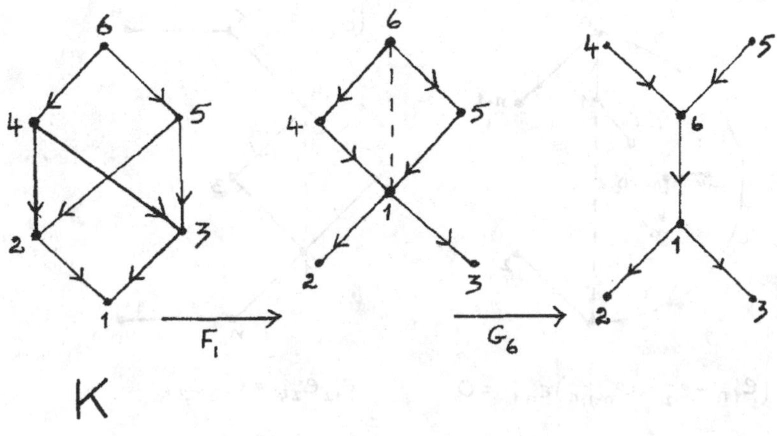

Figure 2

$$e_{12}e_{23} = 0$$

$$e_{12}e_{23}e_{34} = 0$$

$$e_{12}e_{23} = 0$$

Figure 3

$$(e_{1n} - e_{12} \cdots e_{n-1,n})e_{n1'} = 0$$

$$e_{12}e_{24} = e_{13}e_{34}$$

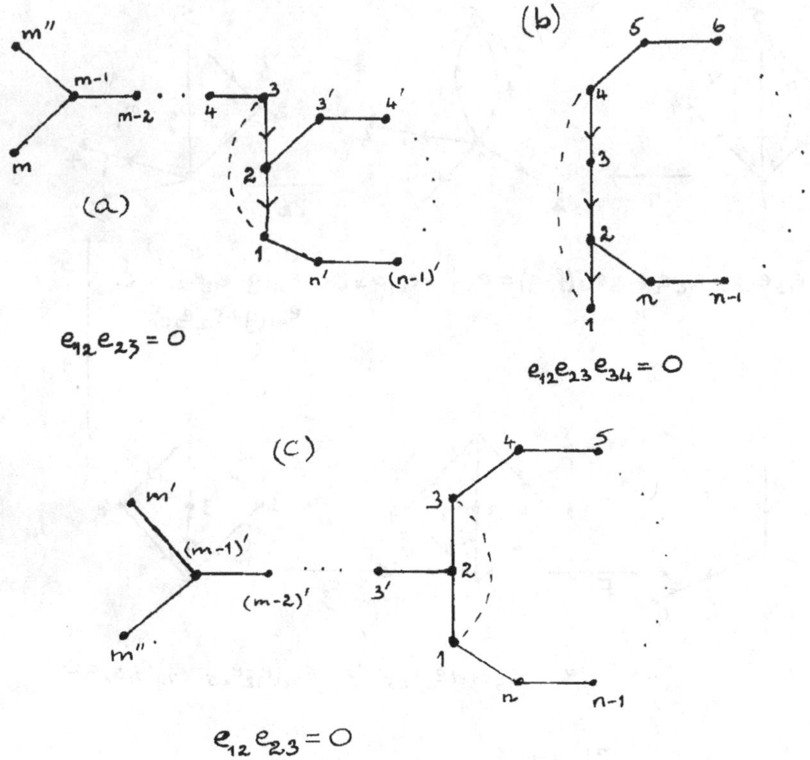

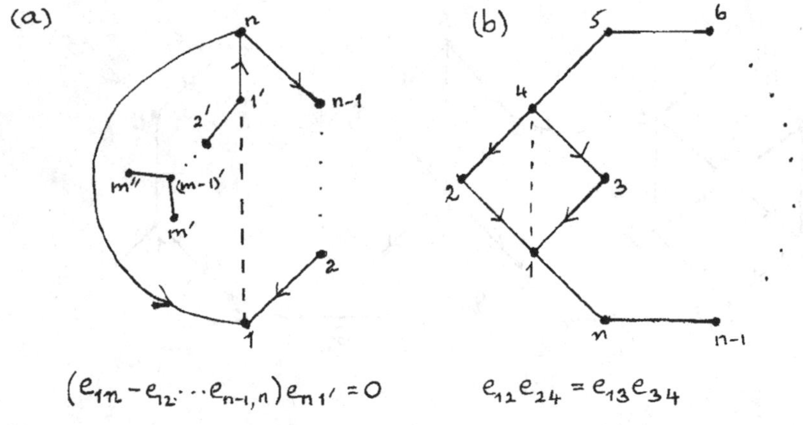

Figure 4

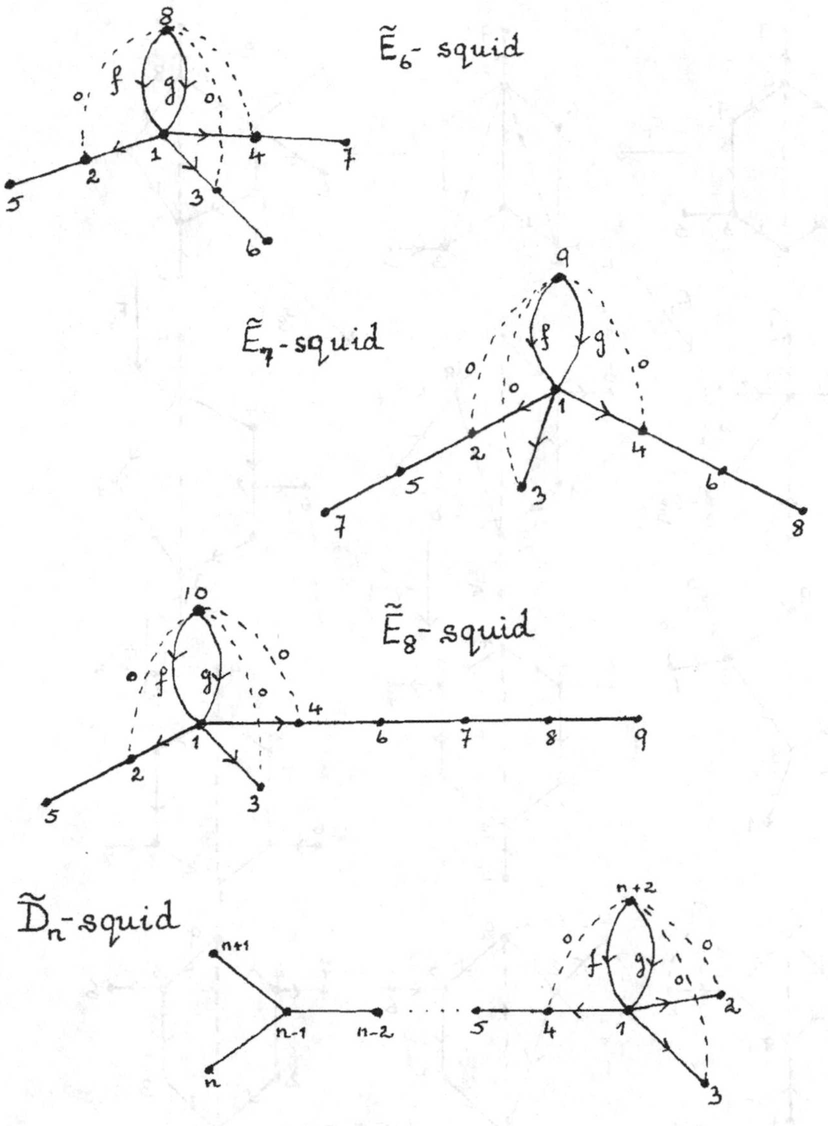

In each case $e_{21}f = e_{31}(f-g) = e_{41}g = 0$

Figure 5

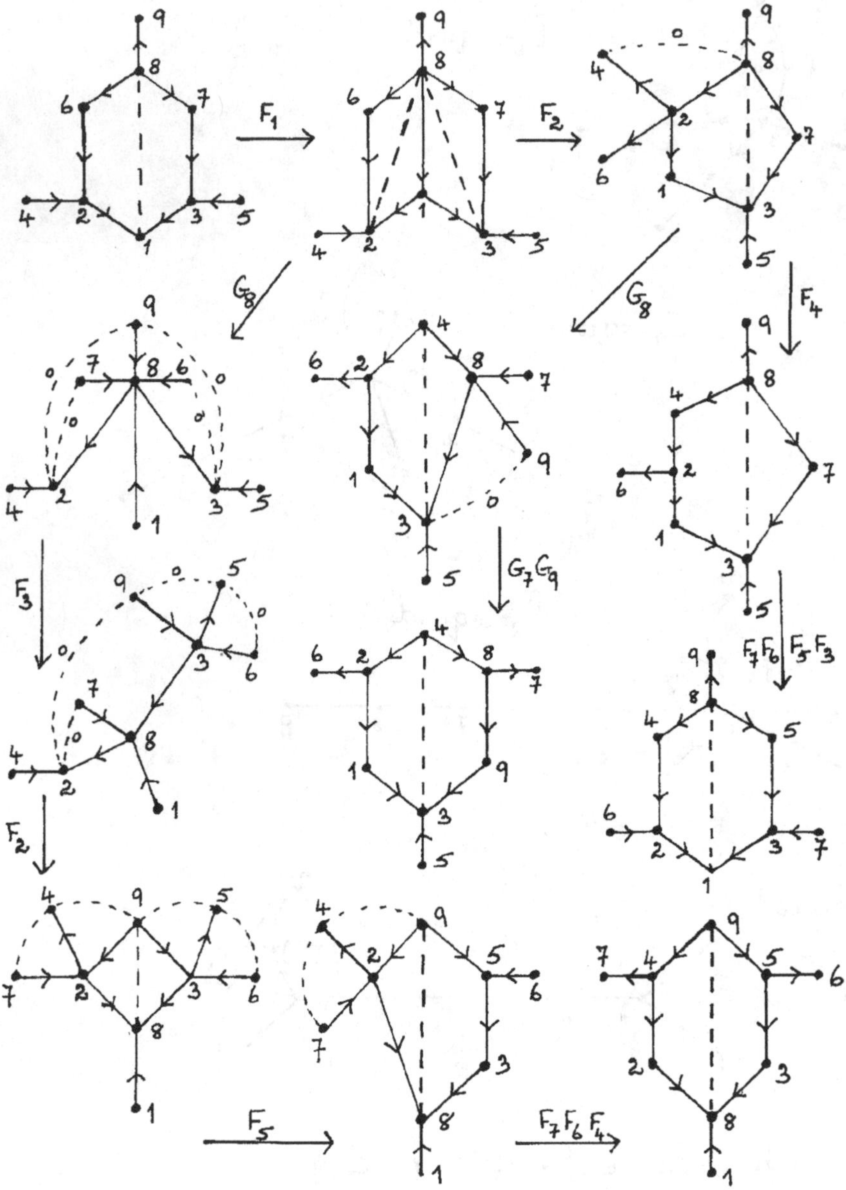

Figure 6

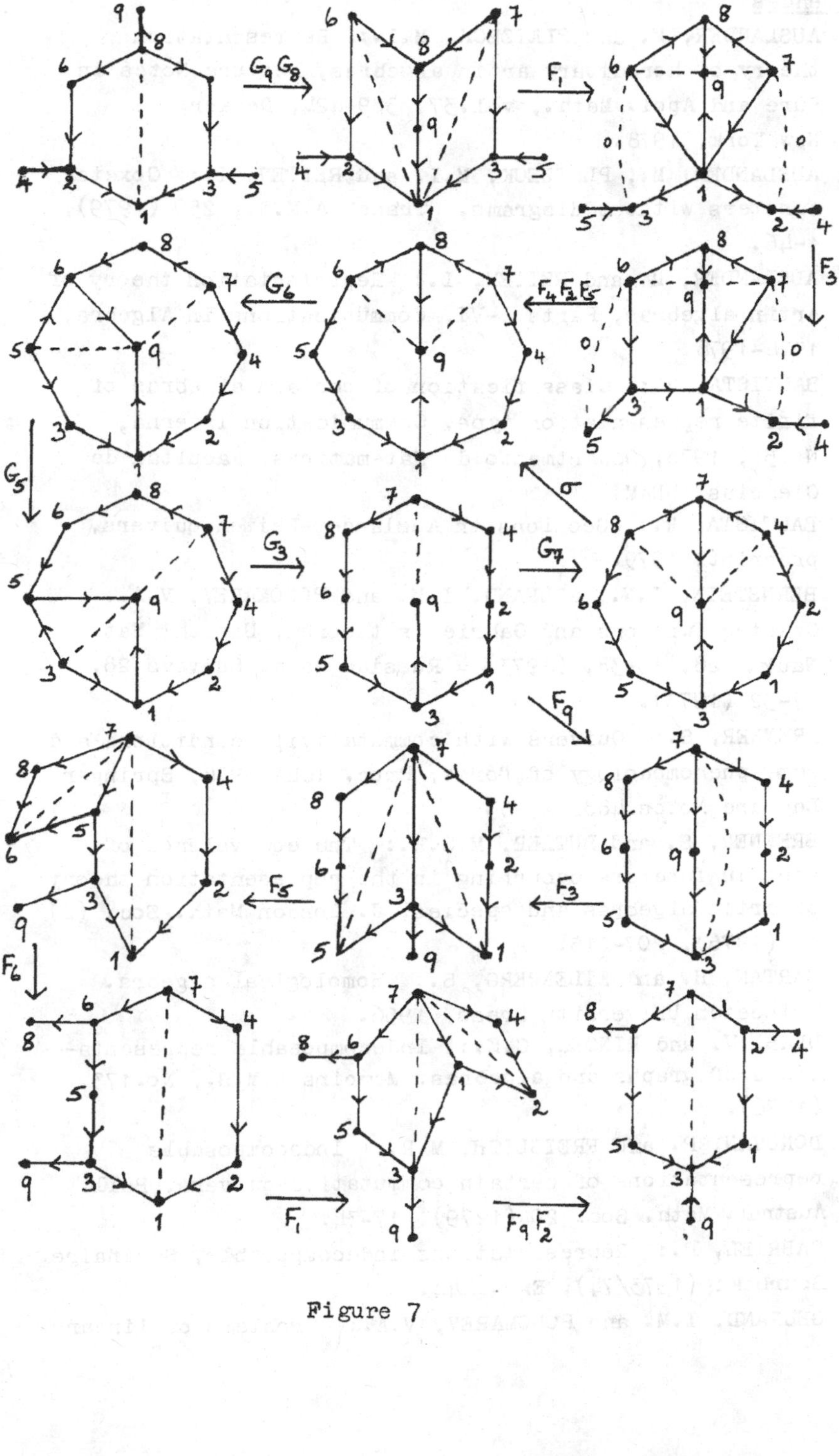

Figure 7

References

1. AUSLANDER, M. and PLATZECK, M.I.: Representation theory of hereditary artin algebras, Lecture Notes in Pure and Appl. Math., vol.37, 389-424, Dekker, New York, 1978.

2. AUSLANDER, M., PLATZECK, M.I. and REITEN, I.: Coxeter functors without diagrams, Trans. A.M.S., 250 (1979), 1-46.

3. AUSLANDER, M. and REITEN, I.: Representation theory of artin algebras, Parts I-VI, Communications in Algebra, 1974-1978.

4. BAUTISTA, R.: Classification of certain algebras of finite representation type, Communication Interna, No.51, 1978, Departmento de Matematicas, Facultad de Ciencias, UNAM.

5. BAUTISTA, R.: Sections in Auslander-Reiten quivers, preprint, 1979.

6. BERNSTEIN, I.N., GELFAND, I.M. and PONOMAREV, V.A.: Coxeter functors and Gabriel's theorem, Uspechi Mat. Nauk. 28, 19-38, (1973) = Russian Math. Surveys 28, 17-32 (1973).

7. BRENNER, S.: Quivers with commutativity conditions and some phenomenology of forms, Proc. ICRA 1974, Springer Lecture Notes 488, 29-53.

8. BRENNER, S. and BUTLER, M.C.R.: The equivalence of certain functors occurring in the representation theory of artin algebras and species, J. London Math. Soc. (2) 14 (1976), 207-215.

9. CARTAN, H. and EILENBERG, S.: Homological algebra. Princeton University Press, 1956.

10. DLAB, V. and RINGEL, C.M.: Indecomposable representations of graphs and algebras, Memoirs A.M.S., No.173, (1976).

11. DONOVAN, P. and FREISLICH, M.R.: Indecomposable representations of certain commutative quivers, Bull. Austral. Math. Soc. 20 (1979), 17-34.

12. GABRIEL, P.: Representations indecomposable, Seminaire Bourbaki (1973/74), Exp. 444.

13. GELFAND, I.M. and PONOMAREV, V.A.: Problems of linear

algebra and the classification of quadruples of subspaces in a finite-dimensional vector space, Coll. Math. Soc. Bolyai 5, Tihany (1970), 163-237.

14. HUGHES, D.W.: Private communication.

15. KAC, V.G.: Infinite root systems, representations of graphs, and invariant theory, preprint, 1979.

16. KAC, V.G.: Infinite root systems, representations of quivers, and invariant theory, This Proceedings.

17. LOUPIAS, M.: Representations indecomposable des ensembles ordonnes finis, These, Universite Tours, 1975. (Summary: Proc. ICRA, 1974, Springer Lecture Notes 488, 201-209.)

18. MARMARIDIS, N.: Darstellungen endlichen Ordnungen, Dissertation, Universität Zürich, 1978.

19. MARMARIDIS, N.: Reflection functors, This Proceedings.

20. NAZAROVA, L.A. and ROITER, A.V.: On a problem of I.M. Gelfand, Funkc. Anal. i Prilozen. 7 (1973), 54-69.

21. NAZAROVA, L.A. and ROITER, A.V.: Categorical matrix problems and the Brauer-Thrall conjecture, Inst. Math. Acad. Sc., Ukr. SSR, Kiev, 1973. = (German translation) Mitteilungen Mathem. Seminar Giessen, Heft 15, 1975.

22. RINGEL, C.M.: Representations of K-species and bimodules, J. Algebra 41 (1976), 269-302.

23. RINGEL, C.M.: This Proceedings.

24. SHKABARA, A.S.: Commutative quivers of tame type, I, preprint, Kiev, 1978.

25. ZAVADSKI, A.G.: Quivers with a distinguished path and no cycles which are of tame type, preprint, Kiev, 1978.

26. ZAVADSKI, A.G. and SHKABARA, A.S.: Commutative quivers and matrix algebras of finite type, preprint, Kiev, 1976.

Departments of Mathematics, The University, P.O. Box 147, LIVERPOOL, L69 3BX, England.

On finite representation type and a theorem of Kulikov

by Hermann Brune

Abstract. For an artin algebra A we show that maximal submodules of pure-projective right A-modules (i.e. of direct sums of finitely generated A-modules) are again pure-projective if and only if A is of finite representation type. More generally we show that a right artinian ring A , such that the injective hulls of finitely generated right A-modules are finitely generated, is right pure-semisimple if and only if maximal submodules of pure-projective right A-modules are pure-projective.

1. Introduction

In 1945 Kulikov [5] proved that subgroups of direct sums of finitely generated abelian groups are again direct sums of finitely generated groups. Since modules over certain finite dimensional hereditary algebras behave rather similar to abelian groups (cf. [10], [9]) we ask whether Kulikov's theorem in some sense carries over to artin algebras. Of course, it cannot carry over in full unless the algebra is of finite representation type (each injective algebra-module being pure-projective). Moreover, Okoh [9] gave a concrete description of a non-pure-projective submodule of a certain pure projective module over an hereditary algebra.

However, in view of the fact that over an artinian ring every submodule of a (usual) projective module P, containing the radical of P, is pure-projective, one may ask whether over an algebra at least "big" (e.g. maximal) submodules always inherit pure-projectivity. The somehow surprising answer is that this property characterizes precisely the algebras of finite representation type (among artin algebras) as well as the full Kulikov-property does. This is a consequence of our more general theorem B (sect. 2).

So infinite representation type for an artin algebra is equivalent to the existence of a non-pure-projective maximal submodule of some pure-projective module. However, we do not know how to describe such a module explicitly.

The proof of theorem B depends mainly on the fact that for a right artinian ring A, whose indecomposable injectives are finitely generated, right pure-semisimple representation type is equivalent to

$$\text{right global dimension } \mathcal{C} \leq 2,$$

where $\mathcal{C}$ is the category of finitely generated right A-modules (theorem A, sect. 2). (Recall, that A is right pure-semisimple if every right A-module is pure-projective.) Theorem A seems to be known for artin algebras (cf. [3], [2]).

2. Categorical context

Throughout let A be a right artinian ring and $\mathcal{C}$ be the category of finitely generated right A-modules.

We want to fix some categorical context for A most of which depends only on the fact that $\mathcal{C}$ is abelian.

Denote by Mod- $\mathcal{C}$ the category of additive abelian group-valued functors on $\mathcal{C}^{op}$ which are also called right $\mathcal{C}$ -modules. In sense of Mitchell's several-object-version of ring theory (cf. [8], [6]) we can work in Mod- $\mathcal{C}$ just as within a usual module category. In particular:

Flat $\mathcal{C}$ -modules are the direct limits of finitely generated projective $\mathcal{C}$ -modules (i.e. of representable functors on $\mathcal{C}^{op}$) and flat dimensions in Mod- $\mathcal{C}$ are defined as usual (cf. [6], [12], [13]). For each finitely presented $\mathcal{C}$ -module E we have projective dimension $E \leq 2$ (cf. [1]), hence weak right global dim $\mathcal{C} \leq 2$.

Now we aim to prove the following

Theorem A. Let the right-artinian ring A be such that injective hulls of finitely generated right A-modules are finitely generated. Then the following are equivalent:

(a) A is right pure-semisimple

(b) Every flat right $\mathcal{C}$ -module is projective (i.e. $\mathcal{C}$ is right-perfect)

(c) right global dim $\mathcal{C} \leq 2$.

The equivalence (a) $\Leftrightarrow$ (b) is well-known (see [12], [11]). The interesting condition is (c).

First let us recall some further consequences from the abelianess of $\mathcal{C}$ (cf. [4], [7]): Left exact functors on $\mathcal{C}^{op}$ coincide with flat $\mathcal{C}$ -modules. The full subcategory $\mathcal{L}$ in Mod- $\mathcal{C}$ of these is abelian and in Mod- $\mathcal{C}$ closed under kernels and injective hulls but not under subobjects and cokernels. Moreover, if Mod-A denotes the category of right A-modules, then there is a full embedding φ : Mod-A $\to$ Mod- $\mathcal{C}$, given by $M \mapsto \mathrm{Hom}_A(-,M)|_{\mathcal{C}}$ and inducing an equivalence Mod-A $\overset{\approx}{\to}$ $\mathcal{L}$ of categories. φ is such that an A-module M is pure-projective if and only if the $\mathcal{C}$ -module $\varphi(M)$ is projective. In particular, finitely generated A-modules correspond to finitely generated projective (representable) $\mathcal{C}$ -modules.

Now we can do the

<u>Proof of theorem A.</u>

(c) $\Rightarrow$ (b) . Let $M \in$ Mod- $\mathcal{C}$ be flat. Then its injective hull E(M) is flat and we consider an exact sequence $0 \to M \to E(M) \xrightarrow{f} F \to 0$ in $\mathcal{L}$. Put C = image f in Mod- $\mathcal{C}$. Then $0 \to M \to E(M) \xrightarrow{f} C \to 0$ is exact in Mod- $\mathcal{C}$ since so is $0 \to M \to E(M) \xrightarrow{f} F$. Observe that the injective hull E(C) of $C \subset F$ is flat being a direct submodule of E(F). So we get an exact sequence.

$$0 \to M \to E(M) \to E(C) \to N \to 0$$

in Mod- $\mathcal{C}$, where E(M) and E(C) are flat and injective.

By asssumption on A every injective right A-module is pure-projective being a direct sum of finitely generated indecomposable modules. So every flat and injective $\mathcal{C}$ -module is projective, via φ (see above). Thus E(C) and E(M) are the first terms of a projective resolution for N. By (c) M is projective.

(b) ⇒ (c). Clear, since weak right gl. dim. $\mathcal{C} \leq 2$.

(a) ⇔ (b) is known (see [12]).

3. Pure semisimple rings

Now we come to our main result.

Theorem B. Let A be a right artinian ring such that the injective hulls of finitely generated right A-modules are finitely generated. If maximal submodules of pure-projective right A-modules are pure-projective, then A is right pure semisimple.

First we consider the context for A developed in section 2. Recall that there is an exact reflector $R : \text{Mod-}\mathcal{C} \to \mathcal{L}$ (cf. [4], [7]). For $M \in \text{Mod-}\mathcal{C}$ let $u(M) : M \to R(M)$ denote the R-universal map. If $f : F' \to F$ is a morphism in $\mathcal{L}$ and $F' \xrightarrow{f} F \xrightarrow{s} C \to 0$ is exact in $\text{Mod-}\mathcal{C}$, then, clearly, $F' \xrightarrow{f} F \xrightarrow{u(C)\circ s} R(C) \to 0$ is exact in $\mathcal{L}$. Now we can do the

Proof of theorem B.

Suppose that maximal submodules inherit pure-projectivity in Mod-A Then also every submodule U of a pure-projective A-module M, such that M/U is of finite length, is pure projective, by induction.

We have to show:

right gl. dim $\mathcal{C} \leq 2$ (by thm. A).

Let $P \in \text{Mod-}\mathcal{C}$ be any small projective (representable functor) and

$H \subset P$ be any $\mathcal{C}$-submodule. It is enough to show: proj. dim. $H \leq 1$ in Mod-$\mathcal{C}$.

At first we get an exact sequence $0 \to F' \to F \xrightarrow{s} H \to 0$ in Mod-$\mathcal{C}$ with projective F and flat F', since weak right gl. dim $\mathcal{C} \leq 2$.
By exactness of R we also have $R(H) \subset P$ (observe $R(P) = P$).
Thus $R(H)$ is small projective (representable), via φ (see sect. 2).
Moreover, $0 \to F' \to F \xrightarrow{u(H)\circ s} R(H) \to 0$ is exact in $\mathcal{L}$, and considering this sequence in Mod-A ($\simeq \mathcal{L}$) we there have F pure-projective and $R(H)$ of finite length.
By assumption F' is pure-projective in Mod-A, thus projective in Mod-$\mathcal{C}$. So proj. dim. $H \leq 1$.

References

[1] M. Auslander, Coherent functors, Proc. Conf. on Categorical Algebra, Springer-Verlag 1966, 189 - 231

[2] M. Auslander, Representation theory of artin algebras II, Comm. Algebra (1974), 269 - 310

[3] P. Gabriel, Algèbres auto-injectives de représentation finie, Séminaire Bourbaki, 32 e année, 1979/80, n° 545

[4] P. Gabriel, Des catégories abéliennes, Bull. Soc. Math. Fr. (1962), 323 - 448

[5] Kulikov, On the theory of abelian groups of arbitrary cardinality [Russian], Mat. Sb, 16 (1945) 129 - 162

[6] D.M. Latch/ B. Mitchell, On the difference between cohomological dimension and homological dimension, J. Pure and Applied Algebra 5 (1974), 333 - 343

[7] B. Mitchell, Theory of Categories, Academic Press, New York (1965)

[8] B. Mitchell, Rings with several objects, Advances in Math. 8 (1972), 1 - 161

[9] F. Okoh, Direct sums and direct products of canonical pencils of matrices, Linear Alg. and Appl. 25 (1979), 1 - 26

[10] C.M. Ringel, Infinite dimensional representations of finite dimensional hereditary algebras, Symposia Math., Vol. 23, Bologna 1979

[11] D. Simson, On pure global dimension of locally finitely presented Grothendieck categories, Fund. Math. 96 (1977), 91 - 116

[12] D. Simson, Functor categories in which every flat object is projective, Bull. Acad. Polon. Sci. 22 (1974), 375 - 380

[13] B. Stenström, Purity in functor categories, J. Alg. 8 (1968), 352 - 361

Hermann Brune

Fachbereich Mathematik, Universität-Gesamthochschule Paderborn

D-4790 Paderborn, GERMANY

HOMOLOGY REPRESENTATIONS OF FINITE GROUPS*

Charles W. Curtis

Actions of finite groups on finite simplicial complexes give rise to representations of the groups on the rational homology groups of the complexes. In §1 a survey is given of general properties of homology representations as they relate various aspects of the geometric realizations of the complexes. Special attention is given to the Lefschetz character, and its relation to fixed point sets and orbit spaces. It is shown that the Lefschetz character plays a role in the theory of homology representations analogous to the permutation character in the theory of permutation representations. It is proved that the endomorphism algebra of a homology representation can be realized on the homology of a certain orbit space, thereby extending a familiar construction of the endomorphism algebra of a permutation representation.

In §2, a sketch is given of an application of the Lefschetz character to the computation of the Steinberg character of a connected reductive algebraic group defined over a finite field. The final section, on homology with a coefficient system, extends the scope from rational homology representations to homology representations of finite groups over a commutative ring. Examples are given of representations of finite groups of Lie type on the homology of the combinatorial building with coefficients, and their connection with a duality operation in the ring of complex-valued characters of a finite group of Lie type.

Most of the results in §1 are no doubt familiar to topologists, and

---

*This research was supported by NSF grant MCS-7801944.

can be presented, with less discussion of subdivision of the underlying complexes, in terms of finite group actions on semi-simplicial complexes. In any case, the main results can be stated in terms of finite group actions on the polyhedra associated with the simplicial complexes, and are therefore independent of the choice of subdivisions of complexes required for the proofs given below. The author wishes to thank G. I. Lehrer, with whom the study resulting in §2 began during his visit to Oregon in 1976, A. Sieradski for contributing a proof of Proposition 1.3, and A. G. Wasserman, who collaborated in the formulation and proof of Propositions 1.7 and 1.8.

1. Homology Representations and the Lefschetz Character.

Let G be a finite group. A G-*complex* is a finite simplicial complex K, with vertices {a} and simplices {σ}, on which G acts as a group of simplicial automorphisms. More precisely, G acts as a permutation group on the vertices of K, and a set of vertices $(a_0,\dots,a_p)$ is a p-simplex σ of K if and only if $g\sigma = (ga_0,\dots,ga_p)$ is a p-simplex, for all g in G, and p = 0, 1,... .

We denote by $|K|$ the underlying topological space of K. We recall that the correspondence $K \to |K|$ is a functor from the category of finite simplicial complexes with simplicial maps to the category of metric spaces, with continuous maps. Therefore if K is a G-complex, there is a corresponding action by G through homeomorphisms of $|K|$.

For the purposes of this section, the *homology* $H_*(K)$ of a simplicial complex K denotes the direct sum of the rational homology groups,

$$H_*(K) = \sum_{p=0}^{\infty} H_p(K).$$

In case K is a G-complex, the G-action on K defines a G-action

$$g_*: H_p(K) \to H_p(K), \quad g \in G,$$

on each rational homology group by linear transformations. Thus $H_*(K)$ and the individual homology groups $\{H_p(K), p = 0,1,\ldots\}$ become modules over the rational group algebra QG of G, and define what we shall call homology representations of G.

We remark that the underlying topological space $|K|$ of K, together with the G-action on $|K|$, and the rational homology groups $\{H_p(K), p = 0,\ldots\}$ with their QG-module structure, remain unchanged (after appropriate identifications) if we replace K by a barycentric subdivision Sd(K) ([4], Ch. VI, §7).

The Lefschetz character Λ of a homology representation of G on K is the virtual character of G defined by

$$\Lambda(g) = \Sigma\,(-1)^i \mathrm{Tr}(g_*, H_i(K)), \quad g \in G.$$

The degree $\Lambda(1)$ of the Lefschetz character is the Euler characteristic $\chi(K)$ of K, and is a topological invariant of $|K|$, given by

$$\chi(K) = \Sigma\,(-1)^i \dim_Q H_i(K).$$

Following Bredon [2], the G-action on K is called regular if, for any subgroup H of G, whenever $(a_0,a_1,\ldots,a_p)$ and $(h_0a_0,h_1a_1,\ldots,h_pa_p)$ are both simplices of K, for $h_0, h_1,\ldots,h_p$ in H, then there exists h in H such that $ha_i = h_ia_i$, $i = 0,\ldots,p$.

There is no loss of generality in restricting our discussion to regular G-actions, as the following result shows.

(1.1) Lemma. *Let K be a G-complex, for a finite group G. Then $Sd^2(K) = Sd(Sd(K))$, with the resulting G-action, is a regular G-complex.*

For the proof, see ([2], Ch. III, Prop. 1.1).

Let K be a regular G-complex. We can then define a simplicial complex K/G as follows. The vertices of K/G are the G-orbits [a] of vertices a of K. A p-simplex of K/G is a set of orbits $([a_0],\ldots,[a_p])$ such that, for some choice of representatives, $(a_0,\ldots,a_p)$ is a p-simplex of K.

The next result is also due to Bredon ([2], Ch. III, §1).

(1.2) Lemma. *Let K be a regular G-complex.*

(i) *The orbit space $|K|/G$, topologized in the usual way, is homeomorphic to the underlying topological space $|K/G|$ of the complex K/G.*

(ii) *The fixed point set* $|K|^G = \{x \in |K| : gx=x \text{ for all } g \in G\}$ *is homeomorphic to* $|K^G|$, *where* K^G *is the subcomplex of K consisting of those simplices whose vertices are fixed under the action of G.*

We now have:

(1.3) Proposition. *Let K be a regular G-complex. The Lefschetz character* Λ *of the homology representation is given by*

$$\Lambda = \Sigma_\sigma (-1)^{\dim \sigma} 1_{G_\sigma}^{\;G} ,$$

where the sum is taken over the G-orbits of simplices of K, and G_σ *denotes the stabilizer of a simplex* σ. *The value of the Lefschetz character at an element* $g \in G$ *satisfies*

$$\Lambda(g) = \chi(|K|^g),$$

where $\chi(|K|^g)$ *is the Euler characteristic of the fixed point set in* $|K|$ *under the action of* g.

Proof. The Hopf Trace Formula asserts that

$$\Lambda(g) = \Sigma (-1)^i \operatorname{Tr}(g_*, C_i(K)),$$

where $C_i(K)$ is the vector space of i-chains, and has a basis consisting of the i-simplices. The group G acts as a permutation group on the i-simplices. It follows that

$$\operatorname{Tr}(g_*, C_i(K)) = \sum_{\dim\sigma = i} 1_{G_\sigma}{}^G(g),$$

where $\{\sigma\}$ ranges over a set of representatives of the G-orbits of i-simplices, and the first part of the proposition is proved.

Applying the first statement, we have

$$\Lambda(g) = \sum_{\dim\sigma=i} (-1)^i 1_{G_\sigma}{}^G g = \Sigma_i (-1)^i n_i(g),$$

where $n_i(g)$ is the number of i-simplices fixed by g. By regularity, the i-simplices fixed by g are the i-simplices of the fixed point complex $K^{<g>}$, where $\langle g \rangle$ is the cyclic group generated by g. Then

$$\Lambda(g) = \chi(K^{<g>}) = \chi(|K|^{<g>}),$$

by the Hopf trace formula applied to the complex $K^{<g>}$, and Lemma 1.2. This completes the proof.

Now let K_1 and K_2 be G-complexes. We may assume, without changing the homology representations or the G-action on the underlying topological spaces, that the following conditions hold.

a) K_1 and K_2 are regular G-complexes;

b) K_1 and K_2 are ordered simplicial complexes, in the sense of [4] ,

p. 67. Thus, the vertices of each complex are partially ordered by a relation which we shall denote by $\leq$ in both cases, with the property that the vertices of each simplex are totally ordered. Moreover, we may assume that for vertices a and a', $a \leq a'$ only when a and a' are both vertices of some simplex.

c) The action of G preserves the order relations on K_1 and K_2, and for vertices a and a' of one of the complexes, we have

i) $a \leq a'$ and $ga' = a'$, for g in G, implies $ga = a$; and

ii) $a \leq a'$ and $ga \leq a'$, for g in G, imply $a = ga$.

In case K_1 and K_2 are regular G-complexes, all these conditions are satisfied for the G-complexes Sd K_1 and Sd K_2, under the order relation of inclusion.

A pair of simplicial complexes satisfying conditions a) - c) will be called a <u>normalized pair</u> of G-complexes.

We now recall the definition of the cartesian product $K_1 \times K_2$ of a pair of ordered simplicial complexes K_1 and K_2 (see [4], p. 67). The set of vertices of $K_1 \times K_2$ is the cartesian product of the vertex set of K_1 with the vertex set of K_2. A p-simplex of $K_1 \times K_2$ is a set of vertices $\{(a_0,b_0),\ldots,(a_p,b_p)\}$, with $a_0,\ldots,a_p$ and $b_0,\ldots,b_p$ vertices (possibly with repetitions) of simplices of K_1 and K_2, which are totally ordered under the partial order relation $(a,b) \leq (a',b')$ defined by the conditions: $a \leq a'$ and $b \leq b'$. If K_1 and K_2 are G-complexes, then their cartesian product becomes a G-complex under diagonal action of G.

We now have:

(1.4) Proposition. Let K_1 and K_2 be a normalized pair of G-complexes.

(i) The G-complex $K_1 \times K_2$ provides a G-equivariant triangulation of the cartesian product space $|K_1| \times |K_2|$, with its diagonal G-action.

(ii) $K_1 \times K_2$ is a regular G-complex.

(iii) There are G-equivariant isomorphisms:

$$H_p(K_1 \times K_2) \cong \bigoplus_{j=0}^{p} H_j(K_1) \otimes H_{p-j}(K_2), \quad p = 0,1,\ldots;$$

and

$$H_*(K_1 \times K_2) \cong H_*(K_1) \otimes H_*(K_2).$$

Proof. The first statement follows from the proof of ([4], Lemma 8.9, p. 68). For the second, let $\{(a_0,b_0),\ldots,(a_p,b_p)\}$ and $\{g_0(a_0,b_0),\ldots,g_p(a_p,b_p)\}$ be simplices of $K_1 \times K_2$, for some elements $g_i \in G$. Then, by induction we may assume $p > 0$, and for some $g \in G$, $g(a_j,b_j) = g_j(a_j,b_j)$, for $j = 1,\ldots,$ $p-1$. Then $(ga_0,\ldots,ga_{p-1},ga_p)$ and $(ga_0,\ldots,ga_{p-1},g_pa_p)$ are vertices of simplices of K_1, where we may assume the vertices of both simplices ordered as indicated. Then $ga_{p-1} \leq ga_p$, and $ga_{p-1} \leq g_pa_p$, hence $a_{p-1} \leq a_p$ and $g_p^{-1}ga_{p-1} \leq a_p$. By the assumption c)ii) above, we have $g_p^{-1}ga_{p-1} = a_{p-1}$, hence by c)i), $g_p^{-1}ga_j = a_j$, for all $j \leq p-1$. It follows that $g_pa_j = g_ja_j$ for $j = 0,\ldots,p$, and similarly $g_pb_j = g_jb_j$ for $j = 0,\ldots,p$, proving that $K_1 \times K_2$ is regular.

The third statement follows by consideration of the complexes of semi-simplicial chains $C_*^+(K_1)$, $C_*^+(K_2)$, $C_*^+(K_1 \times K_2)$ and the Eilenberg-Zilber theorem, as in ([5], Ch. II).

We shall also require the transfer theorem of Conner ([2], Ch. III,

Theorem 2.4).

(1.5) Proposition. *Let K be a regular G-complex. Then*

$$H_*(K/G) \cong \mathrm{inv}_G(H_*(K)),$$

where $\mathrm{inv}_G(H_*(K))$ *is the QG-submodule of* $H_*(K)$ *affording the trivial representation of G.*

We now come to the main results of this section. We shall use the notation (,) to denote the usual scalar product of complex valued class functions on a finite group, and 1_G for the principal character of G.

(1.6) Proposition. (i) *Let K be a regular G-complex, with Lefschetz character* Λ. *Then*

$$(\Lambda, 1_G) = \chi(K/G) = \chi(|K|/G).$$

(ii) *Let* K_1 *and* K_2 *be a normalized pair of G-complexes with Lefschetz characters* Λ_1 *and* Λ_2. *Then*

$$(\Lambda_1, \Lambda_2) = \chi(K_1 \times K_2/G) = \chi(|K_1| \times |K_2|/G).$$

(iii) *Let K be a regular G-complex with Lefschetz character* Λ, *and let H be a subgroup of G. The K is a regular H-complex, and we have*

$$(\Lambda, 1_H{}^G) = \chi(K/H).$$

Proof. By Proposition 1.5 and Lemma 1.2, we have

$$\begin{aligned}(\Lambda, 1_G) &= \Sigma\, (-1)^i \dim_Q \mathrm{inv}_G(H_i(K)) \\ &= \Sigma\, (-1)^i \dim_Q H_i(K/G) = \chi(K/G),\end{aligned}$$

proving the first result. For the second, we have, because the Lefschetz character is integral valued,

$$(\Lambda_1,\Lambda_2) = (\Lambda_1\Lambda_2, 1_G).$$

An easy computation using Proposition 1.4 shows that $\Lambda_1\Lambda_2$ is equal to the Lefschetz character $\Lambda_{K_1 \times K_2}$ of the regular G-complex $K_1 \times K_2$. Thus

$$(\Lambda_1,\Lambda_2) = (\Lambda_{K_1 \times K_2}, 1_G) = \chi(K_1 \times K_2/G),$$

by the first part of the proposition. The last equality follows from Proposition 1.4 (i), (ii), and Lemma 1.2. Finally, it is clear from the definition that K is a regular H-complex, for any subgroup H of G, and that the restriction $\Lambda|_H$ is the Lefschetz character of the regular H-complex K. Then, using Frobenius reciprocity and part (i), we have

$$\chi(K/H) = (\Lambda|_H, 1_H) = (\Lambda, 1_H{}^G),$$

as required.

Remarks. Let X be a finite G-set, with permutation character θ. It is well known that

$$\theta(g) = \mathrm{card}(X^g),$$

where X^g is the set of points of X fixed by g, for g in G, and that

$$(\theta, 1_G) = \mathrm{card}(X/G).$$

Moreover, for two G-sets X_1 and X_2, with permutation characters θ_1 and θ_2, and diagonal action of G on the cartesian product set $X_1 \times X_2$, we have

$$(\theta_1,\theta_2) = \mathrm{card}(X_1 \times X_2/G).$$

These results can be viewed as the zero-dimensional cases of Propositions 1.3 and 1.6. We also note that parts (i) and (ii) of Proposition 1.6 are stated in terms of the Euler characteristics of the underlying topological spaces, and hence are independent of the normalizations of the G-complexes used to prove them.

The next result, obtained jointly with A. Wasserman, shows that the QG-endomorphism algebra of a homology representation $H_*(K)$ can be realized in the homology $H_*(K \times K/G)$.

(1.7) <u>Proposition.</u> <u>Let K be a G-complex such that (K,K) is a normalized pair of G-complexes. Then there exist isomorphisms of vector spaces:</u>

$$H_p(K \times K/G) \cong \bigoplus_{j=0}^{p} \mathrm{Hom}_{QG}(H_j(K), H_{p-j}(K)),$$

<u>for p = 0,1,... . In particular, there is an isomorphism of vector spaces:</u>

$$H_*(K \times K/G) \cong \mathrm{End}_{QG}(H_*(K)).$$

<u>Proof</u>. Because $H_*(K)$ affords a rational representation of G, there exists a G-invariant non-degenerate scalar product:

$$\langle\ ,\ \rangle : H_p(K) \times H_p(K) \to Q,$$

for each p = 0,1,... . It follows that there exist vector space isomorphisms

$$H_p(K \times K) \xrightarrow{\phi} \bigoplus_{j=0}^{p} H_j(K) \otimes H_{p-j}(K) \xrightarrow{\psi} \bigoplus_{j=0}^{p} \mathrm{Hom}_Q(H_j(K), H_{p-j}(K)),$$

for p = 0,1,... , where ϕ is the G-equivariant isomorphism from Proposition 1.4(iii), and ψ is the isomorphism given by:

$$\psi(x \otimes y)(z) = \langle x,z\rangle y, \quad x \in H_j(K),\ y \in H_{p-j}(K),\ z \in H_j(K).$$

It is a standard result that restriction of ψ to the subspace $\mathrm{inv}_G(H_j(K) \otimes H_{p-j}(K))$ defines an isomorphism of vector spaces:

$$\mathrm{inv}_G(H_j(K) \otimes H_{p-j}(K)) \cong \mathrm{Hom}_{QG}(H_j(K), H_{p-j}(K)),$$

for each p, and j = 0,1,... . We also have an isomorphism of vector spaces

$$H_p(K \times K/G) = \mathrm{inv}_G H_p(K \times K),$$

for each p, by Proposition 1.5, since $K \times K$ is a regular G-complex, by Proposition 1.4(ii). Finally, since the isomorphism ϕ: $H_p(K \times K) \to \bigoplus H_j(K) \otimes H_{p-j}(K)$ is G-equivariant, we have, for p = 0,1,...,

$$\mathrm{inv}_G(H_p(K \times K)) \cong \bigoplus_{j=0}^{p} \mathrm{inv}_G(H_j(K) \otimes H_{p-j}(K)).$$

Upon combining the results, we obtain the first isomorphism whose existence is asserted in the statement of the proposition. The second follows from the first, by the additive properties of the HOM functor, and the proof is complete.

Remark. From Proposition 1.7 it follows that, for a given homology representation of G on $H_*(K)$, it is possible to define the structure of a Q-algebra on the vector space $H_*(K \times K/G)$ (or equivalently, on $H_*(|K| \times |K|/G)$ by Proposition 1.4(i)), so that this algebra is isomorphic to the algebra of QG-endomorphisms of $H_*(K)$. The representation theory of this algebra can then be used to investigate the decomposition of the homology representation into its simple components. In the zero-dimensional case, of a finite G-set X, Proposition 1.7 gives the familiar identification of the endomorphism algebra of the permutation representation with an algebra whose basis is indexed by the G-orbits in $X \times X$, and the multiplication of basis elements given by the intersection numbers (see [6]).

The final result of this section is an application of Proposition 1.7 to the interesting question of when there is no cancellation in the Lefschetz character. We have, for a homology representation of G on K with Lefschetz character Λ, $\Lambda = \Lambda_+ - \Lambda_-$, where Λ_+ is the character of the representation of G on the even-dimensional homology groups $\oplus_j H_{2j}$, and Λ_- is afforded by the odd - dimensional homology groups, $\oplus_j H_{2j+1}$. We define the Lefschetz character to be <u>without cancellation</u> if $(\Lambda_+, \Lambda_-) = 0$, i.e. the representations of G on the even and odd dimensional homology have no simple components in common.

(1.8) <u>Proposition. Let K be a regular G-complex, and (K,K) a normalized pair of G-complexes. Then the Lefschetz character of K is without cancellation if and only if the odd dimensional homology of</u> $H_*(K \times K/G)$ <u>vanishes.</u>

<u>Proof</u>. The inner product (Λ_+,Λ_-) is equal to the intertwining number

$$\dim_{Q}(\mathrm{Hom}_{QG}(\oplus_i H_{2i}(K), \oplus_j H_{2j+1}(K)),$$

and is zero if and only if

$$\mathrm{Hom}_{QG}(H_{2i}(K), H_{2j+1}(K)) = 0,$$

for all i and j. The odd-dimensional homology of $K \times K/G$ is the direct sum of the spaces $\mathrm{Hom}_{QG}(H_{2i}(K), H_{2j+1}(K))$, by Proposition 1.7, and the result follows.

2. Application to the Steinberg Character of a Reductive Group over a Finite Field.

In this section, an outline is given, without proofs, of an application of Proposition 1.3 to the computation of the character values of the Steinberg character of a reductive group over a finite field (for details, see[3]). Let k be a field. To every connected reductive algebraic group defined over k, there is associated a simplicial complex $\Delta(G,k)$, the combinatorial building of G, whose simplices are in bijective correspondence with the parabolic k-subgroups of G, ordered by the opposite of the inclusion relation.The group G(k) of k-rational points of G operates as a group of simplicial automorphisms on $\Delta(G,k)$. If n' denotes the k-rank of the derived group of G, $\Delta(G,k)$ has the homotopy type of a bouquet of n'-spheres, and L. Solomon proved that if k is finite, the action of G(k) on the rational homology group $H_{n'-1}(\Delta(G,k))$ affords the Steinberg representation St_G of the finite group G(k) (see [7]).

This geometric interpretation of St_G can be used to compute its character as follows. This is most efficiently done using a different notion of building. To G, k, we associate a spherical building, denoted by B(G,k), which, in contrast to $\Delta(G,k)$, takes into account the center of G. If n is the k-rank of G, the set B(G,k) is, in a G(k)-equivariant way, the (n - n')-fold suspension of $\Delta(G,k)$. Thus, suitably topologized, it has the homotopy type of a bouquet of n-spheres, and in case k is finite, G(k) operates on $H_{n-1}(B(G,k))$ through St_G.

The key property of B(G,k) is its "functorial" behavior: to every k-monomorphism of reductive k-groups $f:G \to H$, there is associated an embedding of topological spaces $B(f):B(G,k) \to B(H,k)$ satisfying the usual conditions. From this remark, it can be proved that if s is a semisimple element of $G(k)$, the fixed point set of s in $B(G,k)$ can be identified with the spherical building of the connected centralizer $Z_G(s)^o$ of s in G (which is also a reductive group defined over k).

We can now sketch the application of these remarks to the computation of St_G, in case k is a finite field. For a suitable triangulation, $B(G,k)$ is the underlying topological space of a finite $G(k)$-complex K, whose rational homology groups are zero except in dimensions 0 and $n-1$, and whose Lefschetz character is given by

$$\Lambda = 1_G + (-1)^{n-1}St_G.$$

For a semisimple element s in $G(k)$, $B(G,k)^s = B(Z_G(s)^o,k)$, hence by Proposition 1.3,

$$\Lambda(s) = \chi(B(G,k)^s) = 1 + (-1)^{m-1}St_{Z_G(s)^o}(1),$$

where m is the k-rank of $Z_G(s)^o$. As a consequence, we obtain

$$St_G(s) = (-1)^{n+m}St_{Z_G(s)^o}(1).$$

If $x \in G(k)$ is not semisimple, then it can be shown that the fixed point set of x on $B(G,k)$ is contractible, so that, by Proposition 1.3 again, we have $St_G(x) = 0$, completing the computation of St_G in all cases.

The construction of $B(G,k)$, and proofs of all the preceding results, are to be found in [3] (cf. also [8], where the result on the values of St_G and a similar approach is announced.)

3. Homology with Coefficients. Duality in the Character Ring of a Reductive Group over a Finite Field

We first recall the definition of homology with coefficients (cf.[5]). Let K be a finite simplicial complex, and let R be a commutative ring. A coefficient system $\mathfrak{M}$ over K is a family of R-modules $\{M_\sigma\}$, indexed by the simplices of K, such that whenever $\sigma' < \sigma$ there exists an R-homomorphism $\phi^{\sigma}_{\sigma'}: M_\sigma \to M_{\sigma'}$, satisfying the conditions that $\phi^{\sigma}_{\sigma} = \mathrm{id}$, and $\phi^{\sigma'}_{\sigma''}\phi^{\sigma}_{\sigma'} = \phi^{\sigma}_{\sigma''}$, whenever $\sigma'' < \sigma' < \sigma$. We define a chain complex $C(K,\mathfrak{M})$, for which the module of p-chains is given by

$$C_p(K,\mathfrak{M}) = \bigoplus_{\dim\sigma = p} M_\sigma, \quad p = 0,1,\ldots ,$$

and a boundary homomorphism $\partial: C_p(K,\mathfrak{M}) \to C_{p-1}(K,\mathfrak{M})$, where

$$\partial(m_\sigma) = \sum_{i=0}^{p} (-1)^i \phi^{\sigma}_{\sigma_i}(m_\sigma), \qquad m_\sigma \in M_\sigma,$$

and σ_i is the ith face of σ, $i = 0,\ldots,p$. It is easily shown that $\partial^2 = 0$, so that we may define the homology of K with coefficients in $\mathfrak{M}$, in the usual way, as $\ker\partial/\mathrm{im}\,\partial$, and denote the resulting graded R-module by

$$H_*(K,\mathfrak{M}) = \oplus_p H_p(K,\mathfrak{M}).$$

We illustrate how this construction can be applied to extend the scope of the homology representations of finite groups considered in §1. Let k be a finite field, G a connected reductive algebraic group defined over k, G(k) the finite group of k-rational points, and $\Delta(G,k)$ the combinatorial building of G, as in §2. We shall define a coefficient system $\mathfrak{M}$ over $\Delta(G,k)$, corresponding to a fixed finitely-generated RG-module

M, for an arbitrary commutative ring R.

For each proper parabolic k-subgroup P of G, let σ_P denote the corresponding simplex in $\Delta(G,k)$, and let

$$M_{\sigma_P} = \mathrm{inv}_{R_u(P)(k)} M,$$

the R-submodule of M affording the trivial representation of the group $R_u(P)(k)$ of k-rational points on the unipotent radical $R_u(P)$ of P. For two simplices, $\sigma_{P'} < \sigma_P$ means that P' contains P, so that $R_u(P') \subset R_u(P)$, and we have an inclusion map

$$\phi^{\sigma_P}_{\sigma_{P'}} : M_{\sigma_P} \to M_{\sigma_{P'}} .$$

We thus obtain a coefficient system $\mathfrak{M}$ over $\Delta(G,k)$ consisting of the R-modules $\{M_{\sigma_P}\}$ and the maps $\{\phi^{\sigma_P}_{\sigma_{P'}}\}$.

There is a natural action of G(k) on the chain complex $C(\Delta(G,k),\mathfrak{M})$, given by

$$m \to xm, \text{ for } x \in G(k),\ m \in M_{\sigma_P} ,$$

where $xm \in M_{\sigma_{xPx^{-1}}}$. The chain groups thereby become RG(k)-modules. It is readily shown that the action of G(k) on the chain complex commutes with the boundardy homomorphism. Thus the homology groups $H_p(\Delta(G,k),\mathfrak{M})$ are all RG(k)-modules, for $p = 0,1,\ldots$.

In case the chain complex and the homology defined above consists of R-free modules (for example when R is a field), the Lefschetz character Λ of the G(k)-action on the coefficient system can be defined by

$$\Lambda(g) = \Sigma\ (-1)^i \mathrm{Tr}(g_*, H_i(\Delta(G,k),\mathfrak{M})), \quad g \in G(k).$$

In this situation, it is not difficult to prove the Hopf trace formula:

$$\Lambda(g) = \Sigma\ (-1)^i \mathrm{Tr}(g_*, C_i(\Delta(G,k),\mathfrak{M})), \quad g \in G(k).$$

Now let $\{P_J\}$ denote a fixed set of standard parabolic k-subgroups, indexed by subsets J of the set S of distinguished generators of the relative Weyl group of G. Then we state as our final result (and leave as an exercise the proof, using the Hopf trace formula):

(3.1) Proposition. The Lefschetz character Λ of $(\Delta(G,k),\mathfrak{M})$ satisfies

$$\Lambda = \mu + (-1)^{|S|-1}\mu^*,$$

where $\mu = \mathrm{Tr}(\ ,M)$, and μ^* is an alternating sum of induced characters,

$$\mu^* = \sum_{J \subseteq S} (-1)^{|J|} \mu_{(P_J)}^{G(k)},$$

and $\mu_{(P_J)} g = \mathrm{Tr}(g, M_{\sigma_{P_J}})$, $g \in G(k)$.

In case R is the complex field C, M is a CG(k)-module, μ is its character, and μ^* is a virtual C-character of G(k) called the dual of μ (see [1]). For example, the dual 1_G^* of the principal character of G(k) is the Steinberg character St_G, as one verifies easily using Proposition 1.3. In fact, Proposition 3.1 can be viewed as an extension of the first part of Proposition 1.3 to the present situation. Alvis has proved [1] that the duality operation $\mu \to \mu^*$ permutes, up to sign, the irreducible complex characters of G(k). We conclude with an unsolved problem, whether there is a vanishing theorem for the homology modules $H_i(\Delta(G,k),\mathfrak{M})$, so that μ^* is afforded by a single homology group, as in the case of St_G.

REFERENCES

1. D. Alvis, The duality operation in the character ring of a finite Chevalley group, to appear.

2. G. E. Bredon, Introduction to compact transformation groups, Academic Press, New York, 1972.

3. C. W. Curtis, G. I. Lehrer and J. Tits, Spherical buildings and the character of the Steinberg representation, to appear.

4. S. Eilenberg and N. Steenrod, Foundations of algebraic topology, Princeton University Press, Princeton, 1952.

5. R. Godement, Théorie des faisceaux, Actualitiés scientifiques et industrielles 1252, Hermann, Paris 1964.

6. L. L. Scott, Modular permutation representations, Trans. Amer. Math. Soc. 175 (1973), 101-121.

7. L. Solomon, The Steinberg character of a finite group with a BN-pair, Theory of finite groups (ed. by R. Brauer and C. H. Sah), W. A. Benjamin, New York, 1969, 213-221.

8. T. A. Springer, Caracteres de groupes de Chevalley finis, Sem. Bourbaki, 1972/73, no. 429.

University of Oregon
Eugene, Oregon 97403

ALGEBRAICALLY RIGID MODULES

Everett C. Dade

These are modules which are isomorphic to all the generic modules in any irreducible algebraic variety of modules in which they lie. Their potential importance comes from the two following facts:

(1) Let A be a finite-dimensional algebra over a field k. Then there are only a countable number of isomorphism classes of finite-dimensional algebraically rigid A-modules. Furthermore, these modules can be classified in some theoretical sense.

(2) Suppose that a block B of a finite group G has a trivial intersection defect group D. Let f be the Green correspondence from B to the corresponding block b of the normalizer $H=N_G(D)$ of D. Then a modular projective-free module $M\varepsilon B$ is algebraically rigid if and only if the corresponding module $f(M)\varepsilon b$ is algebraically rigid. In particular, the Green correspondents f(S) of simple modules $S\varepsilon B$ are all algebraically rigid.

Obviously a classifiable family of modules including all Green correspondents of simple modules could be extremely useful. So it would be worth considerable effort to turn the theoretical classification of (1) into a practical one for at least the H-modules of (2).

In deformation theory Donald and Flanigan[2] have studied rigid modules which are closely related to our algebraically rigid modules. The definition of the former is similar to that of the latter, but uses non-singular one-dimensional local varieties instead of global varieties. In Section 5 below we show that a module is algebraically rigid if and only if it remains rigid under arbitrary algebraic extensions of the ground field. In particular, the two concepts coincide for algebraically closed ground fields.

The advantage in using the global geometry of representations (as in Gabriel [3]) instead of local geometry is that it makes (1) almost obvious. The proof of (1) in Section 1 below, which is due to Alperin, is by a very general argument which applies to many algebraic systems over fields, such as quivers with or without relations, Lie algebras, etc. With slight changes it even applies to algebraic systems over valuation rings.

The only requirement for this argument is that the matrix representations of the system form algebraic varieties in which the equivalence classes are the orbits of some connected algebraic group operating algebraically on these varieties.

Algebraic rigidity does have a local definition, which we give in Section 2 below, in terms of R-forms of modules for suitable valuation rings R. We use this local definition in Section 3 to show that any absolutely simple or projective A-module is algebraically rigid. This parallels similar results for rigid modules in [2]. When A is quasi-Frobenius we also show that the class of algebraically rigid modules is closed under the Heller operators Ω^n and under the direct addition or subtraction of projective modules. Using this last property it is easy to prove (2) in Section 4.

1. Geometry of Representations

We work over a universal domain K satisfying:

(1.1) K is an algebraically closed field of infinite transcendence degree over its prime subfield.

We fix an associative K-algebra A with identity 1_A and finite dimension c. We also fix a K-basis $a_1=1_A, a_2, \ldots, a_c$ of A, and denote by $c_{fgh} \in K$ the corresponding structure constants such that:

(1.2) $\quad a_f a_g = \Sigma_{h=1}^{c} c_{fgh} a_h \in A$, for all $f, g = 1, \ldots, c$.

For any integer $d>0$, let $Mat_d(K)$ be the K-algebra of all $d \times d$ matrices with entries in K. The space $L_d = Hom_K(A, Mat_d(K))$ of all K-linear maps of A into $Mat_d(K)$ is naturally an affine space of dimension cd^2 over K, in which the coordinates $t_{ij}^h(T)$ of any linear map T are given by:

(1.3) $t_{ij}^h(T)$ is the i,j-th entry of the matrix $T(a_h)$, for any $h=1, \ldots, c$ and $i, j = 1, \ldots, d$.

Most of the few facts about algebraic geometry which we need can be found in Sections VI.5bis and VII.3 of Zariski and Samuel [6], which we shall cite as [6,VI] and [6,VII], respectively. If k is any subfield of K, then (1.1) implies that the k-algebra k[t] of functions from L_d to K generated by the coordinate functions t_{ij}^h is a polynomial ring over k in the variables t_{ij}^h. To each subset U of L_d corresponds an ideal $\underline{\underline{I}}_k(U)$ in k[t] consisting of all functions vanishing on U, while to each subset F of k[t] corresponds a subset $\underline{\underline{V}}(F)$ of L_d consisting of all simultaneous zeroes of the functions in F. The subset U is a k-<u>subvariety</u> of L_d if it has the form $\underline{\underline{V}}(F)$ for some subset F of k[t]. In that case U is precisely $\underline{\underline{V}}(\underline{\underline{I}}_k(U))$. For arbitrary subsets U the k-variety $\underline{\underline{V}}(\underline{\underline{I}}_k(U))$ is the <u>closure</u> $cl_k(U)$ of U in the <u>Zariski</u> k-<u>topology</u> for L_d, in which the closed subsets are precisely the k-subvarieties(see [6,VII]). The ascending chain condition for ideals of the polynomial ring k[t] implies the descending chain condition for k-subvarieties of L_d. This, in turn, implies that every k-variety U has a unique decomposition as an irredundant union of a finite number of k-subvarieties $U_1,\ldots,U_r$ which are <u>irreducible</u>, i.e., which cannot be written as the union of two proper k-subvarieties(see Theorem 13 in [6,VII]). The U_i are then called the(<u>irreducible</u>) k-<u>components</u> of U.

A k-subvariety U of L_d is irreducible if and only if its ideal $\underline{\underline{I}}_k(U)$ is prime[6,VII,Theorem 12]. In that case a point $T\varepsilon U$ is <u>generic</u>(or <u>general</u>)for U over k if U is precisely the closure $cl_k(T)$ of the point T, i.e., if:

(1.4) $\underline{\underline{I}}_k(T)=\underline{\underline{I}}_k(U)$.

When k is small enough, e.g., when the algebraically closed field K has infinite transcendence degree over k, every irreducible k-subvariety of L_d has a generic point over k (see page 22 in [6,VI]).

We denote by $Rep_d(A)$ the subset of L_d consisting of all dXd <u>matrix representations</u> of A, i.e., of all identity-preserving K-algebra homomorphisms of A into $Mat_d(K)$. Since a_1 is 1_A, it follows directly from (1.2) and (1.3) that a point $T\varepsilon L_d$ lies in $Rep_d(A)$ if and only if its coordinates $t^h_{ij}(T)$ satisfy:

(1.5a) $t^1_{ij}(T)=\delta_{ij}$,<u>for all</u> i, j=1,...,d,

(1.5b) $\Sigma^d_{e=1}\ t^f_{ie}(T)t^g_{ej}(T)=\Sigma^c_{h=1}\ c_{fgh}t^h_{ij}(T)$,<u>for all</u> f,g=1,..,c, <u>and</u> i, j=1,...,d,

where δ_{ij} is the Kronecker δ-function with values in K. It follows that $Rep_d(A)$ is a k-subvariety of L_d for any subfield k of K containing all the structure constants c_{fgh}.

The unit group $GL(K)=GL_d(K)$ of $Mat_d(K)$ can be made into an irreducible affine variety over any subfield k of K by using the d^2+1 coordinate functions g_{ij} given by:

(1.6a) $g_{ij}(G)$ _is the_ i,j-_th entry of the matrix_ G,_for all_ i,j =1,...,d,

(1.6b) $g_{00}(G)$ _is the inverse_ $\det(G)^{-1}$ _of the determinant of_ G, for any $G\varepsilon GL(K)$. Then GL(K) is an _affine algebraic group_ over k,i.e.,the coordinates of the product G'G or inverse G^{-1} of elements G',$G\varepsilon GL(K)$ are polynomials in the coordinates of those elements with coefficients in k. The group GL(K) acts naturally on L_d,with $G\varepsilon GL(K)$ taking $T\varepsilon L_d$ into the linear map $T^G\varepsilon L_d=$ $Hom_K(A,Mat_d(K))$ defined by:

(1.7) $T^G(a)=G^{-1}T(a)G\varepsilon Mat_d(K)$,_for all_ $a\varepsilon A$.

In view of (1.3) and (1.6) this action is _algebraic over_ k, i.e.,the coordinates of T^G are polynomials in those of T and G with coefficients in k. It follows that this action preserves the Zariski K-topology of L_d. Of course,the subvariety $Rep_d(A)$ is GL(K)-invariant,and the GL(K)-orbit $T^{GL(K)}$ of any $T\varepsilon Rep_d(A)$ is just the class of all matrix representations of A which are equivalent to T.

We shall need the following version of the Closed Orbit Lemma(see page 98 of Borel [1]).

Lemma 1.8. If T _is any point of_ L_d,_then the closure_ $cl_K(T^{GL(K)})$ _of its_ GL(K)-_orbit is an irreducible_ k-_subvariety of_ L_d,_for any subfield_ k _of_ K _containing all the coordinates_ $t^h_{ij}(T)$ _of_ T. _The orbit_ $T^{GL(K)}$ _is a relatively open subset of_ $cl_K(T^{GL(K)})$ _in the Zariski_ k-_topology,i.e.,its complement is a_ k-_subvariety properly contained in_ $cl_K(T^{GL(K)})$.

Proof. Notice that we are taking the closure $C=cl_K(T^{GL(K)})$ of the orbit in the _absolute_(i.e.,K-)topology of L_d,and not just the k-topology. The orbit $T^{GL(K)}$ is the set-theoretic projection on L_d of the subset:

$$U=\{(T^G,G)\mid G\varepsilon GL(K)\}$$

of the product variety $L_d\times GL(K)$. It is evident that U is an irreducible k-subvariety of $L_d\times GL(K)$. It follows that the closure C of its projection is an irreducible k-subvariety of L_d,and that $T^{GL(K)}$ contains some non-empty relatively k-open subset S of C(see page 88 of Lang [5]. Notice that Lang calls 'varieties'what we call'irreducible K-varieties'and uses a more restrictive definition of'defined over k'than we do. None of this detracts from the relevance of his arguments here). Since the action of GL(K) preserves the K-topology of L_d, the closure C of the orbit $T^{GL(K)}$ is GL(K)-invariant. So each S^G, $G\varepsilon GL(K)$, is a relatively K-open subset of C, as is the union $T^{GL(K)}$ of these S^G. Because C and $T^{GL(K)}$ are both invariant under all Galois automorphisms of K over k (applied to the coordinates of points), the K-open subset $T^{GL(K)}$ of C is k-open, and the lemma is proved (see Section III.5 of [5]).

Fix a representation $T\varepsilon Rep_d(A)$. In view of (1.1) there is at least one subfield k of K satisfying:

(1.9a) K _has infinite transcendence degree over_ k,

(1.9b) $c_{fgh}^{h}\varepsilon k$, _for all_ f,g,h=1,...,c,

(1.9c) $t_{ij}(T)\varepsilon k$, _for all_ h=1,...,c _and_ i,j=1,...,d.

We say that T is _algebraically rigid_ with respect to this k if it satisfies the following condition:

(1.10) _If_ V _is any irreducible_ k-_subvariety of_ $Rep_d(A)$ _containing_ T, _then_ $T^{GL(K)}$ _contains every generic point_ T' _of_ V _over_ k.

The following theorem will imply that algebraic rigidity does not depend on the choice of the field k satisfying (1.9).

Theorem 1.11. Suppose that $T\varepsilon Rep_d(A)$ _and a subfield_ k _of_ K _satisfy_ (1.9). _Then_ T _is algebraically rigid with respect to_ k _if and only if the closure_ $cl_K(T^{GL(K)})$ _of its_ GL(K)-_orbit is an irreducible_ K-_component of the variety_ $Rep_d(A)$.

Proof. The K-variety $Rep_d(A)$ contains the orbit $T^{GL(K)}$, and hence contains its closure $C=cl_K(T^{GL(K)})$. We know from (1.9b) and (1.5) that $Rep_d(A)$ is a k-variety, and from (1.9c) and Lemma 1.8 that C is an irreducible k-subvariety. So C is contained in some irreducible k-component V of $Rep_d(A)$. It follows from (1.9a) that V has a generic point T' over k, and from (1.10) that C contains T' and hence V if T is algebraically rigid. Thus C=V is a k-component of $Rep_d(A)$. Because C is K-irreducible by Lemma 1.8, it is also a K-component of $Rep_d(A)$.

Now suppose that C is a K-component of $Rep_d(A)$. The orbit $T^{GL(K)}$, being dense in C, must contain a point T'' lying in no other K-component of $Rep_d(A)$. Since the action of GL(K) leaves invariant the K-topology of $Rep_d(A)$, it follows that no point of $T^{GL(K)}$ can lie in any other K-component of $Rep_d(A)$. In particular, C is the only K-component of $Rep_d(A)$ containing T.

The action of the Galois group Gal(K/k) on the coordinates of points of L_d also leaves invariant the k-subvariety $Rep_d(A)$ and its K-topology. So it permutes among themselves the K-components of $Rep_d(A)$. The union of the Gal(K/k)-conjugates of such a component C' is an irreducible k-subvariety U of $Rep_d(A)$ whose prime ideal is the intersection of $\underline{\underline{I}}_K(C')$ with k[t]. It follows that the k-components of $Rep_d(A)$ are precisely such unions U. Therefore the k-subvariety C is also a k-component of $Rep_d(A)$, and is the only k-component of $Rep_d(A)$ containing T.

Let W be any irreducible k-subvariety of $Rep_d(A)$ containing T. Then W must be contained in the unique k-component C of $Rep_d(A)$ containing T. Since $T^{GL(K)}$ is a k-open subset of C by Lemma 1.8, its non-empty intersection with W is k-open in W. So that intersection contains every generic point of W over k, i.e., (1.10) holds and T is algebraically rigid.

<u>Corollary 1.12.</u> <u>The algebraic rigidity of a representation</u> $T\varepsilon Rep_d(A)$ <u>depends only on the equivalence class</u> $T^{GL(K)}$ <u>of</u> T. <u>It does not depend on the choice of the subfield</u> k <u>of</u> K <u>satisfying</u> (1.9). <u>It does not even depend on the choice of the basis</u> $a_1,\ldots,a_c$ <u>of</u> A.

<u>Proof.</u> The first two statements are obvious consequences of the theorem. Since changing the basis $a_1,\ldots,a_c$ introduces only a K-linear transformation of L_d leaving invariant the Zariski K-topology and the action of GL(K), so is the last statement.

<u>Corollary 1.13.</u> <u>Any irreducible</u> K-<u>component</u> C' <u>of</u> $Rep_d(A)$ <u>is the closure of at most one</u> GL(K)-<u>orbit of algebraically rigid representations. Hence there are at most a finite number of equivalence classes of algebraically rigid representations</u> $T\varepsilon Rep_d(A)$ <u>for any fixed</u> d>0, <u>and at most a countable number of such equivalence classes for all</u> d.

<u>Proof.</u> If C' is the closure of two orbits $T^{GL(K)}$ and $(T')^{GL(K)}$, then both orbits are open subsets of the irreducible K-variety C' by Lemma 1.8, and hence have a non-empty intersection. So they are equal. The rest of the corollary follows from this and the finiteness of the number of K-components of $Rep_d(A)$.

The above theorem gives us the following theoretical method for classifying the algebraically rigid representations in $Rep_d(A)$: First decompose $Rep_d(A)$ into its K-components C_1,

$\ldots, C_q$. Pick a finitely-generated subfield k of K such that each C_i is a k-variety. Choose a general point T_i of C_i over k, for each i. The theorem implies that T_i is algebraically rigid if and only if its orbit $T_i^{GL(K)}$ is an open subset of C_i, i.e., if and only if:

(1.14) $\dim(T_i^{GL(K)}) = \dim(C_i)$,

in the sense of dimensions of varieties. Furthermore, the theorem and Corollary 1.13 imply that every algebraically rigid representation $T \varepsilon \mathrm{Rep}_d(A)$ is equivalent to exactly one of the T_i satisfying this condition. So those T_i form a set of representatives for the equivalence classes of algebraically rigid representations in $\mathrm{Rep}_d(A)$.

2. Specializations of Modules

Geometrically a point $T \varepsilon L_d$ is a *specialization* of a point $T' \varepsilon L_d$ over a subfield k of K if T lies in the unique irreducible k-subvariety $V = \mathrm{cl}_k(T')$ of L_d having T' as a generic point over k. Thus the condition (1.10) for algebraic rigidity of $T \varepsilon \mathrm{Rep}_d(A)$ can be expressed as:

(2.1) $T^{GL(K)}$ *contains every point* $T' \varepsilon \mathrm{Rep}_d(A)$ *specializing to* T *over* k.

Algebraically $T' \varepsilon L_d$ specializes over k to $T \varepsilon L_d$ if and only if the ideal $\underline{\underline{I}}_k(T)$ contains $\underline{\underline{I}}_k(T')$ (see (1.4)). Evaluation of a function $f \varepsilon k[t]$ at the point T is a k-algebra homomorphism of $k[t]$ into K with kernal $\underline{\underline{I}}_k(T)$. The image of this map is the k-subalgebra $k[t(T)]$ of K generated by all the coordinates $t_{ij}^h(T)$ of T. We conclude (see Section II.3 of [5]) that T' specializes over k to T if and only if there is some k-algebra epimorphism ϕ of $k[t(T')]$ onto $k[t(T)]$ such that:

(2.2) $\phi(t_{ij}^h(T')) = t_{ij}^h(T)$, *for all* $h=1,\ldots,c$, *and* $i,j=1,\ldots,d$.

The relation between specializations and GL(K)-orbits is given by:

Lemma 2.3. Suppose that K has infinite transcendence degree over its subfield k, and that T and T_0 lie in L_d. Then T is a specialization over k of some point $T' \varepsilon T_0^{GL(K)}$ if and only if $T^{GL(K)}$ is contained in the k-closure $\mathrm{cl}_k(T_0^{GL(K)})$ of $T_0^{GL(K)}$. In that case every point in $T^{GL(K)}$ is a specialization over k of some point in $T_0^{GL(K)}$.

Proof. The field of fractions k' of $k[t(T_0)]$ in K is finitely-generated over k. So K also has infinite transcendence degree over k', and we may choose a matrix $G' \varepsilon GL(K)$ whose entries $g_{ij}(G')$, $i,j=1,..,d$, are d^2 independent transcendentals over k'. Evidently $T_0^{G'}$ is a generic point for the irreducible k'-variety $cl_K(T_0^{GL(K)})$ of Lemma 1.8, and hence is a generic point over k for $C=cl_k(T_0^{GL(K)})$.

Suppose that T is a specialization over k of some point $T' \varepsilon T_0^{GL(K)}$. We may assume that T' is T_0. If G is any matrix in $GL(K)$, then the epimorphism ϕ of $k[t(T')]=k[t(T_0)]$ onto $k[t(T)]$ satisfying (2.2) can be extended to an epimorphism ϕ' of the k-subalgebra $k[t(T_0),g(G')]$, generated by all the coordinates of both T_0 and G', onto $k[t(T),g(G)]$ sending $g_{ij}(G')$ onto $g_{ij}(G)$ for all i,j (see (1.6)). In view of (1.3), (1.6), and (1.7) the restriction of ϕ' is a k-algebra epimorphism of $k[t(T_0^{G'})]$ onto $k[t(T^G)]$ sending $t_{ij}^h(T_0^{G'})$ onto $t_{ij}^h(T^G)$ for all i,j and h. Thus T^G is a specialization over k of the generic point $T_0^{G'}$ of C, i.e., $T^{GL(K)}$ is contained in C.

If $T^{GL(K)}$ is contained in C, then its point T is a specialization over k of the generic point $T'=T_0^{G'}$ of C. Thus the first conclusion of the lemma is proved. The rest follows immediately from this.

Of course it is not true that every point $T' \varepsilon T_0^{GL(K)}$ specializes over k to a point in $T^{GL(K)}$ under the hypotheses of Lemma 2.3.

We are particularly interested in specializations of T' into T over fields k such that (1.9c) holds, i.e., such that $k[t(T)]$ is just k. When k is algebraically closed the Place Extension Theorem (Theorem 1 in Chapter I of [5] or Theorem 5' in Chapter VI of [6]) says that the epimorphism ϕ of $k[t(T')]$ onto k in (2.2) can be extended to an epimorphism ϕ' onto k of some valuation ring R in K. We may even choose R so that its field of fractions is a given subfield of K containing $k[t(T')]$. Since we can always restrict ϕ' to ϕ, it follows that:

(2.4) *If k is an algebraically closed subfield of K, and if T and T' are points of L_d such that $k[t(T)]=k$, then T' specializes over k to T if and only if there is some valuation subring R of K containing $k[t(T')]$ and some k-algebra epimorphism ϕ of R onto k such that (2.2) holds.*

Notice that the conditions on R in (2.4) imply that ϕ is just the projection onto the first summand in the additive decomposition:

(2.5) $R=k\dotplus J(R)$,

where $J(R)$ is the Jacobson radical of R. So ϕ is uniquely determined by R.

Let Mod(A) be the class of all left, unitary, finite-dimensional A-modules. In view of (1.3) a module $M\varepsilon\text{Mod}(A)$ of non-zero dimension d corresponds to a matrix representation $T\varepsilon\text{Rep}_d(A)$ if and only if there is a K-basis $m_1,\ldots,m_d$ for M such that:

(2.6) $a_h m_j=\Sigma_{i=1}^{d} t_{ij}^{h}(T)m_i$, *for all* $h=1,\ldots,c$, *and* $j=1,\ldots,d$.

We'll say that A, with its basis $a_1,\ldots,a_c$, is *defined over* a subring R of K if the structure constants c_{fgh} of (1.2) are all members of R. In that case the R-*form*:

(2.7) $A_R=Ra_1+\ldots+Ra_c$

is an R-subalgebra and free R-submodule of A from which the latter is obtained by ground ring extension from R to K. An R-*form* M_R for a module $M\varepsilon\text{Mod}(A)$ is then an A_R-submodule of the form:

(2.8) $M_R=Rm_1+\ldots+Rm_d$,

where $m_1,\ldots,m_d$ is a K-basis of M. Thus M_R is a free R-submodule of M of R-rank $d=\dim_K(M)$, as well as an A_R-submodule from which M is obtained by ground ring extension from A_R to A.

We say that a module $M'\varepsilon\text{Mod}(A)$ *specializes* to a module $M\varepsilon\text{Mod}(A)$ over a subfield k of K if M' and M have the same K-dimension d, and either d is zero, or $d>0$ and there exist representations T' and T in $\text{Rep}_d(A)$ corresponding to M' and M, respectively, such that T' specializes over k to T. We shall chiefly be interested in this concept when k satisfies:

(2.9a) k *is an algebraically closed subfield of* K,

(2.9b) K *has infinite transcendence degree over* k,

(2.9c) A *is defined over* k.

Then there is a simple characterization of specialization in terms of R-forms.

Theorem 2.10. Suppose that $M\varepsilon\text{Mod}(A)$ *has a* k-*form* M_k *for some subfield* k *satisfying* (2.9). *Then a module* $M'\varepsilon\text{Mod}(A)$ *specializes over* k *to* M *if and only if there exist* R *and* M'_R *satisfying*:

(2.11a) R *is a valuation subring of* K *such that* (2.5) *holds*,

(2.11b) M'_R *is an* R-*form of* M' *such that the* A_k-*module* $M'_R/J(R)M'_R$ *is isomorphic to* M_k.

Proof. Suppose that M' specializes over k to M. Then M' and M have the same K-dimension d. If d is zero, then (2.11) holds with $R=k$ and $M'_R=0$. So we may assume that $d>0$.

Any k-basis $m_1,\ldots,m_d$ for M_k is also a K-basis for M, and hence determines a representation $T\varepsilon\mathrm{Rep}_d(A)$ by (2.6). By definition some representation $T_0\varepsilon\mathrm{Rep}_d(A)$ corresponding to M is a specialization over k of some representation corresponding to M'. In view of (2.9b) and Lemma 2.3, the equivalent representation T is also a specialization over k of some representation $T'\varepsilon\mathrm{Rep}_d(A)$ corresponding to M'. Since $m_1,\ldots,m_d$ is a k-basis for the A_k-submodule M_k of M, it follows from (2.6) and (2.7) that each $t^h_{ij}(T)$ lies in k, i.e., that $k[t(T)]=k$. So (2.4) gives us a valuation subring R of K containing $k[t(T')]$ such that (2.2) holds for the unique k-algebra epimorphism ϕ of R onto k. If $m'_1,\ldots,m'_d$ is a K-basis for M' giving the representation T' by (2.6), then:

$$M'_R=Rm'_1+\ldots+Rm'_d$$

is an R-form of M', since each $t^h_{ij}(T')$ lies in R and A is defined over the subring k of R. Because ϕ is the projection onto the first summand in (2.5), it follows from (2.2) that there is an A_k-isomorphism of $M'_R/J(R)M'_R$ onto M_k sending $m'_i+J(R)M'_R$ onto m_i for $i=1,\ldots,d$. Therefore (2.11) holds.

Now suppose there exist R and M'_R satisfying (2.11). Then the K-dimension d of M is the k-dimension of the isomorphic modules M_k and $M'_R/J(R)M'_R$. But the dimension of the last module is also the R-rank of the free R-module M'_R, and thus equals the K-dimension of M'. So M' and M have the same K-dimension d.

Since M' specializes to M by definition if d is zero, we may suppose that $d>0$. Let $m_1,\ldots,m_d$ be a k-basis for M_k. Then $m_1,\ldots,m_d$ is a K-basis for M, and the corresponding representation $T\varepsilon\mathrm{Rep}_d(A)$ determined by (2.6) satisfies $k[t(T)]=k$. Because R is a valuation ring there exists an R-basis $m'_1,\ldots,m'_d$ for M'_R such that the isomorphism of (2.11b) sends $m'_i+J(R)M'_R$ onto m_i for all $i=1,\ldots,d$. Using (2.6) we see that (2.2) holds for the projection ϕ of R onto the first summand in (2.5), where $T'\varepsilon\mathrm{Rep}_d(A)$ corresponds to M' and its K-basis $m'_1,\ldots,m'_d$. So T' specializes over k to T by (2.4), and therefore M' specializes over k to M. Thus the theorem is proved.

We say that a module $M\varepsilon\mathrm{Mod}(A)$ is <u>algebraically rigid</u> if it is either zero, or else is non-zero and corresponds to some algebraically rigid matrix representation T via (2.6). In the latter case any matrix representation corresponding to M is algebraically rigid by Corollary 1.12.

Theorem 2.12. For any module $M\varepsilon Mod(A)$ there is some subfield k satisfying (2.9) such that M has a k-form M_k. Fix one such k. Then M is algebraically rigid if and only if it is A-isomorphic to any module $M'\varepsilon Mod(A)$ specializing over k to M.

Proof. When M is zero we let k be the algebraic closure in K of the subfield generated by all the structure constants c_{fgh} of (1.2). Then k satisfies (2.9) by (1.1), and M has the k-form $M_k=0$. The rest of the theorem is trivial in this case. So we may assume that M is non-zero.

Let $m_1,\dots,m_d$ be any K-basis for M, and $T\varepsilon Rep_d(A)$ be the corresponding representation determined by (2.6). Let k be the algebraic closure in K of the subfield generated by all the structure constants c_{fgh} and by all the coordinates $t^h_{ij}(T)$. Then k satisfies (2.9) and M has the k-form:

$M_k=km_1+\dots+km_d$.

Now let k be any subfield of K satisfying (2.9) such that M has a k-form M_k, let $m_1,\dots,m_d$ be a k-basis for M_k, and let T be the corresponding representation in $Rep_d(A)$. Then k and T satisfy (1.9). By Corollary 1.13 the representation T is algebraically rigid if and only if it satisfies (1.10) or the equivalent condition (2.1). In view of Lemma 2.3 the condition (2.1) holds if and only if M is A-isomorphic to every module $M'\varepsilon Mod(A)$ specializing over k to M. So the theorem is proved.

3. Some Rigid Modules

We fix a module $M\varepsilon Mod(A)$, a subfield k satisfying (2.9) such that M has a k-form M_k (see Theorem 2.12), and a module $M'\varepsilon Mod(A)$ specializing over k to M. Then Theorem 2.10 gives us R and M'_R, which we also fix, satisfying (2.11). Theorem 2.12 says that M is algebraically rigid if it is always A-isomorphic to M'.

Any idempotent e in the subalgebra A_k of A_R yields a Peirce decomposition:

$M'_R=eM'_R\dotplus(1-e)M'_R$,

where eM'_R and $(1-e)M'_R$ are also free modules over the valuation ring R. It follows that:

(3.1) The inclusion of eM'_R in M'_R induces a k-isomorphism of $eM'_R/J(R)eM'_R$ onto $e(M'_R/J(R)M'_R)$, which is k-isomorphic to eM_k by (2.11b).

The k-dimension of $eM'_R/J(R)eM'_R$ is the R-rank of the free R-module eM'_R, and hence is the K-dimension of the subspace eM' obtained from eM'_R by ground ring extension. On the other hand, the k-dimension of eM_k is the K-dimension of eM. So (3.1) gives:

(3.2) $\dim_K(eM')=\dim_K(eM)$.

The algebra A_k splits over the algebraically closed field k, and hence contains a primitive idempotent $e=e_S$ of A corresponding to any given simple A-module S. In this case $\dim_K(eM)$ is precisely the multiplicity m(S in M) of S as an A-composition factor of M. So the equality (3.2) gives the well-known result:

(3.3) m(S in M')=m(S in M), *for all simple* A-*modules* S

(see Corollary 1.4 of [3]). As a consequence we have:

Proposition 3.4. Any simple A-*module is algebraically rigid.*

Proof. If M is simple, then (3.3) implies that M', which has the same composition factors as M, is isomorphic to M. So M is algebraically rigid by Theorem 2.12.

Of course, the result corresponding to Proposition 3.4 is known for rigid modules (see Theorem 14 of [2]).

Because k is algebraically closed, the subalgebra A_k contains every central idempotent e of the algebra A. For such e the k-isomorphisms of (3.1) are A_k-isomorphisms. Furthermore eM_k is then a k-form for the A-module eM, and eM'_R is an R-form for eM'. So Theorem 2.10 implies that:

(3.5) eM' *specializes over* k *to* eM, *for any central idempotent* e *of* A.

As an immediate consequence of this remark we have one half of:

Proposition 3.6. Let $e_1,\dots,e_q$ *be the primitive central idempotents of* A. *Then* M *is algebraically rigid if and only if each* $e_iM\in Mod(A)$ *is algebraically rigid, for* $i=1,\dots,q$.

Proof. If each e_iM is algebraically rigid, then (3.5) and Theorem 2.12 imply that e_iM' is A-isomorphic to e_iM, for $i=1,\dots,q$. Therefore $M'=e_1M'\dotplus\dots\dotplus e_qM'$ is A-isomorphic to $M=e_1M\dotplus\dots\dotplus e_qM$, and M is algebraically rigid by Theorem 2.12.

The other half of this proposition follows from:

Lemma 3.7. If M *is algebraically rigid, then so is any* A-*direct summand* N *of* M.

Proof. Let L be a complementary A-submodule to N in $M=N\dotplus L$. By Theorem 2.12 the algebraic rigidity of M is preserved if we replace k by a larger subfield of K satisfying (2.9). So we may assume that k is so large that N and L have k-forms N_k

and L_k, respectively. Then $M_k=N_k \dotplus L_k$ is a k-form for M.

Let N' be any module in Mod(A) specializing over k to N. Theorem 2.10 gives us a valuation subring R of K such that (2.5) holds, and an R-form N'_R of N' such that $N'_R/J(R)N'_R$ is A_k-isomorphic to N_k. From (2.5) it is clear that $L_R=RL_k$ is an R-form for L such that $L_R/J(R)L_R$ is A_k-isomorphic to L_k. Hence the conditions (2.11) are satisfied by the R-form $N'_R \oplus L_R$ of the module $N' \oplus L \in Mod(A)$ and the k-form $N_k \dotplus L_k$ of M. Since M is algebraically rigid, Theorems 2.12 and 2.10 imply that $M=N \dotplus L$ is A-isomorphic to $N' \oplus L$. Therefore N is A-isomorphic to N' by the Krull-Schmidt Theorem in Mod(A), and N is algebraically rigid by Theorem 2.12. This completes the proofs of both the lemma and the proposition.

Since A_k is a split k-algebra, any indecomposable projective A-module is isomorphic to Ae for some primitive idempotent e of A_k. Evidently Ae has the A_k-projective k-form $A_k e$. We conclude that every projective module $P \in Mod(A)$ has a projective k-form P_k.

Suppose the above projective P occurs as the middle term of the exact A-sequence:

(3.8) $0 \longrightarrow N \xrightarrow{f} P \xrightarrow{g} M \longrightarrow 0$,

where M is our usual module. By adjoining a finite number of elements (the coefficients in a basis for the set of K-linear relations among the images $g(p_i)$ for a k-basis $\{p_i\}$ of P_k) to k and taking an algebraic closure in K, we may assume that $M_k = g(P_k)$ is a k-form for M. Then $N_k=f^{-1}(P_k)$ is a k-form for N, and (3.8) comes from the exact A_k-sequence:

(3.9) $0 \longrightarrow N_k \xrightarrow{f} P_k \xrightarrow{g} M_k \longrightarrow 0$

by ground field extension.

From (2.5) it is clear that $P_R=RP_k$ is an A_R-projective R-form for P, and that $P_R/J(R)P_R$ is naturally A_k-isomorphic to P_k. Since $M'_R/J(R)M'_R$ is A_k-isomorphic to M_k by (2.11b) and P_R is A_R-projective, we conclude that the exact A_k-sequence (3.9) is isomorphic to that obtained from an exact A_R-sequence of A_R-lattices:

(3.10) $0 \longrightarrow N'_R \xrightarrow{f'} P_R \xrightarrow{g'} M'_R \longrightarrow 0$

by factoring modulo the radical J(R)X of each lattice X. Extend- the ground ring in (3.10) from R to K we obtain an exact A-sequence:

(3.11) $0 \longrightarrow N' \xrightarrow{f'} P \xrightarrow{g'} M' \longrightarrow 0$

in which N' specializes over k to N by Theorem 2.10.

If the above module P is also A-injective, then we can 'dualize' the above argument, starting from an exact A-sequence (3.8) and a module $N' \in Mod(A)$ specializing over k to N, and constructing the exact A-sequence (3.11) where M' specializes over k to M. Since P is always injective if A is a quasi-Frobenius algebra, these remarks easily imply:

Proposition 3.12. *If A is a quasi-Frobenius algebra, and if (3.8) is an exact A-sequence with P projective, then M is algebraically rigid if and only if N is algebraically rigid.*

Proof. If M is algebraically rigid and $N' \in Mod(A)$ specializes over k to N, then we can construct an exact A-sequence (3.11) where M' specializes over k to M, and hence is isomorphic to M by Theorem 2.12. Since P is projective, it follows from the exactness of both (3.8) and (3.11) that N' is A-isomorphic to N. Therefore N is algebraically rigid by Theorem 2.12, and half the proposition is proved. The other half is proved similarly.

As usual, the A-module M is projective-free if it has no non-trivial projective A-direct summands. In general M is a direct sum of a projective A-submodule and a projective-free A-submodule M_{pf}. We call M_{pf}, which is determined up to A-isomorphisms by M, the projective-free part of M. If (3.8) is an exact A-sequence with P projective, then Shanuel's Lemma and the Krull-Schmidt Theorem imply that the projective-free part N_{pf} of N is determined to within A-isomorphisms by M_{pf} and does not depend upon the choice of (3.8). We call N_{pf} the Heller translate $\Omega(M_{pf})$ of M_{pf}. When A is quasi-Frobenius, then N_{pf} similarly determines $M_{pf}=\Omega^{-1}(N_{pf})$ to within A-isomorphisms. In that case composition gives us well-defined powers Ω^n of Ω for each positive or negative integer n. Since $N=N_{pf}$ whenever $M=M_{pf}$ and P is a projective cover of M in (3.8)(and A is quasi-Frobenius), the above proposition has the

Corollary 3.13. *If A is quasi-Frobenius and M is an algebraically rigid, projective-free module in Mod(A), then $\Omega^n(M)$ is algebraically rigid for all integers n.*

For any class E of simple A-modules there is certainly a smallest A-submodule M^E of M under inclusion such that every A-composition factor of M/M^E is isomorphic to an element of E.

Proposition 3.14. If $P \varepsilon Mod(A)$ *is A-projective and if* E *is any class of simple A-modules, then* $M=P/P^E$ *is an algebraically rigid A-module.*

Proof. We may form an exact A-sequence (3.8) with $N=P^E$. If $M' \varepsilon Mod(A)$ specializes over k to M, then we have an exact A-sequence (3.11). In view of (3.3) the A-modules M and M' have the same A-composition factors. This forces $f'(N')$ to be P^E and M' to be isomorphic to $P/P^E=M$. Therefore M is algebraically rigid by Theorem 2.12.

Corollary 3.15(see Theorem 13 of [2]). *Any projective module* $P \varepsilon Mod(A)$ *is algebraically rigid.*

Proof. Apply the proposition with E the class of all simple A-modules.

Notice that the above proposition and its corollary apply to any algebra A, and not just to quasi-Frobenius ones. The same is not true for the important:

Theorem 3.16. Let A be a quasi-Frobenius algebra. Then a module $M \varepsilon Mod(A)$ *is algebraically rigid if and only if its projective-free part* $N=M_{pf}$ *is algebraically rigid.*

Proof. By definition there is some projective A-submodule P of M such that:

$$M=N \dot{+} P.$$

As usual we may choose k so large that N has a k-form N_k and P has an A_k-projective k-form P_k. Then we can choose:

$$M_k=N_k \dot{+} P_k \tag{3.17}$$

for the k-form of M.

If M is algebraically rigid, then so is N by Lemma 3.7. So we may suppose that N is algebraically rigid. Let M' be any module in Mod(A) specializing over k to M. Theorem 2.10 gives us a valuation subring R of K satisfying (2.5) and an R-form M'_R of M' such that $M'_R/J(R)M'_R$ is A_k-isomorphic to M_k. Evidently $P_R=RP_k$ is an A_R-projective R-form for P such that $P_R/J(R)P_R$ is A_k-isomorphic to P_k. In view of (3.17) we have an A_k-monomorphism Φ of $P_R/J(R)P_R$ into $M'_R/J(R)M'_R$ which comes from some monomorphism ϕ of the A_R-lattice P_R, which is projective, into M'_R. Because k is algebraically closed, the k-form A_k of the quasi-Frobenius algebra A is also quasi-Frobenius. It follows that the R-order $A_R=RA_k$ is also quasi-Frobenius in the sense that the dual of any left projective A_R-lattice is right projective. This implies that the projective A_R-sublattice $\phi(P_R)$

of M_R', which is an R-direct summand since Φ is a monomorphism, has an A_R-complement N_R' in:

(3.18) $M_R' = N_R' \dotplus \phi(P_R)$.

Because ϕ is an A_R-monomorphism and $P_R/J(R)P_R$ is A_k-isomorphic to P_k, the factor module $\phi(P_R)/J(R)\phi(P_R)$ is A_k-isomorphic to P_k. Since $M_R'/J(R)M_R'$ is A_k-isomorphic to M_k, we conclude from this, (3.17), (3.18) and the Krull-Schmidt Theorem for A_k-modules that $N_R'/J(R)N_R'$ is A_k-isomorphic to N_k. So the module $N' = KN_R' \varepsilon$ Mod(A), of which N_R' is an R-form, specializes over k to N by Theorem 2.10, and hence is A-isomorphic to the algebraically rigid module N by Theorem 2.12. It follows from this and (3.18) that $M' = N' \dotplus \phi(P)$ is A-isomorphic to $M = N \dotplus P$, where, of course, we have extended ϕ from P_R to an A-monomorphism of P into M'. Therefore M is algebraically rigid by Theorem 2.12, and the present theorem is proved.

4. Blocks with TI Defect Groups

We assume now that K has prime characteristic p, and that A is the group algebra KG over K of some finite group G. Of course we use the elements of G for our basis $a_1, \ldots, a_c$ of A, so that the R-form A_R of (2.7) is just the group algebra RG over R for any subring R of K.

Fix a block B of KG and a defect group $D \leq G$ of B. We denote by H the normalizer $N_G(D)$ of D in G, and by b the unique block of KH corresponding to B in Brauer's First Main Theorem. We assume that D is a trivial intersection (or TI-) subgroup of G in the usual sense that:

(4.1) $D^\sigma \cap D = 1$, for all $\sigma \varepsilon G-H$.

Let e_b be the primitive central idempotent of KH in the block b. The Green Correspondence f from B to b now has a very simple description.

Lemma 4.2. *If M is a projective-free module in $B \cap$ Mod(KG), then there is a projective-free module $f(M) \varepsilon b \cap$ Mod(KH) determined to within KH-isomorphisms by either of its properties:*

(4.3a) f(M) is isomorphic to the projective-free part $(e_b M_H)_{pf}$ of $e_b M_H$,

(4.3b) M is isomorphic to the projective-free part $(f(M)^G)_{pf}$ of $f(M)^G$,

where, as usual, M_H is the restriction of M to a KH-module, and

$f(M)^G$ <u>is the induction of</u> $f(M)$ <u>to a KG-module. The</u> Green correspondence f <u>is one-to-one between the isomorphism classes of projective-free modules in</u> $B \cap Mod(KG)$ <u>and those in</u> $b \cap Mod(KH)$.

<u>Proof.</u> This is an immediate consequence of Green's paper [4]. Notice that the set $\underline{\underline{X}}$ on page 75 of [4] is now {1} by (4.1), so that the 'error' $O(\underline{\underline{X}})$' in (2.8) of [4] consists only of projective KG-modules. Also the set $\underline{\underline{Y}}$ on page 75 of [4] contains only p-subgroups having trivial intersection with D, so that the 'error' $O(\underline{\underline{Y}})$ in (2.6) of [4] is a direct sum of projective KH-modules and of KH-modules in blocks b' having defect groups $D' \neq D$. So our (4.3) follows from Theorem 2 and the corollary on page 80 of [4].

In the present case the Green Correspondence preserves algebraic rigidity.

<u>Theorem 4.4.</u> <u>Under the above hypotheses a projective-free module</u> $M \in B \cap Mod(KG)$ <u>is algebraically rigid if and only if its Green correspondent</u> $f(M) \in b \cap Mod(KH)$ <u>is algebraically rigid.</u>

<u>Proof.</u> Suppose that M is algebraically rigid. Theorem 2.12 gives us a subfield k satisfying (2.9) such that f(M) has a k-form $f(M)_k$. If $N' \in Mod(KH)$ specializes over k to f(M), then Theorem 2.10 gives us a valuation subring R of K satisfying (2.5) and an R-form N'_R of N' such that $N'_R/J(R)N'_R$ is kH-isomorphic to $f(M)_k$. It follows that $(N'_R)^G$ is an R-form for the induced KG-module $(N')^G$ and that $(N'_R)^G/J(R)(N'_R)^G$ is kG-isomorphic to the k-form $(f(M)_k)^G$ of the KG-module $f(M)^G$. So $(N')^G$ specializes over k to $f(M)^G$ by Theorem 2.10. Since M is algebraically rigid, Theorem 3.16 for the quasi-Frobenius algebra KG, and Property (4.3b) above imply that $f(M)^G$ is algebraically rigid. Therefore $(N')^G$ is KG-isomorphic to $f(M)^G$ by Theorem 2.12. Because N' and f(M) have the same KH-composition factors (see (3.3)), they lie in the same block b. Hence (4.3b) and the isomorphism of $(N')^G$ with $f(M)^G$ imply that f(M) is KH-isomorphic to the projective-free part of N'. Therefore f(M) is isomorphic to N', since their dimensions are the same, and f(M) is algebraically rigid by Theorem 2.12.

Now suppose that f(M) is algebraically rigid. Then (4.3a) and Theorem 3.16 imply that $e_b M_H$ is algebraically rigid. If M has a k-form M_k, for some subfield k satisfying (2.9), and if $M' \in Mod(KG)$ specializes over k to M, then M' lies in the block B by (3.3), and there exist a valuation subring R of K satisfying

(2.5) and an R-form M'_R of M' such that $M'_R/J(R)M'_R$ is kG-isomorphic to M_k (see Theorem 2.10). By restriction the R-form $(M'_R)_H$ of M'_H is kH-isomorphic, modulo its radical $J(R)(M'_R)_H$, to the k-form $(M_k)_H$ of M_H. So M'_H specializes over k to M_H by Theorem 2.10, and $e_bM'_H$ specializes over k to e_bM_H by (3.5). Therefore $e_bM'_H$ is KH-isomorphic to the algebraically rigid module e_bM_H by Theorem 2.12. Since M' lies in B, we conclude from this and (4.3a) that $f(M'_{pf})$ is KH-isomorphic to f(M), and hence that M is KG-isomorphic to the projective-free part M'_{pf} of M'. Because M' and M have the same dimension, they are KG-isomorphic. Therefor M is algebraically rigid by Theorem 2.12, and the present theorem is proved.

Corollary 4.5. *If S is any simple KG-module in B, then* $f(S) \in b \cap \mathrm{Mod}(KH)$ *is algebraically rigid.*

Proof. This follows directly from the theorem and Proposition 3.4 above.

5. Rigidity and Algebraic Rigidity

We return to the situation of the first two sections above, and fix a subfield k of K such that A is defined over k and K has infinite transcendence degree over k. Let k_0 be the algebraic closure of k in K. Since the geometry of k- or k_0-subvarieties of L_d does not change if K is replaced by an extension field, we may suppose that K contains a k_0-subalgebra $k_0[[x]]$, the ring of formal power series in one variable x with coefficients in k_0. Then K contains the subring $k'[[x]]$ and its field of fractions $k'((x))$, for any subfield k' of k_0.

Let M be a module in Mod(A) having a k-form M_k. Donald and Flanigan [2] define a *generic deformation* of the A_k-module M_k to be a $k[[x]]$-form $M'_{k[[x]]}$ of some module $M' \in \mathrm{Mod}(A)$ such that $M'_{k[[x]]}/xM'_{k[[x]]}$ is A_k-isomorphic to M_k. They say that M_k is *rigid* if the $k((x))$-form $k((x))M_k$ of M is $A_{k((x))}$-isomorphic to $k((x))M'_{k[[x]]}$, for any generic deformation $M'_{k[[x]]}$ of M_k. Evidently this occurs if and only if M' is A-isomorphic to M for any such $M'_{k[[x]]}$.

Theorem 5.1. *In the above situation the A-module M is algebraically rigid if and only if its k'-form* $M_{k'}=k'M_k$ *is a rigid* $A_{k'}$*-module for all finite algebraic extension fields k' of k which are subfields of K. In particular, if k is algebraically closed, then M is algebraically rigid if and only if* M_k *is rigid.*

<u>Proof.</u> Suppose that M is algebraically rigid, that k' is a finite algebraic extension field of k in K, and that the k'[[x]]-form $M'_{k'[[x]]}$ of a module M'εMod(A) is a generic deformation of $M_{k'}$. We must show that M' is A-isomorphic to M.

Since k_0 contains k', the product $k_0[[x]]M'_{k'[[x]]}$ is a $k_0[[x]]$-form $M'_{k_0[[x]]}$ of M', and the $A_{k'}$-isomorphism of $M'_{k'[[x]]}/xM'_{k'[[x]]}$ onto $M_{k'}$ given in the definition of generic deformations induces an A_{k_0}-isomorphism of $M'_{k_0[[x]]}/xM'_{k_0[[x]]}$ onto $M_{k_0} = k_0M_k$. Evidently k_0 satisfies (2.9), and $k_0[[x]]$ is a valuation subring of K satisfying (2.5) for k_0 and the radical $xk_0[[x]]$. So M' specializes over k_0 to M by Theorem 2.10, and hence is A-isomorphic to the algebraically rigid module M by Theorem 2.12. Therefore $M_{k'}$ is a rigid $A_{k'}$-module.

Now suppose that M is not algebraically rigid. Then M has non-zero dimension d. If $m_1,\ldots,m_d$ is a k-basis for M_k, then it is a K-basis for M, and the representation TεRep$_d$(A) determined by (2.6) has all its coordinates $t^h_{ij}(T)$ in k. So Lemma 1.8 says that $C=cl_K(T^{GL(K)})$ is simultaneously an irreducible K- and k- subvariety of Rep$_d$(A). Since M is not algebraically rigid, neither is T. By Theorem 1.11 this implies that C is properly contained in some K-component V' of Rep$_d$(A), and hence is properly contained in the k-component V of Rep$_d$(A) containing V'.

The dimension of the irreducible k-variety V is strictly larger than that of its k-subvariety C, and the point TεC has coordinates in k. It follows that there is some irreducible one-dimensional k-subvariety U of V containing T and not contained in C. Then any generic point T' of U over k lies outside $C=cl_K(T^{GL(K)})$, and hence is inequivalent to T. Since T is a specialization of T' over k, there is some k-epimorphism ϕ of k[t(T')] onto k[t(T)]=k such that (2.2) holds. By the Place Extension Theorem ϕ can be extended to a k-homomorphism ϕ'(into the algebraic closure k_0)of some valuation ring R having the same field of fractions k(t(T')) as k[t(T')]. Because k(t(T')) is the function field of a one-dimensional variety over k, the valuations corresponding to R are discrete of rank one, and the residue class field R/J(R) is k-isomorphic to a finite algebraic extension k'=ϕ'(R) of k in K(see the corollary to Theorem 31 in [6,VI]). If we extend the ground field from k to k', and choose for T' a generic point over k' for a k'-component U' of

U containing T, then ϕ' extends to the only k'-epimorphism onto k' of some valuation ring R' having field of fractions k'(t(T')) and satisfying (2.5) for k'. Because R' is a real discrete valuation ring containing its 'residue class field' k', its completion is k'-isomorphic to k'[[x]]. So we can even choose T' in such a way that R' is $k'[[x]] \cap k'(t(T'))$ and J(R')=xR'. If M' is now a module in Mod(A) having a K-basis $m'_1, \ldots, m'_d$ yielding T' in (2.6), then it follows from (2.2) that the k'[[x]]-form

$$M'_{k'[[x]]} = k'[[x]]m'_1 + \ldots + k'[[x]]m'_d$$

of M' has a residue class module $M'_{k'[[x]]}/xM'_{k'[[x]]}$ which is $A_{k'}$-isomorphic to $M_{k'}$, i.e., that it is a generic deformation of $M_{k'}$ over k'. Since M' corresponds to T', which is inequivalent to T, it is not A-isomorphic to M. Therefore $M_{k'}$ is not rigid, and the theorem is proved.

It is easy to construct a module M which is not algebraically rigid yet has a rigid k-form. Suppose that k has a separable extension k' of degree 2 in K. Let A_k be the extension of the k-algebra k' obtained by adjoining an element z satisfying:

(5.2a) $A_k = k' \dotplus k'z$,

(5.2b) $yz = z\bar{y}$, <u>for all</u> $y \in k'$,

(5.2c) $z^2 = 0$,

where $\bar{y}$ is the k-conjugate of y in k'. The K-algebra A obtained from A_k by ground field extension then has two simple modules S and $\bar{S}$ corresponding to the two k-monomorphisms of $A_k/J(A_k) = A_k/zA_k$, which is k-isomorphic to k', into K. The corresponding projective indecomposable A-modules P and $\bar{P}$ both have the same composition factors S, $\bar{S}$ because of (5.2b). Hence they both specialize over k' to the direct sum M of S and $\bar{S}$ (see page 135 of [3]). In particular, M is not algebraically rigid. However, M has the simple A_k-module A_k/zA_k as a k-form, and this module is A_k-rigid by Theorem 14 of [2].

References

[1] A. Borel: <u>Linear Algebraic Groups</u>, W.A. Benjamin, Inc. (New York, Amsterdam) 1969.

[2] J.D. Donald and F.J. Flanigan: Deformations of Algebra Modules, J. Algebra 31 (1974), 245-256.

[3] P. Gabriel: Finite Representation Type is Open, <u>Lecture Notes in Math.</u> 488, Springer Verlag (Berlin, Heidelberg, New

York)1975.

[4] J.A.Green:A Transfer Theorem for Modular Representations, J.Algebra 1(1964),73-84.

[5] S.Lang:Introduction to Algebraic Geometry,Interscience Publishers,Inc.(New York,London)1958.

[6] O.Zariski and P.Samuel:Commutative Algebra II,D.van Nostrand Co.,Inc.(Princeton,Toronto,London,New York)1960.

Department of Mathematics
University of Illinois in Urbana-Champaign
Urbana,IL 61801,U.S.A.

THE PREPROJECTIVE ALGEBRA OF A MODULATED GRAPH

Vlastimil Dlab and Claus Michael Ringel

The present paper generalizes a recent result of I.M. Gelfand and V.A. Ponomarev [4] reported at the Conference by V.A. Rojter.

A *modulated graph* $\mathfrak{M} = (F_i, {}_iM_j, \varepsilon_i^j)_{i,j\in I}$ is given by division rings F_i for all $i \in I$, by bimodules ${}_{F_i}({}_iM_j)_{F_j}$ for all $i \neq j$ in I finitely generated on both sides and by non-degenerate bilinear forms $\varepsilon_i^j : {}_iM_j \otimes {}_jM_i \to F_i$; here, I is a finite index set. Note that the forms ε_i^j give rise to canonical elements $c_j^i \in {}_jM_i \otimes {}_iM_j$. Namely, if $x_1,\dots,x_d$ is a basis of $({}_jM_i)_{F_i}$ and $y_1,\dots,y_d$ the corresponding dual basis of ${}_{F_i}({}_iM_j)$ with respect to ε_i^j , then $c_j^i = \sum_p x_p \otimes y_p$; see section 1.

Define the ring $\Pi(\mathfrak{M})$ as follows. Let $T(\mathfrak{M})$ be the tensor ring of $\mathfrak{M}$: $T(\mathfrak{M}) = \bigoplus_{t\in\mathbb{N}} T_t$, where $T_o = \prod_i F_i$, $T_1 = \bigoplus_{i,j} {}_iM_j$ and

$T_{t+1} = T_1 \otimes_{T_o} T_t$ with the multiplication given by the tensor product.

Then, by definition, $\Pi(\mathfrak{M}) = T(\mathfrak{M})/\langle c\rangle$, where $\langle c\rangle$ is the principal ideal of $T(\mathfrak{M})$ generated by the element $c = \sum_{i,j} c_i^j$.

Let Ω be an (admissible) orientation of $\mathfrak{M}$; thus, for every pair i,j with ${}_iM_j \neq 0$, we prescribe an order indicated by an arrow $i \longrightarrow j$, or $i \longleftarrow j$ in such a way that no oriented cycles occur. Let $R(\mathfrak{M},\Omega)$ be the corresponding tensor ring of $(\mathfrak{M},\Omega)$: $R(\mathfrak{M},\Omega) = \bigoplus_{t \in \mathbb{N}} R_t$ with $R_o = \prod_i F_i$, $R_1 = \bigoplus_{i \to j} {}_iM_j$ and $R_{t+1} = R_1 \otimes_{R_o} R_t$. For the representation theory of $R(\mathfrak{M},\Omega)$ we refer to [3].

<u>THEOREM</u>. *For each orientation* Ω *of* $\mathfrak{M}$, $R(\mathfrak{M},\Omega)$ *is a subring of* $\Pi(\mathfrak{M})$ *and, as a (right)* $R(\mathfrak{M},\Omega)$ *-module,* $\Pi(\mathfrak{M})$ *is the direct sum of all indecomposable preprojective* $R(\mathfrak{M},\Omega)$*-modules (each occurring with multiplicity one).*

This theorem suggests to call $\Pi(\mathfrak{M})$ the preprojective algebra of $\mathfrak{M}$. Recall that an indecomposable $R(\mathfrak{M},\Omega)$-module P is preprojective if and only if there is only a finite number of indecomposable modules X with $\mathrm{Hom}\,(X,P) \neq 0$.

<u>COROLLARY</u>. *The ring* $\Pi(\mathfrak{M})$ *is artinian if and only if the modulated graph is a disjoint union of Dynkin graphs.*

Observe that if $\mathfrak{M}$ is a K-modulation (where K is a commutative field), then $\Pi(\mathfrak{M})$ is a K-algebra. In this case, the corollary may be reformulated as follows: The algebra $\Pi(\mathfrak{M})$ is finite-dimensional if and only if $\mathfrak{M}$ is a disjoint union of Dynkin graphs.

Consider, in particular, the case when $(\mathfrak{M},\Omega)$ is given by a quiver; thus, $F_i = K$ for all i and ${}_iM_j$ is a direct sum of a finite number of copies of ${}_KK_K$. For every arrow x of the quiver, define an "inverse" arrow x^* whose end is the origin of x and whose origin is the end of x . Then $T(\mathfrak{M})$ is the path algebra generated by all arrows x and x^* , and $\Pi(\mathfrak{M})$ is the quotient of $T(\mathfrak{M})$ by the ideal generated by the element $\sum_{\text{all } x} (xx^* + x^*x)$.

<u>COROLLARY</u>. *If* $(\mathfrak{M},\Omega)$ *is given by a quiver, then* $\Pi(\mathfrak{M})$ *is finite-dimensional if and only if the quiver is of finite type.*

For a quiver which is a tree, the last result has been announced by A.V. Rojter [6] in his report on the paper [4]. In contrast to the proofs in [4], our approach avoids use of reflection functors and is based on the explicite description of the category

$P(\mathfrak{M},\Omega)$ of all preprojective $R(\mathfrak{M},\Omega)$-modules. The authors are indebted to P. Gabriel for pointing out that the theorem is, in the case when $(\mathfrak{M},\Omega)$ is given by a quiver, also due to Ch. Riedtmann [7].

1. Preliminaries on dualization

Given a finite-dimensional vector space ${}_FM$, denote by ${}^*M$ its (left) dual space $\mathrm{Hom}({}_FM, {}_FF_F)$. If ${}_FM_G$ is a bimodule and ${}_GX$, ${}_FY$ vector spaces, the adjoint map $\overline{f} : X \to {}^*M \otimes_F Y$ to a map $f : M \otimes_G X \to Y$ is given by $\overline{f}(x) = \sum_{p=1}^{d} \phi_p \otimes f(m_p \otimes x)$, where $x \in X$, $\{m_1, m_2, \ldots, m_d\}$ is a basis of ${}_FM$ and $\{\phi_1, \phi_2, \ldots, \phi_d\}$ is the respective dual basis of $({}^*M)_F$. In particular, if M is an End Y - End X-submodule of the bimodule $\mathrm{Hom}(X,Y)$ and $\chi_M : M \otimes X \to Y$ the evaluation map $\chi_M(m \otimes x) = m(x)$, then $\overline{\chi}_M(x) = \sum_p \phi_p \otimes m_p(x)$. Note that $\overline{\chi}_M$ is a (left) G-homomorphism.

Now, given bimodules ${}_FM_G$, ${}_GN_F$ such that ${}_FM$ and N_F are finite dimensional, let $\varepsilon : M \otimes_G N \to F$ be a non-degenerate bilinear form. Thus, the adjoint $\overline{\varepsilon}$ is an isomorphism $\overline{\varepsilon} : N \to {}^*M$; let $\{n_1, n_2, \ldots, n_d\}$ be a basis of N_F and $\{\phi_1, \phi_2, \ldots, \phi_d\}$ the basis of $({}^*M)_F$ such that $\phi_p = \overline{\varepsilon}(n_p)$ for all $1 \leq p \leq d$. Furthermore, let $\{m_1, m_2, \ldots, m_d\}$ be the dual basis of ${}_FM$. Thus,

$$\varepsilon(m_p \otimes n_q) = (m_p)[\overline{\varepsilon}(n_q)] = (m_p)\,\phi_q = \delta_{pq}\,.$$

Define the canonical element c_ε of $N \otimes_F M$ (with respect to ε) by

$$c_\varepsilon = \sum_{p=1}^{d} n_p \otimes m_p\,.$$

Lemma 1.1. *The element c_ε does not depend on the choice of a basis.*

Proof. Let $\{n_1', n_2', \ldots, n_d'\}$ and $\{m_1', m_2', \ldots, m_d'\}$ be another bases of N_F and ${}_FM$, respectively, so that

$$\varepsilon(m_p' \otimes n_q') = \delta_{pq}\,.$$

Then $n_q' = \sum_j n_j b_{jq}$ and $m_p' = \sum_i a_{pi} m_i$ with b_{jq} and a_{pi} from F.

Since $\delta_{pq} = \varepsilon(m_p' \otimes n_q') = \sum_{i,j} a_{pi}\,\varepsilon(m_i \otimes n_j) b_{jq} = \sum_i a_{pi} b_{iq}$, we have also $\sum_p b_{jp} a_{pi} = \delta_{ji}$.

Thus,

$$\sum_p n'_p \otimes m'_p = \sum_{i,j,p} n_j b_{jp} \otimes a_{pi} m_i$$

$$= \sum_{i,j} n_j \left(\sum_p b_{jp} a_{pi}\right) \otimes m_i = \sum_i n_i \otimes m_i . \bullet$$

If we take, in particular, ${}_G N_F = {}^*({}_F M_G)$ and the evaluation map $\chi : M \otimes_G N \to F$ defined by

$$\chi(m \otimes \phi) = (m)\phi \quad ,$$

we obtain, for every bimodule M, *the canonical element* $c(M) = c_\chi$.

Given a bimodule ${}_F M_G$, define the higher dual spaces ${}^{(t)}{}_F M_G$ inductively by

$${}^{(t+1)}{}_F M_G = {}^*({}^{(t)}{}_F M_G) \ .$$

Thus, ${}^{(t)}M$ is an F-G-bimodule for t even and a G-F-bimodule for t odd.

Lemma 1.2. *Let* ${}_F M_G$ *and* ${}_G N_F$ *be bimodules and* $\varepsilon : {}_F M \otimes_G N_F \to {}_F F_F$ *and* $\delta : {}_G N \otimes_F M_G \to {}_G G_G$ *non-degenerate bilinear forms. Define the maps* ${}^t\eta$ *inductively as follows:*

$${}^0\eta = 1_M : {}_F M_G \to {}^{(0)}M = M \ ;$$

$${}^1\eta = \bar{\varepsilon} : {}_G N_F \to {}^{(1)}M = {}^*M \ ;$$

$${}^{2r}\eta = \overline{\delta[({}^{2r-1}\eta)^{-1} \otimes 1_M]} : {}_F M_G \to {}^{(2r)}M \text{ and}$$

$${}^{2r+1}\eta = \overline{\varepsilon[({}^{2r}\eta)^{-1} \otimes 1_N]} : {}_G N_F \to {}^{(2r+1)}M \ .$$

Then

$$[{}^{2r+1}\eta \otimes {}^{2r+2}\eta](c_\varepsilon) = c({}^{(2r)}M) \quad \textit{and} \quad [{}^{2r}\eta \otimes {}^{2r+1}\eta](c_\delta) = c({}^{(2r+1)}M).$$

Proof. Recall that $c_\varepsilon = \sum_p n_p \otimes m_p$, where $\{m_1, m_2, \ldots, m_d\}$ is a basis of ${}_F M$ and $\{n_1, n_2, \ldots, n_d\}$ the dual basis of N_F with respect to ε. Hence, in order to prove the first equality, it is sufficient to show that, for $m \in M$ and $n \in N$,

$$\delta(n \otimes m) = ({}^{2r+1}\eta(n))[{}^{2r+2}\eta(m)] \ .$$

But, $({}^{2r+1}\eta(n))[{}^{2r+2}\eta(m)] = ({}^{2r+1}\eta(n))[\overline{\delta[({}^{2r+1}\eta)^{-1} \otimes 1_M](m)}] =$

$$= \delta[({}^{2r+1}\eta)^{-1} \otimes 1_M]\,({}^{2r+1}\eta(n)) = \delta[({}^{2r+1}\eta)^{-1}\,{}^{2r+1}\eta(n) \otimes m] =$$

$$= \delta(n \otimes m) .$$

Similarly, since

$$({}^{2r}\eta(m))[{}^{2r+1}\eta(n)] = ({}^{2r}\eta(m))[\overline{\varepsilon[({}^{2r}\eta)^{-1} \otimes 1_N](n)}] =$$

$$= \varepsilon[({}^{2r}\eta)^{-1} \otimes 1_N]({}^{2r}\eta(m)) = \varepsilon[({}^{2r}\eta)^{-1}\,{}^{2r}\eta(m) \otimes n] =$$

$$= \varepsilon(m \otimes n) ,$$

we can derive the second equality for $c({}^{(2r+1)}M)$.

2. Irreducible maps

Recall the definition of an irreducible map [2]: a map $f : X \to Y$ is called irreducible if f is neither a split monomorphism nor a split epimorphism and if, for every factorization $f = f'f''$, either f'' is a split monomorphism or f' is a split epimorphism. Also, recall the definition of the radical of a module category. If X and Y are indecomposable modules, let rad (X,Y) be the set of all non-invertible homomorphisms. If $X = \oplus_p X_p$ and $Y = \oplus_q Y_q$ with indecomposable modules X_p and Y_q , define rad $(X,Y) = \oplus_{p,q}$ rad (X_p,Y_q) , using the identification $\mathrm{Hom}(X,Y) = \oplus_{p,q} \mathrm{Hom}(X_p,Y_q)$. The square $\mathrm{rad}^2(X,Y)$ of the radical is thus the set of all homomorphisms $f : X \to Y$ such that $f = f'f''$, where $f'' \in \mathrm{rad}(X,Z)$ and $f' \in \mathrm{rad}(Z,Y)$ for some module Z . Note that both rad and rad^2 are ideals in our module category; in particular, rad (X,Y) and $\mathrm{rad}^2(X,Y)$ are End Y - End X - submodules of the bimodule ${}_{\mathrm{End}\,Y}\mathrm{Hom}(X,Y)_{\mathrm{End}\,X}$. For indecomposable X and Y, the elements in $\mathrm{rad}\,(X,Y) \setminus \mathrm{rad}^2(X,Y)$ are just the irreducible maps. In this case, we write $\mathrm{Irr}(X,Y) = \mathrm{rad}(X,Y)/\mathrm{rad}^2(X,Y)$, and call $\mathrm{Irr}(X,Y)$ the *bimodule of irreducible maps* (see [5]). In what follows, our main objective is to select a direct complement of $\mathrm{rad}^2(X,Y)$ in $\mathrm{rad}(X,Y)$ which is an EndY-EndX-submodule, and realize in this way

$Irr(X,Y)$ as a subset of $Hom(X,Y)$ rather than just as a factor group. We shall select such complements inductively, using Auslander-Reiten sequences.

Recall that an exact sequence $0 \to X \overset{f}{\to} Y \overset{g}{\to} Z \to 0$ is called an Auslander-Reiten sequence if both maps f and g are irreducible. This implies that both modules X and Z are indecomposable, X is not injective and Z is not projective. Conversely, given an indecomposable non-injective module X, there exists an Auslander-Reiten sequence starting with X, and also dually, given an indecomposable non-projective Z, there is an Auslander-Reiten sequence ending with Z. Moreover, if $0 \to X \overset{f}{\to} Y \to Z \to 0$ is an Auslander-Reiten sequence and $h : X \to X'$ is a map which is not a split monomorphism, then there exists $\alpha : Y \to X'$ such that $h = \alpha f$. (For all these properties, we refer to [2]).

In the sequel, we will consider direct sums of the form $\bigoplus_Y \mathcal{U}(Y)$, where $\mathcal{U}(Y)$ is an abelian group depending on Y, with Y ranging over "all" indecomposable modules. Here, of course, we choose first fixed representatives Y of all isomorphism classes of indecomposable modules and then index the direct sum by these representatives. In fact, all direct sum which will occur in this way will have even only a finite number of non-zero summands.

PROPOSITION 2.1. *Let* X *be an indecomposable non-injective module and* G *be a division ring with*

$$\mathrm{End}\, X = G \oplus \mathrm{rad}\, \mathrm{End}\, X .$$

Assume that, for every indecomposable module Y, *there is given a direct complement* $M(X,Y)$ *of* $\mathrm{rad}^2(X,Y)$ *in* ${}_{\mathrm{End}\,Y}\mathrm{rad}(X,Y)_G$. *Let*

$$0 \longrightarrow X \xrightarrow{(\overline{\chi}_{M(X,Y)})_Y} \bigoplus_Y {}^*M(X,Y) \underset{\mathrm{End}\,Y}{\otimes} Y \xrightarrow{\pi} Z \longrightarrow 0$$

be exact. Then, this is an Auslander-Reiten sequence. Moreover, G *embeds into the endomorphism ring* $\mathrm{End}\, Z$ *of* Z *as a radical complement, and for every* Y, *there is an embedding* σ *of* ${}^*M(X,Y)$ *onto a complement of* $\mathrm{rad}^2(Y,Z)$ *in* ${}_G\mathrm{rad}(Y,Z)_{\mathrm{End}\,Y}$ *such that*

$$\chi_{\sigma^*M(X,Y)} = \pi \mid {}^*M(X,Y) \otimes Y .$$

Proof. Let

$$0 \longrightarrow X \xrightarrow{(f'_{Y,p})_{Y,p}} \bigoplus_{Y} \bigoplus_{p=1}^{d_Y} Y \longrightarrow Z' \longrightarrow 0$$

be an Auslander-Reiten sequence starting with X, where $f'_{Y,p} : X \to Y$ for $1 \leq p \leq d_Y$. Then the residue classes of the elements $f'_{Y,1}$, $f'_{Y,2},\ldots,f'_{Y,d_Y}$ form a basis of the G-vector space $\mathrm{rad}(X,Y)_G/\mathrm{rad}^2(X,Y)_G$ (see Lemma 2.5 of [5]). Let $f_{Y,1}, f_{Y,2},\ldots, f_{Y,d_Y}$ be a G-basis of $M(X,Y)$. By the factorization property of Auslander-Reiten sequences, there is a map

$$\alpha : \bigoplus_{Y} \bigoplus_{p=1}^{d_Y} Y \longrightarrow \bigoplus_{Y} \bigoplus_{p=1}^{d_Y} Y$$

such that $\alpha \circ (f'_{Y,p})_{Y,p} = (f_{Y,p})_{Y,p}$. It follows that α is an automorphism. For, let $E = \mathrm{End}\,(\bigoplus_{Y} \bigoplus_{p=1}^{d_Y} Y)$ and consider the residue class $\bar{\alpha}$ of α in $E/\mathrm{rad}\,E$. Also, consider the factor group

$$M = \mathrm{rad}(X, \bigoplus_{Y} \bigoplus_{p=1}^{d_Y} Y)/\mathrm{rad}^2(X, \bigoplus_{Y} \bigoplus_{p=1}^{d_Y} Y) ,$$

and let $\bar{f}$ and $\bar{f}'$ be the residue classes of $f = (f_{Y,p})_{Y,p}$ and $f' = (f'_{Y,p})_{Y,p}$, respectively. Then $\mathrm{rad}\,E$ annihilates M, and the equality $\bar{\alpha}\,\bar{f}' = \bar{f}$ shows that $\bar{\alpha}$ induces base changes between the bases $(\bar{f}_{Y,p})_p$ and $(f'_{Y,p})_p$ of $\mathrm{Irr}(X,Y)$. This implies that $\bar{\alpha}$ is invertible. Since $\mathrm{rad}\,E$ is nilpotent, α is invertible, as well. Thus, we can form the following commutative diagram

$$\begin{array}{ccccccccc}
0 \longrightarrow & X & \xrightarrow{f'} & \bigoplus_{Y} \bigoplus_{p=1}^{d_Y} Y & \longrightarrow & Z' & \longrightarrow 0 \\
 & \| & & \downarrow \alpha & & \downarrow \beta & \\
0 \longrightarrow & X & \xrightarrow{f} & \bigoplus_{Y} \bigoplus_{p=1}^{d_Y} Y & \xrightarrow{\pi} & Z & \longrightarrow 0 ,
\end{array}$$

where both α and β are isomorphisms. As a consequence, also the lower sequence is an Auslander-Reiten sequence.

Note that we can rewrite $\bigoplus_{p=1}^{d_Y} Y$ as ${}^*M(X,Y) \underset{\text{End } Y}{\otimes} Y$, and then $(f_{Y,p})_p$ becomes $\overline{\chi}_{M(X,Y)}$. For, if $\phi_{Y,1}, \phi_{Y,2}, \ldots, \phi_{Y,d_Y}$ is the dual basis of ${}^*M(X,Y)_{\text{End } Y/\text{rad End } Y}$ with respect to the basis $f_{Y,1}, f_{Y,2}, \ldots, f_{Y,d_Y}$ of ${}_{\text{End } Y/\text{rad End } Y}M(X,Y)$, then we identify

$${}^*M(X,Y) \underset{\text{End} Y}{\otimes} Y = \bigoplus_{p=1}^{d_Y} \phi_{Y,p} \otimes Y \approx \bigoplus_{p=1}^{d_Y} Y ,$$

and

$$\overline{\chi}_{M(X,Y)}(x) = \sum_{p=1}^{d_Y} \phi_{Y,p} \otimes f_{Y,p}(x)$$

is identified with $(f_{Y,p}(x))_p$.

Now, ${}^*M(X,Y)$ is a left G-module, and

$$\overline{\chi}_{M(X,Y)} : X \longrightarrow {}^*M(X,Y) \underset{\text{End } Y}{\otimes} Y$$

is a G-module homomorphism. Hence, under $(\overline{\chi}_{M(X,Y)})_Y$, the module X becomes a G-submodule of $\bigoplus_Y {}^*M(X,Y) \underset{\text{End } Y}{\otimes} Y$, and therefore also the factor module Z has a left G-module structure. Thus, G embeds canonically into End Z and in this way, G becomes a radical complement. This follows from the canonical isomorphism

$$\text{End } X/\text{rad End } X \approx \text{End } Z/\text{rad End } Z ,$$

which is always valid for the outer terms of an Auslander-Reiten sequence.

The restriction of π to ${}^*M(X,Y) \otimes Y$ defines a map σ of ${}^*M(X,Y)$ into $\text{Hom}(Y,Z)$ which is a G-End Y-homomorphism. If we denote again by $\phi_{Y,1}, \phi_{Y,2}, \ldots, \phi_{Y,d_Y}$ an End Y/rad End Y-basis of ${}^*M(X,Y)$, then $\pi \,|\, {}^*M(X,Y) \underset{\text{End } Y}{\otimes} Y \longrightarrow Z$ can be identified with

$$(\phi_{Y,p})_p : \bigoplus_{p=1}^{d_Y} Y \approx \bigoplus_{p=1}^{d_Y} \phi_{Y,p} \otimes Y \longrightarrow Z .$$

Again, using Lemma 2.5 of [5], we see that the residue classes of $\phi_{Y,1}, \phi_{Y,2}, \ldots, \phi_{Y,d_Y}$ in $\mathrm{Irr}(Y,Z)$ form an End Y/rad End Y-basis and that ${}^*M(X,Y)$ is therefore mapped injectively onto a complement of $\mathrm{rad}^2(Y,Z)$ in ${}_G\mathrm{rad}(Y,Z)_{\mathrm{End}\,Y}$. This completes the proof.

Now, assume that X is an indecomposable, non-injective module and that G is a radical complement in End X. If there are given direct complements $M(X,Y)$ of $\mathrm{rad}^2(X,Y)$ in ${}_{\mathrm{End}\,Y}\mathrm{rad}(X,Y)_G$, then the $\sigma^*M(X,Y)$ are direct complements of $\mathrm{rad}^2(Y,Z)$ in ${}_G\mathrm{rad}(Y,Z)_{\mathrm{End}\,Y}$, and the Auslander-Reiten sequence starting with X is of the form

$$0 \longrightarrow X \xrightarrow{(\overline{\chi}_{M(X,Y)})_Y} \oplus\, {}^*M(X,Y) \otimes Y \xrightarrow{(\chi_{\sigma^*M(X,Y)})_Y} Z \longrightarrow 0 .$$

Denote by $c(M(X,Y))$ the canonical element in ${}^*M(X,Y) \otimes M(X,Y)$. Now $\iota : M(X,Y) \hookrightarrow \mathrm{Hom}(X,Y)$ and $\sigma : {}^*M(X,Y) \hookrightarrow \mathrm{Hom}(Y,Z)$, and thus we have a canonical map

$${}^*M(X,Y) \otimes M(X,Y) \longrightarrow \mathrm{Hom}(X,Z) ,$$

namely $\sigma \otimes \iota$ followed by the composition map μ .

<u>PROPOSITION 2.2</u>. *Under the map*

$$\bigoplus_Y {}^*M(X,Y) \otimes M(X,Y) \xrightarrow{\oplus(\sigma\otimes\iota)} \bigoplus_Y \mathrm{Hom}(Y,Z) \otimes \mathrm{Hom}(X,Y) \xrightarrow{(\mu)} \mathrm{Hom}(X,Z) ,$$

the element $\sum_Y c(M(X,Y))$ *goes to zero.*

Observe that, for a fixed module X, there is only a finite number of modules Y such that $M(X,Y) \approx \mathrm{Irr}(X,Y) \neq 0$; therefore, we may form the sum $\sum_Y c(M(X,Y))$.

<u>Proof of Proposition 2.2</u>. First, we are going to show that $c(M(X,Y))$ maps onto $\chi_{\sigma^*M(X,Y)} \circ \overline{\chi}_{M(X,Y)}$. Let $f_1, f_2, \ldots, f_d$ be an End Y/rad End Y-basis of ${}_{\mathrm{End}\,Y/\mathrm{rad}\,\mathrm{End}\,Y}M = M(X,Y)$, and $\phi_1, \phi_2, \ldots, \phi_d$ the corresponding dual basis in ${}^*M_{\mathrm{End}\,Y/\mathrm{rad}\,\mathrm{End}\,Y}$. Then, for $x \in X$, we have

$$\overline{\chi}_M(x) = \sum_p \phi_p \otimes f_p(x) ,$$

and for $\phi \in {}^*M$, $y \in Y$,

$$\chi_{\sigma^*M}(\phi \otimes y) = \sigma(\phi)(y) .$$

Thus,

$$\chi_{\sigma^*M}\, \overline{\chi}_M(x) = \chi_{\sigma^*M}\, (\sum_p \phi_p \otimes f_p(x)) = \sum_p \sigma(\phi_p)(f_p(x)) .$$

This shows that $\chi_{\sigma^*M}\, \overline{\chi}_M$ is equal to $\sum_p \sigma(\phi_p) f_p$, and this is the image of $\sum_p \phi_p \otimes f_p = c(M(X,Y))$ under $\mu(\sigma \otimes 1)$. As a consequence, we conclude that under the map $\bigoplus_Y {}^*M(X,Y) \otimes M(X,Y) \xrightarrow{\oplus(\sigma \otimes 1)} \bigoplus_Y \mathrm{Hom}(Y,Z) \otimes \mathrm{Hom}(X,Y) \xrightarrow{(\mu)} \mathrm{Hom}(X,Z)$, the element $\sum_Y c(M(X,Y))$ goes to $\sum_Y \chi_{\sigma^*M(X,Y)}\, \overline{\chi}_{M(X,Y)}$, which is the composite of the two maps in the corresponding Auslander-Reiten sequence and thus zero. The proof is completed.

Let us point out that, in what follows, we shall not specify any longer the embedding σ of ${}^*M(X,Y)$ into $\mathrm{Hom}(Y,Z)$, but shall simply consider ${}^*M(X,Y)$ to be a subset of $\mathrm{Hom}(Y,Z)$.

REMARK. Let us underline the use of the two distinct tensor products $M(X,Y) \otimes {}^*M(X,Y)$ and ${}^*M(X,Y) \otimes M(X,Y)$. Whereas the first one is used for the ordinary evaluation map

$$\chi : M(X,Y) \otimes {}^*M(X,Y) \longrightarrow \mathrm{End}\, Y/\mathrm{rad}\, \mathrm{End}\, Y$$

given by $\chi(f \otimes \phi) = f(\phi)$, it is the second one which has to be used for the composition map μ . Namely, using the above embedding ${}^*M(X,Y) \hookrightarrow \mathrm{Hom}(Y,Z)$, we can consider

$${}^*M(X,Y) \otimes M(X,Y) \hookrightarrow \mathrm{Hom}(Y,Z) \otimes \mathrm{Hom}(X,Y) \xrightarrow{\mu} \mathrm{Hom}(X,Z) ,$$

and $\mu(\phi \otimes f) = \phi \circ f$.

3. The preprojective modules

Now, let us consider the particular case of the irreducible maps between indecomposable preprojective $R(\mathfrak{M},\Omega)$-modules. First, recall the way in which these modules can be inductively obtained from the indecomposable projective ones.

For each $i \in I$, there is an indecomposable projective $R(\mathfrak{M},\Omega)$-module $P(i)$. Indeed, denoting by e_i the primitive idempotent of $R(\mathfrak{M},\Omega)$ corresponding to the identity element of the ith

factor F_i in $R_o = \prod_i F_i$, $P(i) = e_i R(\mathfrak{M},\Omega)$. Note that $P(i)/\mathrm{rad}\,P(i)$ is the simple $R(\mathfrak{M},\Omega)$-module corresponding to the vertex i which defines $P(i)$ uniquely up to an isomorphism. Moreover, note that $\mathrm{End}\,P(i) = F_i$, and thus it is a division ring. The irreducible maps between projective modules are always rather easy to determine. Here, for $R(\mathfrak{M},\Omega)$, there are irreducible maps from $P(j)$ to $P(i)$ if and only if $i \to j$ in Ω. In fact, ${}_iM_j$ can be easily embedded in $\mathrm{Hom}\,(P(j), P(i))$ in such a way that

$${}_iM_j \oplus \mathrm{rad}^2(P(j), P(i)) = \mathrm{rad}\,(P(j), P(i))$$

as F_i-F_j-bimodules. This follows either from the explicit description of the modules $P(i)$ given in [3], or from the fact that $\oplus\,{}_iM_j$ is a direct complement of $\mathrm{rad}^2 R(\mathfrak{M},\Omega)$ in $\mathrm{rad}\,R(\mathfrak{M},\Omega)$. As a result, given two indecomposable projective $R(\mathfrak{M},\Omega)$-modules P and P', we can always choose a direct complement $M(P,P')$ of $\mathrm{rad}^2(P,P')$ in ${}_{\mathrm{End}\,P'}\,\mathrm{rad}(P,P')_{\mathrm{End}\,P}$, and we can identify these $M(P,P')$ with the given bimodules ${}_iM_j$, where $i \to j$.

Now, the indecomposable preprojective modules can be derived from the projective ones by using powers of the Coxeter functor C^- (as defined in [3]) or of the Auslander-Reiten translation $A^- = \mathrm{Tr}\,D$ ("transpose of dual" of [2], and also [1]). Thus, we denote by $P(i,r)$ the module obtained from $P(i)$ by applying the rth power of one of the mentioned constructions. (It is clear from the uniqueness result in [3] that $C^{-r}\,P(i) \approx A^{-r}\,P(i)$.)

LEMMA 3.1. *Assume that* X *and* Y *are indecomposable modules and that there exists an irreducible map* $X \to Y$. *If one of the modules* X, Y *is preprojective, then both are. Furthermore, if* $X = P(i,r)$ *and* $Y = P(j,s)$, *then either* $s = r$ *and* $i \leftarrow j$, *or* $s = r+1$ *and* $i \to j$.

Proof. This lemma is well-known, so let us just outline a proof. Using shifts by powers of the Coxeter functors C^+ and C^- (see [3]) or of the Auslander-Reiten translations $A = D\,\mathrm{Tr}$ and $A^- = \mathrm{Tr}\,D$ (see [2] and [1]), we can assume that X is projective. If Y is not projective, then we get from the Auslander-Reiten sequence ending with Y, an irreducible map from AY to X.

Since X is projective, this map cannot be an epimorphism and thus it has to be a monomorphism. Consequently, AY is projective.

Now, in view of Proposition 2.1, we obtain by induction on the "layer" r of the indecomposable preprojective $R(\mathfrak{M},\Omega)$-modules $P(i,r)$ the following result.

PROPOSITION 3.2. a) *If we choose, for any two indecomposable projective modules* P *and* P', *a direct complement* $M(P,P')$ *of* $\mathrm{rad}^2(P,P')$ *in* ${}_{\mathrm{End}\,P}\mathrm{rad}(P,P')_{\mathrm{End}\,P'}$, *then this determines a direct complement* $M(P,P')$ *of* $\mathrm{rad}^2(P,P')$ *in* $\mathrm{rad}\,(P,P')$ *for any indecomposable preprojective modules* P, P'.

b) *If we identify, for any arrow* $i \to j$ *the bimodule* $M(P(j), P(i))$ *with* ${}_iM_j$, *then this yields an identification of any* $M(P(j,r), P(i,r))$ *with* ${}^{(2r)}{}_iM_j$ *and any* $M(P(i,r), P(j,r+1))$ *with* ${}^{(2r+1)}{}_iM_j$ *for* $i \to j$.

PROPOSITION 3.3. *Every map between two indecomposable preprojective modules is a sum of composites of maps from the various* $M(P,P')$.

Proof. Let Y be an indecomposable preprojective module, say $Y = P(i,r)$. Then the radical of the endomorphism ring E of $\bigoplus_{\substack{j \in I \\ 0 \le s \le r}} P(j,s)$ is generated (by using the addition and multiplication) by an arbitrary complement of $\mathrm{Rad}^2 E$ in $\mathrm{Rad}\,E$. So we may choose as a complement the direct sum of $M(P(j,s), P(j',s'))$.

4. Abstract definition of the full subcategory of the preprojective modules

First, let us introduce the following notation indicating the operation of the division rings F_i and F_j: For $i \to j$, put

$$ {}^{2r}{}_iM_j = {}^{(2r)}({}_iM_j) \quad \text{and} \quad {}^{2r+1}{}_jM_i = {}^{(2r+1)}({}_iM_j) . $$

Now, define the category $P(\mathfrak{M},\Omega)$ as follows: The objects of $P(\mathfrak{M},\Omega)$ are pairs (i,r), $i \in I$, $r \ge 0$ with the endomorphism rings F_i. For $i \to j$,

$$ M((j,r),(i,r)) = {}^{2r}{}_iM_j $$

and

$$M((i,r),(j,r+1)) = {}^{2r+1}_{\ \ j}M_i \ .$$

Denote by $F(\mathfrak{M},\Omega)$ the free category generated by these morphisms using the tensor products over F_i. Furthermore, for every (j,r), take

$$c(j,r) = \sum_{i\to j} c({}^{2r}_{\ i}M_j) + \sum_{j\to k} c({}^{2r+1}_{\ \ k}M_j) \in$$

$$\bigoplus_{i\to j} ({}^{2r+1}_{\ \ j}M_i \otimes {}^{2r}_{\ i}M_j) \oplus \bigoplus_{j\to k} ({}^{2r+2}_{\ \ j}M_k \otimes {}^{2r+1}_{\ \ k}M_j) \ ,$$

and denote by J the category ideal generated by all elements $c(j,r)$. The category $P(\mathfrak{M},\Omega)$ is then defined as the factor category of $F(\mathfrak{M},\Omega)$ by the ideal J.

Observe that the definition of $P(\mathfrak{M},\Omega)$ requires only the knowledge of the bimodules ${}_iM_j$ for $i \to j$ (and neither the corresponding bimodules ${}_jM_i$, nor the bilinear forms ε_i^j and ε_j^i).

<u>PROPOSITION 4.1</u>. *The full subcategory of the preprojective modules of the category of all* $T(\mathfrak{M},\Omega)$*-modules is equivalent to* $P(\mathfrak{M},\Omega)$.

<u>Proof</u>. Using Proposition 3.2, there is a canonical functor Γ from $F(\mathfrak{M},\Omega)$ to the subcategory of preprojective $T(\mathfrak{M},\Omega)$-modules given by the choice of $M(P(i),P(j)) = {}_jM_i$ for projective modules $P(i),P(j)$ where $j \to i$. Also by Proposition 3.3, Γ is surjective. Moreover, according to Proposition 2.2, the elements $c(j,r)$ are mapped to zero.

Conversely, let a morphism $f : (j,r) \to (j',r')$ from $F(\mathfrak{M},\Omega)$ be mapped under Γ to zero. We are going to show that f must lie in the ideal J. This is clear if $r = r'$; for, then $f = 0$. Thus, assume that $f \neq 0$ and proceed by induction on $r' - r$. Now j and r are fixed; let $\{\ldots g_p \ldots\}$ be the union of bases of all vector spaces ${}_{F_i}({}^{2r}_{\ i}M_j)$ for all i with $i \to j$ and ${}_{F_k}({}^{2r+1}_{\ \ k}M_j)$ for all k with $j \to k$, and let $\{\ldots g'_p \ldots\}$ be the union of the corresponding dual bases of $({}^{2r+1}_{\ \ j}M_i)_{F_i}$ and $({}^{2r+2}_{\ \ j}M_k)_{F_k}$.

Thus, $c(j,r) = \sum_p g_p' \otimes g_p$. Now, $f = \sum_p h_p \otimes g_p$, where h_p is a morphism of $F(\mathcal{M},\Omega)$ either from (i,r) or $(k,r+1)$ to (j',r'). Since there is an Auslander-Reiten sequence

$$0 \longrightarrow P(j,r) \xrightarrow{(\Gamma(g_p))_p} Q \xrightarrow{(\Gamma(g_p'))_p} P(j,r+1) \longrightarrow 0$$

and since

$$0 = \Gamma(f) = \sum_p \Gamma(h_p)\, \Gamma(g_p) ,$$

we can factor $(\Gamma(h_p))_p : Q \to P(j',r')$ through $(\Gamma(g_p'))_p$. Hence, there is a homomorphism $\tilde{u} : P(j,r+1) \to P(j',r')$ such that

$$\Gamma(h_p) = \tilde{u}\, \Gamma(g_p') .$$

And, since Γ is surjective, we can find $u : (j,r+1) \to (j',r')$ in $F(\mathcal{M},\Omega)$ such that $\Gamma(u) = \tilde{u}$. Obviously, the elements $h_p - u \otimes g_p'$ lie in the kernel of Γ , and therefore, by induction, they belong to J . Consequently,

$$f = \sum_p h_p \otimes g_p = \sum_p (h_p - u \otimes g_p') \otimes g_p + \sum_p u \otimes g_p' \otimes g_p$$

also belongs to J ; for, $\sum_p u \otimes g_p' \otimes g_p = u \otimes c(j,r)$.

5. Proof of the theorem

The proof of the theorem consists in identifying the additive structure of $\Pi(\mathcal{M})$ with a factor of a subcategory of $F(\mathcal{M},\Omega)$. Indeed, we may consider both $F(\mathcal{M},\Omega)$ and $P(\mathcal{M},\Omega)$ defined in section 4 as abelian groups forming the direct sum of all $\mathrm{Hom}((i,r),(j,s))$. Denote by $\Phi(\mathcal{M},\Omega)$ and $\Pi(\mathcal{M},\Omega)$ the respective subgroups of all $\mathrm{Hom}((i,0),(j,s))$. Then, both $\Phi(\mathcal{M},\Omega)$ and $\Pi(\mathcal{M},\Omega)$ contain a subring $R = \bigoplus_{i,j} \mathrm{Hom}((i,0),(j,0))$ which is obviously isomorphic to $R(\mathcal{M},\Omega)$. Furthermore, under the composition in $\Pi(\mathcal{M},\Omega)$, $\Pi(\mathcal{M},\Omega)$ is a right $R(\mathcal{M},\Omega)$-module; for, if $f : (i,0) \to (j,s)$ and $a : (k,0) \to (i,0)$ from R , then $fa : (k,0) \to (j,s)$ in $\Pi(\mathcal{M},\Omega)$.

PROPOSITION 5.1. $\Pi(\mathcal{M},\Omega)_{R(\mathcal{M},\Omega)}$ *is isomorphic to the direct sum of all* ∨ *preprojective* $R(\mathcal{M},\Omega)$*-modules (each occurring with multiplicity one).*

∨ = indecomposable

Proof. Using the notation of section 3, the indecomposable preprojective R-modules are $P(j,s)$, $j \in I$, $s \geq 0$. In particular, $P(j,0)$ are the indecomposable projective R-modules and thus

$R_R = \bigoplus_{i \in I} P(i,0)$. For every R-module X_R,

$$X_R \approx \mathrm{Hom}({}_R R_R, X_R) = \mathrm{Hom}({}_R[\bigoplus_i P(i,0)], X_R) =$$

$$= [\mathrm{Hom}(\bigoplus_i P(i,0)_R, X_R)]_R = [\bigoplus_i \mathrm{Hom}(P(i,0)_R, X_R)]_R .$$

Hence,

$$P(j,s) = [\bigoplus_i \mathrm{Hom}(P(i,0), P(j,s))]_R$$

and thus under the identification of $P(j,s)$ with (j,s) and $\mathrm{Hom}(P(i,0), P(j,s))$ with the maps in $\Pi(\mathfrak{M},\Omega)$, we get the statement.

Now, define the map $\Delta : T(\mathfrak{M}) \to T(\mathfrak{M},\Omega)$ as follows. First, the morphisms in $F(\mathfrak{M},\Omega)$ can be described in the following way: For an (unoriented path) $w = i_{n+1} - i_n - \ldots - i_2 - i_1$ of $\mathfrak{M}$, call the number of arrows $i_{t+1} \leftarrow i_t$, $1 \leq t \leq n$, in Ω the layer $\lambda(w)$ of w. Then, the morphisms in $F(\mathfrak{M},\Omega)$ are the elements of the tensor products

$$ {}^{r_n}_{i_{n+1}}M_{i_n} \otimes \ldots \otimes {}^{r_2}_{i_3}M_{i_2} \otimes {}^{r_1}_{i_2}M_{i_1} ,$$

where $r_t = 2\lambda(i_t - i_{t-1} - \ldots - i_2 - i_1) + \begin{cases} 0 & \text{if } i_{t+1} \to i_t \\ 1 & \text{if } i_{t+1} \leftarrow i_t \end{cases}$.

Now, the map Δ is defined by

$$ {}_{i_{n+1}}M_{i_n} \otimes \ldots \otimes {}_{i_3}M_{i_2} \otimes {}_{i_2}M_{i_1} \xrightarrow{{}^{r_n}\eta \otimes \ldots \otimes {}^{r_2}\eta \otimes {}^{r_1}\eta} {}^{r_n}_{i_{n+1}}M_{i_n} \otimes \ldots \otimes {}^{r_2}_{i_3}M_{i_2} \otimes {}^{r_1}_{i_2}M_{i_1} ,$$

where ${}^{r}\eta$ are the maps of Lemma 1.2 for $M = {}_iM_j$ and $N = {}_jM_i$.

From the definition of $\Phi(\mathfrak{M},\Omega)$, it is clear that $\Phi(\mathfrak{M},\Omega)$ is just the image of $T(\mathfrak{M})$ under Δ. Also, Δ is obviously $R(\mathfrak{M},\Omega)$-linear.

LEMMA 5.2. $\Delta(\langle c \rangle) = J \cap \Phi(\mathfrak{M},\Omega)$.

Proof. By definition, $c = \sum_j (\sum_i c_j^i) = \sum_j c(j)$; note that $c(j) = e_j c e_j$, where e_j is the idempotent of $T(\mathfrak{M})$ corresponding to the identity of F_j ; thus <c> is the ideal generated by all c(j)'s. Hence, the statement follows from Lemma 1.2 taking into account that, by definition,

$$\Delta(1 \otimes 1 \otimes \ldots \otimes c(j) \otimes \ldots \otimes 1) = 1 \otimes 1 \otimes \ldots \otimes c({}^{r}M) \otimes \ldots \otimes 1 .$$

Now, from Lemma 5.2, it follows that Δ defines an isomorphism of $\Pi(\mathfrak{M}) = T(\mathfrak{M})/\langle c\rangle$ onto $\Pi(\mathfrak{M},\Omega) = \Phi(\mathfrak{M},\Omega)/J \cap \Phi(\mathfrak{M},\Omega)$. This completes the proof of the theorem.

The corollaries follow from the results in [2].

REFERENCES

[1] Auslander, M., Platzeck, M.I. and Reiten, I.: Coxeter functors without diagrams. Trans. Amer. Math. Soc. 250 (1979), 1-46.

[2] Auslander, M. and Reiten, I.: Representation theory of artin algebras III. Comm. Algebra 3 (1975), 239-294; V. Comm. Algebra 5 (1977), 519-554.

[3] Dlab, V. and Ringel, C.M.: Indecomposable representations of graphs and algebras. Memoirs Amer. Math. Soc. No.173 (Providence, 1976).

[4] Gelfand, I.M. and Ponomarev, V.A.: Model algebras and representations of graphs. Funkc. anal. i prilož. 13 (1979), 1-12.

[5] Ringel, C.M.: Report on the Brauer-Thrall conjectures: Rojter's theorem and the theorem of Nazarova and Rojter. These Lecture Notes.

[6] Rojter, A.V.: Gelfand-Ponomarev algebra of a quiver. Abstract, 2nd ICRA (Ottawa, 1979).

[7] Riedtmann, Ch.: Algebren, Darstellungsköcher, Überlagerungen und zurück. Comment. Math. Helv., to appear.

Department of Mathematics
Carleton University
Ottawa, Ontario K1S 5B6
Canada

Fakultät für Mathematik
Universität
D-4800 Bielefeld
West Germany

HEREDITARY ARTINIAN RINGS OF FINITE REPRESENTATION TYPE

P.Dowbor, C.M.Ringel, D.Simson

Recall [11,6] that a hereditary finite dimensional algebra is of finite type if and only if a corresponding diagram is a disjoint union of the Dynkin diagrams A_n, B_n, C_n, D_n, E_6, E_7, E_8, F_4, G_2 occouring in Lie theory. Here, we will consider the general case of a hereditary artinian ring A and associate to it a diagram $\Gamma(A)$. It turns out that A is of finite type if and only if $\Gamma(A)$ is the disjoint union of the Coxeter diagrams A_n, B_n ($= C_n$), D_n, E_6, E_7, E_8, F_4, G_2, H_3, H_4, $I_2(p)$ which classify the irreducible Coxeter groups [3] . However, the existence of rings of type H_3, H_4 and $I_2(p)$ with $p = 5$ or $p \geqslant 7$ remains open: it depends on rather difficult questions concerning division rings. On the other hand, we define for any Coxeter diagram "branch system" which generalize the root systems of the Dynkin diagrams. The dimension types of a hereditary artinian ring of finite representation type just form such a branch system.

The results of sections 1, 2 and 5 were obtained by P.Dowbor and D.Simson who announced part of them in the papers [9, 10] and at the Ottawa Conference 1979. The results were obtained independently by C.M.Ringel who announced them at the Oberwolfach meeting on division rings in 1978.

1. Bimodules of finite representation type

Let F and G be division rings, and ${}_FM_G$ a bimodule. Denote by $\mathcal{L}({}_FM_G)$ the category of finite dimensional representations of ${}_FM_G$, a representation being of the form $V = (X_F, Y_G, \varphi: X_F \otimes {}_FM_G \longrightarrow Y_G)$ with dimension type $\underline{\dim} V = (\dim X_F, \dim Y_G)$. Given ${}_FM_G$, let $M^R = \mathrm{Hom}_G({}_FM_G, {}_GG_G)$, $M^L = \mathrm{Hom}_F({}_FM_G, {}_FF_F)$, and $M^{R(i+1)} = (M^{Ri})^R$, $M^{L(i+1)} = (M^{Li})^L$, with $M^{Ro} = M = M^{Lo}$. Note that if $\dim M_G$ is finite,

$M^{RL} \cong M$, whereas if $\dim_F M$ is finite, then $M^{LR} \cong M$. We say that M is a bimodule with <u>finite dualisation</u> if all bimodules M^{Ri} and M^{Li} are finite dimensional on either side.

<u>Proposition 1. Assume $_F M_G$ is of finite representation type, with indecomposable representations $P_1, \ldots, P_m$. Let $\dim P_i = (x_i, y_i)$, and assume we have choosen an orderig with $x_i/y_i \leq x_{i+1}/y_{i+1}$. Then this ordering is unique, and there are Auslander-Reiten sequences</u>

$$0 \longrightarrow P_{i-1} \longrightarrow M^{L(i-1)} \otimes P_i \longrightarrow P_{i+1} \longrightarrow 0$$

<u>The bimodule M has finite dualisation; M^{Li} is one-dimensional as a right vector space, for some i, and the bimodules M and M^{Lm} are (semilinear) isomorphic.</u>

Let outline the main steps of the proof:

If $\dim M_G$ is finite, we define a functor $C_1^+ : \mathcal{L}(_F M_G) \longrightarrow \mathcal{L}(M^L)$ as usual [2,7]: given $(X_F \otimes _F M_G \xrightarrow{\varphi} Y_G)$, let $A_G = \ker \varphi$. Using $X_F \otimes _F M_G \cong \mathrm{Hom}(M^L, X_F)$, we get from the kernel map $A_G \longrightarrow \mathrm{Hom}(M^L, X_F)$ as adjoint a map of the form $A_G \otimes M^L \longrightarrow X_F$. Clearly, under C_1^+, the full subcategory $\mathcal{L}_1(_F M_G)$ of $\mathcal{L}(_F M_G)$ of objects without simple projective direct summands is equivalent to the full subcategory $\mathcal{L}_2(M^L)$ of $\mathcal{L}(M^L)$ of objects without simple injective direct summands. In particular, $\mathcal{L}(M^L)$ and $\mathcal{L}(M)$ have the same representation type.

Similary, if $\dim(M^R)_F$ is finite, define $C_1^- : \mathcal{L}(_F M_G) \longrightarrow \mathcal{L}(M^R)$ mapping $(X_F \otimes _F M_G \xrightarrow{\varphi} Y_G)$ onto the cokernel of the adjoint map $X_F \longrightarrow \mathrm{Hom}_G(_F M_G, Y_G) \cong Y_G \otimes M^R$, and thus establishing an equivalance between $\mathcal{L}_2(M)$ and $\mathcal{L}_1(M^R)$.

If $\dim_F M$ or $\dim M_G$ is infinite, then clearly $\mathcal{L}(M)$ cannot be of bounded representation type. Thus, the functors C_i^+ and C_i^- show that M has to be a bimodule with finite dualisation if $\mathcal{L}(M)$ is of finite representation type.

The functors C_1^+, C_1^- have nice properties: If (X_F, Y_G, φ) is indecomposable in $\mathcal{L}(M)$, then its image in $\mathcal{L}(M^R)$ is either zero, in which case $(X_F, Y_G, \varphi)=(F_F, 0, o)$ is simple injective, or else $\underline{\dim}\ C_1^-(X, Y, \varphi) = (y, yb-x)$, where $\underline{\dim}\ (X,Y,\varphi) = (x, y)$ and $b = \dim(M^R)_F$. In particular, there is a unique representation in $\mathcal{L}(M)$ of type $(x,y) \neq (1,0)$ if and only if there is a unique one in $\mathcal{L}(M^R)$ of type $(y, yb-x) \neq (0,1)$.

Let $C_i^+: \mathcal{L}(M) \longrightarrow \mathcal{L}(M^{Li})$, $C_i^-: \mathcal{L}(M) \longrightarrow \mathcal{L}(M^{Ri})$ be the iterated functors Call (X, Y, φ) in $\mathcal{L}(M)$ <u>preprojective</u> provided $C_i^+(X,Y,\varphi) = 0$ for some i, and <u>preinjective</u> provided $C_i^-(X,Y,\varphi) = 0$ for some i. Obviously, these modules are characterized by their dimension types. Let P_i be the indecomposable module (if it exists) with $C_i^+ P_i = 0$ and $C_{i-1}^+ P_i \neq 0$. Being determined by their dimension types, P_i, P_j can be ismorphic only for $i=j$. Let P_m be the last such module which exists (assuming $\mathcal{L}(M)$ of finite representation type). Then all P_i with $1 \leqslant i \leqslant m$ exist, and P_{m-1}, P_m have to be injective. Since the set of preprojective modules is closed under indecomposable submodules of direct sums, and contains the indecomposable injective modules, it contains all indecomposable modules. Similary, all indecomposable modules are also preinjective. Next, one proves that $\mathrm{Hom}(P_1,P_2) \cong {}_FM_G$, and therefore $\mathrm{Hom}(P_i,P_{i+1}) \cong M^{L(i-1)}$. Also, there is an obvious exact secuence

$$0 \longrightarrow P_{i-1} \longrightarrow \mathrm{Hom}(P_{i-1},P_i)^L \otimes P_i \longrightarrow P_{i+1} \longrightarrow 0 ,$$

and it is left almost split [1] (prove it by induction using the functors C_i^-).

Similary, let I_i be the indecomposable module with $C_i^- I_i = 0$ and $C_{i-1}^- I_i \neq 0$. Then $\mathrm{Hom}(I_2,I_1) \cong M^{R2}$. Since $P_{m-1} = I_2$, $P_m = I_1$, we conclude that $M^{L(m-2)} \cong M^{R2}$, thus $M^{Lm} \cong M$.

Finally, if the right dimension of all M^{Li} would be $\geqslant 2$, then one proves by induction on i, that P_i exists (for any $i \geq 1$) and that $\underline{\dim}\, P_i = (x_i, y_i)$ with $0 \leq x_i < y_i$. This produces a countable number of isomorphism classes of indecomposable modules.

2. Dimension sequences

A sequence $a = (a_1, \ldots, a_m)$ of length $|a| = m \geq 2$ with $a_i \in \mathbb{N}$ is called a <u>dimension sequence</u> provided there exists $x_i, y_i \in \mathbb{N}$ ($1 \leq i \leq m$), with

$$a_i x_i = x_{i-1} + x_{i+1}, \quad a_i y_i = y_{i-1} + y_{i+1} \quad (1 \leq i < m),$$

$x_o = -1$, $x_1 = 0$, $x_m = 1$, and $y_o = 0$, $y_1 = 1$, $y_m = 0$. The set of vectors $\binom{x_i}{y_i}$, with $1 \leq i \leq m$, is called the <u>branch system</u> defined by a. Note that it is uniquely determined by a, and conversly, it determines a. This generalizes the positive part of the usual rank 2 root systems: for the dimension sequences (0,0), (1,1,1), (1,2,1,2) and (1,3,1,3,1,3) the branch systems are just the positive roots of $A_1 \times A_1$, A_2, B_2 and G_2, respectively.

<u>Proposition 2. The branch system in $\mathbb{R}^2$ can be constructed as follows: $\{(1,0), (0,1)\}$ is a branch system. If $\mathcal{B}$ is a branch system, and p, q are neighbors in $\mathcal{B}$, then $\mathcal{B} \cup \{p+q\}$ is a branch system.</u>

Here, in a finite subset $\mathcal{B}$ of $\mathbb{R}^2$ consisting of pairwise linearly intependent elements, we call two elements neighbors in case the lines through these elements are neighbors in the set of all lines through elements of $\mathcal{B}$. G. Bergman has pointed out that the branch systems $\{(x_i, y_i) \mid 1 \leq i \leq m\}$ correspond just to the Farey sequences $\frac{x_i}{y_i}$ (see [16]). In particular, the numbers x_i, y_i always are without common divisor.

The proposition above can be reformulated as follows.

<u>Proposition 2'. The set $\mathcal{D}$ of dimension sequences can be obtained as follows:</u> $(0,0)\in\mathcal{D}$, <u>and if</u> $(a_1,\dots,a_m)\in\mathcal{D}$, <u>then all the sequences</u> $(a_1,\dots,a_{i-1}, a_i+1, 1, a_{i+1}+1, a_{i+2},\dots,a_m)$ <u>for</u> $1\leqslant i<m$ <u>belong to</u> $\mathcal{D}$.

Note that for any other dimension sequence a, we have $a_i = 1$ for some $i \geqslant 2$. An easy consequence is the following

<u>Corollary. The set $\mathcal{D}$ of dimension sequences is closed under cyclic permutations.</u>

Given a dimension sequence $a = (a_1,\dots,a_m)$, let $a^+ = (a_m,a_1,\dots,a_{m-1})$. Let us give the list of all dimension sequences of length $\leqslant 7$, up to the cyclic permutations and reversion:

$(0,0)$; $(1,1,1)$; $(1,2,1,2)$; $(1,2,2,1,3)$;
$(1,2,2,2,1,4)$, $(1,2,3,1,2,3)$, $(1,3,1,3,1,3)$;
$(1,2,2,2,2,1,5)$, $(1,2,2,3,1,2,4)$, $(1,2,3,2,1,3,3)$,
$(1,4,1,2,3,1,3)$.

3. Coxeter diagrams

Let $\Gamma = \{1,\dots,n\}$ and assume there is given a set map $d: \Gamma\times\Gamma \longrightarrow \mathcal{D}$. Note that d defines on Γ the structure of an oriented graph, if we draw an arrow $i \longrightarrow j$ in case $d(i,j) \neq (0,0)$. With (Γ,d) we also will consider the unoriented graph given by $(\Gamma,|d|)$, where two different points are connected by at most two edges, and any edge has assigned a number $\geqslant 3$. If i is a sink for (the orientation defined by) d, define $(\pi_i d)(i,j) = d(j,i)^+$, $(\pi_i d)(j,i) = (0,0)$, and $(\pi_i d)(j,k) = d(j,k)$ for all $j, k \neq i$. For any sink i, define on $\mathbb{R}^n$ a linear transformation $\sigma_i = \sigma(d)_i$ as follows: if $x = (x_j) \in \mathbb{R}^n$, let

$(\sigma_i x)_j = x_j$ for $j \neq i$, and $(\sigma_i x)_i = -x_i + \sum_j d(i,j)_1\, x_j$.

If $i_1,\dots,i_t$ is a $(+)$-admissible sequence (thus i_s is a sink for $\pi_{s-1}\cdots\pi_1 d$, for all $1 \leqslant s \leqslant t$), define

$$\sigma_{i_t \ldots i_1} = \sigma(\pi_{t-1} \ldots \pi_1 d)_{i_t}\, \sigma_{i_{t-1} \ldots i_1} .$$

We call $y \in \mathbb{N}^n$ <u>preprojective</u> provided there exists a (+)-admissible sequence $i_1, \ldots, i_t$ such that $\sigma_{i_t \ldots i_1}(x)$ is one of the canonical base vectors $(0, \ldots, 1, 0, \ldots, 0)$. We say that (Γ, d) is of <u>finite type</u> provided there are only finitely many preprojective vectors, and then we call the set of preprojective vectors the <u>branch system</u> of (Γ, d).

<u>Theorem 1.</u> (Γ, d) <u>is of finite type if and only if</u> $(\Gamma, |d|)$ <u>is the disjoint union of Coxeter diagrams</u> A_n, $B_n (=C_n)$, D_n, E_6, E_7, E_8, F_4, G_2, H_3, H_4, $I_2(p)$ $(p=5$ or $p \geq 7)$. <u>If</u> $(\Gamma, |d|)$ <u>is one of the Coxeter diagrams of rank</u> n <u>and with Coxeter number</u> h, <u>then the branch system for</u> (Γ, d) <u>has precisely</u> $\frac{1}{2} nh$ <u>elements</u>.

Of course, the branch systems for the diagrams A_n, D_n, E_6, E_7, E_8, F_4, G_2 are precisely the positive roots of the corresponding root system; there are two possible branch systems for B_n, namely the positive roots of B_n or C_n. The branch systems of the type $I_2(p)$, H_3, and H_4 have p, 15, and 60 elements, respectively.

As an example, let us write down the branch system for $\bullet \longrightarrow \bullet \xrightarrow{5} \bullet$, with second dimension sequence $(2,1,3,1,2)$. It contains the following 15 vectors

$$\begin{array}{l}(1\,0\,0),\ (0\,1\,0),\ (0\,0\,1),\ (1\,1\,0),\ (0\,1\,1),\\ (1\,1\,1),\ (0\,1\,2),\ (0\,2\,1),\ (1\,1\,2),\ (1\,2\,1),\\ (1\,2\,2),\ (2\,2\,1),\ (1\,3\,2),\ (1\,3\,3),\ (1\,4\,2),\end{array}$$

and is dependent on the orientation and the given dimension sequence.

The proof of the theorem is rather technical. First, one constructs explicity for every pair (Γ, d) with $(\Gamma, |d|)$ a Coxeter diagram the corresponding branch system. For the converse, one only has to consider the pairs (Γ, d) for wich all proper subdiagrams of $(\Gamma, |d|)$ are Coxeter diagrams, and again an explicit calculation shows that in these cases there are infinitely many preprojective vectors.

4. Hereditary artinian rings

Given an artinian ring A, let A^o be its basic ring, $A^o/\mathrm{rad}A^o = \prod_{i=1}^{n} F_i$, and $\mathrm{rad}A^o/(\mathrm{rad}A^o)^2 = \bigoplus_{i,j} {}_iM_j$ as $\prod_{i=1}^{n} F_i$-bimodule. Then $\mathfrak{S} = (F_i, {}_iM_j)$ is called the species of A. Let $d(i,j)$ be the dimension sequence of the bimodule ${}_iM_j$, thus we have associated to A a pair (Γ, d) and we denote by $\Gamma(A)$ the pair $(\Gamma, |d|)$.

Theorem 2. The hereditary artinian ring A is of finite representation type if and only if $\Gamma(A)$ is disjoint union of Coxeter diagrams A_n, B_n $(=C_n)$, D_n, E_6, E_7, E_8, F_4, G_2, H_3, H_4, $I_2(p)$ ($p=5$ or $p \geq 7$). In this case, the dimension types of the indecomposable A-modules form the branch system for (Γ, d).

As in the case of an algebra [6], one sees that for a hereditary artinian ring A of finite representation type, the basic ring A^o always is the tensor ring over its species.

5. Problems on division rings

The main problem which arises and which we are not able to answer is the question about the possible dimension sequences a of bimodules ${}_FM_G$ of finite representation type. We may assume that $a_2 = 1$ (using a cyclic permutation of a, thus replacing M by some M^{Li}). Then there is given a division ring inclusion $G \hookrightarrow F$, and ${}_FM_G = {}_FF_G$. Since $M^L = {}_GF_F$, the dimension sequence of ${}_FF_G$ starts as follows:

$$a_1 = \dim F_G \ , \ a_2 = 1 \ , \ a_3 = \dim {}_GF \ .$$

Thus, immediatly, we are confronted with the question of differrent left or right index of a division subring G of F, (see [4]).

Of particular interest is the question whether there exists a bimodule with the dimension sequence of type $I_2(5)$, say $(2,1,3,1,2)$, since it would give rise to rings of type H_3 and H_4. In fact, it is easy to see that the only dimension sequence starting with $(2,1,3,1,\ldots)$ is $(2,1,3,1,2)$. Thus, we would need $G \subseteq F$ with $\dim F_G = 2$, $\dim {}_G F = 3$, and $\dim_F \mathrm{Hom}({}_G F_F, {}_G G) = a_4 = 1$. Let us point out certain consequences of these conditions. Since the length of $(2,1,3,1,2)$ is 5, it follows from $M^{L5} \cong M$ that F, G have to be isomorphic. Also F contains a division subring H again isomorphic to F with $\dim F_H = 3$, $\dim_H F = 2$, since $({}_F F_G)^{L3}$ has the dimension sequence $(3,1,2,2,1)$.

Bimodules with suitable dimension sequences would produce interesting examples of artinian rings. For example, we may ask whether there exists a bimodule ${}_F M_G$ of finite representation type with $\dim_F M = 2 = \dim M_G$, thus dimension sequence $(2,2,a_3,\ldots,a_m)$. Of course, according to [13] such a bimodule has to be simple as a bimodule (this also follows from proposition 1, since dualisation leads to a bimodule which is one-dimensional on one side). Note that in case of odd m, the division rings F, G have to be isomorphic, so that we also can form the trivial ring extension $R = F \ltimes M$, and this then would be a local ring with $(\mathrm{rad} R)^2 = 0$, left length 3, right length 3, and with precisely $m-1$ indecomposable modules. In particular, the dimension sequences of type $I_2(5)$ would give a local ring R with $(\mathrm{rad} R)^2 = 0$, left length 3, right length 3, and precisely 4 indecomposable modules.

Similary, a bimodule with dimension sequence $(a_1,\ldots,a_{u+v+1})$ where $a_1 = u$, $a_2 = 1$, $a_{u+1} = 3$, $a_{u+v} = 1$, $a_{u+v+1} = v$, and the remaining $a_i = 2$ would be of local-colocal representation type (any indecomposable modules has a unique minimal submodule or unique maximal submodule). In contrast, the finite dimensional algebras with this property all have been classified in [15].

This shows very clearly the dependence of the representation theory of artinian rings on questions concerning division rings, a fact which to have been exhibited for the first time in [14].

In particular, under the assumption that there are no pairs of division rings $F \supsetneq G$ with $\dim_G F = 2$, and $\dim F_G = 2$, but finite, the Coxeter diagrams H_3, H_4, and $I_2(p)$ cannot be realized as $\Gamma(A)$ for any artinian ring A, so they could be excluded from theorem 2.

Of course, in case we assume that the division rings F are finitely generated over their centers, then we exclude immediately the cases H_3, H_4, and $I_2(p)$, thus one has the following (see also [8]):

<u>Theorem 3.</u> <u>The hereditary artinian ring A with $A/\text{rad}A$ finitely generated over its center is of finite representation type if and only if $\Gamma(A)$ is the disjoint union of Coxeter diagrams A_n, B_n $(=C_n)$, D_n, E_6, E_7, E_8, F_4, G_2.</u>

<u>References</u>

[1] Auslander, M., Reiten, I.: Representation theory of artin algebras III: Almost split sequences. Comm. Algebra 3 (1975), 239-294.

[2] Berstein, I. N., Gelfand, I. M. Ponomarev, V. A.: Coxeter functors and Gabriel's theorem. Uspiechi Mat. Nauk 28, 19-33 (1973), translation: Russian Math. Surveys 28 (1973), 17-32.

[3] Bourbaki, N.: Groupes et algèbres de Lie, Ch. 4,5,6. Paris: Hermann 1968.

[4] Cohn, P. M.: Skew field constructions. London Math. Soc. Lecture Note Series 27, Cambridge 1977.

[5] Dlab, V., Ringel, C. M.: Decomposition of modules over right uniserial rings. Math. Z. 129 (1972), 207-230.

[6] -,- : On algebras of finite representation type. J. Algebra 33 (1975), 306-394.

[7] -,- : Indecomposable representations of graphs and algebras. Memoirs Amer. Math. Soc. 173 (1976).

[8] Dowbor, P., Simson, D,: Quasi-artin species and rings of finite representation type. J. Algebra, to appear.

[9] -,- : A characterization of hereditary rings of finite representation type. to appear.

[10] -,- : On bimodules of finite representation type. preprint.

[11] Gabriel, P.: Unzerlegbare Darstellungen I. Manuscripta Math. 6 (1972), 71-103.

[12] - : Indecomposable representations II. Symposia Math. Ist. Naz. Alta Mat. Vol. XI (1973), 81-104.

[13] Ringel, C. M.: Representations of K-species and bimodules. J. Algebra 41 (1976), 296-302.

[14] Rosenberg, A., Zelinsky, D.: On the finiteness of the injective hull. Math. Z. 70 (1959), 372-380.

[15] Tachikawa, H.: On algebras of which every indecomposable representation has an irreducible one as the top or the bottom Loewy constituent. Math. Z. 75 (1961), 215-227.

[16] Vinogradov, I. M.: Elements of number theory. Dover Publications (1954).

P. Dowbor, D. Simson
Institute of Mathematics,
Nicholas Copernicus University,
87-100 Toruń

C. M. Ringel
Fakultät für Mathematik,
Universität,
D-4800 Bielefeld

TAME AND WILD MATRIX PROBLEMS

Ju.A. Drozd

In [13] Nazarova and Roiter proving the famous Brauer-Thrall conjecture showed that if Λ is a finite-dimensional algebra over an algebraically closed field then either Λ is of finite type, i.e. has only a finite number of non-isomorphic indecomposable representations, or the classification of its representations includes the problem on the canonical form of matrices with respect to conjugacy. In the last case Λ is of strictly unbounded type, i.e. there is an infinite number of dimensions each possessing infinitely many non-isomorphic indecomposable representations. Numerous examples (see [2-5,10-12] etc.) show that algebras of infinite type split in turn into two disjoint classes: "tame" algebras whose indecomposable representations may be parametrized by several discrete and one continuos parameters and "wild" algebras for which the classification of representations includes the classical unsolved problem on the canonical form of pairs of matrices with respect to conjugacy. The last problem is apparently of extreme difficulty; at all events, as is well known, it includes the classification of representations of any algebra.

Freislich and Donovan [7] proposed an explicit definition of the terms "tame" and "wild" and conjectured that any algebra of infinite type is either tame or wild. We prove this conjecture for algebras over an algebraically closed field and in a weakened form for algebras over a perfect field. Just as in [13] the natural scope for the proof is a rather wide class of "matrix problems" containing in particular the problems appearing in the classification of representations of algebras (a method for reducing the classification of representations to such problems is propesed in §4). A suitable language fór their formulation is that of differential graded categories (DGC) proposed by Roiter and Kleiner in [14,9].

The proof method is an usual inductive method of solving matrix problems: we reduce a matrix A, obtain a new problem but with matrices of lower dimension and then the proof can be easily accomplished by induction. If the matrix A may be reduced by elementary transformations, we use the reduction algorithm constructed by Roiter and Kleiner in [14,9]. But if A may be reduced only by conjugacy, one cannot in general construct an analogous algorithm since there exists infinitely many non-conjugate matrices of fixed dimension. However, for non-wild problems it turns out to exist such polynomial f(x) that in any exact indecomposable representation the matrix f(A) has to be nilpotent. It makes possible to construct some reduction algorithm for such problems (§2) and thus to obtain a foundation for an inductive proof which is given in §3.

1. Tame and wild DGC

We shall use terminology and results of the paper [9] (or [14]) All categories and algebras considered are supposed to be categories and algebras over a fixed field K. For a graded category U the symbol $U_m(X,Y)$ will denote the set of morphisms of degree m from X to Y. All graded categories are supposed to be semi-free, i.e. such that $U_m=U_1\otimes_{U_0}\dots\otimes_{U_0}U_1$ (m times). For a differential graded category (DGC) U R(U,C) denotes the category of representations of DGC U in a (non-graded) category C. If C is the category of free modules of finite rank over an algebra Λ, we shall write R(U,Λ) instead of R(U,C) and the representations of U in C will be called the representations of U over Λ . We define the <u>dimension</u> of a representation $F \in R(U,\Lambda)$ as the function dim F defined on the set of objects of U and putting in correspondence to an object X the rank of the free module F(X). The set of representations of U over Λ of dimension d is denoted by $R_d(U,\Lambda)$.

If F is a representation of U over Λ and M is a representation of Λ over an algebra Γ , one can define the representation $F\otimes_\Lambda M \in R(U,\Gamma)$ setting $(F\otimes_\Lambda M)(X)=F(X)\otimes_\Lambda M$. Then $\dim(F\otimes_\Lambda M)=\dim F.\dim M$ where $\dim M=\mathrm{rank}_\Gamma M$. Thus fixing M we obtain a functor $R(U,\Lambda)\dashrightarrow R(U,\Gamma)$ and fixing F we obtain a functor $R(\Lambda,\Gamma)\dashrightarrow R(U,\Gamma)$.

Let now $\hat{U}$ be the completion of U (cf. [9,14]), $\tilde{U}$ be the category of matrices over $\hat{U}$ (the smallest additive category containing $\hat{U}$). Any differential graded functor (DGF) $T:\hat{U}\dashrightarrow\tilde{U}'$ induces the functor $T^*_C:R(U',C)\dashrightarrow R(U,C)$ for any additive category C in the following way. A representation $F:U'_0\dashrightarrow C$ can be uniquely prolonged to the representation $\tilde{F}:\tilde{U}'_0\dashrightarrow C$ and one can set $T^*_C F(X)=\tilde{F}(TX)$; similarly T^*_C may be defined on morphisms of representations. If C is the category of free Λ-modules of finite rank, we shall write T^*_Λ instead of T^*_C; the functor T^*_K will be denoted simply T^*. If $M\in R(\Lambda,\Gamma)$ then it is easy to see that $T^*_\Gamma(F\otimes_\Lambda M)=T^*_\Lambda F\otimes_\Lambda M$.

Remind the definition of tame and wild DGC given in [6] . A representation $F\in R(U,\Lambda)$ will be called <u>strict</u> if $F\otimes_\Lambda M\cong F\otimes_\Lambda N$ implies $M\cong N$ for any algebra Γ and any $M,N\in R(\Lambda,\Gamma)$. A DGC U is said to be <u>wild</u> if it possess a strict representation over the free algebra $\Sigma = K\{x,y\}$ with two generators. Freely said, U has a "family of representations depending on 2 matrix parameters" and two representations of this family are isomorphic if and only if the corresponding pairs of matrices are conjugate. Remark that in [6] a weaker condition was demanded: only the field K itself was taken as the "test" algebra Γ, but it follows from the subsequent that, at least for free DGC and for finite-dimensional algebras, these conditions are equivalent.

A <u>rational</u> algebra over the field K is a finitely generated algebra Γ such that $K[x] \subset \Gamma \subset K(x)$ or, the same, Spec Γ is a smooth rational curve over K. Denote K_s a separable closure of K. A DGC U is said to be <u>tame</u> if for any dimension d there exists a finite family of representations $\{F_\alpha\} \subset R_d(U,\Gamma)$ where Γ is a rational algebra over K_s such that any indecomposable representation $G \in R_d(U,K_s)$ is isomorphic to $F_\alpha \otimes_\Gamma M$ for some $M \in R_1(\Gamma, K_s)$. The set $\{F_\alpha\}$ is called a <u>parametrizing family</u> of representations of DGC U of dimension d. Remark that a DGC of finite type is automatically tame. Moreover, this definition will obviously not change if one admits each F_α to be a representation over its own rational algebra Γ_α.

From now on U will be expected to be finitely generated, i.e. such that its set of objects is finite: $\text{ObU} = \{X_1, \dots, X_n\}$, and U is generated by a finite set S of morphisms (in view of semi-freedom one can expect that $S = S_0 \cup S_1$ where $S_0 \subset U_0$ and $S_1 \subset U_1$). In this case a dimension d of representations of U may be identified with a vector $(x_1, \dots, x_n)$ where $x_i = d(X_i)$. Denote $|d| = x_1 + \dots + x_n$. If we write representations of U of dimension d in a matrix form, we can consider $R_d = R_d(U,K_s)$ as an algebraic variety over K_s. If U is tame, it follows from the definition that there is subvariety in R_d of a dimension $m \leq |d|$ which intersects all isomorphism classes. On the other hand, if $F \in R_{d_0}(U,\Sigma)$ is a strict representation, the representations of the form $F\otimes_{\Sigma} M$ where $M \in R_k(\Sigma, K_s)$ form a subvariety in R_{kd_0} of the dimension $2k^2$ whose intersection with any isomorphism class has a dimension not greater than k^2-1. If we suppose U to be both tame and wild then, for $k > |d_0|$, we easily obtain a contradiction [6] and arrive to the following result.

Proposition 1. No finitely generated DGC can be both tame and wild.

Pass to the formulation of the main theorem. Remind [9,14] that a free DGC U is said to be triangular provided some system S of its free generators can be numbered in such a way $S=\{a_1,\dots,a_s\}$ that Da_i is contained in the subcategory generated by $\{a_1,\dots,a_{i-1}\}$ (in particular $Da_1=0$). The set S will be called a free triangular system of generators. For technical purposes we need some broader class of DGC than free triangular ones. Namely, let U be a DGC, $a \in U_0(X,X)$ such morphism that $Da=0$. Then one can consider the category of fractions $U[a^{-1}]$ (cf. [8]) and on the other hand a-adique completion of the category U denoted by $U[[a]]$. The differential D may be extended to both these categories so that the natural functors $U \dashrightarrow U[a^{-1}]$ and $U \dashrightarrow U[[a]]$ become DGF and hence induce the functors $R(U[a^{-1}],C) \dashrightarrow R(U,C)$ and $R(U[[a]],C) \dashrightarrow R(U,C)$. The first of them establishes an equivalence between $R(U[a^{-1}],C)$ and the full subcategory of $R(U,C)$ consisting of such representations F that $F(a)$ is invertible. If, moreover, all spaces of morphisms in the category C are finite-dimensional, the second one establishes an equivalence between $R(U[[a]],C)$ and the full subcategory of $R(U,C)$ consisting of such representations F that $F(a)$ is nilpotent. Let now S be a free triangular system of generators of a DGC U. An element $a \in S_0$ will be called minimal if $Da=0$. In addition, if $a:X \dashrightarrow X$, it is said to be a minimal loop and if $a:X \dashrightarrow Y$ ($X \neq Y$), it is said to be a minimal edge. If $\{a_1,\dots,a_k\}$ is a set of minimal loops from S_0 and $\{f_1,\dots,f_k\}$ is a set of polynomials from $K[x]$, one may construct the DGC $U'=U[f_1(a_1)^{-1},\dots,f_l(a_l)^{-1}][[f_{l+1}(a_{l+1}),\dots,f_k(a_k)]]$. Such DGC will be called almost free and the image of S under the natural inclusion $U \dashrightarrow U'$ will be called an almost free system of generators of U' (remark that an almost free system of generators is necessarily triangular). The images in U' of the minimal loops $a_1,\dots,a_k$ will be caleed marked loops.

Now the main theorem may be formulated as follows.

THEOREM 1. Any almost free DGC over an algebraically closed field is either tame or wild.

Going over to non-closed fields, we obtain some weakened result. Namely, let F be a representation of U over an algebra Λ. Call F semi-strict if for any representation $M \in R(\Lambda,K_s)$ there exists only a finite number (up to isomorphism) of such representations $N \in R(\Lambda,K_s)$ that $F\otimes_\Lambda M \cong F\otimes_\Lambda N$. Call a DGC U semi-wild provided it possesses a semi-strict representatin over Σ. Simple arguments on field extensions give the following result.

Proposition 2. Let L be an algebraic extension of the field K, U a DGC over K, $U_L=U\otimes_K L$. Then if U is semi-wild, such is also U_L and if U_L is tame, such is also U. If, moreover, L is separable, U and U_L are simultaneously tame or semi-wild.

Corollary 1. Any almost free DGC over a perfect field is either tame or semi-wild.

Remark that the author knows no examples of semi-wild but not wild DGC. It seems very probable that any semi-wild DGC over arbitrary field is really wild.

Theorem 1 implies another important result concerning the Tits form [9,14]. Remind that the Tits form of a free DGC U is the quadratic form $Q=Q_U$ whose value on a vector $d=(x_1,\dots,x_n)$ is

$$Q(d)=\sum_{i=1}^{n} x_i^2 + \sum_{i,j} t_{ij}x_ix_j-\sum_{i,j} s_{ij}x_ix_j$$

where s_{ij} is the number of free generators contained in $U_0(X_i,X_j)$ and t_{ij} is the number of free generators contained in $U_1(X_i,X_j)$. It is easy to check [6] that for a tame DGC its Tits form is weakly non-negative, i.e. $Q(d)\geqslant 0$ for any vector with non-negative coordinates.

Corollary 2. Let U be a free triangular DGC over an algebraically closed (perfect) field. If its Tits form is not weakly non-negative then U is wild (semi-wild).

For a lot of important examples of DGC (cf..[3,4,10]) but unfortunately not always the converse is true: if Q_U is non-negative then U is tame. It would be very important and interesting to distinguish a rather wide class of DGC for which it is always the case by virtue of some "natural" causes (an analogue of Shurian DGC of finite type [9,14]).

2. Reduction of minimal loops

Theorem 1 may be, of cuorse, formulated in the following way:

(*) If an almost free DGC U (over an algebraically closed field K)is not wild then for any dimension d there exists a parametrizing family of representations of U of this dimension.

It is natural to attempt to prove this statement inductively by the dimension as it was done in [9,14] for the I Brauer-Thrall conjecture. More precise, consider for an almost free DGC U the quadratic form $Q^0=Q^0_U$ (negative part of the Tits form) whose value on $d=(x_1,\ldots,x_n)$ is $Q^0(d)=\sum_{i,j}s_{ij}x_ix_j$ where s_{ij} is the number of almost free generators in $U_0(X_i,X_j)$. In other words, if d is a dimension of a representation then $Q^0(d)$ is the "number of places" in matrices defining this representation. Our aim is to "reduce" one matrix after what we should obtain a new matrix problem, i.e. a new DGC U' and a new dimension d' of representations such that $Q^0_{U'}(d')<Q^0_U(d)$. For such reduction we have to choose $a\in S_0$ (S being an almost free system of generators of U) for which Da already contains no generators of degree 0. Now there are three possible cases:

1) a is a minimal loop, i.e. Da=0 and $a:X\dashrightarrow X$;

2) a is a minimal edge, i.e. Da=0 and $a:X\dashrightarrow Y$ $(X\neq Y)$;

3) $Da\neq\emptyset$.

In the last case $Da=\sum\alpha_i\varphi_i$ where $\varphi_i\in S_1$, $\alpha_i\in K$ and one can assume, rechoosing the generators from S_1, that $Da\in S_1$. Such generator will be called <u>irregular</u>.

A reduction algorithm for minimal edges and irregular generators was construted in [9] (§8) or [14] (§3). We shall formulate the appropriate result in a convenient form.

<u>Proposition</u> 3. Let U be an almost free DGC, S its almost free system of generators, $a\in S\cap U(X,Y)$ be either a minimal edge or an irregular generator. Then there exists a DGC U' and a DGF $T:\hat{U}\dashrightarrow\tilde{U}'$ having the follwing properties:

(1) for any category C the functor T^*_C is strict and full;

(2) $T^*:R(U',K)\dashrightarrow R(U,K)$ is an equivalence of categories;

(3) if $T^*F(X)\neq 0$ and $T^*F(Y)\neq 0$ where $F\in R(U',K)$ then $Q^0_{U'}(\dim F)<$ $<Q^0_U(\dim T^*F)$;

(4) if either a is irregular or there are no marked loops in $U(X,X) \cup U(Y,Y)$ then U' is almost free too.

This proposition is sufficient to proove the I Brauer-Thrall conjecture since a DGC of finite type cannot possess any minimal loop. But tame DGC may still possess minimal loops so we have yet to consider this case too. The reduction of minimal loops is complicated by the fact that there are infinitely many non-conjugate matrices of fixed dimension. That is why we cannot obtain a complete analogue of proposition 3. Nevertheless we shall construct some algorithm which turns out to be sufficient for problems which are not a priori wild.

The first part of this algorithm is based on the following observation. If A is a linear operator in a finite-dimensional space V then V splits into a direct sum of invariant subspaces $V=V_0 \oplus V_1$ so that the restriction of A on V_0 is nilpotent and its restriction on V_1 is non-singular. Moreover, if B is an operator in V commuting with A then V_0 and V_1 are invariant under B. It is not difficult to deduce from here the following result.

<u>Proposition</u> 4. Let U be an almost free DGC, $a \in U(X,X)$ a minimal loop and f(x) a non-zero polynomial. Then there exist a DGC U' and a DGF $T:\hat{U} \dashrightarrow \tilde{U}'$ having the following properties:

(1) for any category C the functor T_C^* is strict and full;

(2) $T^*:R(U',K) \dashrightarrow R(U,K)$ is an equivalence of categories;

(3) $Q^0_{U'}(\dim F) \leq Q^0_U(\dim T^*F)$ and if the operator $T^*F(f(a))$ is neither non-singular nor nilpotent then this inequality is strict;

(4) if U(X,X) possesses no marked loops except maybe a then U' is almost free too.

The second part of the algorithm is based on the fact that if A is such operator in a space V that $A^m=0$ then V splits into a direct sum of invariant subspaces $V=V_1\oplus\ldots\oplus V_m$ such that if A_i is the restriction of A on V_i then the matrix of A_i in some basis has the form

$$\begin{bmatrix} 0 & E & 0 & \ldots & 0 \\ 0 & 0 & E & \ldots & 0 \\ \ldots & \ldots & \ldots & \ldots & \ldots \\ 0 & 0 & 0 & \ldots & E \\ 0 & 0 & 0 & \ldots & 0 \end{bmatrix}$$

(i horisontal and vertical bands). By formalizing the reduction of A to this form we obtain the following statement.

<u>Proposition</u> 5. Let U be an almost free DGC, $a\in U(X,X)$ a minimal loop and m a positive integer. Then there exist a DGC U' and a DGF $T:\hat{U}\to\widetilde{U}'$ having the following properties:

(1) for any category C the functor T_C^* is strict and full;

(2) The image of the functor $T^*:R(U',K)\to R(U,K)$ consists of such representations G that $G(a)^m=0$;

(3) if $T^*F(X)\neq 0$, then $Q^0_{U'}(\dim F)<Q^0_U(\dim T^*F)$;

(4) if U(X,X) possesses no marked loops except maybe a then U' is almost free too.

The construction of the category U' and the functor T is in both cases analogous to that of [9] (§8) or [14] (§3). We obtain one more corollary from propositions 3-5.

<u>Corollary</u> 3. Let U' be the DGC constructed according to one of propositions 3-5. If U' is wild then U is also wild.

3. Proof of the main theorem

As was marked above, theorem 1 in the form (*) will be proved by means of induction by $Q^0_U(d)$ for all DGC U and dimensions d.

Surely, we can assume d to be a strict dimension, i.e. $d(X)\neq 0$ for $X\in \mathrm{Ob}U$. If $Q_U^0(d)=1$, there is only one element a of degree 0 in the almost free system of generators and also if $a:X \dashrightarrow Y$ then $d(X)=d(Y)=1$. Im this case the existence of a parametrizing family is evident. Hence, from now on we can assume the theorem valid for all U' and d' such that $Q_{U'}^0(d')<m$ and prove it for $Q_U^0(d)=m$.

First of all we shall mark some known examples of wild DGC. As naturally, it is convenient to represent the objects of a DGC by points and the elements of an almost free system of generators by arrows (entire for those of degree 0 and dash for those of degree 1).

Proposition 6. The following almost free DGC are wild:

W_1: 

; $Da=Db=0$;

W_2: 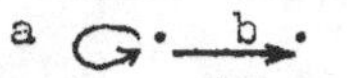

; $Da=Db=0$;

W_3: 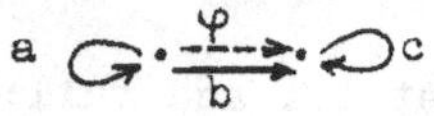

; $Da=Dc=D\varphi=0$;

W_4: 

; $Da=D\varphi=0$,

in addition, in cases W_3 and W_4 b is a regular generator, i.e. Db cannot be substituted for φ .

It is not difficult to construct explicitely a strict representation over Σ for all these DGC.

Now let U be a non-wild almost free DGC with an almost free system of generators S, d a such dimension of its representations that $Q_U^0(d)=m$. If U possess a minimal edge or an irregular generator then proposition 3 is applicable (example W_2 of propositicn 6 guarantees the conditios of item (4) of this proposition). Let U' and T be the DGC and DGF mentioned in proposition 3. Consider all dimensions d' of representations of U' such that $Q_{U'}^0(d')<m$ (there is,

of course, a finite number of them). By the inductive supposition, for each of them there exists a parametrizing family of representations of U' over some rational algebra Γ. Let $\{F_\alpha\}$ be the set of such representations of these families that $\dim T^*_\Gamma F_\alpha = d$. If $G \in$ $\in R_d(U,K)$ is an indecomposable representation then $G \cong T^*F$ for some indecomposable representation $F \in R_{d'}(U',K)$ and also $Q^0_{U'}(d') < m$, so $F \cong F_\alpha \otimes_\Gamma M$ for some F_α and $M \in R_1(\Gamma, K)$. But then $G \cong T^*F \cong T^*(F_\alpha \otimes_\Gamma M) \cong$ $T^*_\Gamma F_\alpha \otimes_\Gamma M$. Hence, $\{T^*_\Gamma F_\alpha\}$ is a parametrizing family of representations of U of dimension d.

Thus, we may henceforth assume U to have no minimal edge or irregular generator. Now let $a \in S$ be a minimal loop,, $f \in K[x]$ a non-zero polynomial. Then, using proposition 4 as above, we can construct such set of representations $\{G_\alpha\}$ of dimension d that any indecomposable representation $G \in R_d(U,K)$ for which $G(f(a))$ is neither invertible nor nilpotent is isomorphic to $G_\alpha \otimes_\Gamma M$ for some G_α and M. Therefore, it is sufficient to construct parametrizing families of representations of dimension d for the categories $U[f(a)^{-1}]$ and $U[[f(a)]]$. Let $f(x) = (x-\alpha)^k g(x)$ where $\alpha \in K$, $g(\alpha) \neq 0$. Applying proposition 4 to the category $U[[f(a)]]$ and the polynomial $x-\alpha$, we reduce analogously the second problem to the construction of parametrizing families for the categories $U[[a-\alpha]]$ and $U[[g(a)]]$. If we continue this procedure, we arrive to the necessity to construct parametrizing families for $U[f(a)^{-1}]$ and $U[[a-\alpha_i]]$ where α_i are the roots of $f(x)$. But we can replace in the category $U[[a-\alpha_i]]$ the generator a by $a-\alpha_i$ and also if $a \in$ $\in U(X,X)$ and $d(X)=1$ then $F(a-\alpha_i)^L = 0$ for any representation F of this category of dimension d. Hence, we can apply proposition 5 and using the inductive assumption construct a parametrizing family of representations of dimension d for each DGC $U[[a-\alpha_i]]$.

Therefore, it remains to construct a parametrizing family of representations of dimension d for $U[f(a)^{-1}]$ where a is any minimal loop and f(x) a non-zero polynomial. Remark that in view of example W_1 of proposition 6 there exists at most one minimal loop in U(X,X) for each object X. If S_0 consists only of minimal loops and there are several of them, there are no strict representations of dimension d, and if there is a unique minimal loop, the parametrizing family is given by the Jordan normal form. Suppose that there are generators in S_0 besides minimal loops. Then in view of triangularity such $b \in S_0$ can be found that Db includes only minimal loops. If $b: X \dashrightarrow Y$, there must be a minimal loop either in U(X,X) or in U(Y,Y). For definiteness, let it be $a \in U(X,X)$. Futher two cases are possible.

Case 1. There are no minimal loops in U(Y,Y). Then Db= $=\sum_{i=1}^{t} p_i(a)\varphi_i$ where $\varphi_i \in S_1$ and $p_i(x)$ are some polynomials. Set $f(x)=p_1(x)\dots p_t(x)$. In DGC $U[f(a)^{-1}]$ all $p_i(a)$ are invertible so all φ_i may be replaced in the system of generators by $p_i(a)\varphi_i$. But then the generator b is irregular in $U[f(a)^{-1}]$ and we arrive to the variant for which the theorem has already been proved.

Case 2. U(Y,Y) possesses a minimal loop c (maybe X=Y and a=c). Then $Db=\sum_{i,j} p_{i,j}(a)\varphi_i q_{ij}(c)$ for some $\varphi_i \in S_1$ and some polynomials p_{ij}, q_{ij} and also $\varphi_i \in U_1(X,Y)$. We shall consider $U_1(X,Y)$ as a module over the polynomial ring $K[x,y]$ setting xu=au and yu=uc for $u \in U_1(X,Y)$. Then $Db=\sum_i r_i(x,y)\varphi_i$ where $r_i(x,y)=$ $=\sum_j p_{ij}(x)q_{ij}(y)$. Denote h(x,y) the greatest common divisor of $r_i(x,y)$. Then there exist polynomials f(x) and $g_i(x,y)$ such that

$f(x)h(x,y)=\sum_i g_i(x,y)r_i(x,y)$ or

$\sum_i f(x)^{-1}g_i(x,y)r_i(x,y)/h(x,y)=1$.

Therefore there exists an invertible matrix over the ring $K[x,y,f(x)^{-1}]$ whose first row is $(r_1/h,\dots,r_k/h)$ and so we are able to rechoose the system of generators of DGC $U[f(a)^{-1}]$ and to insert in it the element $\psi=\sum_i (r_i(x,y)/h(x,y))\varphi_i$. Then $Db=h(x,y)\psi$ and, if $\deg h>0$, U contains a sub-DGC of the form W_3 or W_4 from proposition 6 what is impossible as these DGC are wild. Hence $\deg h = 0$ and the generator b is irregular in $U[f(a)^{-1}]$, so we again arrive to the already examined variant and the theorem is completely proved.

Remark that just the same demonstration also gives the following result.

Proposition 7. Let U be a tame DGC, $a:X\to X$ a minimal loop from U. Then for each dimension d there exists such polynomial f(x) that for each indecomposable representation F $R_d(U,K)$ in which the operator $F(f(a))$ is not nilpotent $F(b)$ for any generator $b\neq a$, $b:X\to Y$ or $b:Y\to X$ for some Y. In particular, if $d(Y)\neq 0$ for some $Y\neq X$ then the operator $F(f(a))$ is nilpotent in any indecomposable representation of dimension d.

In all known examples the polynomial f(x) may be chosen independent on dimension d. It seems very probable that it is always the case.

4. Representations of algebras

Now we shall propose a method which allows to reduce the calculation of representations of algebras to that of DGC. This method is based on the following known fact [1]. Let Λ be a finite-dimensional algebra, J its radical. Consider the category $P(\Lambda)$ whose objects are homomorphisms $\varphi: Q \to P$, Q and P being projective finite-dimensional Λ-modules, such that $\operatorname{Im}\varphi \subset PJ$ and $\operatorname{Ker}\varphi \subset QJ$. If $\varphi': Q' \to P'$ is another object of $P(\Lambda)$ then a morphism from φ to φ' is defined as a pair of homomorphisms (f,g), where $f: Q \to Q'$, $g: P \to P'$ and also $\varphi g = f\varphi'$.

Proposition 8. The functor $C: P(\Lambda) \to R(\Lambda, K)$ associating to a homomorphism φ its cokernel is a representation equivalence (i.e. each object of $R(\Lambda,K)$ is isomorphic to $C\varphi$ for some φ and $C\varphi \cong C\psi$ implies $\varphi \cong \psi$).

Let $P_1, \dots, P_n$ be all non-isomorphic indecomposable projective Λ-modules. Set $H_{ij} = \operatorname{Hom}_\Lambda(P_i, P_jJ)$. The multiplication of homomorphisms induces the maps $\mu_{ij}: \oplus_k H_{ik} \otimes H_{kj} \to H_{ij}$. We shall construct a free DGC $U = U_\Lambda$ as follows. Set $\operatorname{Ob} U = \{X_1, \dots, X_n, Y_1, \dots, Y_n\}$ and consider the family of graded K-modules $\underline{M}$ where

$$\underline{M}_0(X_i, Y_j) = \underline{M}_1(X_i, X_j) = \underline{M}_1(Y_i, Y_j) = H^*_{ij}$$

(the dual space to H_{ij}) and all other modules of the family are zero. Consider the graded category U generated by this family of modules [14]. Determine a differential D in U whose value on $\underline{M}_1(X_i, X_j)$ and $\underline{M}_1(Y_i, Y_j)$ coincides with the homomorphism $\mu^*_{ij}: H^*_{ij} \to \oplus_k H^*_{ik} \otimes H_{kj}$ dual to μ_{ij} and whose value on $\underline{M}_0(X_i, Y_j)$ coincides with the difference of the homomorphisms

$\underline{M}_0(X_i, Y_j) \to \oplus_k \underline{M}_0(X_i, Y_k) \otimes M_1(Y_k, Y_j)$ and

$\underline{M}_o(X_i,Y_j) \dashrightarrow \oplus_k M_1(X_i,X_k) \otimes \underline{M}_O(X_k,Y_j)$ both dual to $\mathcal{M}_{ij}$.
It is easy to check that U turns in such way into a free triangular DGC.

Proposition 9. The category $P(\Lambda)$ is equivalent to the full subcategory of $R(U_\Lambda,K)$ consisting of the representations which have no direct summands of the form E_i where $E_i(X_i)=K$, $E_i(X_j)=0$ for $i\neq j$ and $E_i(Y_j)=0$ for all j.

We shall demonstrate how one can construct the homomorphism $\varphi:Q \dashrightarrow PJ$ corresponding to a representation $F\in R(U,K)$. Denote $V_i=F(X_i)$, $W_j=F(Y_j)$. Then F determines a homomorphism $H_{ij}^* \dashrightarrow \mathrm{Hom}(V_i,W_j)$; But $\mathrm{Hom}(H_{ij}^*,\mathrm{Hom}(V_i,W_j)) \cong V_i^* \otimes H_{ij} \otimes W_j \cong \mathrm{Hom}_\Lambda(Q,PJ)$ where $Q=\oplus_i P_i \otimes V_i$ and $P=\oplus_j P_j \otimes W_j$. In addition, it is easy to verify that if F has no direct summands of the form E_i then $\mathrm{Ker}\,\varphi \subset QJ$ for the corresponding homomorphism $\varphi:Q \dashrightarrow PJ$, so $\varphi \in P(\Lambda)$. Put $\varphi = \Phi(F)$. Analogously the value of Φ on morphisms of representations is determined and thus we obtain a functor $\Phi:R(U,K) \dashrightarrow P(\Lambda)$ which is easy to prove to be an equivalence of categories.

Propositions 8,9 and theorem 1 implie the main theorem for representations of algebras.

Theorem 2. Any finite-dimensional algebra over an algebraically closed field is either tame or wild.

Corollary 4. Any finite-dimensional algebra over a perfect field is either tame or semi-wild.

REFERENCES

1. Auslender M. Representation dimension of Artin Algebras. Queen Mary College Math.Notes, 1971.
2. Bondarenko V.M.,Drozd Ju.A. Representation type of finite groups. Zapiski Nauchn.Semin.LOMI, 71(1977),24-41.
3. Dlab V.,Ringel C.M. Indecomposable representations of graphs and algebras. Mem.Amer.Math.Soc.173(1976).
4. Donovan P.,Freislich M.-R. The representation theory of finite graphs and assosiated algebras. Carleton Math.Lecture Notes, No.5,1973.
5. Drozd Ju.A. Representations of commutative algebras. Func. Analiz i Prilozen. 6,No.4(1972),41-43.0
6. Drozd Ju.A. On tame and wild matrix problems. "Matrix problems" Kiev (1977), 104-114.
7. Freislich M.-R.,Donovan P. Some evidence for an extension of the Brauer-Thrall conjecture. Sonderforschungsbereich Theor. Math.,40,Bonn,1973.
8. Gabriel P.,Zisman M. Calculus of fractions and homotopy theory. Springer-V.,1967.
9. Kleiner M.M.,Roiter A.V. Representations of differential graded categories, "Matrix problems",Kiev (1977),5-70.
10. Nazarova L.A. Representations of quivers of infinite type. Izv.Akad.Nauk SSSR.Ser.mat.,37(1973),752-791.
11. Nazarova L.A. Partially ordered sets of infinite type. Izv. Akad.Nauk SSSR.Ser.mat.,39(1975),963-991.
12. Nazarova L.A. Polyquivers of infinite type. Trudy Matem.Inst. V.A.Steklova Akad.Nauk SSSR,148(1978),175-189.
13. Nazarova L.A.,Roiter A.V. Categorical matrix problems and the Brauer-Thrall conjecture. Preprint Inst.Mat.Akad.Nauk USSR,1973.
14. Roiter A.V.,Kleiner M.M. Representations of differential graded categories. Lecture Notes in Math.,488(1975),316-339.

REMARKS ON PROJECTIVE RESOLUTIONS

Edward L. Green*

In this paper we study information which is contained in projective resolutions of modules over left Artin rings. We begin by showing that given a left Artin ring of finite left global dimension there are partial orders on the isomorphism classes of indecomposable projective left modules in such a way that the partial orderings reflect the finiteness of the left global dimension. Although there may be a finite number of such orderings there is one which is most natural which is discussed at the end of the first section. The second section contains a number of examples showing that the results of section 1 are the best possible. We also apply the results of section 1 to the study Artin algebras of finite representation type (i.e., algebras with only a finite number of nonisomorphic indecomposable finitely generated left modules).

The last section of the paper deals with finding the relation ideal of an algebra. More precisely, if k is an algebraically closed field and R is a basic finite dimensional k-algebra, it is well-known that R is a homomorphic image of a special tensor k-algebra T having the same quiver as R. Thus there is an algebra

---

* This research was partially supported by a grant from the National Science Foundation.

surjection $f:T \to R$. We develop techniques for finding both $f:T \to R$ and generators for the ideal ker(f) from the minimal projective R-resolutions of the simple left R-modules.

We conclude this introduction with notations which will be used throughout this paper. All modules will be left modules. If R is a ring then mod(R) and p-mod(R) will denote the category of finitely generated R-modules and the full subcategory of mod(R) consisting of the finitely generated projective R-modules respectively. We let $\underline{P}(R)$ denote the set of isomorphism classes of indecomposable projective R-modules. If P is an indecomposable projective R-module we will, by abuse of notation, denote its isomorphism class in $\underline{P}(R)$ by P also.

If R is a left Artin ring and $M \in \text{mod}(R)$, let

$$\cdots \to P_n^* \xrightarrow{f_n} P_{n-1}^* \xrightarrow{f_{n-1}} \cdots \to P_1^* \xrightarrow{f_1} P_0^* \xrightarrow{f_0} M \to 0$$

be a minimal projective R-resolution of M. We denote the nth-syzygy of M (i.e. ker (f_{n-1})) by $\Omega_n(M)$. We set $\Omega_0(M) = M$. If $M \in \text{mod}(R)$, we let proj(M) denote the projective cover of M.

Finally, if X is a set and $<_a$, $<_b$ are two partial orderings on X we say $<_a$ is <u>weaker than</u> $<_b$ if for all x, $y \in X$ whenever $x <_b y$ then $x <_a y$. If S is a set of partial orderings on X we say a partial ordering in S is <u>minimal in S</u> if there is no partial ordering in S weaker than it. If $<_a$ is a partial ordering on X we let $w(<_a)$, the width of $<_a$, denote the largest number of elements in X in a linearly ordered chain. We say

$<_a$ is nontrivial if $w(<_a) \geq 2$.

§1. Partial orderings on $\underline{P}(R)$

Let R be a left Artin ring. We say a partial ordering $<_a$ on $\underline{P}$ (R) is _of type r for R_ if for all $t \geq 0$ and for all simple R-modules S we have the following property:

given an indecomposable summand M of $\Omega_{r+t}(S)$ and given an indecomposable projective summand P of $\text{proj}(\Omega_1(M))$ then there exists an indecomposable projective summand Q of proj(M) such that $P <_a Q$.

Basically, the definition says that $<_a$ is a partial ordering on $\underline{P}(R)$ of type r if given a minimal projective R-resolution of a simple R-module S:

$$\cdots \to P^*_{r+1} \to P^*_r \cdots\cdots \to P^*_1 \to P^*_0 \to S \to 0$$

then if Q is an indecomposable summand of P^*_{r+t+1} there is an indecomposable summand Q' of P^*_{r+t} so that $Q <_a Q'$ for $t \geq 0$. Although the definition is a bit stronger, the above weaker statement is useful at times.

We say R admits a nontrivial partial ordering on $\underline{P}(R)$ if there is an $r \geq 0$ such that there is nontrivial partial ordering of type r minimal in the set of all partial orderings of type r for R.

The next result proves the existence of nontrivial partial orderings for left Artin rings of finite left global dimension and also shows that the existence characterizes such rings. We note first that if the left global dimension of the left Artin ring R is $n < \infty$ then the existence of a partial ordering of type $r < n$

implies R admits a nontrivial partial ordering. This follows because all partial orderings on $\underline{P}(R)$ of type less than n must have width greater than 1.

Proposition 1.1. Let R be a left Artin ring. Then the left global dimension of R is finite if and only if R admits a nontrivial partial ordering. If $<_a$ is a partial ordering of type r for R then

(*) left global dimension of $R \leq r + w(<_a) - 1$.

Proof. If $<_a$ is a partial ordering of type r for R then (*) follows directly from the definition of a partial ordering of type r. Thus we have the "if" part of the proposition. Now suppose that the global dimension of R is $n < \infty$. Let A be the subset of $\underline{P}(R)$ consisting of those isomorphism classes of indecomposable projective R-modules which occur as n^{th}-syzygies. Partition A into B_1, B_2, ... by Loewy length, i.e., B_1 consists of the isomorphism classes of indecomposable projective R-modules in A of smallest Loewy length, B_2 consists of the classes in A of the next smallest Loewy length occurring, etc. Since A is a finite set this process ends at say B_s. We set $B_{s+1} = \underline{P}(R) - (\cup_{i=1}^{s} B_1) = \underline{P}(R) - A$. Let $<_a$ be the partial order on $\underline{P}(R)$ generated as follows:

if $P, Q \in \underline{P}(R)$ then $P <_a Q$ if the following two conditions hold:

(i) if $P \in B_i$ then $Q \in B_{i+j}$ for some $j > 0$

(ii) there is a simple R-module S and an indecomposable summand M of $\Omega_{n-1}(S)$ so that if $0 \to P^{**} \to P^* \to M \to 0$ is a minimal R-projective resolution of M then P is isomorphic to a summand of

P^{**} and Q is isomorphic to a summand of P^{*}.

We claim that $<_a$ is a partial ordering of type $n - 1$ for R. (If so, we are done.) Let M be an indecomposable summand of $\Omega_{n-1}(S)$ for some simple R-module S. Let P be an indecomposable summand of $\mathrm{proj}(\Omega_1(M))$. We want to show there is an indecomposable summand Q of $\mathrm{proj}(M)$ such that $P <_a Q$. Since the left global dimension of R is n, it follows that $\mathrm{proj}(\Omega_1(M)) = \Omega_1(M)$. By definition of the partial ordering, we need only show that there is an indecomposable summand of $\mathrm{proj}(M)$ of larger Loewy length than the Loewy length of P. But $0 \to \Omega_1(M) \to \mathrm{proj}(M) \to M \to 0$ is a minimal R-projective resolution of M. Thus $\Omega_1(M)$ is contained in $\mathrm{rad}(R) \cdot \mathrm{proj}(M)$ where $\mathrm{rad}(R)$ denotes the Jacobson radical of R. The result immediately follows. □

The construction given in the proposition is the one on which the intrinsic partial ordering is based. Before getting to that ordering we remark that it is of interest how small we may take r, the type of the ordering, to be since it is an invariant of the ring. As shown in the proposition, we may always take r to be one less than the left global dimension of the ring. Example 2.1 shows that this is the best possible. Many times the smallest type turns out to be 0 or 1. The following result classifies when $r = 0$ occurs.

<u>Lemma 1.2</u>. Let R be a left Artin ring. The following statements are equivalent:

(a) the left global dimension of $R/(\mathrm{rad}(R))^2$

is finite.

(b) the left quiver of R has no oriented cycles.

(c) there is a partial ordering of type 0 for R.

Proof. Let $\overline{R} = R/(\text{rad}(R))^2$. By definition, (see [6, Appendix]) the left quiver of R is the same as the left quiver of $\overline{R}$. By [6, Appendix], (a) and (b) are equivalent. Furthermore, if the left global dimension of $\overline{R}$ is finite, then it equals the length of the longest (oriented) path in the quiver. If the quiver of R has no oriented cycles then it induces a natural partial ordering on a full set of primitive nonisomorphic idempotents and hence on $\underline{P}(R)$. It is easy to check that this partial ordering is of type 0. Thus (b) implies (c). Finally, if $<_a$ is a partial ordering of type 0 for R consider the projective resolutions of the simple R-modules:

$$P_1^* \to P_0^* \to S \to 0$$

If S is a simple R-module then $P_0^* = \text{proj}(S)$ and $P_1^* = \text{proj}(\text{rad}(R) \cdot P_0^*)$. Since the partial ordering is type 0 the definitions imply that there are no oriented cycles in the quiver of R and we are done. □

The above proof shows that when there is a partial ordering of type 0 it is by necessity unique and constructable from the quiver of R. Unfortunately, when the quiver of R has oriented cycles, there may be different minimal partial orderings of type r where r is as small as possible -- see example 2.3. This leads us to the intrinsic partial ordering described below:

We find the generators of the partial ordering $<_a$ in a step by step manner. First we partition $\underline{P}(R)$ into $B_1, \cdots, B_s, B_{s+1}$ as in Proposition 1.1.

Step 0: Give $\underline{P}(R)$ the trivial (i.e., width 1) partial order.

If we can complete step (i) we proceed to step (i + 1):

if $P, Q \in \underline{P}(Q)$ then let $S = \{P <_a Q$: if

(a) $P \not<_a Q$ by steps (0) - (i)

(b) if $P \in B_i$, $i \leq s$ then $Q \in B_{i+j}$ for some $j > 0$

(c) if $P \in B_{s+1}$ then so is Q

(d) there is a simple R-module

S and an indecomposable summand M of $\Omega_{n-i}(S)$ so that if $P^{**} \to P^* \to M \to 0$ is a minimal projective presentation of M then P is isomorphic to a summand of P^{**} and Q is isomorphic to a summand of P^*.$\}$

We say step (i + 1) is complete if when one adds the relations in S to the relations of steps (0) - (i) they generate a partial ordering. The process then goes to step (i + 2). If they do not, the process stops at step (i) and we have constructed the intrinsic partial ordering of type n - i for R.

We conclude this section with a number of remarks.

(1) The partial orderings on $\underline{P}(R)$ are defined using only the projective resolutions of the simple R-modules. If one used the projective resolutions of all finitely generated R-modules in the definition one would also get partial orderings on $\underline{P}(R)$. Note that 1.1 and 1.2 remain valid. More generally, one may define partial orderings on $\underline{P}(R)$ for any set of finitely

generated R-modules provided that there is an upper bound to the left projective dimensions of the modules in the set.

(2) In a dual fashion one may define partial orderings on the isomorphism classes of indecomposable injective R-modules in the case of finite injective dimension.

(3) At this time we are unable to prove that the intrinsic partial ordering is of type r_0 where r_0 is the smallest r such that there is a partial ordering of type r for R when $r_0 > 0$. If $r_0 = 0$ then the intrinsic partial ordering can be of type $r > 0$.

(4) As example 2.3 shows, the intrinsic partial ordering need not be minimal.

(5) The inequality given in 1.1(*) is usually not equality. Of course, equality can occur (e.g., example 2.1).

(6) The partition $B_1, \cdots, B_s, B_{s+1}$, and property (i) in the definition of $<_a$ in Proposition 1.1 cannot be weakened to "$j \geq 0$" instead of "$j > 0$" as shown by examples 2.5 and 2.6.

§2. Examples and an application.

Throughout this section k denotes a fixed field. We freely use the notation of [6, Appendix].

Example 2.1. Given $n \geq 2$, we construct a finite dimensional k-algebra of global dimension n having two nonisomorphic simple modules.

Let $A = k \times k$ be the product ring of two copies of k. Let $e_1 = (1,0)$ and $e_2 = (0,1)$. Let V be the direct sum of α copies of ${}_2k_1 = Ae_2 \otimes_k e_1A$ and let W be the direct sum of β copies of ${}_1k_2$. Let $X = V \oplus W$ viewed as an A-A bimodule. Let T be the special tensor k-algebra

$$A \oplus X \oplus (\otimes_A^2 X) \oplus (\otimes_A^3 X) \oplus \cdots$$

Let $v_1, \cdots, v_\alpha$ and $w_1, \cdots, w_\beta$ be fixed k-base of V and W respectively. Assume $\beta \le \alpha \le \beta + 1$. Finally let $\underline{I}$ be the two-sided T-ideal generated by

$$w_jv_i \quad \text{for} \quad i \le j \le \beta \quad \text{and}$$

$$v_jw_i \quad \text{for} \quad i+1 \le j \le \alpha$$

Setting $R = T/I$ we have

Proposition 2.2.

(a) R is a finite dimensional k-algebra.

(b) R/rad(R) is isomorphic to A as k-algebras.

(c) The left global dimension of R is $\alpha + \beta$.

Proof. Left to the reader. □

We also have the following facts whose proofs we omit. Since $R/\text{rad}(R) = A$ we view Ae_i, $i = 1, 2$ as simple R-modules. We have that the left projective dimensions of these modules are given by

$$\text{p. dim}_R\, Ae_i = \begin{cases} \alpha + \beta & \text{if } i = 1 \\ \alpha + \beta - 1 & \text{if } i = 2 \end{cases}.$$

Furthermore, $\Omega_{\alpha+\beta}(Ae_1) = \begin{cases} Re_1 & \text{if } \alpha = \beta \\ Re_2 & \text{if } \alpha = \beta + 1. \end{cases}$

Moreover if $\alpha = \beta$, $Re_1 \subseteq Re_2$ and if $\alpha = \beta + 1$ then $Re_2 \subseteq Re_1$. Thus the intrinsic partial ordering is

$Re_1 <_a Re_2$ if $\alpha = \beta$ and $Re_2 <_a Re_1$ if $\alpha = \beta + 1$. Finally, we claim that the intrinsic partial ordering is of type $\alpha + \beta - 1$. This follows from the fact that each projective cover of $\Omega_i(Re_1)$ consists either of a direct sum of copies of Re_1 or a direct sum of copies of Re_2 but not both and alternates between these two possibilities as i increases. □

Example 2.3. This example shows the nonuniqueness of the partial orderings on $\underline{P}(R)$.

Let T be the special tensor k-algebra with quiver

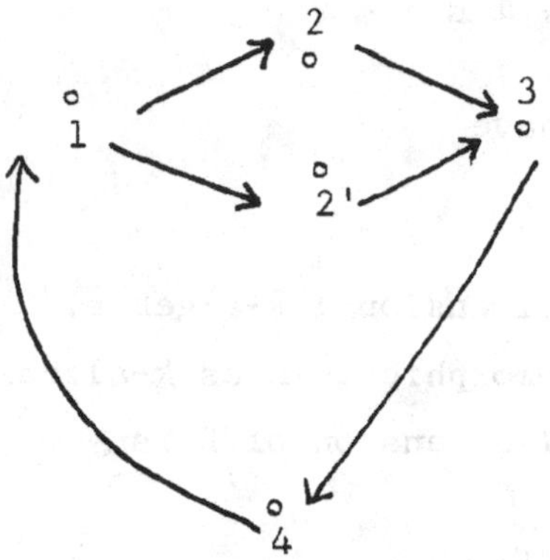

Let $T = A \oplus M \oplus (\otimes^2 M) \oplus \cdots$ where $A = k_1 \times k_2 \times k_{2'} \times k_3 \times k_4$ with $k_i \cong k$ and $M = {}_2k_1 \oplus {}_{2'}k_1 \oplus {}_3k_2 \oplus {}_3k_{2'} \oplus {}_4k_3 \oplus {}_1k_4$. Let ${}_jx_i$ be a fixed nonzero element of the summand ${}_jk_i$ of M. Let I be the two-sided ideal generated by ${}_1x_{44}x_3$ and $({}_3x_{22}x_1 - {}_3x_{2'2'}x_1)$. Set $R = T/I$.

Let S_i and P_i denote the simple R-module and indecomposable projective R-module corresponding to the ith vertex of the quiver of R. Then the projective R-resolutions of the S_i's are given by

$$0 \longrightarrow P_3 \longrightarrow P_2 \oplus P_{2'} \longrightarrow P_1 \longrightarrow S_1 \longrightarrow 0$$
$$0 \longrightarrow P_3 \longrightarrow P_2 \longrightarrow S_2 \longrightarrow 0$$
$$0 \longrightarrow P_3 \longrightarrow P_{2'} \longrightarrow S_{2'} \longrightarrow 0$$
$$0 \longrightarrow P_1 \longrightarrow P_4 \longrightarrow P_3 \longrightarrow S_3 \longrightarrow 0$$
$$0 \longrightarrow P_1 \longrightarrow P_4 \longrightarrow S_4 \longrightarrow 0$$

Since the quiver of R has oriented cycles and the left global dimension of R is 2, the results of section 1 imply the only nontrivial partial orders on $\underline{P}(R)$ are of type 1. The intrinsic one is generated by

$$\{P_3 <_a P_2,\ P_3 <_a P_{2'},\ P_1 <_a P_4\}$$

But $\{P_3 <_b P_2,\ P_1 <_b P_4\}$ and $\{P_3 <_c P_{2'},\ P_1 <_c P_4\}$ both also generate partial orderings of type 1. The latter two are the minimal ones. □

Application to algebras of finite representation type.

Let R* be an Artin algebra of finite representation type and let $M_1, \cdots, M_r$ be a full set of non-isomorphic indecomposable finitely generated R*-modules. Let $R = \text{End}_{R^*}(\coprod_{j=1}^{r} M_j)$. The left global dimension of R is ≤ 2 (with equality if and only if R* is not a semi-simple ring) [1]. It is well-known that the quiver of R is the same as the Auslander-Reiten graph of R*. Recall that Auslander-Reiten graph of R* (sometimes called the irreducible map graph of R*) has as vertices the isomorphism classes of the indecomposable finitely generated left R*-modules and an arrow from vertex M to vertex N if and only if there is an irreducible map from M to N (see [3, 4] for definitions). Note that the category mod(R*) is equivalent to the category p-mod(R). Thus a partial order on $\underline{P}(R)$ induces a partial order on the isomorphism classes of indecomposable R*-modules. Since the projective R-resolutions of simple R-modules are constructed from almost split sequences of R*-modules and from maps rad(R*) P → P where P is an indecomposable projective R*-module, the intrinsic partial ordering induced on R* is intimately related to the Auslander-Reiten graph of R*. We now describe how to translate the intrinsic partial order on R to R*.

Assume the Auslander-Reiten graph of R* has oriented cycles. Then the partial ordering on the isomorphism classes of indecomposable R*-modules is generated as follows:

if M, N are indecomposable R*-modules then $M <_a N$ if the following conditions are satisfied

(1) there is an irreducible map $f: M \to N$

(2) M is not injective

(3) if N is not injective then the R-Loewy length of the R-module $\mathrm{Hom}_{R*}(\ ,M)$ is less than the R-Loewy length of the R-module $\mathrm{Hom}_{R*}(\ ,N)$.

We note that the R-Loewy length of the R-module $\mathrm{Hom}_{R*}(\ ,M)$ is the largest n so that there is a sequence of R*-modules $N_{n-1}, \cdots, N_1$ and irreducible R*-module maps

$$N_{n-1} \xrightarrow{f_{n-1}} N_{n-2} \to \cdots \to N_1 \xrightarrow{f_1} M$$

with nonzero composition.

We remark that there seems to be a close connection between the preinjective partition of mod(R*), [2], and the induced order described above. The next example though shows that, in general, there does not seem to be much connection between the preinjective partition of p-mod(R) and the intrinsic partial ordering on $\underline{P}(R)$. Before proceeding to the examples we should also remark that it would be nice to know necessary and sufficient conditions on R* so that the Auslander-Reiten graph has no oriented cycles.

<u>Example 2.4</u>. Let T be the special tensor k-algebra with quiver:

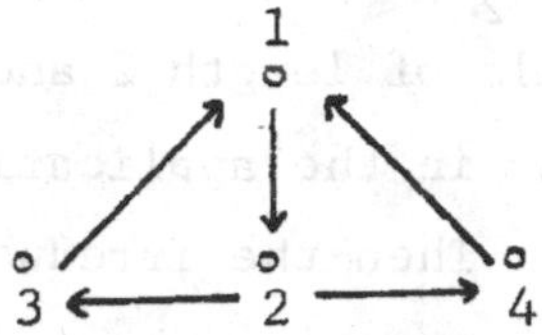

As in example 2.3, we let $T = A \oplus M \oplus (\otimes_A^2 M) \oplus \cdots$ and fix nonzero elements ${}_ix_j$ in the various ${}_ik_j$'s occurring as summands of M. Let I be the two-sided ideal in T generated by ${}_4x_2\,{}_2x_1$, ${}_2x_1\,{}_1x_4$ and ${}_1x_3\,{}_3x_2$.

Let $R = T/I$ and S_i (respectively P_i) be the simple (resp. projective) R-module corresponding to the ith vertex of the quiver. Then the minimal projective R-resolutions of the simple R-modules are given below:

$$0 \longrightarrow P_4 \longrightarrow P_2 \longrightarrow P_1 \longrightarrow S_1 \longrightarrow 0$$

$$0 \longrightarrow P_1 \longrightarrow P_3 \oplus P_4 \longrightarrow P_2 \longrightarrow S_2 \longrightarrow 0$$

$$0 \longrightarrow P_1 \longrightarrow P_3 \longrightarrow S_3 \longrightarrow 0$$

$$0 \longrightarrow P_4 \longrightarrow P_2 \longrightarrow P_1 \longrightarrow P_4 \longrightarrow S_4 \longrightarrow 0$$

The map $P_1 \to P_3 \oplus P_4$ in line 2 is given by $P_1 \subseteq P_3$ and the zero map $P_1 \to P_4$. Thus we find the intrinsic partial order on $\underline{P}(R)$ is generated by

$$P_4 <_a P_2,\ P_2 <_a P_1 \text{ and } P_1 <_a P_3$$

and is of type 1. On the other hand the preinjective partition of $\{P_1, P_2, P_3, P_4\}$ is given by $I_0 = \{P_2, P_3\}$ and $I_1 = \{P_1, P_4\}$. □

The next two examples show that we may not change the definition of the intrinsic partial ordering for R.

Example 2.5. Let R* be the ring $k[x]/(x^3)$. Let M_1 be the simple R*-module, M_2 the unique (up to isomorphism) indecomposable R*-module of length 2 and let M_3 be R* as a left S-module. As in the application we set $R = \mathrm{End}_{R^*}(M_1 \oplus M_2 \oplus M_3)$. Then the irreducible map graph of R* is

$$M_1 \rightleftarrows M_2 \rightleftarrows M_3$$

Let S_1, S_2 and S_3 denote the three nonisomorphic simple R-modules and let $P_i = \mathrm{Hom}_{R^*}(\ ,M_i)$. Then the minimal R-resolutions of the S_i are given by

$$\begin{array}{ccccccccccc} 0 & \longrightarrow & P_1 & \longrightarrow & P_2 & \longrightarrow & P_1 & \longrightarrow & S_1 & \longrightarrow & 0 \\ 0 & \longrightarrow & P_2 & \longrightarrow & P_1 \oplus P_3 & \longrightarrow & P_2 & \longrightarrow & S_2 & \longrightarrow & 0 \\ & & 0 & \longrightarrow & P_2 & \longrightarrow & P_3 & \longrightarrow & S_3 & \longrightarrow & 0 \end{array}$$

Now the R-Loewy length of P_2 is greater than that of P_1. Thus the intrinsic partial order on $\underline{P}(R)$ is generated by $P_1 <_a P_2$, $P_2 <_a P_3$. Note that this is the only possible partial ordering for R of type 1.

Example 2.6. Let T be the special tensor k-algebra associated to the quiver.

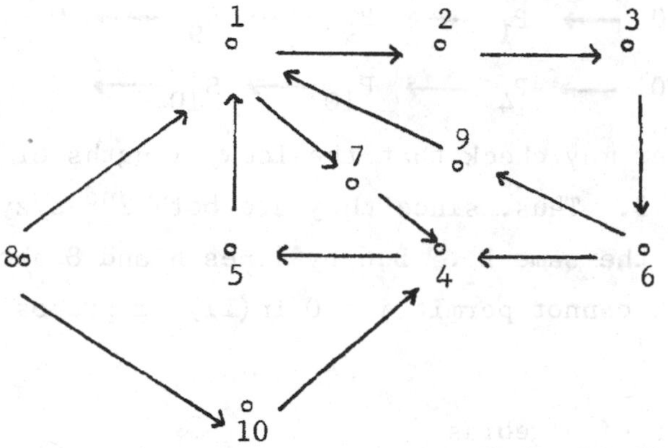

As in 2.4, 2.5 let ${}_ix_j$ be nonzero elements in the appropriate ${}_ik_j$. Let I be the two-sided ideal generated by

$$ {}_6x_3\,{}_3x_2\,,\ {}_2x_1\,{}_1x_5\,,\ {}_7x_1\,{}_1x_5\,,\ {}_5x_4\,{}_4x_7\,, $$

$$ {}_1x_9\,{}_9x_6 - {}_1x_5\,{}_5x_4\,{}_4x_6 \text{ and } {}_4x_{10}\,{}_{10}x_8 - {}_4x_7\,{}_7x_1\,{}_1x_8 $$

Using the same notations as in the earlier examples, the minimal projective resolutions of the simple R-modules is given below:

$$ 0 \to P_2 \oplus P_7 \to P_1 \to S_1 \to 0 $$

$$ 0 \to P_6 \to P_3 \to P_2 \to S_2 \to 0 $$

$$ 0 \to P_6 \to S_3 \to 0 $$

$$ 0 \to P_5 \to P_4 \to S_4 \to 0 $$

$$ 0 \to P_2 \oplus P_7 \to P_1 \to P_5 \to S_5 \to 0 $$

$$ 0 \to P_1 \to P_4 \oplus P_9 \to P_6 \to S_6 \to 0 $$

$$ 0 \to P_5 \to P_4 \to P_7 \to S_7 \to 0 $$

$$ 0 \to P_4 \to P_1 \oplus P_{10} \to P_8 \to S_8 \to 0 $$

$$0 \longrightarrow P_1 \longrightarrow P_9 \longrightarrow S_9 \longrightarrow 0$$

$$0 \longrightarrow P_4 \longrightarrow P_{10} \longrightarrow S_{10} \longrightarrow 0$$

The reader may check that the Loewy lengths of both P_1 and P_4 are 3. Thus, since they are both 2nd-syzygies they occur in the same B_i. But by lines 6 and 8 above we see that we cannot permit $j = 0$ in(ii) of proposition 1.1.

§3. Relations of algebras.

Let k be a fixed field. Let R be a finite dimensional k-algebra with $R/\text{rad}(R) = \prod_{i=1}^{n} k$. Then by [6, Appendix, 7] R is a homomorphic image of a special tensor k-algebra T with the same quiver as R. Say $0 \to I \to T \overset{\pi}{\to} R \to 0$ where π is a k-algebra surjection. Note that if we set $A = \prod_{i=1}^{n} k$ and $M = \text{rad}(R)/(\text{rad}/(R))^2$ viewed as an A - A bimodule, we have $T \cong T_A(M) = A \oplus M \oplus (\otimes_A^2 M) \oplus \cdots$. Both R and T have the same quiver which we denote by $\underset{\sim}{Q}$. Let $J = \pi^{-1}(\text{rad}(R))$ then $J = \coprod_{i\geq 1}(\otimes^i M)$. In a sense, all information about R is contained in T and the ideal I. Equivalently, all information about R is contained in knowledge of $\underset{\sim}{Q}$ and the relations on $\underset{\sim}{Q}$ which generate I. For example, the left global dimension of R is finite if and only if there exists n such that I^n/I^nJ is a projective left R-module. (In this case, the left global dimension of R is $\leq 2n$).

In this section we give a method of finding generators of I as a left T-module. Of course, one way of doing this is to choose appropriate generators of M, formally multiply them and see which relations occur.

This at times is awkward and given the projective R-resolutions of the simple R-modules we describe another method of finding the generators of I.

Let $S_1, \ldots, S_n$ be a full set of nonisomorphic simple R-modules. Let $P_i = \mathrm{proj}(S_i)$ for $i = 1,\ldots,n$. For $i = 1, \ldots, n$, let

(*) $P_2^{(i)} \xrightarrow{A^i} P_1^{(i)} \xrightarrow{B^i} P_0^{(i)} \to S_i \to 0$ be a projective R-resolution of S_i. For each i, we fix direct sum decompositions

$$P_0^{(i)} = \Lambda e_i,\quad P_1^{(i)} = \coprod_{j=1}^{n_1} \Lambda e_{1j}^{(i)},\quad P_2^{(i)} = \coprod_{j=2}^{n_2} \Lambda e_{2j}^{(i)}$$

where the e_i and $e_{u,v}^{(i)}$ are from a fixed set of primitive orthogonal idempotents. Using the above decompositions we view the maps $B^{(i)}$ and $A^{(i)}$ in (*) as matrices, i.e.

$$B^{(i)} = \begin{bmatrix} u_{11}^i \\ \vdots \\ u_{1n_1}^i \end{bmatrix}_{n_1 \times 1}, \qquad A^{(i)} = \begin{bmatrix} v_{j\ell}^{(i)} \end{bmatrix}_{n_2 \times n_1}$$

where $u_{1j}^i \in e_{1j}^{(i)} Re_i$, $v_{j\ell}^{(i)} \in e_{2j}^{(i)} Re_{1\ell}^{(i)}$. Note that the u_{1j}^i generate $\mathrm{rad}(R)e_i$ as a left R-module. Thus this choice of $B^{(i)}$ as a matrix determines both the quiver $\underline{Q}$ and the map $\pi: T \to R$. Next we note that the $v_{j\ell}^{(i)}$ can be written as polynomials with coefficients in k and "variables" u_{1j}^i's There is no unique way of doing this but for each $v_{j\ell}^{(i)}$ choose one such polynomial $F_{j\ell}^{(i)}(u_{1\alpha}^{(\beta)}) = v_{j\ell}^{(i)}$.

Now $\pi: T \to R$ induces an isomorphism

$\bar{\pi}: M \to rad(R)/(rad(R))^2$. Let the image of $u_{1\alpha}^{(\beta)}$ in $rad(R)/(rad(R))^2$ be denoted by $\bar{u}_{1\alpha}^{(\beta)}$. Set $x_{1\alpha}^{(\beta)} = \bar{\pi}^{-1}(\bar{u}_{1\alpha}^{(\beta)})$. Then we have

Proposition 3.1. The ideal I is generated as a left T-module by all elements of the form, for $\ell = 1,\cdots,n$

$$\sum_{j=1}^{n_1} F_{ij}^{(\ell)}(x_{1\alpha}^{(\beta)}) \cdot x_{1j}^{(\ell)},\ i = 1, \cdots, n_2$$

together with all products of M of the $x_{1\alpha}^{(\beta)}$'s where $M \geq$ index of nilpotency of R.

Before proving 3.1 we remark that basically 3.1 says that I is generated by multiplying the matrices given by lifting $A^{(i)}$ and $B^{(i)}$ to matrices in T, taking the entries, together with elements of J^M for large M.

Proof of Proposition 3.1. Let I* be the left ideal generated by the $\sum F_{ij}^{(\ell)}(x_{1\alpha}^{(\beta)} \cdot x_{ij}^{(\ell)}$ and the products. Since (*) is exact $I^* \subseteq I$. Let $t \in I$. Without loss of generality we may assume $t = te^{(\ell)}$. Since $I \subseteq J$ by construction, and since the $x_{1j}^{(\ell)}$ generate $Je^{(\ell)}$ as a left T-module, we may write

(3.2) $\quad t = \Sigma_{j=1}^{n_1} t_j x_{1j}^{(\ell)}$, where $B^{(\ell)}$ is size $n_1 \times 1$.

Then we have

(3.3) $$t = [t_1,\cdots,t_n] \cdot \begin{bmatrix} x_{11}^{(\ell)} \\ \vdots \\ x_{1n_1}^{(\ell)} \end{bmatrix}$$

Now $Z = [\pi(t_1), \cdots, \pi(t_{n_1})]$ can be viewed as an element of $P_1^{(\ell)}$. But $(Z)B^{(\ell)} = 0$ since $\pi(t) = 0$. Thus $Z \in \ker(B^{(\ell)}) = \text{image } (A^{(\ell)})$. Thus there is $[r_1, \cdots, r_{n_2}] \in P_2^{(\ell)}$, $r_i \in Re_{2i}^{(\ell)}$ so that $[r_1, \cdots, r_{n_2}]B^{(\ell)} = A$, where $A^{(\ell)}$ is size $n_2 \times n_1$. Let $s_i \in T$ such that $\pi(s_i) = r_i$ for $i = 1, \cdots, n_2$. Then

if $\quad [s_1, \cdots, s_{n_2}]\,[F_{ij}^{(\ell)}(x_{1\alpha}^{(s)})] = [t_1', \cdots, t_{n_1}']$

we have

(3.4) $$t_i - t_i' \in T$$

Thus $$t = \sum_{i=1}^{n_2} s_i \Big(\sum_{j=1}^{n_1} F_{ij}^{(\ell)}(x_{1\alpha}^{(\beta)})\Big) \cdot x_{1j}^{(\ell)} - \Big(\sum_{j=1}^{n_1} (t_j' - t_j)\, x_{1j}^{(\ell)}\Big)$$

The first part of the right hand side is in I* and the second part is in IJ. Apply the above procedure to each $t_i' - t_i$ we get $t \in I^* + IJ^2$. Continuing, we get $t \in I^* + IJ^M$, but $J^M \subseteq I^*$ and we are done. □

We conclude with some remarks. If we continue the projective resolutions of the simple R-modules and lift the matrices to entries in T and multiply them we get entries in $IJ, I^2, I^2J, \cdots$ successfully. In general though, the entries do not generate the ideals. It would be useful to know when they do. More precisely, if

$$\cdots \longrightarrow P_4 \xrightarrow{D^{(i)}} P_3^{(i)} \xrightarrow{C^{(i)}} P_2^{(i)} \xrightarrow{B^{(i)}} P_1^{(i)} \xrightarrow{A^{(i)}} P_0^{(i)} \longrightarrow S_i \longrightarrow 0$$

Lifting the matrices $A^{(i)}$, $B^{(i)}$, $\cdots$ to A*, B*, ..

with entries in T then the entries of

$$C^* B^* A^* \text{ are in } IJ$$
$$D^* C^* B^* A^* \text{ are in } I^2$$
$$\text{etc.}$$

Adding large products of $x_{1\alpha}^{(\beta)}$ to these generators at times yield left generating sets of IJ, I^2, etc. General conditions when this occurs at this time are open.

<u>References</u>:

[1] Auslander, M., Representation dimension of Artin algebras, Queen Mary College Math. Notes, London.

[2] ____________, Preprojective modules over artin algebras, Ring Theory:Proceedings of the 1978 Antwerp Conference, M. Dekker, NY 1979.

[3] Auslander, M., and Reiten, I., Representation theory of artin algebras III. Almost split sequences, Comm. in Algebra, <u>3</u> (1975) 239-294.

[4] ____________, Representation theory of artin algebras IV:Invariants given by almost split sequences, Comm. in Algebra <u>5</u> (1977), 441-578.

[5] Gabriel, P., Indecomposable representations II, Symposia Mathematica, XI, Academic Press, London (1973), 81-104.

[6] Gordon, R. and Green, E. L., Modules with cores and amalgamations of indecomposable modules, Memoirs of the Am. Math. Society, <u>10</u> 187 (1977), 1-145

[7] Jans, J. P. and Nakayama, T., On the dimension

of modules and algebras, VII, Nagoya Math. J. 11 (1957), 67-76.

Virginia Polytechnic Institute
and State University
Blacksburg, Virginia 24061

VINBERG'S CHARACTERIZATION OF DYNKIN DIAGRAMS USING SUBADDITIVE FUNCTIONS WITH APPLICATION TO DTr-PERIODIC MODULES

Dieter Happel, Udo Preiser, Claus Michael Ringel

Let R be an Artin algebra. Given two indecomposable modules M,N, let $\mathrm{Irr}(M,N) = \mathrm{rad}(M,N)/\mathrm{rad}^2(M,N)$ be the bimodule of irreducible maps [5] and denote by a_{MN} the length of $\mathrm{Irr}(M,N)$ as an $\mathrm{End}(N)$-module, by a'_{MN} its length as an $\mathrm{End}(M)$-module. Note that in case M is not injective, then a_{MN} is equal to the multiplicity of N occuring in the middle term of the Auslander-Reiten sequence starting with M, whereas if N is not projective, then a'_{MN} is equal to the multiplicity of M occurring in the middle term of the Auslander-Reiten sequence ending with N. The <u>Auslander-Reiten quiver</u> $A(R)$ has as vertices the isomorphism classes of the indecomposable R-modules, and there is an arrow $[M] \to [N]$ provided $\mathrm{Irr}(M,N) \neq o$. We endow this arrow with the valuation (a_{MN}, a'_{MN}), and, in this way we obtain a valuated quiver. We denote the Auslander-Reiten translations by $A = \mathrm{DTr}$, $A^- = \mathrm{TrD}$. An indecomposable module is called <u>stable</u> provided $A^n M \neq o, A^{-n} M \neq o$ for all $n \in \mathbb{N}$. The full subquiver $A_s(R)$ of $A(R)$ consisting of the isomorphism classes of stable modules is called the stable <u>Auslander-Reiten-quiver</u>. Any component of the stable Auslander-Reiten quiver determines (uniquely) a Cartan matrix, and we call it its <u>Cartan class</u>. Also, a module M is called <u>periodic</u> provided $A^p M \approx M$ for some $p \in \mathbb{N}$.

<u>Theorem.</u> <u>The Cartan class of a component of the stable Auslander-Reiten quiver of an Artin algebra containing periodic modules is either a Dynkin diagram or</u> A_∞.

In the case of R being an algebra of finite representation type over an algebraically closed field, this is the famous result of Riedtmann [5], the extension to arbitrary Artin algebras of finite representation type being due to Todorov [9]. Todorov also has considered the general case of components of $A_s(R)$ containing periodic modules and reduced their Cartan classes to Dynkin diagrams or A_∞, A_∞^∞, B_∞, C_∞, D_∞. Thus, our only contribution is the elimination of the possibilities A_∞^∞, B_∞, C_∞, D_∞ (lemma 3). Note that the other cases actually do occur.

We will provide a rather elementary self-contained proof of the theorem using only the structure theorem for Riedtmann quivers and Auslander's theorem on the existence of indecomposable modules of arbitrarily large length in any infinite component of an Auslander-Reiten quiver. It was the technique of Todorov which motivated the present presentation: her sole use of length functions and inequalities seemed to ask for an axiomatic treatment using additive and subadditive functions (copying the additivity property of the ordinary length function on Auslander-Reiten sequences). This notion of an additive function was introduced by Bautista [2]. It was M. Auslander who pointed out during his visit to Bielefeld in June 1979 that the methods of Todorov should furnish an interesting combinatorial characterization of the Dynkin diagrams. In fact, such a characterization follows from the investigations of Vinberg in [10]: namely, the Dynkin diagrams are the only finite Cartan matrices with subadditive functions which are not additive. We will need an extension of this result to Cartan matrices which are not necessarily finite and provide a direct proof of the general result. In the same way, one also characterizes the Cartan matrices with additive functions; in the finite case, this result again is due to Vinberg [10], and also to Berman, Moody and Wonenburger [3]; it will be used in a forthcoming paper [4] to deal with binary polyhedral groups.

The authors are indebted to many participants of the Ottawa conference 1979, in particular P. Gabriel, M.I. Platzeck and I. Reiten, for stimulating discussions on this topic and helpful remarks concerning the final form of the manuscript.

1. A characterization of Dynkin diagrams

Let I be an index set. A Cartan matrix C on I is a function $C : I \times I \longrightarrow \mathbb{Z}$ satisfying the following properties

(1) $C_{ii} = 2$ for all $i \in I$.

(2) $C_{ij} \leq 0$ for all $i \neq j$ in I.

(3) $C_{ij} = 0$ if and only if $C_{ji} = 0$.

Note that we write C_{ij} instead of $C(i,j)$. The underlying graph of C has as vertices the elements of I, and edges $\{i,j\}$ for all pairs $i \neq j$ with $C_{ij} \neq 0$.

Of course, the easiest way to write down those Cartan matrices we will be interested in, is to start with the underlying graph and add to the edges pairs of numbers $\underset{i\qquad\qquad j}{\circ\overset{(|C_{ij}|,|C_{ji}|)}{\rule{3em}{0.4pt}}\circ}$ in case $C_{ij}C_{ji} \neq 1$, the "valuation".

The Cartan matrix C will be called connected in case the underlying graph is connected. In particular, we are interested in the Dynkin diagrams

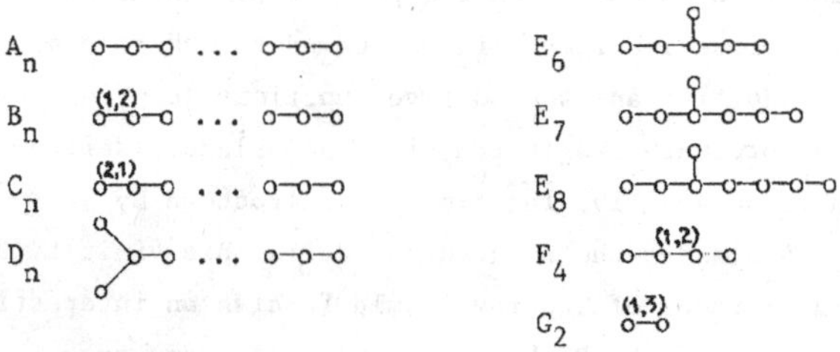

the Euclidean diagrams

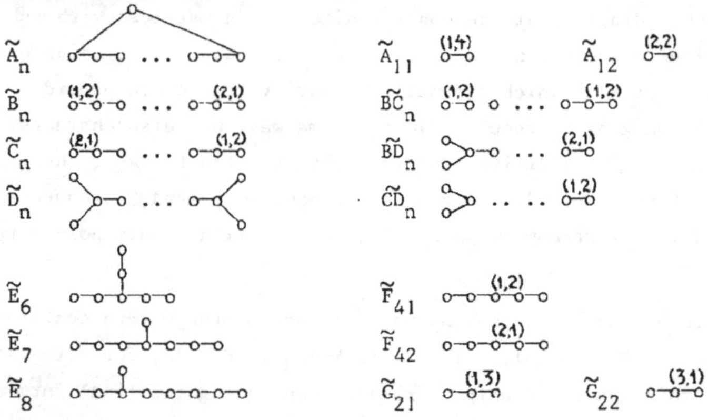

and the following infinite diagrams

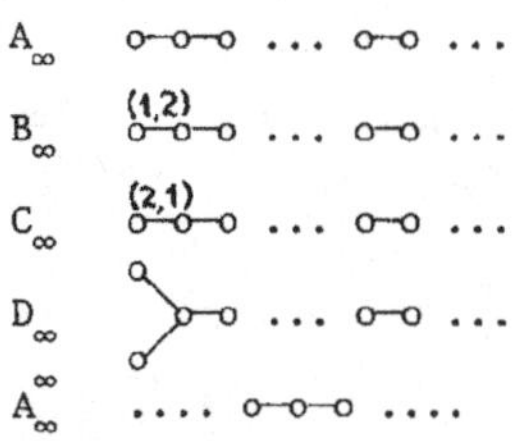

Let C be a Cartan matrix on I. By a <u>subadditive function for</u> C we will mean a function $d : I \rightarrow \mathbb{N} = \{1,2,3,\ldots\}$ satisfying $\sum_{i \in I} d_i C_{ij} \geq 0$ for all $j \in I$. Again, we write d_i instead of $d(i)$.
Such a function is called <u>additive</u> provided we even have $\sum_{i \in I} d_i C_{ij} = 0$ for all $j \in I$. (In case I is finite, an additive function is also called a null root [2]. Note that in case I is infinite, the existence of a subadditive function immediately implies that for fixed j, all but a finite number of C_{ij} are zero.)

<u>Lemma 1</u>. Let C be a Euclidean diagram. Then any subadditive function for C is additive.

<u>Proof</u>: Let C^t be the transpose of C, thus $C^t_{ij} = C_{ji}$ for all $i,j \in I$. With C also C^t is a Euclidean diagram. Now for every Euclidean diagram, there is an additive funtion h, see the table below. Given the Euclidean diagram C, let us denote by ∂ a fixed additive function for C^t, thus $\partial C^t = 0$. Let d be a subadditive function for C. Then $(dC)\partial^t = d(C\partial^t) = 0$. By assumption, the components of dC are ≥ 0, those of ∂ are > 0. Therefore the equality $(dC)\partial^t = 0$ implies that all components of dC are zero, which means that d is additive.

In the tables below, we have listed for every Euclidean diagram C an additive function h for C.
[Note that any other additive function for C is an integral multiple of this h. Namely, given a second additive function h' for C, we can form a non-trivial linear combination of h and h' which vanishes for some $i \in I$. However, it is well-known (and easy to see) that the Cartan matrices of Dynkin diagrams are regular. Thus the linear combination has to be the zero function, and therefore h' is a $\mathbb{Q}$-multiple of h. Since $h_i = 1$ for some $i \in I$, we see that h' even has to be an $\mathbb{N}$-multiple of h.]

Type	diagram	h
$\tilde{A}_{11}$		21
$\tilde{A}_{12}$		11
$\tilde{B}_n$		11...11
$\tilde{C}_n$		12...21
$\tilde{BC}_n$		22...21
$\tilde{BD}_n$	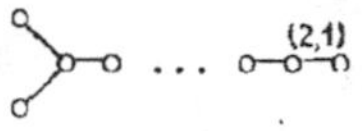	$^1_1$2...22

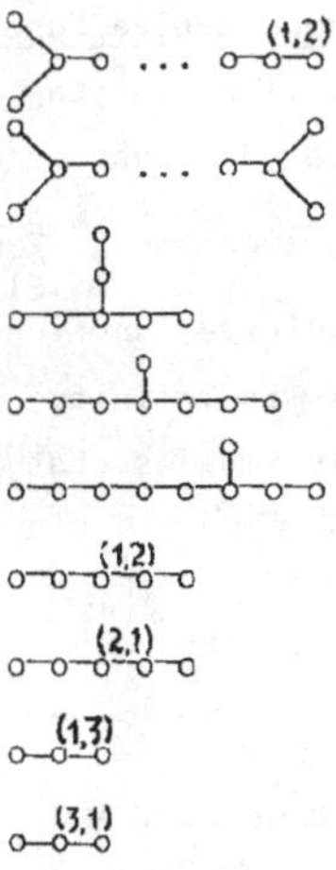

$\widetilde{CD}_n$	$\begin{smallmatrix}1\\1\end{smallmatrix}2\ldots 21$
$\widetilde{D}_n$	$\begin{smallmatrix}1\\1\end{smallmatrix}2\ldots 2\begin{smallmatrix}1\\1\end{smallmatrix}$
$\widetilde{E}_6$	$\begin{smallmatrix}1\\2\\12321\end{smallmatrix}$
$\widetilde{E}_7$	$\begin{smallmatrix}2\\1234321\end{smallmatrix}$
$\widetilde{E}_8$	$\begin{smallmatrix}3\\12345642\end{smallmatrix}$
$\widetilde{F}_{41}$	12321
$\widetilde{F}_{42}$	12342
$\widetilde{G}_{21}$	121
$\widetilde{G}_{22}$	123

Given two Cartan matrices C on I and C' on I', then we call C' <u>smaller</u> than C provided $I' \subseteq I$ and $|C'_{ij}| \le |C_{ij}|$ for all i,j in I'.

<u>Lemma 2</u>. Let C, C' be two different Cartan matrices, with C' smaller than C. Let d be a subadditive function for C. Then $d|I'$ is a subadditive function for C' which is not additive.

<u>Proof</u>: Let $j \in I'$, then

$$2d_j \ge \sum_{\substack{i\in I\\ i\neq j}} d_i|C_{ij}| \ge \sum_{\substack{i\in I'\\ i\neq j}} d_i|C_{ij}| \ge \sum_{\substack{i\in I'\\ i\neq j}} d_i|C'_{ij}|$$

shows that $d|C'$ is subadditive, again. If I' is a proper subset of I, choose $j \in I'$, $i \in I\setminus I'$ which are neighbors, then

$$\sum_{\substack{i\in I'\\ i\neq j}} d_i|C_{ij}| > \sum_{\substack{i\in I'\\ i\neq j}} d_i|C_{ij}|,$$

thus $d|C'$ is not additive. If $|C'_{ij}| < |C_{ij}|$ for some i,j in I', then

$$\sum_{\substack{i\in I'\\ i\neq j}} d_i|C_{ij}| > \sum_{\substack{i\in I'\\ i\neq j}} d_i|C'_{ij}|,$$

thus again, $d|C'$ is not additive.

Lemma 3. Every subadditive function for any one of A_∞^∞, B_∞, C_∞ or D_∞ is additive and bounded.

Proof: Consider first A_∞^∞. We may assume $I = \mathbb{Z}$, with edges $\{i,i+1\}$. Given $d : \mathbb{Z} \to \mathbb{N}$, there is some $i \in \mathbb{Z}$, where d takes its minimum. But the subadditivity means $2d_i \geq d_{i-1}+d_{i+1}$, which combined with $d_{i-1} \geq d_i$, $d_{i+1} \geq d_i$ gives $d_{i-1} = d_i = d_{i+1}$. By induction, we see that d is constant.

In writing down a subadditive function d, we will use the valued graph and attach to each vertex i the numbers d_i. In case B_∞,

$$d_0 \overset{(1,2)}{\text{——}} d_1 \text{——} d_2 \text{——} d_3 \cdots$$

we obtain from d a subaddtive function on A_∞^∞, namely

$$\cdots d_2 \text{——} d_1 \text{——} d_0 \text{——} d_1 \text{——} d_2 \cdots .$$

In case C_∞, we obtain from

$$d_0 \overset{(2,1)}{\text{——}} d_1 \text{——} d_2 \text{——} d_3 \cdots$$

a subadditive function on A_∞^∞, namely

$$\cdots d_2 \text{——} d_1 \text{——} 2d_0 \text{——} d_1 \text{——} d_2 \cdots .$$

In case D_∞, we obtain from

$$\begin{array}{l} d_0 \\ \quad\searrow \\ \quad\quad d_1 \text{——} d_2 \text{——} d_3 \cdots \\ \quad\nearrow \\ d_{0'} \end{array}$$

a subadditive function on A_∞^∞, namely

$$\cdots d_2 \text{——} d_1 \text{——} d_0 + d_{0'} \text{——} d_1 \text{——} d_2 \cdots .$$

In all three cases, the obtained function on A_∞^∞ has to be constant, thus d is additive and bounded.

Theorem. *Let* C *be a connected Cartan matrix and* d *a subadditive function for* C.

(a) C *is either a Dynkin diagram, a Euclidean diagram or one of* A_∞^∞, A_∞, B_∞, C_∞, D_∞.

(b) *If* d *is not additive, then* C *is a Dynkin diagram or* A_∞.

(c) *If* d *is unbounded, then* C *is* A_∞.

Proof: If C is neither a Dynkin diagram nor one of A_∞, A_∞^∞, B_∞, C_∞, D_∞, then there exists a Euclidean diagram C' which is smaller than C (an easy verification).
Now if $C' \neq C$, then $d|C'$ cannot be additive, according to lemma 2. This is a contradiction, since $d|C'$ must be additive, according to lemma 1. This proves (a). If d is not additive, then Euclidean diagrams and A_∞^∞, B_∞, C_∞ and D_∞ cannot occur according to lemma 1 and lemma 3, this proves (b). If d is unbounded, then I has to be infinite, and only A_∞ remains according to lemma 3.

Remarks. For any Euclidean diagram, we have seen in the table of lemma 1 an additive function. Restricting these functions to proper subdiagrams, we obtain for all Dynkin diagrams subadditive functions which then cannot be additive. Thus, the Dynkin diagrams are the only Cartan matrices on a finite index set for which there exist subadditive functions which are not additive. This characterization of the Dynkin diagrams is due to Vinberg [10]. Also, there are the obvious additive functions on A_∞^∞, B_∞, C_∞, D_∞ (see lemma 3), and for A_∞, there are both additive functions, and subadditive functions which are not additive, for example

1 - 2 - 3 - 4 - 5 ...

2 - 4 - 5 - 6 - 7 ...

Finally, there are no additive functions for a Dynkin diagram C (since C is a regular matrix). Thus, the Euclidean diagrams are the only Cartan matrices on finite index sets with additive functions. This characterization of the Euclidean diagrams is due to Vinberg [10] and Berman-Moody-Wonenburger [3].

2. The application

For a quiver $\Gamma = (\Gamma_o,\Gamma_1)$ with Γ_o the set of vertices and Γ_1 the set of arrows, we always will assume that it does not have loops or double arrows. If x is a vertex, we denote by x^+ the set of endpoints of arrows with starting point x, and by x^- the set of starting points of arrows with endpoint x. In case the sets x^+ and x^- are finite for all x, we will call the quiver locally finite.

A Riedtmann quiver $\Delta = (\Delta_o,\Delta_1,\tau)$ is given by a quiver (Δ_o,Δ_1), together with an injective function $\tau : \Delta'_o \to \Delta_o$ defined on a subset Δ'_o of Δ_o satisfying $(\tau x)^+ = x^-$. Given an arrow $\alpha : y \to x$, there is a unique arrow $\tau x \to y$ and this arrow will be denoted by $\sigma\alpha$. A Riedtmann quiver is called stable provided τ is defined on all of Δ_o and is also surjective. Of course, any Riedtmann quiver has a unique maximal stable Riedtmann subquiver. (These concepts have been introduced in [5], there, a Riedtmann quiver is called "Darstellungsköcher".) A vertex x of a Riedtmann quiver Δ will be called periodic provided $\tau^p(x) = x$ for some $p \in \mathbb{N}$. We will be interested in stable Riedtmann quivers containing periodic elements.

An important example of a Riedtmann quiver is the following: let Γ be an oriented tree (a quiver with underlying graph a tree), and define $\mathbb{Z}\Gamma$ as follows: its vertices are the elements of $\mathbb{Z} \times \Gamma_o$, and given an arrow $\alpha : x \to y$, there are arrows $(n,\alpha) : (n,x) \to (n,y)$ and $\sigma(n,\alpha) : (n+1,y) \to (n,x)$ for all $n \in \mathbb{Z}$. Finally, let $\tau(n,\alpha) = (n+1,\alpha)$. Note that in this way, we obtain a stable Riedtmann quiver.

Given a quiver (Γ_o,Γ_1), a function $a : \Gamma_1 \to \mathbb{N} \times \mathbb{N}$ will be called a valuation, and $\Gamma = (\Gamma_o,\Gamma_1,a)$ a valued quiver. The image of $\alpha : x \to y$ will be denoted by (a_α,a'_α), or also (a_{xy},a'_{xy}). If Γ is a valued quiver, we can associate with it a Cartan-matrix $C = C(\Gamma)$ on the index set Γ_o as follows: for $x \in \Gamma_o$, let $C_{xx} = 2$, for $x \neq y$ in Γ_o, let $C_{xy} = -a_{xy} - a'_{yx}$, where $a_{xy} = o = a'_{xy}$ in case there is no arrow with starting point x and endpoint y. In case we deal with a valued oriented tree Γ, then (Γ_o,Γ_1) and C together determine the valuation.

A valued Riedtmann quiver $\Delta = (\Delta_o,\Delta_1,\tau,a)$ is given by a Riedtmann quiver (Δ_o,Δ_1,τ) and a valuation a for (Δ_o,Δ_1) such that $a_{\sigma\alpha} = a'_\alpha$,

$a'_{\sigma\alpha} = a_\alpha$ for all $\alpha : y \to x$ with $x \in \Delta'_o$. A typical example is again the following: let (Γ_o,Γ_1,a) be a valued oriented tree, and define on $\mathbb{Z}(\Gamma_o,\Gamma_1)$ a valuation by $a_{(n,\alpha)} = a_\alpha = a'_{\sigma(n,\alpha)}$ and $a'_{(n,\alpha)} = a'_\alpha = a_{\sigma(n,\alpha)}$. This valued Riedtmann quiver is denoted by $\mathbb{Z}(\Gamma_o,\Gamma_1,a)$.

Proposition. Let Γ,Γ' be valued oriented trees. Then $\mathbb{Z}\Gamma$ and $\mathbb{Z}\Gamma'$ are isomorphic if and only if the Cartan matrices $C(\Gamma)$ and $C(\Gamma')$ are isomorphic. Given any stable valued Riedtmann quiver Δ, there is a valued oriented tree Γ and a group G of automorphisms of $\mathbb{Z}\Gamma$ such that Δ is isomorphic to $\mathbb{Z}\Gamma/G$.

In case Δ is isomorphic to $\mathbb{Z}\Gamma/G$ for some valued oriented tree Γ, we call $C(\Gamma)$ the Cartan class of Δ; it is uniquely determined by Δ.

The proof of the proposition follows immediately from the corresponding result on Riedtmann quivers without valuations [5]. Namely, if $\Delta = (\Delta_o,\Delta_1,\tau,a)$ is a valued Riedtmann quiver, and $(\Delta_o,\Delta_1,\tau) = \mathbb{Z}(\Gamma_o,\Gamma_1)/G$ for some oriented tree (Γ_o,Γ_1), then using the projection from $\mathbb{Z}(\Gamma_o,\Gamma_1)$ onto (Δ_o,Δ_1,τ), the valuation a of Δ gives rise to a valuation on $\mathbb{Z}(\Gamma_o,\Gamma_1)$, also denoted by a, in such a way that $\mathbb{Z}(\Gamma_o,\Gamma_1)$ becomes a valued Riedtmann quiver. The canonical embedding of (Γ_o,Γ_1) into $\mathbb{Z}(\Gamma_o,\Gamma_1)$ given by $x \mapsto (o,x)$ endows (Γ_o,Γ_1) with a valuation, again denoted by a, and clearly $(\Delta_o,\Delta_1,\tau,a) = \mathbb{Z}(\Gamma_o,\Gamma_1,a)/G$. If x is a sink in (Γ_o,Γ_1), denote by $\sigma_x(\Gamma_o,\Gamma_1,a)$ the full valued subquiver of $\mathbb{Z}(\Gamma_o,\Gamma_1,a)$ with vertices (o,y) for $y \neq x$, and $(1,x)$. It is obvious that the Cartan matrices $C(\Gamma_o,\Gamma_1,a)$ and $C(\sigma_x(\Gamma_o,\Gamma_1,a))$ are isomorphic. This shows the unicity of the Cartan matrix associated to $\mathbb{Z}(\Gamma_o,\Gamma_1,a)$.

If $\Delta = (\Delta_o,\Delta_1,\tau,a)$ is a valued Riedtmann quiver, a subadditive function ℓ for Δ is, by definition, a function $\ell : \Delta_o \to \mathbb{N}$ satisfying

$$\ell(x) + \ell(\tau x) \geq \sum_{y \in x^-} \ell(y)a'_{yx} ,$$

for all $x \in \Delta'_o$. Such a function is called additive, provided we always have equality (for all $x \in \Delta'_o$).

Theorem. Let $\Delta = (\Delta_0, \Delta_1, \tau, a)$ be a stable valued Riedtmann quiver which is connected, and contains a periodic vertex. Assume there is a subadditive function ℓ for Δ.

(a) The Cartan class of Δ is either a Dynkin diagram, a Euclidean diagram, or one of A_∞, A_∞^∞, B_∞, C_∞, D_∞.

(b) If ℓ is not additive, then the Cartan class of Δ is a Dynkin diagram, or A_∞.

(c) If ℓ is unbounded, then the Cartan class of Δ is A_∞.

Proof: First note that the existence of a subadditive function implies that Δ is locally finite.

Let us show that any vertex of Δ has to be periodic. For, let x be periodic, say $\tau^p x = x$. Now,

$\tau^p(x^+) = (\tau^p(x))^+ = x^+$ shows that τ^p induces a permutation on the finite set x^+, and therefore τ^{pm} the identity on x^+, for some $m \in \mathbb{N}$. Thus any $y \in x^+$ also is periodic. Similarly, any $y \in x^-$ is periodic. But in this way, using in addition τ, we can reach any other vertex of Δ, since we assume that Δ is connected.

Let Δ be a quotient of $\mathbb{Z}\Gamma$, with Γ a valuated oriented tree with Cartan matrix C. We can assume that $\Gamma = \{0\} \times \Gamma$ is embedded into $\mathbb{Z}\Gamma$, and denote the corresponding map $\Gamma \longrightarrow \mathbb{Z}\Gamma \longrightarrow \Delta$ just by $u \mapsto \tilde{u}$. By definition of C, we have

$$C_{uv} = \begin{cases} 2 & u = v \\ -a_{\tilde{u}\tilde{v}} & u \longrightarrow v \\ & \text{in case} \\ -a'_{\tilde{v}\tilde{u}} & v \longrightarrow u \\ 0 & \text{otherwise} \end{cases}$$

Assume now there is given a subadditive function ℓ for Δ. We consider first the case where there exists a fixed number p with $\tau^p x = x$ for all vertices x of Δ. For example, this clearly is true in case Γ is finite. From ℓ we obviously obtain a τ-invariant subadditive function d for Δ, by

$$d(x) = \sum_{i=0}^{p-1} \ell(\tau^i x),$$

and d is additive if and only if ℓ is. Namely, $\tau^p x = x$ shows that $d(x) = d(\tau x)$, thus

$$2d(x) = d(x) + d(\tau x) = \sum_{i=0}^{p-1} \ell(\tau^i x) + \sum_{i=1}^{p} \ell(\tau^i x)$$

$$= \sum_{i=0}^{p-1} [\ell(\tau^i x) + \ell(\tau(\tau^i x))]$$

$$\geq \sum_{i=0}^{p-1} \sum_{y\in(\tau^i x)^-} \ell(y)\, a'_{y,\tau^i x}$$

$$= \sum_{i=0}^{p-1} \sum_{z\in x^-} \ell(\tau^i z)\, a'_{\tau^i z,\tau^i x}$$

$$= \sum_{z\in x^-} \sum_{i=0}^{p-1} \ell(\tau^i z)\, a'_{z,x}$$

$$= \sum_{z\in x^-} d(z) a'_{z,x}$$

where we have written $y \in (\tau^i x)^- = \tau^i(x^-)$ in the form $y = \tau^i z$, and used that $a'_{\tau^i z,\tau^i x} = a'_{z,x}$ for all x,z. Thus d is a τ-invariant subadditive funtion for Δ which is additve iff ℓ is additive. We consider now the composed map $\Gamma \longrightarrow \Delta \longrightarrow \mathbb{N}$, given by $u \mapsto d(\tilde{u})$.

Note that $\tilde{u}^-$ is the disjoint union of $\{\tilde{v} \mid v \in u^-\}$ and $\{\tau\tilde{v} \mid v \in u^+\}$, thus

$$2d(\tilde{u}) \geq \sum_{z\in\tilde{u}^-} d(z)\, a'_{z,\tilde{u}}$$

$$= \sum_{v\in u^-} d(\tilde{v})\, a'_{\tilde{v}\tilde{u}} + \sum_{v\in u^+} d(\tilde{v})\, a'_{\tau\tilde{v},\tilde{u}}$$

$$= \sum_{v\in u^-} d(\tilde{v})\, a'_{\tilde{v}\tilde{u}} + \sum_{v\in u^+} d(\tilde{v})\, a_{\tilde{u}\tilde{v}}$$

$$= -\sum_{v\in u^-} d(\tilde{v})\, c^t_{vu} - \sum_{v\in u^+} d(\tilde{v})\, c^t_{vu}$$

$$= -\sum_{v\neq u} d(\tilde{v})\, c^t_{vu}\, .$$

This shows that we obtain in this way a subadditive function for C^t, which is additive or unbounded iff ℓ is additive, or unbounded, respectively. Thus, the existence of a subadditive function ℓ on Δ implies that C has to be a Dynkin or Euclidean diagram or one of A_∞, A_∞^∞, B_∞, C_∞, D_∞. In case ℓ is additive, C must be Dynkin or A_∞, and in case ℓ is unbonded, C must be of the form A_∞.

Finally, consider the case where we only have for every vertex x of Δ a number $p(x)$ depending on x with $\tau^{p(x)}(x) = x$. In particular, Γ is infinite. Choosing a finite subdiagram Γ' of Γ, and Δ' the stable Riedtmann quiver generated by Γ', we see that Γ' has to be a Dynkin diagram or a Euclidean diagram. As a consequence, Γ only can be one of A_∞, A_∞^∞, B_∞, C_∞, or D_∞.

We claim that for Γ of type A_∞, A_∞^∞, B_∞, C_∞, or D_∞, any automorphism group G of $\mathbb{Z}\Gamma$ containing an element g with $g(n,x) = (n+p,x)$ for some $(n,x) \in \mathbb{Z}\Gamma$ and some $p \geq 1$, must contain a translation (an automorphism of the form $(m,y) \mapsto (m+q,y)$ for all $(m,y) \in \mathbb{Z}\Gamma$). Namely, in the cases A_∞, B_∞, C_∞, we use the following numbering

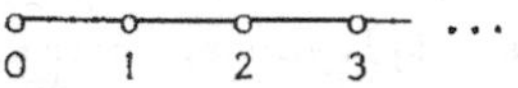

of the vertices of Γ. Any automorphism of $\mathbb{Z}\Gamma$ maps a subset of the form $\mathbb{Z}\times\{x\}$ into itself (this is clear for $x = 0$, since $\mathbb{Z}\times\{0\} = \{(n,x) \mid |(n,x)^+| = 1\}$, and follows by induction for the remaining x). If now $g(n,x) = (n+p,x)$ for some (n,x), then also all neighbours (m,y) of (n,x) will satisfy $g(m,y) = (m+p,y)$. Similarly, for D_∞, use the numbering

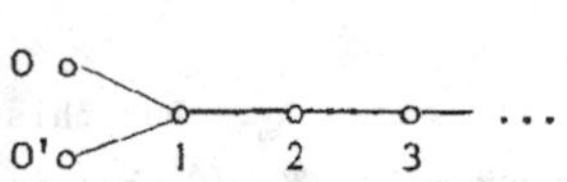

Then the subsets $\mathbb{Z}\times\{0,0'\}$, and $\mathbb{Z}\times\{x\}$ with $x \geq 1$ are mapped into themselves by any automorphism. If $g(n,x) = (n+p,x)$ for some (n,x), then also $g(m,y) = (m+p,y)$ for all neighbours with $y \geq 1$. If $(m,0)$ is a neighbour of $(n,1)$, and $g(n,1) = (n+p,1)$, then we only can conclude that $g^2(m,0) = (m+2p,0)$, however this then implies that g^2 is a translation. Finally consider the case A

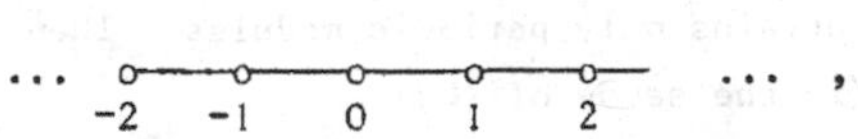

where we may assume that $g(n,0) = (n+p,0)$, for some n,p. If $(m,1)$ is a neighbour of $(n,0)$, then either $g(m,1) = (m+p,1)$, and then g is a translation, or else $g(m,1) = (m+p,-1)$, and then at least g^2 is a translation.

As a consequence, we see that in all cases there is a fixed number q with $\tau^q(z) = z$ for all vertices z in Δ, thus we are in the previous case, and the theorem is proved.

An immediate consequence of this result is the theorem stated in the introduction: Note that the Auslander-Reiten quiver is always locally finite. Consider a component C of $A_s(R)$ containing a periodic module, and let ℓ be the ordinary length function, it clearly is subadditive. Note that ℓ is additive on C if and only if C is even a component of the complete Auslander-Reiten quiver $A(R)$. We may assume that R is connected. Now, if ℓ is not additive on C, then the Cartan class of C can only be a Dynkin diagram or A_∞, by part (b). If, on the other hand, ℓ is additive, then R cannot be of finite representation type, since there exists a component of the Auslander-Reiten quiver without projective modules, namely C. But then ℓ cannot be bounded on C, by a theorem of Auslander [1], see also [7]. Thus, we can apply (c) and see that the Cartan class of C is A_∞.

As a first application, we obtain Riedtmann's theorem [5], and its generalisation to arbitrary Artin algebras due to Todorov [9]:

Corollary 1. Let R be an Artin algebra of finite representation type. Let C be a connected component of $A_s(R)$. Then the Cartan class of C is a Dynkin diagram.

Proof: We only have to exclude the case A_∞. But this case is impossible since for any automorphism group G, $\mathbb{Z}A_\infty/G$ has infinitely many points.

As a second application, we can describe completely those components of the Auslander-Reiten quiver which contain a periodic module but no projective ones.

Corollary 2. Let R be an Artin algebra and C a connected component of $A(R)$ which contains only periodic modules. Then C is a quasi-serial component (in the sense of [6]).

Proof: Since we deal with a component of $A(R)$, the ordinary length function is additive. Thus, the Cartan class is A_∞. But this then implies that C is quasi-serial.

3. Example

We have seen that a component of the Auslander-Reiten quiver with only periodic modules is quasi-serial. Let us exhibit the example of a component with stable part of Cartan class A_∞ containing periodic modules which is not quasi-serial.

Consider the Artin algebra R defined by the following quiver with relations

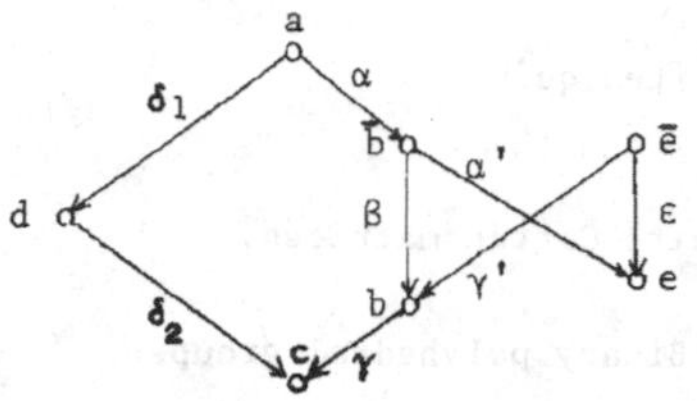

$$\gamma\beta\alpha = \delta_2\delta_1$$

$$\alpha'\alpha = o$$

$$\gamma\gamma' = o$$

and its component C containing the simple module corresponding to the vertex d. Then C has the following form (We denote any module by its composition factors in a suggestive way, the dotted lines have to be identified in order to form a cylinder):

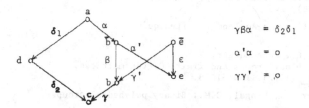

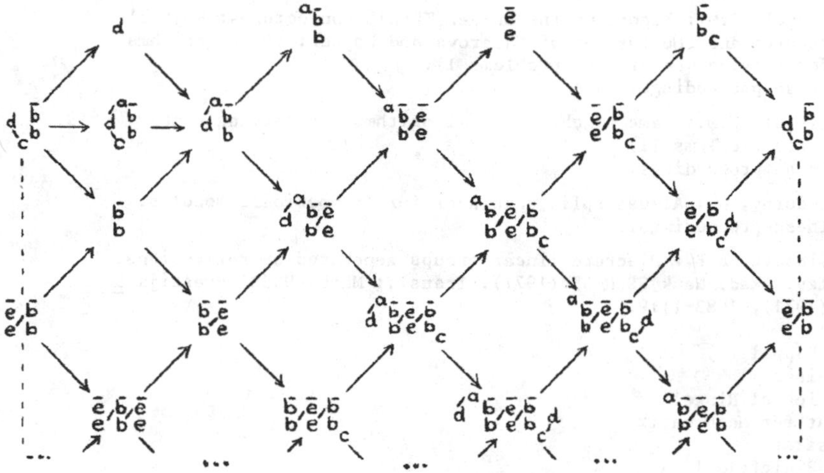

Further examples can be built by using suitable regular enlargements and regular co-enlargements of tame quivers, see [8].

Remark. Note that the example above gives an algebra with infinitely many indecomposables which are both preprojective and preinjective in the sense of Auslander and Smalø. Namely, in C all modules containing the composition factor corresponding to the vertex a are preprojective, those containing the composition factor corresponding to the vertex c are preinjective.

References

[1] Auslander, M.: Applications of morphisms determined by objects. Proc. Conf. on Representation Theory, Philadelphia (1976). Marcel Dekker (1978), 245-327.

[2] Bautista, R.: Sections in Auslander-Reiten quivers. These proceedings.

[3] Berman, S., Moody, R., Wonenburger, M.: Cartan matrices with null roots and finite Cartan matrices. Indiana Math. J. 21 (1972), 1091-1099.

[4] Happel, D., Preiser, U., Ringel, C.M.: Binary polyhedral groups and Euclidean diagrams. To appear in Manuscripta Math.

[5] Riedtmann, Chr.: Algebren, Darstellungsköcher, Überlagerungen und zurück. To appear in Comm. Helv.

[6] Ringel, C.M.: Finite dimensional hereditary algebras of wild representation type. Math. Z. 161 (1978), 235-255.

[7] Ringel, C.M.: Report on the Brauer Thrall conjectures: Rojter's theorem and the theorem of Nazarova and Rojter. (On algorithms for solving vectorspace problems I). These proceedings.

[8] Ringel, C.M.: Tame algebras (On algorithms for solving vectorspace problems II) These proceedings.

[9] Todorov, G.: Almost split sequences for TrD-periodic modules. These proceedings.

[10] Vinberg, E.B.: Discrete linear groups generated by reflections. Izv. Akad. Nauk SSSR 35 (1971). Transl.: Math. USSR Izvestija 5 (1971), 1083-1119.

Dieter Happel
Udo Preiser
Claus Michael Ringel
Fakultät für Mathematik
Universität
D-4800 Bielefeld 1
West-Germany

TRIVIAL EXTENSION OF ARTIN ALGEBRAS

Yasuo Iwanaga and Takayoshi Wakamatsu

I. Throughout this paper, we fix the following notations:

Λ = an artin algebra over a commutative artin ring R,

D = the self-duality: $\mathrm{mod}\ \Lambda \to \mathrm{mod}\ \Lambda^{op}$ of a category of finitely generated left Λ-modules defined by $D(X) = \mathrm{Hom}_R(X, E(R/J(R)))$ for $X \in \mathrm{mod}\ \Lambda$, where $J(*)$ is the radical of a ring $*$ and $E(M)$ is an injective hull of a module M,

$T(\Lambda) = \Lambda \ltimes D(\Lambda)$ is a trivial extension of Λ by a Λ-bimodule $D(\Lambda)$. (See Fossum-Griffith-Reiten [4] for details.)

We recall that an artin algebra Λ is a symmetric artin algebra if $\Lambda \cong D(\Lambda)$ as Λ-bimodules. (Auslander-Platzeck-Reiten [1])

Then we have

Proposition 1. $T(\Lambda)$ is a symmetric artin algebra.

Thus, every artin algebra is a homomorphic image of some symmetric artin algebra.

Now, we present the following problem.

PROBLEM. When is $T(\Lambda)$ of finite representation type? Especially, if $T(\Lambda)$ is of finite representation type, then $g\ell.\dim \Lambda < \infty$?

In this paper, we will give some partial answers of this problem, which covers Green-Reiten [6, Proposition 3.2]. Moreover, Tachikawa [8] and Yamagata [9] also have the different partial answers, and we should refer the Müller's work [7], on which we will note later.

First, we investigate in case of Λ having the square-zero radical. Assume $J(\Lambda)^2 = 0$, and let $\mathcal{G}(\Lambda)$ be a graph of Λ defined as follows: if $\{e_1, \ldots, e_m\}$ is the complete set of orthogonal primitive idempotents in Λ, then the set of vertices of $\mathcal{G}(\Lambda)$ is $\{1, \ldots, m\}$ and an edge with a value $\underset{i}{\cdot}\overset{(d_{ij},\, d_{ji})}{\overline{\qquad\qquad}}\underset{j}{\cdot}$ means $e_j\, J(\Lambda)\, e_i \neq 0$, $d_{ij} = |{}_{e_j\Lambda e_j} e_j J(\Lambda) e_i|$ and $d_{ij} = |e_j J(\Lambda) e_{i\, e_i \Lambda e_i}|$, where $|M|$ denotes the composition length of a module M. Further $\mathcal{D}(T(\Lambda))$ denotes the separated diagram of $T(\Lambda)/J(T(\Lambda))^2$ in the sense of Gabriel [5] and Dlab-Ringel [3]. Then we have

<u>Theorem 2</u>. Assume $J(\Lambda)^2 = 0$. $T(\Lambda)$ is of finite representation type if and only if $\mathcal{G}(\Lambda)$ is disjoint union of a Dynkin graph, i.e. valued graphs: A_n $(n \geqq 1)$, $B_n (n \geqq 2)$, $C_n (n \geqq 3)$, $D_n (n \geqq 4)$, $E_n (n=6,7,8)$, F_4 and G_2. In this case, $\mathcal{D}(T(\Lambda))$ is a disjoint union of two $\mathcal{G}(\Lambda)$'s, the number of indecomposable left (or right) $T(\Lambda)$-modules is equal to the twice of the number of indecomposable representations of $\mathcal{G}(\Lambda)$ and $g\ell.\dim \Lambda < \infty$.

<u>Corollary 3</u>. Let Λ be an artin algebra with $\mathcal{G}(\Lambda/J(\Lambda)^2)$ a disjoint union of a Dynkin graph, then Λ is a homomorphic image of a hereditary ring.

Next, in connection with the latter question in PROBLEM, we obtain

<u>Proposition 4</u>. Let Λ be a quasi-Frobenius artin algebra (i.e. Λ is self-injective) and non-semisimple, then $T(\Lambda)$ is of infinite representation type.

Here we note that a quasi-Frobenius ring has an infinite global dimension unless it is semisimple.

II. First we state

<u>Proof of Proposition 1</u>. Define $\varphi\colon T(\Lambda) \to \mathrm{Hom}_R(T(\Lambda), E(R/J(R)))$ by $[\varphi(\lambda,f)](\lambda',f') = f(\lambda') + f'(\lambda)$ for $\lambda, \lambda' \in \Lambda$ and $f, f' \in D(\Lambda)$, then φ is an isomorphism of $T(\Lambda)$-bimodule.

The proof of Theorem 2 and Proposition 4 will be done by claiming the following lemmas. From now on we put $\Gamma = T(\Lambda)$.

<u>Lemma 1</u>. Assume $J(\Lambda)^2 = 0$. For a primitive idempotent e in Λ, $J(\Gamma)e/J(\Gamma)^2 e \cong J(\Lambda)e \oplus D(eJ(\Lambda))$.

<u>Proof</u>. Since $J(\Gamma)e/J(\Gamma)^2 e \cong J(\Lambda)e/J(\Lambda)^2 e \oplus D(\Lambda)e/(J(\Lambda)D(\Lambda) + D(\Lambda)J(\Lambda))e$ in general, it is enough to show $D(\Lambda)e/(J(\Lambda)D(\Lambda)+D(\Lambda)J(\Lambda))e \cong D(eJ(\Lambda))$. In case of ${}_\Lambda D(\Lambda)e$ simple, $J(\Lambda)D(\Lambda)e = 0$ and so $D(\Lambda)e = D(\Lambda)J(\Lambda)e$. On

the other hand, if ${}_{\Lambda}D(\Lambda)e$ is not simple $J(\Lambda)D(\Lambda)e \neq 0$ and hence $J(\Lambda)D(\Lambda)e$ contains $D(\Lambda)J(\Lambda)e$ because $(J(\Lambda)D(\Lambda)+D(\Lambda)J(\Lambda))e \cong \mathrm{Soc}\,(\Gamma e)$ is simple by Proposition 1. Therefore $D(\Lambda)e/(J(\Lambda)D(\Lambda)+D(\Lambda)J(\Lambda))e$ is zero or $D(\Lambda)e/J(\Lambda)D(\Lambda)e \cong D(\mathrm{Soc}(e\Lambda))$ according to ${}_{\Lambda}D(\Lambda)e$ simple or not and, in either case, it is isomorphic to $D(eJ(\Lambda))$ in our assumption $J(\Lambda)^2 = 0$.

Lemma 2. Assume $J(\Lambda)^2 = 0$, then $\mathfrak{G}(\Gamma)$ is a disjoint union of $\mathfrak{G}(\Lambda)$ and $\mathfrak{G}(\Lambda)^*$ where $\mathfrak{G}(\Lambda)^*$ is obtained by for each edge $i \overset{(dij,dji)}{\longrightarrow} j'$ between i and j' at the top and bottom row, respectively in $\mathfrak{G}(\Lambda)$, taking an edge $i' \overset{(dij,dji)}{\longrightarrow} j$ between i' and j.

Proof. Clear by $f(J(\Gamma)e/J(\Gamma)^2e) \cong fJ(\Lambda)e \oplus D(eJ(\Lambda)f)$ for any primitive idempotents e and f in Λ, which follows from Lemma 1.

Lemma 3. Assume $J(\Lambda)^2 = 0$. If $\mathfrak{G}(\Lambda)$ contains no cycle, then $\mathfrak{G}(\Gamma)$ is a disjoint union of two $\mathfrak{G}(\Lambda)$'s.

Proof. $\mathfrak{G}(\Lambda)$ with an orientation corresponds one to one with $\mathfrak{G}(\Lambda)$, and $\mathfrak{G}(\Lambda)$ is independent to an orientation of $\mathfrak{G}(\Lambda)$ from Lemma 2. Further, it is possible to give $\mathfrak{G}(\Lambda)$ an orientation such that each vertex in $\mathfrak{G}(\Lambda)$ is either sink or source by the assumption. Then we get $\mathfrak{G}(\Lambda) = \mathfrak{G}(\Lambda)$ and $\mathfrak{G}(\Lambda)^* = \mathfrak{G}(\Lambda)$.

In general, let n be the nilpotency index of Λ, i.e. $J(\Lambda)^n = 0$ and $J(\Lambda)^{n-1} \neq 0$, then the nilpotency index of $T(\Lambda)$ is $n+1$. Hence the next lemma plays an important role.

Lemma 4. Let Λ be a quasi-Frobenius ring, n the nilpotency index of Λ and M an indecomposable Λ-module with $J(\Lambda)^{n-1}M \neq 0$, then M is projective.

Proof. See Müller [7, Lemma 3.4].

For an artin ring Λ, $a(\Lambda)$ and $n(\Lambda)$ denote the numbers of non-isomorphic indecomposable and simple Λ-modules, respectively. Now let Λ is a quasi-Frobenius ring and n the nilpotency index of Λ, then Λ is of finite representation type if and only if so is $\Lambda/J(\Lambda)^{n-1}$, and further assume the Loewy length of any projective indecomposable Λ-module is n, we have an equation $a(\Lambda) = a(\Lambda/J(\Lambda)^{n-1}) + n(\Lambda)$.

Lemma 5. Let Λ_1 and Λ_2 be artin algebras over a commutative artin ring R, and assume Λ_1 and Λ_2 are Morta equivalent then $T(\Lambda_1)$ and $T(\Lambda_2)$ are so.

Proof. Let ${}_{\Lambda_1}P$ be a progenerator with $\mathrm{End}_{\Lambda_1}(P) \cong \Lambda_2$, then P can be regarded as a $T(\Lambda_1)$-module through a trivial mapping $P \to D(\Lambda_1)\otimes_{\Lambda_1}P$ which we denote by $\tilde{P}$. $\tilde{P}$ is a progenerator as a $T(\Lambda_1)$-module and $\mathrm{End}_{T(\Lambda_1)}(\tilde{P}) \cong \mathrm{End}_{\Lambda_1}(P) \alpha \mathrm{Hom}_{\Lambda_1}(P, D(\Lambda_1)\otimes_{\Lambda_1}P)$, so we show ${}_{\Lambda_2}\mathrm{Hom}_{\Lambda_1}({}_{\Lambda_1}P_{\Lambda_2}, {}_{\Lambda_1}D(\Lambda)\otimes_{\Lambda_1}P_{\Lambda_2})_{\Lambda_2} \cong {}_{\Lambda_2}D(\Lambda_2)_{\Lambda_2}$ as Λ_2-bimodules. Now,

$$\begin{aligned}
{}_{\Lambda_2}\mathrm{Hom}_{\Lambda_1}({}_{\Lambda_1}P_{\Lambda_2}, {}_{\Lambda_1}D(\Lambda_1)\otimes_{\Lambda_1}P_{\Lambda_2})_{\Lambda_2} &\cong {}_{\Lambda_2}\mathrm{Hom}_{\Lambda_1}({}_{\Lambda_1}P_{\Lambda_2}, {}_{\Lambda_1}\Lambda_{1\Lambda_1})\otimes_{\Lambda_1}D(\Lambda_1)\otimes_{\Lambda_1}P_{\Lambda_2}\\
&\cong {}_{\Lambda_2}\mathrm{Hom}_{\Lambda_1}({}_{\Lambda_1}D(\Lambda_1)_{\Lambda_1}, {}_{\Lambda_2}D(P)_{\Lambda_1})\otimes_{\Lambda_1}D(\Lambda_1)\otimes_{\Lambda_1}P_{\Lambda_2}\\
&\cong {}_{\Lambda_2}D(P)\otimes_{\Lambda_1}P_{\Lambda_2}\\
&\cong {}_{\Lambda_2}\mathrm{Hom}_{\Lambda_1}(\mathrm{Hom}_{\Lambda_1}({}_{\Lambda_1}P_{\Lambda_2}, {}_{\Lambda_1}\Lambda_{1\Lambda_1}), \mathrm{Hom}_R({}_{\Lambda_1}P_{\Lambda_2}, I))_{\Lambda_2}\\
&\cong {}_{\Lambda_2}\mathrm{Hom}_R(\mathrm{Hom}_{\Lambda_1}({}_{\Lambda_1}P_{\Lambda_2}, {}_{\Lambda_1}\Lambda_{1\Lambda_1})\otimes {}_{\Lambda_1}P_{\Lambda_2}, I)_{\Lambda_2}\\
&\cong {}_{\Lambda_2}\mathrm{Hom}_R({}_{\Lambda_2}\Lambda_{2\Lambda_2}, I)_{\Lambda_2}\\
&\cong {}_{\Lambda_2}D(\Lambda_2)_{\Lambda_2},
\end{aligned}$$

where $I = E(R/J(R))$.

Proof of Theorem 2. "Only if": Since Γ is of finite representation type, $\mathcal{S}(\Gamma)$ contains no cycle and so does $\mathcal{G}(\Lambda)$. Hence $\mathcal{G}(\Lambda)$ is a disjoint union of a Dynkin graph by Lemma 3, Gabriel [5] and Dlab-Ringel [3]. (Note: $\mathcal{G}(\Lambda)$ is always connected if we assume Λ is twosided indecomposable.)

"If": Since $\mathcal{G}(\Lambda)$ contains no cycle, $\mathcal{S}(\Gamma)$ is a disjoint union of two $\mathcal{G}(\Lambda)$'s by Lemma 3 and now $\mathcal{G}(\Lambda)$ is a disjoint union of a Dynkin graph by the assumption. Hence $\Gamma/J(\Gamma)^2$ is of finite representation type again by [5] and [3], which shows that Γ is also of finite representation type by Lemma 4.

Assume now Γ is of finite representation type. It is easy to see the Loewy length of any projective indecomposable Γ-module is three and hence $a(\Gamma) = a(\Gamma/J(\Gamma)^2) + n(\Gamma)$. Here $a(\Gamma/J(\Gamma)^2) = a(\mathcal{S}(\Gamma)) - n(\Gamma/J(\Gamma)^2) = 2\cdot a(\mathcal{G}(\Lambda)) - n(\Gamma)$ by Lemma 3, [5] and [3], where $a(\mathcal{S}(\Gamma))$ and $a(\mathcal{G}(\Lambda))$ are the numbers of non-isomorphic indecomposable representations of $\mathfrak{D}(\Gamma)$ and $\mathfrak{G}(\Lambda)$, respectively. Therefore we get $a(\Gamma) = 2 \cdot a(\mathfrak{G}(\Lambda))$.

Proof of Corollary 3. Since $\mathfrak{G}(\Lambda/J(\Lambda)^2)$ is a disjoint union of a Dynkin graph, $gl.\dim \Lambda/J(\Lambda)^2 < \infty$ by Green-Reiten [6] and hence Λ is a homomorphic image of a hereditary ring by Chase [2, Theorem 4.1].

Proof of Proposition 4. By Lemma 5, we may suppose Λ is self-basic and then Λ is a Frobenius algebra, that is, ${}_\Lambda D(\Lambda) \cong {}_\Lambda\Lambda$ and $D(\Lambda)_\Lambda \cong \Lambda_\Lambda$ as left and right Λ-modules, respectively. Hence, by similar argument as Lemma 1, we have $J(\Gamma)e/J(\Gamma)^2e \cong J(\Lambda)e/J(\Lambda)^2e \oplus D(\mathrm{Soc}(e\Lambda))$ for a primitive idempotent e in Λ. Now, $\mathrm{Soc}(e\Lambda)_\Lambda$ is simple and so $D(\mathrm{Soc}(e\Lambda)) \cong \Lambda f/J(\Lambda)f$ for some primitive idempotent f in Λ, where e and f are corresponding each other by Nakayama permutation of Λ. Thus, we assume Λ is twosided indecomposable and let $e_1, \dots, e_m$ be a complete set of orthogonal primitive idempotents in Λ and σ Nakayama permutation of Λ. Then $\mathfrak{D}(\Gamma)$ contains edges $(1, \sigma(i_1))$, $(i_1, \sigma(i_2))$, ..., $(i_{n-1}, \sigma(i_n))$, ... with all i_j's different, but this can not continue infinitely, so $i_k = i_j$ for some $k > j$, which means that there is a cycle in $\mathfrak{D}(\Gamma)$. Therefore we get the desired result.

III. Finally we give some examples and remarks.

(1) An example of $\mathfrak{G}(\Lambda) = C_n$: let $K \supsetneqq k$ be fields with $(K:k) = 2$ and

$$\Lambda = \left\{ \begin{pmatrix} \alpha & & \\ \beta & x_2 & \\ & x_2 & \\ & y & x_3 \end{pmatrix} ; \alpha, \beta \in K \text{ and } x_2, x_3, y \in k \right\}$$

a k-algebra with an usual matrix multiplication. Then

$$\mathfrak{G}(\Lambda) = \underset{1}{\cdot} \overset{(2,1)}{\text{———}} \underset{2}{\cdot} \text{———} \underset{3}{\cdot} = C_3 ;$$

$\mathfrak{D}(\Lambda) =$ 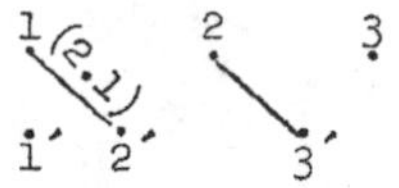 $= A_1 \cup B_2 \cup A_2 \cup A_1$;

$$\mathfrak{D}(T(\Lambda)) = \underset{1}{\cdot} \overset{(2,1)}{\text{———}} \underset{2'}{\cdot} \text{———} \underset{3}{\cdot} \cup \underset{1'}{\cdot} \overset{(2,1)}{\text{———}} \underset{2}{\cdot} \text{———} \underset{3'}{\cdot} = C_3 \cup C_3 .$$

(2) Let Λ_1 and Λ_2 be finite dimensional algebras constructed from the quivers 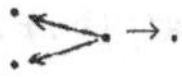 and 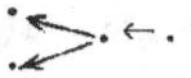, respectively. Then we have

$$a(\Lambda_1) = 12,\ a(\Lambda_2) = 8 \quad \text{and} \quad a(T(\Lambda_1)) = a(T(\Lambda_2)) = 24 .$$

Furthermore, Tachikawa [8] showed that if Λ is a hereditary artin algebra of finite representation type, then $a(T(\Lambda)) = 2 \times a(\Lambda)$. However, in non-hereditary case, it happens that $a(T(\Lambda))$ is no multiple of $a(\Lambda)$. For example, let Λ be a serial artin algebra with admissible sequence 1, 2, 2, then $a(\Lambda) = 5$ and $a(T(\Lambda)) = 12$.

(3) It is easy to see that even if artin algebras Λ_1 and Λ_2 are stably equivalent, $T(\Lambda_1)$ and $T(\Lambda_2)$ are not in general. For, every artin algebra with square-zero radical is stably equivalent to a hereditary artin algebra and the trivial extension of a hereditary artin algebra of finite representation type by an injective cogenerator is always of finite representation type by [8], but it does not hold for an artin algebra with square-zero radical by Theorem 2.

(4) Müller's construction [7] of a weakly symmetric ring Γ from an artin ring Λ with square-zero radical can be done by using a trivial extension provided Λ is an artin algebra, but Γ is different from our $T(\Lambda)$. In case of Müller's one, $a(\Gamma) = 2(a(\Lambda) + n(\Lambda))$ but it does not hold in our case as seen in (2).

References

1. AUSLANDER, M., PLATZECK, M.I. and REITEN, I.: Periodic modules over weakly symmetric algebras, J. Pure Appl. Alg. 11 (1977) 279-291.

2. CHASE, S.U.: A generalization of the ring of triangular matrices, Nagoya Math. J. 18 (1961) 13-25.

3. DLAB, V. and RINGEL, C.M.: Indecomposable representations of graphs and algebras, Mem. Amer. Math. Soc. 173 (1976).

4. FOSSUM, R.M., GRIFFITH, P.A. and REITEN, I.: Trivial extnesions of Abelian categories, Lect. Notes in Math. 456 (1975), Springer-Verlag, Berlin-Heidelberg-New York.

5. GABRIEL, P.: Unzerlegbare Darstellungen I, Manuscripta math. 6 (1972) 71-103.

6. GREEN, E.L. and REITEN, I.: On the construction of ring extensions, Glasgow Math. J. 17 (1976) 1-11.

7. MÜLLER, W.: Unzerlegbare Moduln uber artinschen Ringen, Math. Z. 137 (1974) 197-226.

8. TACHIKAWA, H.: Proceedings of ICRA II, Ottawa, 1979.

9. YAMAGATA, K.: Proceedings of ICRA II, Ottawa, 1979.

MODEL THEORY AND REPRESENTATIONS OF ALGEBRAS

C.U. Jensen and H. Lenzing

In this paper we give a preliminary account of some results concerning the number of indecomposable representations of Artinian rings, in particular in the case of finite-dimensional algebras over a field.

The results are obtained from model theoretical considerations of Artinian rings, resp. finite-dimensional algebras, and their indecomposable modules. Detailed proofs will be published in [4].

We first recall some basic notions and facts from model theory. Let L be the first order language of associative rings with unity. Two rings R and S are called *elementarily equivalent* (notation: $R \equiv S$) if R and S satisfy the same first order sentences in L. By a theorem of Keisler-Shelah [2] the rings R and S are elementarily equivalent if and only if there exists a set I and an ultrafilter F on I such that the corresponding ultrapowers R^I/F and S^I/F are isomorphic rings. A class C of rings is called *elementarily closed* if $R \equiv S$, $R \in C$, implies $S \in C$. Further a class C of rings is called *axiomatizable* [*finitely axiomatizable*] if C can be defined by a family of first order sentences [one first order sentence] in L. By [2] a class C is axiomatizable if and only if C is elementarily closed and closed under formation of ultraproducts. Further C is finitely axiomatizable if and only if C is axiomatizable and the class $\overline{C}$ of rings not in C is closed under formation of ultraproducts.

It is not hard to see that the left Artinian rings of specified length form a finitely axiomatizable class. Hence the class of all left Artinian rings is elementarily closed. Similarly the Artin algebras (i.e. the left Artinian rings that are finitely generated as modules over their center) form an elementarily closed class. If R and S are elementarily equivalent left Artinian rings, they have the same left global dimension and the same left self-injective dimension. Further, elementarily equivalent finite-dimensional algebras over a field have the same

Hochschild dimension.

If R is left Artinian and t is a positive integer we denote by $\mathrm{Ind}_t(R)$ the *set of isomorphism classes of indecomposable left R-modules of length* t. The corresponding number (cardinality) is denoted $|\mathrm{Ind}_t(R)|$.

The following is a key result for our model theoretical study of indecomposable representations.

Proposition 1. *Let* (R_α), $\alpha \in I$, *be a family of left Artinian rings of fixed length* d *and let* F *be a (non-principal) ultrafilter on* I. *Then the ultraproduct* $R^* = \prod_\alpha R_\alpha/F$ *is a left Artinian ring of length d. Further*

i) If (M_α), $\alpha \in I$, *is a family of left* R_α*-modules of fixed length* t, *the ultraproduct* $M^* = \prod_\alpha M_\alpha/F$ *is a left* R^**-module of length* t.

ii) If (M_α), $\alpha \in I$ *and* (N_α), $\alpha \in I$, *are two families of left* R_α*-modules of constant lengths, then* $M^* = \prod_\alpha M_\alpha/F$ *and* $N^* = \prod_\alpha N_\alpha/F$ *are isomorphic* R^**-modules if and only if* M_α *and* N_α *are isomorphic* R_α*-modules for almost all* α, *i.e. for all* α *in some subset of* I *belonging to the ultrafilter* F.

iii) For any left R^**-module* M *of finite length* t *there exists a family* (M_α), $\alpha \in I$, *of left* R_α*-modules of length* t *such that* M *and* $\prod_\alpha M_\alpha/F$ *are isomorphic* R^* *- modules.*

iv) If (M_α), $\alpha \in I$, *is a family of left* R_α*-modules of constant length, then* $M^* = \prod_\alpha M_\alpha/F$ *is an indecomposable* R^**-module if and only if* M_α *is an indecomposable* R_α*-module for almost all* α.

v) If (M_α), $\alpha \in I$, *is a family of left* R_α*-modules of constant length and* s *is a given integer* ≥ 0, *then the projective (injective) dimension of the* R^**-module* $M^* = \prod M_\alpha/F$ *is* s *if and only if the projective (injective) dimension of the* R_α*-module* M_α *is* s *for almost all* α.

vi) Let us assume that for all $\alpha \in I$,

$$0 \to M_1 \xrightarrow{f_1} M_2 \xrightarrow{f_2} M_3 \to 0 \qquad (*)$$

is a sequence of left R-*modules of finite length and* R-*homomorphisms. The corresponding sequence of ultrapowers*

$$0 \to M_1^* \xrightarrow[f_1^*]{} M_2^* \xrightarrow[f_2^*]{} M_3^* \to 0$$

where $M_i^* = M_i^I/F$, $(i = 1,2,3)$ *and* $f_i^* = f_i^I/F$, $(i = 1,2)$, *is an almost split exact sequence of* R^* *-modules if and only if* (*) *is an almost split exact sequence of* R-*modules.*

vii) Retain the notation of vi). An R-*homomorphism*

$$M_1 \xrightarrow[f_1]{} M_2$$

between R-*modules of finite length is irreducible in the sense of* [1] *if and only if the corresponding homomorphism of ultrapowers*

$$M_1^* \xrightarrow[f_1^*]{} M_2^*$$

is an irreducible homomorphism of R^* *-modules.*

The next result is merely a useful reformulation of the assertions i)-iv) of proposition 1.

<u>Proposition 2.</u> *Let* (R_α), $\alpha \in I$, *be a family of left Artinian rings of constant length and let* F *be a (non-principal) ultrafilter on* I. *If* $R^* = \Pi R_\alpha/F$ *is the corresponding ultraproduct, then for any integer* t *the "canonical" mapping*

$$\prod_\alpha \mathrm{Ind}_t(R_\alpha)/F \to \mathrm{Ind}_t(R^*)$$

is bijective.

The propositions 1 and 2 have a number of consequences among which we mention the following

<u>Theorem 3.</u> *Let* R *and* S *be elementarily equivalent left Artinian rings. If* R *is of finite representation type, so is* S. *Moreover, when* R *and* S *are of finite*

representation type, there is a bijective mapping between the left R-modules of finite length and the left S-modules of finite length that preserves the length, the projective dimension, the injective dimension and the indecomposability of modules. Further, this mapping yields an isomorphism between the corresponding Auslander-Reiten graphs.

Remark. By the Löwenheim-Skolem theorem any ring is elementarily equivalent to a countable ring; hence theorem 3 shows that for the study of most module theoretic properties of a ring of finite representation type it suffices to consider the case when R is countable.

Theorem 4. *For fixed integers* d *and* L *the left Artinian rings of length* d *for which any indecomposable left module has length* $\leq L$ *form an axiomatizable class* C *; there exists a universal bound* $a(d,L) \in \mathbb{N}$ *such that the total number of indecomposable left* R*-modules is* $\leq a(d,L)$ *for every* $R \in C$.

In the rest of this paper we consider Artinian rings that are finite-dimensional algebras over an *infinite field*.

The next result is obtained by a combination of proposition 2 and the Ringel version of the Nazarova-Rojter result concerning the second Brauer-Thrall conjecture [5], i.e. for an algebra R of infinite representation type there is an integer p such that $|\mathrm{Ind}_{pn}(R)| = |R|$ for all $n \in \mathbb{N}$.

Theorem 5. *For any pair of integers* (d,t) *there is an integer* $b = b(d,t)$ *such that for any infinite field* K *and any* d*-dimensional* K*-algebra* R *of finite representation type we have*

$$|\mathrm{Ind}_t(R)| \leq b \ .$$

Corollary 6. *For a fixed* $d \in \mathbb{N}$ *the* d*-dimensional* K*-algebras of finite representation type,* K *being an unspecified infinite field, form an axiomatizable class.*

Using a result of Gabriel [3] we obtain

Theorem 7. *For a fixed* $d \in \mathbb{N}$ *and a fixed characteristic* p , (p = *a prime or* 0), *the* d*-dimensional* K*-algebras of finite representation type,* K *being an arbitrary algebraically closed field of characteristic* p*, form a finitely axiomatizable class.*

Theorem 8. *For a fixed* $d \in \mathbb{N}$ *and a fixed characteristic* p, (p = a *prime or* 0), *there is an increasing sequence of integers*

$$0 = g_0(p) < g_1(p) < g_2(p) < \dots$$

with the property that for any d*-dimensional* K*-algebra* R *of infinite representation type,* K *being an arbitrary algebraically closed field of characteristic* p*, each of the intervals* $]g_{n-1}(p), g_n(p)]$ *contains a number* t *for which* $\mathrm{Ind}_t(R)$ *is infinite.*

Remark. As usual d is a fixed integer. For the characteristic p let g(p) be a smallest possible $g_1(p)$. There are at most finitely many prime numbers p for which $g(p) < g(0)$. We do not know whether the numbers g(p) are bounded or not. If they are bounded, then $g(p) = g(0)$ for all but a finite number of primes p. If they are not bounded, the class of d-dimensional algebras of finite representation type over algebraically closed fields of unspecified characteristic is not finitely axiomatizable.

For algebras over arbitrary algebraically closed fields we prove the following

Theorem 9. *For any pair of integers* (d,t) *there is an integer* $c = c(d,t)$ *such that for any* d*-dimensional algebra* R *over any algebraically closed field* K *one has either* $|\mathrm{Ind}_t(R)| = |K|$ *or* $|\mathrm{Ind}_t(R)| \leq c$.

We finish this paper by some results concerning the number of finite-dimensional algebras of finite representation type. For an integer d and a field K let $\mathrm{Fin}_d(K)$ be the *set of isomorphism classes of* d*-dimensional* K*-algebras of finite representation type.*

Theorem 10. *Let* (K_α), $\alpha \in I$, *be a family of infinite fields and let* F *be a (non-principal) ultrafilter on* I. *If* $K^* = \Pi K_\alpha / F$ *is the corresponding ultraproduct and* d *is a fixed integer there is a canonical injective mapping*

$$\varphi : \prod_{\alpha} \mathrm{Fin}_d(K_\alpha)/F \longrightarrow \mathrm{Fin}_d(K^*)$$

Moreover, if the fields K_α *are algebraically closed of the same characteristic the mapping* φ *is bijective.*

Theorem 10 has several corollaries among which we mention the following.

Corollary 11. *If* d *is an integer and* K *an algebraically closed field then* $\mathrm{Fin}_d(K)$ *is either finite or* $|\mathrm{Fin}_d(K)| = |K|$. *If* $\mathrm{Fin}_d(K)$ *is finite, then* $|\mathrm{Fin}_d(K)| = |\mathrm{Fin}_d(L)|$ *for any algebraically closed field* L *of the same characteristic as* K.

Corollary 12. *Assume* d *is an integer for which* $|\mathrm{Fin}_d(K)|$ *is a finite number* n *for one and hence all algebraically closed fields* K *of characteristic* 0. *Then there exists a finite set* P *of exceptional prime numbers such that* $|\mathrm{Fin}_d(L)| \leq n$ *for any algebraically closed field* L *of characteristic* p, $p \notin P$.

Corollary 13. *Let* p *be a prime number or* 0, *and let* d *be an integer for which* $\mathrm{Fin}_d(K)$ *is finite for one and hence all algebraically closed fields* K *of characteristic* p. *Then there exists a finite algebraic extension* F *of the prime field of characteristic* p *with the property that for any algebraically closed field* K *of characteristic* p *every algebra* $R \in \mathrm{Fin}_d(K)$ *is defined over* F.

Remark. Theorem 10 and its corollaries also hold true if we replace $\mathrm{Fin}_d(K)$ by the set of isomorphism classes of d-dimensional K-algebras of finite representation type having prescribed global dimension, self-injective dimension or Hochschild dimension.

Note, added in proof (June 1980). With the aid of Theorem 5 it was shown by C. Herrmann [Axioms for algebras of finite representation type, preprint 1980] that finite representation type always forms a finitely axiomatizable class, provided the base fields considered are infinite. In the following we describe the principle results, which arise in the combination of [4] and this result. Detailed proofs will appear in a joint publication with Herrmann, which replaces [4].

1. For a fixed $d \in \mathbb{N}$, the d-dimensional K-algebras of finite representation type, K being an arbitrary infinite field, form a finitely axiomatizable class F_d. This result replaces Theorem 7. We note that the proof does not depend on Gabriel's theorem [3].

2. By elimination of quantifiers [see for example: Kreisel, G. and J.L. Krivine: Elements of mathematical Logic (Model theory). North-Holland Publishing Company, Amsterdam 1971] the finite axiom system for F_d implies the existence of a finite number of polynomials $f_1,\ldots,f_s$ with *integer coefficients*, such that a d-dimensional algebra over an arbitrary algebraically closed field is of finite representation type if and only if the structure constants of R satisfy a prescribed system (a Boolean combination) of equations and inequations in $f_1,\ldots,f_s$. In other words: finite representation type is *constructible over* $\mathbb{Z}$. It is obvious that this result parallels and in fact complements Gabriel's theorem [3], that finite representation type is *open over* K, for K algebraically closed.

3. There is a general principle of *characteristic transfer* for finite representation type. More precisely: For any first order statement ϕ (in the language of d-dimensional algebras over algebraically closed fields) there is a finite set P of exceptional prime numbers, such that the following are equivalent:

(i) ϕ holds true for F_d in characteristic 0,

(ii) ϕ holds true for F_d in characteristic p,

for every $p \notin P$.

Particular cases of this principle are:

a) The "Gabriel numbers" $g(p)$ are bounded (see Remark following Theorem 8), and in fact $g(0) = g(p)$ for all but a finite number of prime numbers p.

b) If $|\mathrm{Fin}_d(K)| = n$ is finite in characteristic 0, then $|\mathrm{Fin}_d(L)| = n$ in characteristic p for all but a finite number of primes p. (K and L are supposed to be algebraically closed).

4. There is a primitive recursive procedure which (in the algebraically closed case) allows to determine - at least in principle - the constants $c(d,t)$ (Theorem 9) and $g(p)$ (Theorem 8). Provided these constants are known,

$$\phi : \quad |\mathrm{Ind}_t(R)| \leq c(d,t) \quad \text{for} \quad 1 \leq t \leq g ,$$

where $g = \max_p g(p)$, is an *explicit first order axiom for* F_d. Again, via elimination of quantifiers, ϕ allows - in principle - to determine the defining polynomials $f_1,\ldots,f_s$ of finite representation type.

BIBLIOGRAPHY

[1] Auslander, M. and I. Reiten: Representation theory of Artin algebras IV., Comm. Algebra 5 (1977), 443 - 518.

[2] Chang, C.C. and H.J. Keisler: Model theory. North Holland Publ. Comp., Amsterdam 1977.

[3] Gabriel, P.: Finite representation type is open, Representations of algebras. Lecture Notes in Mathematics 488, Springer 1975.

[4] Jensen, C.U. and H. Lenzing: Applications of model theory to representations of algebras, to appear.

[5] Ringel, C.M.: Report on the Brauer-Thrall conjectures: Rojter's theorem and the theorem of Nazarova and Rojter. These Proceedings.

Christian U. Jensen
Københavns Universitets
Matematiske Institut
Universitetsparken 5
DK-2100 København
Denmark

Helmut Lenzing
Fachbereich Mathematik der
Universität - Gesamthochschule
D-4790 Paderborn
W.-Germany

SOME REMARKS ON REPRESENTATIONS OF QUIVERS AND INFINITE ROOT SYSTEMS

VICTOR G. KAČ

This is an addendum to my paper [4]. The purpose of it is to give simpler proofs of the main results of [4] in a more general situation. In [4] properties of the infinite root systems are used in the representation theory of quivers. Here properties of the root systems (and their existence, which in [4] is deduced from the theory of the Kac-Moody Lie algebras) are obtained in the framework of the representation theory of quivers. We do not exclude edges-loops from our consideration. This makes us introduce a more general notion of infinite root system than the one in [4].

In the remainder of the article some remarks on related topics are made and some open problems are discussed. They include:

a) an "abstract" definition of an infinite root system (i.e., a definition which does not depend on the basis);

b) multiplicities of roots and ζ-functions of quivers;

c) a connection with the problem of classification of prehomogeneous linear groups.

We keep the notations of [4]. The base field $\mathbb{F}$ is arbitrary unless otherwise stated.

I am grateful to P. Gabriel for the remark that my proof can be extended to the quivers with edges-loops and to C.M. Ringel for giving me some interesting examples of representations of quivers.

1. (Generalized) infinite root systems.

An (n x n) square matrix $A = (a_{ij})$ with integral entries is called a (generalized) Cartan matrix if

(C1) $a_{ii} \leq 2$ and even;

(C2) $a_{ij} \leq 0$ for $i \neq j$;

(C3) $a_{ij} = 0$ implies $a_{ji} = 0$, $i,j = 1,\ldots,n$.

Notice that Lemmas 1.2 and 1.3 of [4] hold in this more general situation. The lists of Cartan matrices of positive and zero type is almost the same as Tables P and Z in [4]: one should only add to Table Z the (1 x 1) zero matrix which we denote by $A_0^{(1)}$. The Dynkin diagram of a Cartan matrix A is defined in the same way as in [4] with additional $\frac{1}{2}(2 - a_{ii})$ edges-loops to a vertex p_i.

Let A be a (generalized) Cartan (n x n)-matrix, let Γ be a free abelian group with free generators $\alpha_1,\ldots,\alpha_n$ and let Γ_+ be the set of all non-zero elements in Γ of the form $\alpha = k_1\alpha_1 + \ldots + k_n\alpha_n$ with $k_i \geq 0$, $i = 1,\ldots,n$. For $\alpha = \Sigma k_i\alpha_i \in \Gamma$ we call the support of α the subdiagram of the Dynkin diagram of A, consisting of those vertices p_i, for which $k_i \neq 0$, and all the edges joining these vertices.

The set $\Pi = \{\alpha_i | a_{ii} = 2\}$ is called the set of simple roots. We define the positive root system $\Delta_+ = \Delta_+(A)$, associated with A, by the properties:

(R1) $\{\alpha_1,\ldots,\alpha_n\} \subset \Delta_+ \subset \Gamma_+$; $2\alpha_i \notin \Gamma_+$ if $\alpha_i \in \Pi$;

(R2) if $\alpha = \Sigma k_j\alpha_j \in \Delta_+$, $\alpha_i \in \Pi$ and $\alpha \neq \alpha_i$, then $\alpha + k\alpha_i \in \Delta_+$ if and only if $-p \leq k \leq q$, $k \in \mathbb{Z}$, where p and q are some non-negative integers satisfying $p - q = \sum_j a_{ij}k_j$;

(R3) if $\alpha \in \Delta_+$, $\alpha_i \notin \Pi$ and the vertex p_i is joined by an edge with a vertex from the support of α, then $\alpha + \alpha_i \in \Delta_+$.

The set $\Delta = \Delta_+ \cup (-\Delta_+)$ is called the root system. For $\alpha_i \in \Pi$ we define a reflection r_i by

$$r_i(\alpha_j) = \alpha_j - a_{ij}\alpha_i, \quad j = 1,\ldots,n,$$

and call the group generated by all these reflections the Weyl group. We call the fundamental set the following subset in Γ_+:

$$K = \{\alpha = \sum_j k_j\alpha_j \in \Gamma_+ | \sum_j a_{ij}k_j \leq 0 \text{ if } \alpha_i \in \Pi;$$
$$\text{support } \alpha \text{ is connected}\}.$$

Notice that properties (R1) - (R3) define Δ_+ uniquely; the existence and other properties of Δ_+ will be deduced from the representation theory of quivers.

We call $\alpha \in \Delta$ a nil root if the support of α is one of the diagrams of zero type and $\alpha = k\Sigma a_i\alpha_i$, a_i's being the labels of the Dynkin diagram ($a_1 = 1$ for $A_0^{(1)}$), and $k \in \mathbb{Z}\setminus\{0\}$.

Note that the set Δ is W-invariant. The roots from $\Delta^{re} = \bigcup_{w \in W} w(\Pi)$ are called real roots and from $\Delta^{im} = \Delta \setminus \Delta^{re}$ are called imaginary roots.

2. Dimensions of indecomposable representations of quivers.

We recall that a quiver is an oriented graph (S,Ω) (we admit edges-loops), where S is a connected graph with n vertices $S_o = \{p_1,\ldots,p_n\}$ and Ω is an orientation of S. Denote by S_1 the set of edges of S. We associate with S a symmetric Cartan matrix $A = (a_{ij})$ as follows: $-a_{ij}$ is the number of edges, connecting p_i and p_j in S if $i \neq j$ and $a_{ii} = 2 - 2\#$ (loops-edges in p_i), $i,j = 1,\ldots,n$. This is a bijection between the finite connected graphs and the indecomposable symmetric (generalized) Cartan matrices, S being the Dynkin diagram of A. We define a bilinear form (,) on Γ by $(\alpha_i,\alpha_j) = \frac{1}{2}a_{ij}$. This form is W-invariant. It is also clear that $(\alpha,\alpha) \leq 0$ for $\alpha \in K$.

We recall the definition of the category $\mathcal{M}(S,\Omega)$. An object is a collection (U,φ) of finite-dimensional vector spaces U_p, $p \in S_o$, and linear maps $\varphi_\ell: U_{i(\ell)} \to U_{f(\ell)}$ for any size $\ell \in S_1$ ($i(\ell)$ and $f(\ell)$ denote the initial and finite vertices of the oriented edge ℓ). A morphism $\Psi: (U,\varphi) \to (U',\varphi')$ is a collection of linear maps $\Psi_p: U_p \to U_p'$, $p \in S_o$, such that $\Psi_{f(\ell)}\varphi_\ell = \varphi_\ell'\Psi_{i(\ell)}$. A class of equivalence of isomorphic objects of $\mathcal{M}(S,\Omega)$ is called a representation of the quiver (S,Ω). The element $\sum_i (\dim U_{p_i})\alpha_i \in \Gamma_+$ is called the dimension of the representation.

Denote by $d(S,\Omega)$ the set of dimensions of indecomposable representations of the quiver (S,Ω). The problem we are concerned with is to describe this set.

The following lemma is trivial.

Lemma 1. The set $d(S,\Omega)$ satisfies the properties (R1) and (R3) of a positive root system. Any $\alpha \in d(S,\Omega)$ has a connected support.

Lemma 2. Suppose that $\mathbb{F}$ is infinite. Then the set $d(S,\Omega)$ contains the fundamental set K. Moreover, if $\alpha \in K$ is not a nil root and U is a representation of dimension α with minimal possible dimension of End U, then U is absolutely indecomposable; if char $\mathbb{F} = 0$, then End U = $\mathbb{F}$. In particular,[1] $\mu_\alpha \geq 1 - (\alpha,\alpha)$.

Proof is exactly the same as that of Lemmas 2.5 and 2.7 in [4]. The only additional remark we need is that $\sum_j a_{ij}k_j \leq 0$ if $\alpha_i \notin \Pi$ and $\alpha = \sum_j k_j\alpha_j \in \Gamma_+$.

The following lemma follows from the existence of a reflection functor in the case of an admissible vertex p_i of (S,Ω) (i.e., a source or a sink).

Lemma 3. If p_i is an admissible vertex of the quiver (S,Ω) and $\alpha \in d(S,\Omega)$, $\alpha \neq \alpha_i$, then[2] $r_i(\alpha) \in d(S,\tilde{r}_i(\Omega))$. Moreover, $\mu_\alpha = \mu_{r_i(\alpha)}$ and in the case of a finite base field $\mathbb{F}$ the numbers of indecomposable (or absolutely indecomposable) representations of dimensions α and $r_i(\alpha)$ are equal.

---

[1] μ_α is the "number of parameters" of the set of indecomposable representations of dimension α of the quiver (S,Ω)(see [4] for a precise definition).

[2] $\tilde{r}_i(\Omega)$ is an orientation of the graph S obtained from Ω by reversing the direction of arrows along all the edges containing p_i.

Lemma 4. Provided that $\mathbb{F}$ is algebraically closed, the set $d(S,\Omega)$ does not depend on the orientation Ω of the graph S; moreover, μ_α does not depend on Ω. In the case of a finite base field $\mathbb{F}$ the number of indecomposable (or absolutely indecomposable) representations of dimension α does not depend on the orientation Ω.

Proof. Let $\alpha = \sum_{i=1}^{n} k_i \alpha_i \in \Gamma_+$ and $V_1, \ldots, V_n$ be vector spaces of dimensions $k_1, \ldots, k_n$. Recall that the classification of the representations of a quiver (S,Ω) is equivalent to the classification of the orbits of the linear group $G^\alpha(\mathbb{F}) = GL_{k_1}(\mathbb{F}) \times \ldots \times GL_{k_n}(\mathbb{F})$ operating in the space

$$(1) \qquad \mathcal{M}^\alpha(S,\Omega) = \bigoplus_{\ell \in S_1} \mathrm{Hom}_{\mathbb{F}}(V_{i(\ell)}, V_{f(\ell)}).$$

The reversing of the direction of an arrow of the quiver (S,Ω) gives a new quiver (S,Ω_1) and is equivalent to the replacement of the corresponding summand in (1) by a contragredient representation of the group G^α.

Suppose now that $\mathbb{F}$ is a finite field. Recall that by a theorem of Brauer, for any linear finite group G operating in a vector space $V \simeq \mathbb{F}^k$ the numbers of orbits in V and V* are equal (see [4], Lemma 2.10 for the proof). This implies that if $U \simeq \mathbb{F}^m$ is the space of another representation of G, then the numbers of orbits in $U \oplus V$ and $U \oplus V^*$ are equal (one should apply the Brauer theorem to all the linear groups G_x, $x \in U$, operating in V and V*).

These two remarks imply immediately that the number of all representations of dimension α does not depend on the orientation of the quiver.

Now we obtain immediately by induction on the height α that the number of indecomposable (over the finite field $\mathbb{F}$) representations of dimension α does not depend on Ω (we use the uniqueness of the decomposition of a representation into direct sum of indecomposable representations).

The fact that the number of absolutely indecomposable representations of dimension α does not depend on Ω is also proven by induction on height α for any finite field $\mathbb{F}$. The proof is more delicate. It uses the fact that any indecomposable representation over $\mathbb{F}$ is an essentially unique absolutely indecomposable representation over a bigger finite field $\mathbb{F}' \supset \mathbb{F}$, considered over $\mathbb{F}$. The details can be found in Appendix to [4].

The fact that $d(S,\Omega)$ and μ_α do not depend on Ω follows from the preceding result by the following

Proposition 1. Let A be a finite dimensional algebra and α be an element from the Grotendique ring $K_o(A)$. If the base field is $\mathbb{F}_q$, $q = p^s$, then the number $m_t^\alpha(A)$ of absolutely indecomposable representations of A of "dimension" α over field $\mathbb{F}_{q^t}$ is given by the following formula:

$$(2) \quad m_t^\alpha(A) = rq^{Nt} + \lambda_2^t + \ldots + \lambda_k^t - \mu_1^t - \ldots - \mu_s^t,$$

where r and N are positive integers and $\lambda_2,\ldots,\mu_s$ are complex numbers (not depending on t) such that $|\lambda_i|,|\mu_j|$ are non-negative half-integral powers of q smaller than q^N. The number N is equal to the number of parameters and r to the number of irreducible components of maximal dimension of the set of indecomposable representations of A over an algebraically closed field of characteristic p.

If the base field $\mathbb{F}$ is algebraically closed and of characteristic 0, then for all but a finite number of primes p for a reduction mod P the numbers N and r are again the number of parameters and number of irreducible components of maximal dimension of the set of indecomposable representations of A.

Proof. The set of representations of A of dimension α is the set of orbits of an algebraic group G operating on an algebraic variety M, the subset of absolutely indecomposable representations being a constructible G-invariant subset $X \subset M$.

By Rosenblicht's theorem, we can represent X as a union of G-invariant algebraic varieties $X = \bigcup_{i=1}^{s} X_i$, such that each X_i/G is again an algebraic variety.

Since G_x is connected for any $x \in M$ (as the group of units in the endomorphism ring), we obtain bijections between the set of $G(\mathbb{F}_q)$-rational orbits on $M(\mathbb{F}_q)$, the set of $\mathbb{F}_q$-rational points on $\bigcup_i X_i/G$ and the set of absolutely indecomposable representations defined over $\mathbb{F}_q$ (see Appendix to [4] for details).

Recent general results of Deligne [9] give now formula (2). A standard reduction mod P argument proves the last statement.

An immediate consequence of Lemmas 3 and 4 is:

Lemma 5. Suppose that the base field $\mathbb{F}$ is finite or algebraically closed. Then the set $d(S,\Omega)\setminus\{\alpha_i\}$ is r_i-invariant (and, therefore, $d(S,\Omega) \cup (-d(S,\Omega))$ is W-invariant). Moreover, over a finite base field the numbers of indecomposable (or absolutely indecomposable) representations of dimension α and $w(\alpha)$, $w \in W$, are equal;

over an algebraically closed field one has: $\mu_\alpha = \mu_{w(\alpha)}$, $w \in W$.

Now we are able to prove the final:

Lemma 6. For an algebraically closed base field, the set $d(S,\Omega)$ is exactly the set of positive roots $\Delta_+(A)$, where A is the Cartan matrix of the graph S.

Proof. We will prove that the set $d = d(S,\Omega)$ satisfies properties (R1)-(R3) of $\Delta_+ = \Delta_+(A)$. The properties (R1) and (R3) of Δ_+ are satisfied by Lemma 1. By Lemma 5, $\Delta_+^{re} \subset d$ and since the support of any $\alpha \in d$ is connected we obtain that $d = \Delta_+^{re} \cup (\bigcup_{w \in W} w(K))$, where K is the fundamental set (since for any $\alpha \in d \setminus (K \cup \{\alpha_1, \ldots, \alpha_n\})$ there is a reflection r_i such that height $r_i(\alpha) <$ height α).

Now we prove (R2) for any $\alpha \in d$. If $\alpha = \alpha_j$, this property obviously holds. Therefore, this property holds also for any root $\alpha \in \Delta_+^{re} \subset d$. If $\alpha \in d \setminus \Delta_+^{re}$, then $\alpha \in M = \bigcup_{w \in W} w(K)$. I claim that the set M is convex (i.e., if $\beta, \gamma \in M$, then any $\delta \in [\beta,\gamma] \cap \Gamma$ also lies in M). Indeed, let $\hat{M}$ and $\hat{K}$ be the open kernels of the convex hulls of M and K in the space $V = \Gamma \otimes_{\mathbb{Z}} R$; $\hat{M}$ is a convex cone. We introduce the canonical Riemanian metric on $\hat{M}$ (see e.g. [10]). This metric is W-invariant and W operates discretely on the Riemanian manifold M since W is a discrete subgroup in GL(V)). Therefore, any segment $[\alpha, w(\alpha)]$, $\alpha \in \hat{M}$, $w \in W$, intersects only a finite number of hyperplanes of reflections, say, $r_{\beta_1}, \ldots, r_{\beta_s} \in W$. But then $[\alpha, w(\alpha)] \subset \bigcup_i r_{\beta_i} \ldots r_{\beta_1} \hat{K}$. Clearly, this implies that M is convex.

So (R2) is satisfied for any $\alpha \in M$, which completes the proof of the Lemma.

We summarize the obtained results in the following two theorems (cf. [4]).

Theorem 1. *Let* (S,Ω) *be a quiver and let the base field* $\mathbb{F}$ *be a finite field* $\mathbb{F}_q$. *For* $\alpha \in \Gamma_+$ *let* $m_t^\alpha(S,\Omega)$ *denote the number of absolutely indecomposable and* $\overline{m}_t^\alpha(S,\Omega)$ *denote the number of indecomposable representations of* (S,Ω) *of dimension* α *defined over* $\mathbb{F}_{q^t}$. *Then*

a) $m_t^\alpha(S,\Omega)$ *and* $\overline{m}_t^\alpha(S,\Omega)$ *do not depend on the orientation* Ω *of* S *and the action of* W *on* α.

b) *For* $\alpha \notin \Delta_+$ *there is no indecomposable representations of* (S,Ω) *of dimension* α.

c) *For* $\alpha \in \Delta_+^{re}$ *there exists a unique indecomposable representation of* (S,Ω) *of dimension* α *which is absolutely indecomposable and is defined over the prime field.*

d) *For* $\alpha \in \Delta_+^{im}$ *there exists complex numbers* $\lambda_2,\dots,\lambda_k,\mu_1,\dots,\mu_s$ *(depending on* α *but not on t) and positive integers* N *and* r *such that* $|\lambda_i|,|\mu_j|$ *are non-negative half-integral powers of* q *smaller than* q^N, $N \geq 1-(\alpha,\alpha)$ *and*

$$(3)\quad m_t^\alpha(S,\Omega) = rq^{Nt} + \lambda_2^t + \dots + \lambda_k^t - \mu_1^t - \dots - \mu_s^t .$$

Analogous formula takes place for $\overline{m}_t^\alpha(S,\Omega)$. *One has:* $m_t^\alpha(S,\Omega) = \overline{m}_t^\alpha(S,\Omega)$ *for a non-divisible* α.

Theorem 2. *Let* (S,Ω) *be a quiver and let the base field* $\mathbb{F}$ *be algebraically closed. Let* $\Delta_+ = \Delta_+(A)$ *be the positive root system, where* A *is the Cartan matrix of the graph* S. *Then*

a) *For* $\alpha \in \Gamma_+$, α *is a dimension of an indecomposable representation of the quiver* (S,Ω) *if and only if* $\alpha \in \Delta_+$.

b) For $\alpha \in \Delta_+^{re}$ there exists a unique indecomposable representation of (S,Ω) of dimension α.

c) For $\alpha \in \Delta_+^{im}$ there exists an infinite number of indecomposable representations of (S,Ω) of dimension α. Moreover, the number of parameters of the set of indecomposable representations of dimension α is at least $1 - (\alpha,\alpha) > 0$ and does not depend on Ω and the action of W.

3. Further remarks.

a) Infinite root systems. An immediate consequence of the results of sec. 2 is

Proposition 2 (cf. [4]). Let A be a symmetric square matrix with integral entries, satisfying condition (C1)-(C3) of sec. 1. Then the associated positive root system Δ_+ (satisfying the properties (R1)-(R3)) exists. Moreover, $\Delta_+ = \Delta_+^{re} \cup \Delta_+^{im}$, where $\Delta_+^{re} = \bigcup_{w \in W} (w(\Pi) \cap \Gamma_+)$ and $\Delta_+^{im} = \bigcup_{w \in W} w(K)$.

Remark. The statement that in the case of a Cartan matrix, associated with a graph without loops, any element from K is a root appears in [5] (see Theorem ; however, it seems that there is a gap in the proof of the crucial Proposition 3 - in the case k = 1).

The results of sec. 2 can be extended to the case of species (see [2], [1] for definitions) when the base field is finite. In particular, this gives a generalization of Proposition 2 for a symmetrisable A. For an arbitrary field the reduction mod p argument does not work and I can extend the results of sec. 2 only modulo the following conjecture (cf. [4]).

Conjecture (*). Let G be a linear algebraic group operating in a vector space V defined over a field $\mathbb{F}$ of characteristic 0. Then the cardinalities of the sets of the orbits with a unipotent stabilizer (or with a stabilizer such that its maximal split torus is trivial) of the group G in V and V* and the number of parameters of these sets are equal.

Now I would like to give an "abstract" definition of an (ordinary) infinite root system. Let Γ be a full lattice in a real vector space V. We recall that a reflection in a vector $\alpha \in V$ is an automorphism Γ_α of V such that its fixed point set has codimension 1, $\Gamma_\alpha(\alpha) = -\alpha$ and $\Gamma_\alpha(\Gamma) = \Gamma$.

Let Δ be a subset in Γ { }; we denote by Δ^{re} the set of vectors from Δ in which there exists a reflection preserving Δ and by W the group generated by all the reflections in vectors from Δ. The set Δ is called a root system (in general infinite) if the following conditions are satisfied:

(i) Γ is the $\mathbb{Z}$-span of Δ^{re};

(ii) For any $\beta \in \Delta$ and $w \in W$ all the points of Γ which lie on the segment $[\beta, w(\beta)]$ belong to Δ;

(iii) For $\beta \in \Delta \setminus \Delta^{re}$ the set $W(\beta)$ lies in an open half-space.

This definition includes non-reduced root systems (i.e., some of $2\alpha_i$'s may lie in Δ) which naturally appear in Lie superalgebras (see [3]), but I do not know whether they are related to representations of graphs.

Note also that one can easily show that for a finite Δ this definition is equivalent to a usual definition of a finite root system [8].

For simplicity we excluded from the abstract definition of root systems the case when the graph contains an edge-loop (see sec. 1). One can see from sec. 1 and 2 that they are also important. One can define infinite dimensional Lie algebras $\mathfrak{g}(A)$, associated with Cartan matrices introduced in sec. 1. The root system of $\mathfrak{g}(A)$ is then the system Δ. One can also define highest weight representations for these Lie algebras and prove the character formula (cf. [3]). In the simplest new case of the (1 x 1) zero matrix A the Lie algebra $\mathfrak{g}(A)$ is the infinite Heisenberg algebra.

b) Representations of quivers over non-closed fields.

As was mentioned in a), all the results of sec. 2 can be proven for an arbitrary base field $\mathbb{F}$ modulo conjecture (*).

The first open question is: for a root $\alpha \in \Delta_+^{re}$ is it true that the unique indecomposable representation of dimension α is defined over the prime field (this is proven in sec. 2 only in the case of fields of non-zero characteristic). It would be also interesting to give an explicit construction of these representations. Ringel has done it in [6] in the rank 2 case in terms of some generalized reflection functions.

It is easy to show that if there exists an indecomposable representation over $\mathbb{F}$ of dimension α, then either $\alpha \in \Delta_+^{im}$, or $\alpha = k\beta$, where $\beta \in \Delta_+^{re}$; if, moreover, the Brauer group of $\mathbb{F}$ is trivial, then $\alpha \in \Delta_+$.

Of course, all the results of sec. 2 would be extended to an arbitrary field $\mathbb{F}$ if one proves that the set $d(S,\Omega)$ does not depend on Ω over $\mathbb{F}$.

c) <u>ζ-function of a finite dimensional algebra.</u>

Let A be a finite-dimensional algebra over $\mathbb{F}_q$ and α be an element from $K_o(A)$. Denote by $m_n^\alpha(A)$ the number of absolutely indecomposable representations of A of "dimension" α defined over field $\mathbb{F}_{q^n}$. We set

$$\Phi_{A,\alpha}(z) = \sum_{n\geq 1} \frac{1}{n} m_n^\alpha(A)\, z^n$$

and define a ζ-function

$$\zeta_{A,\alpha}(z) = \exp \Phi_{A,\alpha}(z).$$

From (2) we obtain that

$$\zeta_{A,\alpha}(z) = \frac{\prod_{i=1}^{s} (1-\mu_i z)}{(1-q^N z)^r \prod_{i=2}^{k} (1-\lambda_i z)} .$$

In the case of a quiver S conjecture 1 from Appendix in [4] about the multiplicity m_α of a root α can be stated as follows:

$$m_\alpha = \oint \Phi_{s,\alpha}(z)\, dz$$

where the contour of integration is any circle with the radius less than 1 and the center in 0. If Conjecture 3 from [4] is true, then Conjecture 1 can be stated as follows: m_α = multiplicity of the pole of $\zeta_{S,\alpha}(z)$ in $z = 1$.

d) <u>A connection with prehomogeneous linear groups.</u>

A *prehomogeneous* linear algebraic group G operating in a vector space V is a linear group, admitting dense orbit in V. For irreducible representations these groups have been classified in [7]. An essential (and the most difficult) part of the case of general reductive groups is to classify the linear groups $G^{\alpha} = GL_{k_1} \times \ldots \times GL_{k_n}$ operating in $\mathcal{M}^{\alpha}(S,\Omega) = \bigoplus_{\ell \in S_1} \mathrm{Hom}_{\mathbb{F}}(V_{i(\ell)}, V_{f(\ell)})$, associated with a quiver (S,Ω) and $\alpha = \Sigma\, k_i \alpha_i \in \Gamma_+$, which are prehomogeneous. Of course, a necessary condition is that $(\alpha,\alpha) \geq 1$.

Let S be a connected graph. Let $\alpha \in \Gamma_+$ and let Ω be an orientation of S. Denote by (a) the following procedure: we take an admissible vertex $p_i \in S_o$ and replace α by $r_i(\alpha) + s\alpha_i$, where s is the minimal non-negative integer such that $r_i(\alpha) + s\alpha_i \in \Gamma_+$, and replace Ω by $\tilde{r}_i(\Omega)$. Denote by (b) the following procedure: we take $\ell_o \in S_1$ such that for the "generic" stabilizer H of G^{α} in $\bigoplus_{\ell \neq \ell_o} \mathrm{Hom}_{\mathbb{F}}(V_{i(\ell)}, V_{f(\ell)})$ the maximal dimensions of H-orbits in $\mathrm{Hom}_{\mathbb{F}}(V_{i(\ell_o)}, V_{f(\ell_o)})$ and the dual are equal, and reverse the direction of the edge ℓ_o (one has this situation, for instance, when H is reductive). Denote by $D(S,\Omega)$ (or $D_1(S,\Omega)$) the subset of those $\alpha \in \Gamma_+$ which can be transformed to 0 by iteration of the procedures (a) and (b) (resp. (a)). Clearly, if $\alpha \in D(S,\Omega)$, then G^{α} has a dense orbit in $\mathcal{M}^{\alpha}(S,\Omega)$. It seems that the following should be true.

<u>Conjecture.</u> G^{α} <u>has a dense orbit in</u> $\mathcal{M}^{\alpha}(S,\Omega)$ <u>if and only if</u> $\alpha \in D(S,\Omega)$.

<u>Remark.</u> I have conjectured in [4] that if G^{α} has a dense orbit in $\mathcal{M}^{\alpha}(S,\Omega)$, then $\alpha \in D_1(S,\Omega)$. Ringel has constructed a counterexample to this conjecture. His quiver is: $0 \rightrightarrows 0 \rightarrow 0$ and $\alpha = 3\alpha_1 + 6\alpha_2 + \alpha_3$. It is easy to see that $\alpha \in D(S,\Omega)$ but $\alpha \notin D_1(S,\Omega)$.

References

[1] DLAB, V., RINGEL, C.M.: Indecomposable representations of graphs and algebras. Memoirs of Amer. Math. Soc. 6, 173, 1-57 (1976).

[2] GABRIEL, P.: Indecomposable representations II. Symposia Math. Inst. Naz. Alta Mat. XI, 81-104 (1973).

[3] KAC, V.G.: Infinite dimensional algebras, Dedekind's η-function, classical Möbius function and the very strange formula. Adv. in Math. 30, 85-136 (1978).

[4] KAC, V.G.: Infinite root systems, representations of graphs and invariant theory. Inv. Math. 56(1980), 57-92

[5] OVSIENKO, S.A.: On the root systems for arbitrary graphs, Matrix Problems, 81-87 (1977).

[6] RINGEL, C.M.: Reflection functors for hereditary algebras, preprint (1979).

[7] SATO, M., KIMURA, T.: A classification of irreducible prehomogeneous vector spaces and their relative invariants. Nagoya Math. J. 65, 1-155 (1977).

[8] SERRE, J.-P.: Algèbres de Lie semi-simples complexes, New York-Amsterdam: Benjamin 1966.

[9] DELIGNE, P.: La conjecture de Weil II, Publ. Math. IHES, to appear.

[10] KOECHER, M.: Positivitätsbereiche im R^n, Amer. J. Math. 79, 3, 575-596 (1957).

Massachusetts Institute of Technology
Department of Mathematics
Room 2-178
Cambridge, Massachusetts
02139
U.S.A.

Correction

To the axioms of the positive root system on p.3 of this paper, and also on pp. 58,63 and 69 of [4], one should add one more axiom:

If $\alpha \in \Delta_+ \setminus \Pi$, then $\alpha - \alpha_i \in \Delta_+$ for some $\alpha_i \in \Pi$.

I am grateful to J. Morita who pointed out this to me.

SYMMETRIC ALGEBRAS OF FINITE REPRESENTATION TYPE

Herbert Kupisch and Eberhard Scherzler

Let K be an algebraically closed field. It is a well-known conjecture that for a given natural number d there are only finitely many algebras of finite representation type, having dimension d over K. The conjecture is true for generalized uniserial algebras [7], group algebras [6], symmetric algebras [10,11] and algebras without cycles [1].
In this paper we show that it is true for quasi-Frobenius-algebras.

If R has finite representation type, with R is associated the following algebra A_R. Let N be the radical of R and $e_1,\dots,e_n$ a maximal system of primitive orthogonal idempotents, such that $Re_i \not\cong Re_j$ for $i \neq j$. Then a K-basis B of A_R is the collection of subsets $e_iN^\rho e_j \neq 0$ of R,

$$B = \{e_iN^\rho e_j,\ i,j=1,\dots,n;\ \rho=0,1,2,\dots\},$$

and the multiplication on B is given by the formula

$$e_iN^\rho e_j e_k N^\mu e_q = \delta_{jk} e_i N^\nu e_q ,$$

where ν is determined by the property that the product on the left, if not 0, is in N^ν but not in $N^{\nu+1}$ (see [8]).

If R is a symmetric algebra of finite representation type, with R is associated a second algebra A'_R which is defined in the same way as A_R except for the following modification:

$$e_j N e_i e_i N e_k = 0, \text{ if } c_{ii} = 4 \text{ and } c_{jj} = c_{kk} = c_{ji} = c_{ki} = 2,$$

where (c_{ji}) denotes the Cartan-matrix of R.

Then, in case of a symmetric algebra, the basic algebra R^o of R is isomorphic to A_R, if char $K \neq 2$, and to A_R, or A'_R, if char $K = 2$ [10,11].

Our objective is to prove that in case R is a non symmetric quasi-Frobenius-algebra, the basic algebra R^o is isomorphic to A_R.

The two cases combined yield

Theorem C: Let R be an indecomposable quasi-Frobenius-algebra of finite representation type.

a) If R is not a symmetric algebra, then the basic algebra R^o of R is isomorphic to A_R.

b) If R is a symmetric algebra, then

$R \cong A_R \cong A'_R$, if char $K \neq 2$;

$R \cong A_R$ or $R \cong A'_R$, if char $K = 2$.

Since A_R and A'_R either has a K-basis which together with 0 form a semigroup, we have

Corollary: For a given natural number d there are only finitely many quasi-Frobenius-algebras of finite representation type, having dimension d over K.

In case of a symmetric algebra, the fact that R^o is isomorphic to A_R or A'_R rests on the following

Theorem A [10]: Let R be a weakly symmetric algebra satisfying the following condition (P): For every ideal I of R the separated quiver Q'(I) contains no subquiver of type $\tilde{A}_m$ or $\tilde{D}_m$, the extended Dynkin-diagrams. Then R has a uniserial projective module.

In the same way part a) of Theorem C follows from the corresponding result for quasi-Frobenius-algebras which we state in

Theorem B: Let R be a quasi-Frobenius-algebra satisfying condition (P). Then R has a uniserial projective module.

We give a brief outline of the paper.
Section 1 is a list of notations and of some known facts about algebras of finite representation type.
Sections 2 and 3 contain the proof of Theorem B. To prove it we shall reduce it to the weakly symmetric case. For this purpose we first pass from R to $A = A_R$ (clearly, R has a uniserial projective module, if A has), and then from A to a weakly symmetric algebra $\bar{A}$; roughly speaking, $\bar{A}$ is the quotient algebra with respect to the action of the group $G = \langle \sigma \rangle$, generated by a Nakayama-automorphism σ of A which induces a permutation on the basis B of A. The precise definition of $\bar{A}$ will be given in section 2, where also some

results concerning the relationship between A and $\bar{A}$ are proved. In particular, $\bar{A}$ is a weakly symmetric algebra which satisfies condition (P), so that Theorem A applies, providing us with uniserial projective $\bar{A}$-modules.

The remaining task then is to show that there is at least one uniserial projective $\bar{A}$-module which lifts to a uniserial (projective) A-module. This will be done in section 3 by studying V-sequences [10] in $\bar{A}$ and the corresponding sequences in A.

Sections 4 and 5 are devoted to the proof of Theorem C, where, in order to make the paper self-contained, the symmetric case will be included.

In section 4 we prove that, if R is not weakly symmetric, then R is regular in the sense of [8] , i.e. $e_iNe_iNe_j = e_iNe_jNe_j$ for all $i,j = 1,...,n$. This will be needed in section 5 to show that Theorem C follows from Theorem B.

We want to point out that Ch. Riedtmann has announced [14] a full classification of all quasi-Frobenius-algebras of finite representation type over an algebraically closed field, which in particular, gives an independent proof of the conjecture for these algebras. She also observed that for a symmetric algebra the statement $R^o \cong A_R$ of [10, Satz 2] is in general not true, if char $K = 2$. A counterexample of hers was communicated to the first author by Gabriel. The correct statement [11] is that of part b) in Theorem C.

1.

In this section we fix the notation and collect some facts from [3], [5] and [8] about algebras of finite representation type.

Throughout the paper K is an algebraically closed field and R is an indecomposable basic quasi-Frobenius-algebra over K. N denotes the radical of R, n the number of simple R-modules.

$$R = \bigoplus_{i=1}^{n} Re_i$$

is a direct decomposition of R, the e_i being primitive orthogonal idempotents.

$F_i = Re_i/Ne_i$, $i = 1,\ldots,n$ then represent the non isomorphic simple (left) R-modules.

c_{ji} , $i,j = 1,\ldots,n$, are the Cartan-invariants of R, i.e. c_{ji} is the number of composition factors of Re_i which are isomorphic to F_j. Since K is a splitting field for R, we have

$$c_{ji} = \dim_K e_j Re_i \ .$$

It is well known ([3],[5]) that, if R has finite representation type, the following condition (P) holds:

(P) For every ideal I of R the separated quiver Q'(I) has

(1) no subquiver of type $\tilde{A}_m$: 

(2) no subquiver of type $\tilde{D}_m$: 

Here Q'(I) denotes the separated quiver $S'_{\bar{I}}(\bar{R})$ in the sense of [3] with

$$\bar{R} = R/U \ , \ U = NI + IN \ , \ \bar{I} = I/U.$$

In addition to the above notation R will be assumed to satisfy condition (P).

(P1) implies that the algebra $A = A_R$ exists [8]. We denote by

$$B = \{u^{\rho}_{ji}\}$$

as defined in [8,(1.13),(1.14),(2,3)] the <u>standard basis</u> of A.

We recall that $u_{ji} = u^{1}_{ji}$ corresponds to $e_j N e_i$ and that, according to [8, Satz 1.1, Folgerung 1.1 - 1.2], we have

$$u^{\rho}_{ii} = u_{ii} \cdot u_{ii} \cdot \ldots \cdot u_{ii} \qquad (\rho \text{ times}),$$

and for $i \neq j$

(a) $$u^{\rho}_{ji} = u_{ji} \cdot u^{\rho-1}_{ii} \quad , \quad \rho = 1,\ldots\ldots,c_{ji} \quad ,$$

or

(b) $$u^{\rho}_{ji} = u^{\rho-1}_{jj} \cdot u_{ji} \quad , \quad \rho = 1,\ldots\ldots,c_{ji} \quad ,$$

holds. u_{ji} is <u>right regular</u> in case (a), <u>left regular</u> in case (b), <u>regular</u>, if (a) and (b) hold. Clearly, u_{ji} regular means

(c) $$u_{ji}u_{ii} = u_{jj}u_{ji} \ .$$

In particular, u_{ji} is regular, if $c_{ji} = 1$.

If there is no danger of confusion, we denote by e_i the primitive orthogonal idempotents u^{o}_{ii}, $i = 1,\ldots,n$, of A, by N the radical and by c_{ji}, $i,j = 1,\ldots,n$, the Cartan-invariants of A, which coincide with the Cartan-invariants of R [8].

Two elements $u,v \in B$ are <u>independent</u> (see [8,§2]), if $u \notin (v)$ and $v \notin (u)$. A non empty subset $B_o \subset B$ is <u>independent</u>, if any two elements of B_o are independent.

The separated quiver $Q'(B_o)$ of an independent subset B_o of B is the separated quiver $Q'(\langle B_o \rangle)$ of the ideal $\langle B_o \rangle$ generated by B_o.

The definition implies that $Q'(B_o)$ is the quiver which has $\{1,\dots,n\} \times \{0,1\}$ as set of points, and there is an arrow from $(i,0)$ to $(j,1)$ if and only if $e_j B_o e_i \neq 0$ (i.e. $u^{\rho}_{ji} \in B_o$ for some ρ).

Finally, an independent subset B_o of B is called a B-chain, if $Q'(B_o)$ is a Dynkin-diagram or an extended Dynkin-diagram. The type of a B-chain is the type of the diagram.

We usually write a B-chain as a sequence of its elements.

In this terminology (P1) and (P2) are equivalent to the following conditions (P1') and (P2') respectively:

(P1') There is no B-chain of type $\tilde{A}_m$.

(P2') There is no B-chain of type $\tilde{D}_m$.

The above definition also implies

(1.1) Let B_o be an independent subset of B with the following property: For every e_i either $e_i B_o$, resp. $B_o e_i$, is empty or contains at least two elements. Then B_o contains a B-chain of type $\tilde{A}_m$.

Let s_i denote the maximal number of independent elements in Be_i. As a consequence of (P2') we have

(1.2) $$s_i \leq 3 \text{ for every } i.$$

If R is a weakly symmetric algebra, i.e.

(1.3) $\operatorname{soc}(Re_i) \cong Re_i/Ne_i$ for every i,

a sequence $J = (j_1, \ldots, j_s)$, $j_\nu \in \{1, \ldots, n\}$, is a <u>V-sequence</u>, if one of the following equivalent conditions (see [10,(1.1)]) holds:

(i) $c_{j_\nu j_\mu} \neq 0$ if and only if $|\mu-\nu| \leq 1$

(ii) $c_{j_\nu j_{\nu+1}} \neq 0$ and $c_{j_\nu j_{\nu+2}} = 0$.

Since (c_{ji}) is a symmetric matrix in a weakly symmetric algebra, the definition implies

(1.4) If R is an indecomposable weakly symmetric algebra, then for any $i,j \in \{1, \ldots, n\}$ there exists a V-sequence $J = (i, \ldots, j)$.

Also from (i) and [8,§2] it follows

(1.5) To every V-sequence $J = (j_1, \ldots, j_s)$ correspond two B-chains:

$$u_{j_1 j_2}, u_{j_3 j_2}, u_{j_3 j_4}, \ldots, u_{j_{s-1} j_s} \quad \text{and}$$
$$u_{j_2 j_1}, u_{j_2 j_3}, u_{j_4 j_3}, \ldots, u_{j_s j_{s-1}} \quad .$$

Here we have assumed that s is even. If not, $u_{j_{s-1} j_s}$ and $u_{j_s j_{s-1}}$ have to be interchanged. This convention will be used throughout.

2.

In this section we introduce the algebra $\bar{A}$ and deduce some basic properties.

By the assumption on R we know from [8,§1] that A is a Frobenius-algebra. We choose a Nakayama-automorphism σ (see [12]) of A which permutes the idempotents e_i. The construction of A then implies that σ induces a permutation on B. Hence

$$G = \langle\sigma\rangle$$

is a finite group of automorphisms of A.
We consider G also as a group acting on $\{1,\ldots\ldots,n\}$ by $\tau(i) = j$, if $\tau(e_i) = e_j$ for $\tau \in G$. Then, again by the construction of A,

$$\tau(u^m_{pq}) = u^m_{\tau(p)\tau(q)} \ .$$

We write $\bar{x}$, resp. $\bar{i}$ for the orbit of $x \in B$, resp. $i \in \{1,\ldots\ldots,n\}$ and $\alpha(i)$ for the length $|\bar{e}_i|$ of the orbit $\bar{e}_i = \bar{u}^o_{ii}$.

Because of (P) we have (see [9,§1, Lemma])

(2.1) For a fixed $i \in \{1,\ldots,n\}$ let $e = \sum_{j\in\bar{i}} e_j$. Then eAe is a generalized uniserial quasi-Frobenius-algebra.

Also (P) implies

(2.2) Let $c_{ik} \neq 0$. Then the following holds

a) If $\alpha(i) > \alpha(k)$, then $\alpha(k)$ divides $\alpha(i)$.

b) $\alpha(i) > \alpha(k)$ if and only if there exist different representatives u_{jk}, u_{tk} in $\bar{u}_{ik}$.

Proof. a) is a consequence of (1.1) and (P). Using a) we see that $\alpha(i) > \alpha(k)$ if and only if

$$\tau(e_k) = e_k \quad \text{and} \quad \tau(e_i) \neq e_i \quad \text{for some} \quad \tau \in G,$$

which proves b).

In particular we obtain from (2.2 b)

(2.3) If $c_{ik} \neq 0$ and $\alpha(i) > \alpha(k)$, then Ae_k is not uniserial.

To define $\bar{A}$ we want to show that the multiplication on B naturally induces a multiplication on the set $\bar{B} = \{\bar{x} \mid x \in B\}$ of orbits. For this sake we need the following lemma.

(2.4) <u>Lemma</u>: Let $u, v \in B$ and $x, x' \in \bar{u}$, $y, y' \in \bar{v}$, and assume that $xy \neq 0$ and $x'y' \neq 0$. Then

$$x'y' = \tau(xy) \quad \text{for some} \quad \tau \in G.$$

<u>Proof</u>. Without restriction we can assume that $y' = y$. Let

$$x = u^r_{jl}\ , \ y = u^s_{lt}\ , \ x' = \pi(x) \quad \text{for some} \quad \pi \in G.$$

We consider the following two cases according to whether (2.5) holds for $q = t,j$ or not.

(2.5) There exists some $\tau \in G$ such that $u_{\tau(q)q} \notin \operatorname{Soc} Ae_q$.

First assume that (2.5) does not hold for $q = t,j$, which means

(2.5a) $u_{\tau(q)q} \in \operatorname{soc} A$ for every $\tau \in G$, $q = t,j$.

Since

$$a = xy \neq 0\ , \ b = x'y = \pi(u^r_{jl})u^s_{lt} \neq 0\ ,$$

we have $\pi(1) = 1$. If $\pi(j) = j$, we are done. If $\pi(j) \neq j$,

$$B_o = \{\tau(a) \ , \tau(b) \mid \tau \in G\}$$

is because of (2.5a) an independent subset of B satisfying the hypothesis of (1.1). Hence B_o contains a B-chain of type $\tilde{A}_m$, a contradiction to (P) which proves the lemma in case that (2.5) does not hold for $q = j,t$.

Now, assume that (2.5) holds for $q = j$ or t. Then we show

(2.6) $$\alpha(j) = \alpha(1) = \alpha(t) \ ,$$

from which the assertion of the lemma follows immediately.

Since c_{j1}, c_{1t}, $c_{jt} \neq 0$, to prove (2.6) it is enough to prove

(2.7) If $c_{ki} \neq 0$ and (2.5) holds for $q = i$ or $q = k$, then $\alpha(i) = \alpha(k)$.

Proof. Let

$$E = \sum_{j\in\bar{i}} e_j + \sum_{j\in\bar{k}} e_j \quad \text{and} \quad A' = EAE \ .$$

Then A' is an indecomposable Frobenius-algebra satisfying (P). Therefore we can assume without restriction that A' = A. Suppose $\alpha(i) > \alpha(k)$. From $c_{ki} \neq 0$ and $A = A'$ we conclude that Ne_i/N^2e_i has a composition factor $F_{k'}$, $k' \in \bar{k}$, and Ne_k/N^2e_k has a composition factor $F_{i'}$, $i' \in \bar{i}$. By (1.1) Ne_i/Ne_i^2 or Ne_k/N^2e_k is simple. Hence (2.2) shows that Ne_i/N^2e_i is simple and, replacing k by k',

(2.8) $$Ne_i/N^2e_i \cong F_k \ .$$

Moreover, (2.2) shows that Ne_k/N^2e_k has at least two composition factors F_d and F_t , $d,t \in \bar{i}$. Now, assume that besides it has a composition factor F_q, $q \in \bar{k}$. Then, by the preceeding argument, e_qN/e_qN^2 has two composition factors corresponding to some d' and t' $d', t' \in \bar{i}$ which gives a B-chain

$$u_{dk} , u_{tk} , u_{qk} , u_{qt'} , u_{qd'}$$

of type $\tilde{D}_m$, a contradiction to (P). This proves the first part of

(2.9) a) Every composition factor of Ne_k/N^2e_k is isomorphic to some F_j, $j \in \bar{i}$.

b) $N^2e_k/N^3e_k \cong F_p$ for some $p \in \bar{k}$.

The second follows in view of (2.1) and (2.8) from the first.

Next we show that Ae_i is uniserial. From (2.8) and (2.9a) we know that every composition factor of N^2e_i/N^3e_i is isomorphic to some F_j, $j \in \bar{i}$. This and (2.1) imply that N^2e_i/N^3e_i is simple,

$$N^2e_i/N^3e_i \cong F_j \quad \text{for some} \quad j \in \bar{i} .$$

Hence we see that $N^me_i/N^{m+1}e_i$ is simple for $m = 1,2,3,\ldots$, if $N^me_i \neq 0$, and so Ae_i is uniserial.

Now, let us assume that (2.5) holds for $q = i$. Then (2.8) and (2.9) imply

$$N^3e_i \neq 0 \quad \text{and} \quad N^3e_i/N^4e_i \cong F_{k'} \quad \text{for some } k' \in \bar{k} ,$$

and, since soc $Ae_i \cong F_{\sigma(i)}$,

$$N^4 e_i / N^5 e_i \cong F_j \quad , \text{ for some } j \in \bar{i} .$$

This by (2.8) implies that (2.5) also holds for $q = k$. Therefore, to prove (2.7) we can assume that (2.5) holds for $q = k$. Then it follows from (2.9) that $N^3 e_k \neq 0$ and there exist $u_{k'j}$, $u_{jk} \in B$ with $j \in \bar{i}$, $k' \in \bar{k}$, such that for $u = u_{k'j} u_{jk}$ we obtain

(2.10) $Nu/N^2 u \cong F_{j'}$ for some $j' \in \bar{i}$.

Since we have assumed $\alpha(i) > \alpha(k)$, we can choose $\tau \in G$ such that $\tau(k') = k'$, $\tau(k) = k$, $\tau(j') \neq j'$. Then $\tau(u) = u$ by (2.1). Applying τ to Au, we get from (2.10) that Au is not uniserial and so Ae_j is not, a contradiction. This proves (2.7) and completes the proof of the lemma.

Because of the lemma we can define $\bar{A}$ by

(2.11) <u>Definition</u>: $\bar{A}$ is the algebra having as K-basis the set of orbits

$$\bar{B} = \{\bar{x} \mid x \in B\} ,$$

and the multiplication on $\bar{B}$ given by

$$\bar{u}\bar{v} = \begin{cases} \overline{xy} \text{ , if } xy \neq 0 \text{ for some } x \in \bar{u},\ y \in \bar{v} \\ 0 \text{ else .} \end{cases}$$

The construction shows that $A_{\bar{A}}$ exists and coincides with $\bar{A}$.

Also, if $\left\{v^{\mu}_{\ell j}\right\}$ is the standard basis of $\bar{A}$, we can identify its elements with the orbits $\bar{x} \in \bar{B}$,

$$\bar{B} = \left\{v^{\mu}_{\ell j}\right\} .$$

From the definition we easily deduce some basic properties of $\bar{A}$.

(2.12) If $\bar{B}_o$ is an independent subset of $\bar{B}$ and $B_o = \{x \in B | \bar{x} \in \bar{B}_o\}$, then B_o is independent.

Proof. Clearly, x and y belonging to different orbits are independent. However, if $\bar{x} = \bar{y}$, then x and y are on the same level in the radical, since G is a group of automorphisms. Hence x and y are independent.

In particular, because of (1.1), this implies

(2.13) Every $\bar{B}$-chain of type $\tilde{A}_m$, resp. $\tilde{D}_m$, can be lifted to a B-chain of the same type.

Since A is a quasi-Frobenius-algebra, we know that

(2.14) $\quad \operatorname{soc} Ae_i \cong A\sigma(e_i)/N\sigma(e_i)$ for every i.

This implies

(2.15) $\quad \operatorname{soc} \bar{A}\bar{e}_i \cong \bar{A}\bar{e}_i/\bar{N}\bar{e}_i$ for every i $\quad (\bar{N} = \operatorname{rad} \bar{A})$

which says, that $\bar{A}$ is a weakly symmetric algebra. Hence the Cartan-matrix $(\bar{c}_{ji})$ of $\bar{A}$ is symmetric.

Here we write

$$\bar{c}_{ji} \text{ instead of } c_{\bar{j}\bar{i}} \,.$$

(2.16) a) If $c_{ji} \neq 0$, then $\bar{c}_{ji} \neq 0$.

b) If $\bar{c}_{ji} \neq 0$, then $c_{i'j'} \neq 0$ for some $i' \in \bar{i}$, $j' \in \bar{j}$.

<u>Proof</u>: Obvious.

(2.13), (2.15) and (2.16) in particular give the following

(2.17) <u>Proposition</u>: $\bar{A}$ is an indecomposable weakly symmetric algebra satisfying condition (P).

3.

In this section we prove Theorem B. Our purpose is to show that among the uniserial projective $\bar{A}$-modules which by Theorem A exist, there is at least one $\bar{A}\bar{e}_i$, such that Ae_i is also uniserial. The starting point is the following

(3.1) <u>Lemma</u>: Let $\bar{A}\bar{e}_i$ be uniserial. If Ae_i is not uniserial, then $\alpha(i) < \alpha(k)$ for some k with $c_{ki} \neq 0$.

<u>Proof</u> follows from (2.2).

We put

$$\bar{c}_i = \bar{c}_{ii} - 1 \text{ for } i = 1, \ldots, n \,.$$

(3.2) Lemma: If $\bar{c}_i > 1$ for some i, then $\alpha(i) = \alpha(j)$ for every j.

Proof. From (2.7) we know

(i) Let $\bar{c}_i, \bar{c}_k > 1$ and $\bar{c}_{ki} \neq 0$. Then $\alpha(i) = \alpha(k)$.

Suppose $\alpha(i) \neq \alpha(k)$ for some k. Since $\bar{A}$ is indecomposable, there is a V-sequence $J = (j_1, \ldots, j_s)$ of minimal length satisfying

$$\bar{c}_{j_1} > 1, \quad \alpha(j_1) = \ldots = \alpha(j_{s-1}) \neq \alpha(j_s)$$

By [10,(1.4)], if $\bar{c}_{j_s} > 1$, then $s = 2$ and thus (i) applies. If $\bar{c}_{j_s} = 1$, the minimal length of J implies

(ii) $\bar{c}_{j_1} > 1, \ \bar{c}_{j_2} = \ldots = \bar{c}_{j_s} = 1.$

Then, because of $[8, F_7]$, the two $\bar{B}$-chains associated with J together with $v_{j_1 j_1}$ form a $\bar{B}$-chain

$$v_{j_s j_{s-1}}, \ldots, v_{j_2 j_1}, v_{j_1 j_1}, v_{j_1 j_2}, \ldots, v_{j_{s-1} j_s} .$$

By (2.2) and (2.12) this $\bar{B}$-chain can be lifted to a B-chain of type $\tilde{A}_m$ or $\tilde{D}_m$ according to whether $\alpha(j_s) < \alpha(j_{s-1})$ or $\alpha(j_s) > \alpha(j_{s-1})$, a contradiction to (P) which proves (3.2).

In the remaining case, when $\bar{c}_j = 1$ for all j, we need some further information on the behavior of α on a V-sequence.

(3.3) Lemma: If $J = (j_1, \ldots, j_s)$ is a V-sequence in $\bar{A}$, then

$\alpha(j_\nu)$ can change at most once, i.e. $\alpha(j_\lambda) \neq \alpha(j_{\lambda+1})$ for some λ implies:

$$\alpha(j_1) = \ldots = \alpha(j_\lambda) \quad \text{and} \quad \alpha(j_{\lambda+1}) = \ldots = \alpha(j_s).$$

Proof. Suppose (3.3) is not true and $J = (j_1, \ldots, j_s)$ is a minimal counter-example. Let

$$v_{j_1 j_2}, \ldots, v_{j_{s-1} j_s}$$

be one of the $\bar{B}$-chains associated with J. By (1.1), (2.2) and (2.12) it can be lifted to a B-chain of type $\tilde{A}_m$ in case

$$\alpha(j_1) < \alpha(j_2) = \ldots = \alpha(j_{s-1}) > \alpha(j_s),$$

and of type $\tilde{D}_m$ in case

$$\alpha(j_1) > \alpha(j_2) = \ldots = \alpha(j_{s-1}) > \alpha(j_s)$$

or

$$\alpha(j_1) > \alpha(j_s) = \ldots = \alpha(j_{s-1}) < \alpha(j_s) .$$

This contradiction proves the lemma.

We set

$$d = \max(\alpha(j), j = 1, \ldots, n)$$

(3.4) Lemma: Let $\bar{c}_j = 1$ for all j and let $J = (j_1, \ldots, j_r)$ be a V-sequence in $\bar{A}$. If $\alpha(j_1) = d$ and $\bar{s}_{j_1} = 3$, then

$$\alpha(j_1) = \ldots = \alpha(j_r).$$

Proof. Suppose not and let $J = (j_1, \ldots, j_r)$ by a counter-example of minimal length. Then by (3.3)

(i) $\alpha(j_1) = \ldots = \alpha(j_{r-1}) > \alpha(j_r)$.

Put $i = j_1$, $k = j_2$. Since $\bar{s}_i = 3$, there are independent elements

(ii) v_{pi}, v_{qi}, v_{ti} .

Case 1: v_{ki} and at least 2 further elements in (ii), say v_{pi} and v_{qi}, are independent. By (i), for some $\tau \in G$ we have

$$\tau(e_{j_r}) = e_{j_r}, \quad \tau(e_{j_\nu}) \neq e_{j_\nu} \quad \text{for } \nu < r.$$

This implies that, if the e_j are enumerated suitably,

$$u_{pi}, u_{qi}, u_{ki}, u_{kj_3}, \ldots, u_{j_{r-1}j_r}, \tau(u_{j_{r-1}j_r}), \ldots, \tau(u_{pi})$$

is a B-chain of type $\tilde{D}_m$, which contradicts (P).
In particular, this case applies to the V-sequences (i,p), (i,q), (i,t). Consequently we have

(iii) $\alpha(i) = \alpha(p) = \alpha(q) = \alpha(t)$

Case 2: v_{ki} and at most one element in (ii), say v_{ti}, are independent. Then v_{ki} and v_{pi} as well as v_{ki} and v_{qi} are not independent. This implies $\bar{c}_{pk} \neq 0$ and $\bar{c}_{qk} \neq 0$. If $\bar{s}_k = 3$, by the minimal length of J we have $r = 2$ and, using (iii),

$$\alpha(k) < \alpha(i) = \alpha(p) = \alpha(q).$$

Thus, for a suitable $\tau \in G$,

$$u_{pk}, u_{qk}, \tau(u_{pk}), \tau(u_{qk})$$

is a B-chain of type $\tilde{D}_m$. We therefore may assume $\bar{s}_k \leq 2$, $\alpha(k) = \alpha(i)$ and $r \geq 3$. Accordingly, v_{j_3k} and v_{pk} (or v_{qk})

are not independent, hence $\bar{c}_{pj_3} \neq 0$. By the minimal length of J this means that $J' = (i,p,j_3,\dots,j_r)$ is a V-sequence of the type considered in case 1, proving the lemma.

<u>Proof of Theorem B</u>. Since Re_i is uniserial if and only if Ae_i is uniserial, it is enough to show that a uniserial projective A-module exists.

If $\bar{c}_i > 1$ for some i, then the assertion follows from Theorem A, (3.1) and (3.2).

Now assume that $\bar{c}_j = 1$ for all j and that no uniserial Ae_j exists. Choose $\bar{A}\bar{e}_i$ such that $\alpha(i) = d$ is maximal. By (3.1) we have $\bar{s}_i = 2$ or $\bar{s}_i = 3$. If $\bar{s}_i = 3$, let $J = (i,\dots,q)$ be a V-sequence with $\bar{A}\bar{e}_q$ uniserial. Since $\alpha(q) < d$, we can apply (3.4) to get a contradiction.

Thus we may assume that $\bar{s}_i = 2$ and that there is no j with $\alpha(j) = d$ and $\bar{s}_j = 3$. By [10,(2.8),(2.9)] we can construct a V-sequence $J = (j_1,\dots,i,\dots,j_r)$ such that $\bar{s}_{j_1}$ and $\bar{s}_{j_r}$ are 1 or 3. By our assumption, this implies $\alpha(j_1) < \alpha(i)$ and $\alpha(j_r) < \alpha(i)$, a contradiction to (3.3).

4.

In this section we prove that, if R is not weakly symmetric, then every u_{ji} is regular (see §1). This will be needed in the proof of Theorem C.

We recall from [8] that Ae_i and Re_i are <u>right regular</u> (resp. left regular), if every $u_{ji} \in Be_i$ is right regular (resp. left regular). A and R are <u>regular</u>, if u_{ji} is regular for all $i,j \in \{1,\dots,n\}$.

In case of a weakly symmetric algebra, Ae_i is right regular or left regular [8,II,Folgerung 1].

(4.1) <u>Proposition</u>: If R satisfies (P) and R is not weakly symmetric, then R is regular.

<u>Proof</u>. Suppose not. Then, since u_{ki} is left regular if and only if $u_{\sigma(i)k}$ is right regular (2.14), there exist i and k such that u_{ki} is not left regular. First we observe

(4.2) a) $0 \neq u_{ki}^2 = u_{ki}u_{ii}$ and $u_{kk}u_{ki} = u_{ki}^t$ or 0, $t > 2$

b) $c_{ki} \geq 2$.

c) $u_{ki} \nmid u_{ii}$.

Here, $u|v$, resp. $u \nmid v$, denotes $v \in (u)$, resp. $v \notin (u)$ (see [8,§2]).

Clearly a) and b) hold. If c) does not hold, then

$$u_{ik}u_{ki} = u_{ii}$$

This implies

$$u_{ki}^2 = u_{ki}u_{ii} = u_{ki}u_{ik}u_{ki} = u_{kk}^t u_{ki} \quad , \quad t \geq 1,$$

a contradiction to a).

Let σ be as in §2 and

$$p = \sigma(i) .$$

Then, according to (2.14)

$$\operatorname{soc} Ae_i = K \cdot u_{pi}^d \cong F_p \quad \text{for some } d \geq 1 .$$

This means that for every $u_{ji}^{\rho} \in Be_i$ there exists an u_{pj}^{ν} such that

(4.3) $$u_{pj}^{\nu} u_{ji}^{\rho} = u_{pi}^{d} .$$

Also, (2.1) and [7,Satz 5] imply

(4.4) If $l,j \in \bar{i}$ and $c_{lj} \neq 0$, then

a) u_{lj} is regular

b) $u_{lj} \mid u_{jj}$, if $c_{jj} > 1$

c) $c_{jj} = c_{ii}$.

From (4.4b) and (4.2c) it follows that $u_{ki} \nmid u_{pi}$, hence, by (4.3)

$$u_{pk} u_{ki} = u_{pi}^{r} = u_{pi} u_{ii}^{r-1} \quad , \quad r > 1$$

and

$$u_{pk} u_{ki}^{2} = u_{pk} u_{ki} u_{ii} = u_{pi} u_{ii}^{r} \neq 0.$$

This, in particular tells us that $c_{ii} > 2$. Passing to $\bar{A}$ we have $\bar{c}_{ii} > 2$. Then (3.2) implies

(4.5) $$\alpha(j) = \alpha(i) > 1 \quad \text{for every } j,$$

since we have assumed that R is not weakly symmetric.

Finally, we show that there exists a B-chain of type $\tilde{D}_m$. This follows from

(4.6) Let $c_{ii} > 2$, u_{ki} not left regular and $\alpha(i) = \alpha(k) > 1$. Then

a) $c_{kj} \neq 0$ for all $j \in \bar{i}$

b) $c_{qi} \neq 0$ for all $q \in \bar{k}$

c) $u_{qi} \nmid u_{ii}$ for all $q \in \bar{k}$

d) Ne_j/N^2e_j is not simple for all $j \in \bar{i}$

e) Ne_q/N^2e_q is simple for all $q \in \bar{k}$

f) u_{qi} and u_{ti} are independent for all $q,t \in \bar{k}$, $q \neq t$.

Proof. We put

$$e = \sum_{j\in\bar{i}} e_j + \sum_{j\in\bar{k}} e_j \text{ and } A' = eAe.$$

Then A' is also a quasi-Frobenius-algebra which satisfies (P). Hence, without restriction we can assume that $A = A'$ in (4.6).

a) By (4.2) and (4.4) we have

$$0 \neq u_{ki}^2 = u_{ki}u_{ii} = u_{ki}u_{ij}u_{ji} \text{ for every } j \in \bar{i},$$

which implies a).

b) follows from a) and (4.5).

c) Let $q \in \bar{k}$. Since u_{ki} is not left regular, we can choose $j \in \bar{i}$ such that u_{qj} is not left regular. Suppose, c) does not hold, i.e.

$$u_{iq}u_{qi} = u_{ii}.$$

Then from $c_{ii} > 2$ and (4.4) it follows

(i) $\quad u_{iq}u_{qi}u_{ij} = u_{ii}u_{ij} = u_{ij}u_{jj} \neq 0.$

Also, by (4.2) and (4.4) we know $u_{ij} | u_{jj}$ and $u_{qj} \nmid u_{jj}$. Hence $u_{qj} \nmid u_{ij}$. Therefore (i) and the fact that u_{qj} is right regular imply

$$u_{qi}u_{ij} = u_{qj}$$

and so

$$u_{qj}^2 = u_{qi}u_{ij}u_{jj} = u_{qi}u_{ii}u_{ij} = u_{qi}u_{iq}u_{qi}u_{ij} = u_{qq}^t u_{qj}, \quad t \geq 1,$$

a contradiction, since u_{qj} is not left regular.

d) Suppose Ne_j/N^2e_j is simple, say

$$Ne_j/N^2e_j \cong F_t.$$

Then c) implies $t \in \bar{i}$. Repeating this argument shows that Ae_j has no composition factor F_q with $q \in \bar{k}$, a contradiction.

e) follows from d) and (1.1).

f) For the same reason as in d) we have by e)

(ii) $\qquad Ne_k/N^2e_k \cong F_j, \quad j \in \bar{i}.$

Suppose f) does not hold. Then there exist $q,t \in \bar{k}$ such that

$$u_{qi} = u_{qt}u_{ti}.$$

Since (ii) implies

$$u_{qt} \in N^2 \text{ for all } q,t \in \bar{k},$$

we have

$$u_{qt} = u_{ql}u_{lt} \quad \text{for some } l \in \bar{i},$$

hence, using c) and (4.4),

$$u_{qi} = u_{ql}u_{lt}u_{ti} = u_{ql}u_{li}^m = u_{ql}u_{li}u_{ii}^{m-1}, \quad m > 1.$$

This, however, gives the contradiction $u_{qi} \in u_{qi}N$.

Now, we apply (4.5), (4.6)b),c) and f) to obtain a B-chain

$$u_{qi}, u_{ti}, u_{ii}, u_{it}, u_{iq}$$

of type $\tilde{D}_m$. In view of (P) this completes the proof of the proposition.

(4.7) Remark. It is easy to see that (4.1) does not hold if R is weakly symmetric.

5.

In this section we prove that Theorem C follows from Theorem B.

If R is a weakly symmetric algebra which satisfies condition (P), we denote by A_R' the following algebra: A K-basis of A_R' is the standard basis B of A_R, denoted by $B' = \{u'^{\rho}_{ji}\}$, and the multiplication on B' is defined by

$$u'^{\mu}_{kj} \cdot u'^{\rho}_{ji} = \begin{cases} 0, \text{ if } u_{kj} \text{ and } u_{ij} \text{ are not left regular and } \rho + \mu \neq c_{ij} + 1 \\ u^{\mu}_{kj} \cdot u^{\rho}_{ji} \text{ else.} \end{cases}$$

It follows from (P) that this is an associative product; also it is easy to see that A_R' is a symmetric algebra [11].
We sometimes ommit the ' and write B and u^{ρ}_{ji} instead of B' and u'^{ρ}_{ji}, if it is clear from the context which of the two algebras is considered.
From [8] we recall that u_{ji} is not left regular if and only if u_{ij} is not right regular and that in this case we have

(i) $c_{ii} \geq 4$, $c_{jj} = 2$ and $1 < c_{ji} < c_{ii} - 1$.

Moreover, finite representation type even implies [15, Satz 6]:

(ii) $c_{ii} = 4$, if u_{ji} is not left regular for some j.

Hence the definition of A_R' given here coincides with that given in the introduction.

In view of (4.1), Theorem C is thus an immediate consequence of the following

(5.1) <u>Proposition</u>: Let $R = R^{\circ}$ be an indecomposable basic quasi-Frobenius-algebra satisfying condition (P).

a) If R is regular, then $R \cong A_R$.

b) If R is not regular (hence weakly symmetric) and satisfies (ii), then

$R \cong A_R \cong A_R'$, if char $K \neq 2$;

$R \cong A_R$ or $R \cong A_R'$, if char $K = 2$.

<u>Proof</u>. To make the paper selfcontained we include the proof of statement b) [11]. Therefore it will be assumed that, in addition to (P), R satisfies (ii). Then we have by [8,I,F_{10}, (2.12) and 8,II,Satz 1].

(5.2) u_{ji} is not left regular if and only if

$c_{ii} = 4$ and $c_{ji} = c_{jj} = 2$.

To prove the proposition we shall construct a basis of R which has the same multiplication as B has.

We use induction on $g(R)$ = number of simple R-modules.

For $g(R) = 1$ there is nothing to prove, since (P) implies that R is a local uniserial algebra.

Assume that $g(R) = n > 1$. By Theorem B, R has a uniserial projective module Re_i. We fix Re_i and put

$$e = 1 - \sum_{j \in I} e_j \quad \text{and} \quad S = eRe.$$

Then $S = 0$ or S is also a quasi-Frobenius-algebra which

satisfies (P).

In case $S = 0$ we know from (2.1) that R is a generalized uniserial algebra, consequently the assertion holds [7,Satz 6]. Therefore we can assume that $0 \neq g(S) < g(R)$. Hence by our induction hypothesis $S \cong A_S$ or $S \cong A'_S$. Identifying S with A_S or A'_S and A_S with eA_Re or eA'_Re we have

(5.3a) $eA_Re = A_S = S \subset R$ or

(5.3b) $eA'_Re = A'_S = S \subset R$,

according as $S \cong A_S$ or $S \cong A'_S$.

Since Re_i is uniserial, Re_j is uniserial for every $j \in \bar{i}$. This implies that for every $j \in \bar{i}$ there correspond $p(j)$ and $q(j)$ such that

$$Ne_j/N^2e_j \cong F_{p(j)}, \quad e_jN/e_jN^2 \cong F^*_{q(j)} \quad (F^*_K = e_KR/e_KN)$$

and, since R is indecomposable and $S \neq 0$,

(5.4) $p(j), q(j) \notin \bar{i}$ for every $j \in \bar{i}$.

Accordingly, there exists a set of elements

(5.5) $T = \{x_{p(j)j}, y_{jq(j)} \mid j \in \bar{i}\}$ with

$$x_{p(j)j} \in e_{p(j)}Ne_j \quad \text{and} \quad Rx_{p(j)j} = Ne_j .$$

$$y_{jq(j)} \in e_jNe_{q(j)} \quad \text{and} \quad y_{jq(j)}R = e_jN .$$

To construct from T and eBe the basis we need, we first show that we can assume

$$x_{p(j)j}\, y_{jq(j)} \in eBe \quad \text{for every} \quad j \in \bar{i} .$$

For this purpose, let $j \in \bar{i}$. We write

$$p = p(j) \quad \text{and} \quad q = q(j)$$

if there is no danger of confusion.

According to [8, Folgerung 1.1 - 1.2 and (1.13)] we have

(5.6) $$e_pNe_je_jNe_q = e_pN^te_q = e_pNe_q(e_qNe_q)^r \quad \text{or}$$
$$= (e_pNe_p)^r e_pNe_q$$

for some r and t.

Now, by (5.3) and (5.4), $e_pNe_q \subset eA_Re$. Hence $e_pN^te_q$ has a basis B_{pq} consisting of a subset of the standard basis B of A_R, resp. A_R',

$$B_{pq} = \left\{u_{pq}u_{qq}^r, u_{pq}u_{qq}^{r+1}, \ldots\ldots, u_{pq}u_{qq}^d\right\} \quad \text{or}$$
$$= \left\{u_{pp}^r u_{pq}, u_{pp}^{r+1}u_{pq}, \ldots\ldots, u_{pp}^d u_{pq}\right\},$$

where $u_{pq}u_{qq}^{d+1} = 0$, resp. $u_{pp}^{d+1}u_{pq} = 0$.

(5.5) and (5.6) thus imply

$$x_{pj}y_{jq} = \sum_{\nu=r}^{d} \alpha_\nu u_{pq}u_{qq}^\nu, \quad \alpha_\nu \in K, \quad \alpha_r \neq 0, \quad \text{or}$$

$$= \sum_{\nu=r}^{d} \beta_\nu u_{pp}^\nu u_{pq}, \quad \beta_\nu \in K, \quad \beta_r \neq 0.$$

Replacing y_{jq} by a suitable element of e_pNe_j (or similar x_{pj}), we see that we can assume

(5.7) $$x_{pj}y_{jp} = u_{pq}u_{qq}^r \quad \text{or}$$
$$= u_{pp}^r u_{pq} \in eBe.$$

Since Re_j and e_jR are uniserial, the order of the composition factors of Re_j, resp. e_jR is uniquely determined. We denote this order by

$F_{\nu_1}, F_{\nu_2}, \ldots\ldots, F_{\nu_m}$ and $F^*_{\mu_1}, F^*_{\mu_2}, \ldots\ldots, F^*_{\mu_m}$

respectively, where $\nu_k = \nu_k(j)$, $\mu_k = \mu_k(j)$ and

$\nu_1 = \mu_1 = j$ and $\nu_2 = p$, $\mu_2 = q$.

Then (5.3),(5.4) and (5.5) imply

$$N^2 e_j = Nx_{pj} = Re_{\nu_3} Ne_p x_{pj} = \begin{cases} Ru_{\nu_3 p} x_{pj}, & \text{if } \nu_3 \notin \bar{I} \\ Ry_{\nu_3 p} x_{pj}, & \text{if } \nu_3 \in \bar{I} \end{cases}$$

We put $w_{2j} = x_{pj}$,

$$w_{3j} = \begin{cases} u_{\nu_3 p} x_{pj}, & \text{if } \nu_3 \notin \bar{I} \\ y_{\nu_3 p} x_{pj}, & \text{if } \nu_3 \in \bar{I}, \end{cases}$$

and define w_{k+1j} inductively by

$$\text{(5.8a)} \quad w_{k+1j} = \begin{cases} u_{\nu_{k+1}\nu_k} w_{kj}, & \text{if } \nu_k, \nu_{k+1} \notin \bar{I} \\ x_{\nu_{k+1}\nu_k} w_{kj}, & \text{if } \nu_k \in \bar{I} \\ y_{\nu_{k+1}\nu_k} w_{kj}, & \text{if } \nu_{k+1} \in \bar{I} \end{cases},$$

where

$$\text{(5.8b)} \quad u_{\nu_{k+1}\nu_k} \in eBe, \; u_{\nu_{k+1}\nu_k} \notin N^2, \quad x_{\nu_{k+1}\nu_k}, \; y_{\nu_{k+1}\nu_k} \in T.$$

Because of (P) the elements $u_{\nu_{k+1}\nu_k}$, $x_{\nu_{k+1}\nu_k}$, $y_{\nu_{k+1}\nu_k}$ are uniquely determined by ν_k and ν_{k+1} and (5.8b). We call them the <u>irreducible factors</u> of $w_{\ell j}$.

Thus we have

(5.9) a) $w_{\ell j} \in N^{\ell-1} e_j$, $w_{\ell j} \notin N^{\ell}$.

b) $w_{\ell j}$ is uniquely a product of irreducible factors

given by (5.8 b) and the composition series of Re_j.
This product is called the <u>normal factorization</u> of $w_{\ell j}$ with respect to eBe and T.

The construction of the $w_{\ell j}$ in particular shows that for every $j \in \bar{I}$

$$B_j = \{e_j \ , w_{\ell j} \ , \quad \ell = 2, \dots m\}$$

is a K-basis of Re_j.
In the same way we obtain a K-basis B_j^* of e_jR,

$$B_j^* = \{e_j \ , w_{j\ell} \ , \ \ell = 2 , \dots , m\} .$$

For every $j \in \bar{I}$ we put

$$\hat{u}^{\rho}_{\nu_k(j)j} = w_{kj} \quad ,$$

if $\nu_k(j)$ appears for the ρ-th time in the sequence $v_2(j), \dots, v_m(j)$, and define $\hat{u}^{\rho}_{j\mu_k(j)}$ analogously. Also, we write $\hat{u}_{t\ell}$ instead of $\hat{u}^1_{t\ell}$.

Now, since R is selfinjective, the normal factorization of an element $\hat{u}^{\rho}_{t\ell}$ with $t, \ell \in \bar{I}$ is the same whether we start with ℓ or with t (see (5.12 b)). In particular this implies

(5.10) If $\nu_k(j) = j$ for the ρ-th time, then
$$\hat{u}^{\rho}_{\nu_k(j)j} = \hat{u}^{\rho}_{j\mu_k(j)} = \hat{u}_{jj} \cdot \ldots \cdot \hat{u}_{jj} \qquad (\rho \text{ times})$$

and

(5.11) $\hat{B} = eBe \bigcup_{j \in \bar{I}} (B_j \cup B_j^*)$ is a K-basis of R.

Here, as in the construction of B_j and B_j^*, eBe is considered as a subset of $S \subset R$. For convenience, we write $\hat{x}$ instead of x for every $x \in eBe$.

Before proceeding to the final step of the proof of (5.1) we note some further properties of $\hat{B}$.

(5.12) a) If Re_i is regular, then the composition factors of Re_j, $j \in \bar{I}$, have cyclic order.

b) If $\nu_k \in \bar{I}$, then $F^*_{\nu_k}, F^*_{\nu_{k-1}}, \ldots, F^*_{\nu_1}$ is the order of the first k composition factors of $e_{\nu_k}R$.

Proof. a) is obvious, b) follows from the fact that R is self-injective.

(5.13) Let $u = v_1 \ldots v_r$ with $v_\ell \in eBe \cup T$ and $v_1 = ev_1$, $v_r = v_r e$. Then $u = 0$ or $u \in eBe$. Moreover, u is completely determined by the multiplication in S and by (5.6).

Proof follows by an induction argument on the number r of factors v_ℓ from (5.4), (5.7) and (5.9).

(5.14) If Re_j is regular, $j \in \bar{I}$, then

$$\hat{u}_{kk}\hat{u}_{kj} = \hat{u}_{kj}\hat{u}_{jj} \quad \text{for every} \quad k = 1,\ldots,n .$$

Proof. First assume, that $\hat{u}_{kk}\hat{u}_{kj} \neq 0$. Then F_k appears at least twice as a composition factor of Re_j. This, by (5.9) and (5.12a), implies $\hat{u}_{kj}\hat{u}_{jj} \neq 0$. So, without restriction we can assume that $\hat{u}_{kj}\hat{u}_{jj} \neq 0$. In this case, (5.9), (5.10) and (5.12a) show

$$0 \neq \hat{u}_{kj}\hat{u}_{jj} = \hat{u}_{kj}v_1 \ldots v_s \quad \text{with} \quad v_1 = y_{jq}, \quad v_s = x_{pj},$$

and

$$v_r v_{r+1} \dots v_s = \hat{u}_{kj} \text{ for some } r \le s,$$

where $v_1, \dots, v_s$ are irreducible factors. This implies

$$\hat{u}_{kj} v_1 \dots\dots v_{r-1} = \hat{u}^{\rho}_{kk} \text{ for some } \rho,$$

hence

$$\hat{u}_{kj}\hat{u}_{jj} = \hat{u}^{\rho}_{kk}\hat{u}_{kj}$$

Using (5.12 a) again we obtain $\rho = 1$ which proves (5.14).

If R is regular, so is S. Therefore, as a consequence of (5.14) we have

(5.15) If R is regular, then

$$\hat{u}_{kk}\hat{u}_{k\ell} = \hat{u}_{k\ell}\hat{u}_{\ell\ell} \text{ for all } k,\ell = 1,\dots,n.$$

To complete the proof of (5.1) we now treat the two parts of the proposition separately.

Part a) is an immediate consequence of

(5.16) Let Re_i be regular. Then the map

$$\varphi\colon \hat{B} \longrightarrow B\ ,\ \hat{u}^{\rho}_{kt} \longrightarrow u^{\rho}_{kt}$$

induces an isomorphism between R and A_R or R and A'_R according as $S = A_S$ or $S = A'_S$.

Proof. It is enough to prove

(i) $$\varphi(\hat{u}^{\rho}_{kt}\hat{u}^{\mu}_{t\ell}) = u^{\rho}_{kt}u^{\mu}_{t\ell} \quad .$$

(i) is obvious, if $k,\ell,t \notin \bar{i}$. If $t \in \bar{i}, k,\ell \notin \bar{i}$ it follows from (5.6), (5.7), (5.9), (5.13) and the fact that, since Re_j is left regular for $j \in \bar{i}$, we have

$$u_{pj}u_{jq} = u^{r+1}_{pq}$$

not only in A_R, but also in A'_R.

Thus we can assume without restriction that $\ell \in \bar{i}$ and, besides, F_k is a composition factor of $R\hat{u}^{\mu}_{t\ell}$. Then the construction of B_ℓ implies that there exists some ν such that

$$\hat{u}^{\nu}_{k\ell} = v_1 \cdot \ldots \cdot v_r \cdot \hat{u}^{\mu}_{t\ell} \in B_\ell \, ,$$

where $v_1, \ldots, v_r$ are irreducible factors of $\hat{u}^{\nu}_{k\ell}$.

Their product is in $\hat{B}$,

$$v_1 \cdot \ldots \cdot v_r = \hat{u}^{\delta}_{kt} \quad \text{for some } \delta \, ,$$

which follows from (5.13) or from (5.11) according as $k,t \notin \bar{i}$ or not. Moreover, choosing ν minimal, we have

$$\delta = 1.$$

This is obvious, if u_{kt} is left regular, and is implied by (5.14), if u_{kt} is right regular, since $\ell \in \bar{i}$ and so $u_{t\ell}$ is regular.

Hence

(ii) $$\hat{u}_{kt}\hat{u}^{\mu}_{t\ell} = \hat{u}^{\nu}_{k\ell} \quad .$$

By the choice of ν and by the fact that $u_{t\ell}$ is regular we also have

(iii) $$u_{kt}u^{\mu}_{t\ell} = u^{\nu}_{k\ell}$$

in A_R, resp. A'_R.
Using the same argument as above (to show that $\delta = 1$) we now obtain (i) from (ii) and (iii).

Part b). We put

$$c_k = c_{kk} - 1 \quad \text{for} \quad k = 1,\dots,n.$$

In view of (5.16) we can assume that Re_i is not regular. Since Re_i is uniserial and so left regular, this means that Re_i is not right regular. In particular we then have

$$c_i = 1 .$$

Note that, since R is weakly symmetric, we also have $\bar{i} = \{i\}$ and $c_{j\ell} = c_{\ell j}$ for all $\ell, j \in \{1,2,\dots,n\}$.

Case 1. There exists an $Re_q \neq Re_i$ which is not right regular. First we prove

(5.17) Let u_{ji} be not right regular for some j. Then

a) $c_j = 3$, $c_{ji} = 2$.

b) $c_{qi} = c_q = 1$.

c) either $u_{qi} \mid u_{ji}$ or $u^2_{ji} \mid u_{qi}$ holds.

d) $c_{qj} = 2$. In particular, u_{qj} is not left regular.

Proof. a) follows from (5.2), since u_{ji} is not right regular if and only if u_{ij} is not left regular.

b) From the assumption that Re_q is not right regular we know that u_{qk} is not left regular for some k. Hence by (5.2),

$$c_k = 3 \ , \ c_{qk} = 2 \ .$$

This implies by [8,II,Folgerung 2,Satz 6] and [9,Theorem 2,(2.3)]

(5.18) $c_{q\ell} \neq 0$ for every ℓ with $c_\ell > 1$.

In particular $c_{qj} \neq 0$. From this we conclude that $c_{qi} \neq 0$, since otherwise,

$$u_{ij} \ , \ u_{qj} \ , \ u_{jj} \ , \ u_{ji} \ , \ u_{jq}$$

would be a B-chain of type $\tilde{D}_m$. On the other hand, $c_i = c_q = 1$ implies $c_{qi} \leq 1$, hence $c_{qi} = 1$.

c) Since Re_i is uniserial, one of the following relations holds

(i) $u_{qi} | u_{ji}$, (ii) $u_{ji}^2 | u_{qi}$ or

(iii) $u_{ji} | u_{qi}$ and $u_{qi} | u_{ji}^2$.

However, (iii) implies

$$u_{jq} u_{qj} = u_{jj} \ ,$$

which contradicts the fact that, $u_{jj}^3 \neq 0$ and $u_{qq}^2 = 0$.

This proves c), and moreover, shows that

$$u_{jq} u_{qj} = u_{jj}^r \quad \text{with} \quad r > 1 \ .$$

By [8,F_6], this gives $c_{qj} \leq 2$. Since (i) as well as (ii) implies that $c_{qj} \geq 2$, we have $c_{qj} = 2$ which, because of (5.2), proves d).

We now show that in the present case (5.16) also holds, which is equivalent to

$$(5.19) \qquad \varphi(\hat{u}^{\rho}_{kt}\hat{u}^{\mu}_{t\ell}) = u^{\rho}_{kt}u^{\mu}_{t\ell} .$$

For $t = i$ (5.19) follows from (5.6), (5.7) and (5.13), since $c_i = 1$ implies that in the normal factorization of $\hat{u}^{\rho}_{ki}$, $k \neq i$, the only factor which is not in S, is $\hat{u}_{pi} = x_{pi}$. Hence we can assume that $t \neq i$, $k = i$ or $\ell = i$, say $\ell = i$.

If u_{kt} is left regular or u_{ti} is right regular (hence regular), the argument in the proof of (5.16) applies.

Thus we can restrict to the case that u_{kt} is not left regular and u_{ti} is not right regular. Then we have

$$u^{\rho}_{kt}u^{\mu}_{ti} = u_{kt}u^{\rho+\mu-1}_{ti} \quad \text{and} \quad \hat{u}^{\rho}_{kt}\hat{u}^{\mu}_{ti} = \hat{u}_{kt}\hat{u}^{\rho+\mu-1}_{ti}$$

Moreover, $\mu, \rho \leq 2$, by (5.2) and (5.17), and so $\rho + \mu \leq 4$.

Now, $\rho + \mu = 4$ is impossible, as $\hat{u}^{3}_{ti} = 0$. For the same reason, if $\rho + \mu = 3$, the argument in (5.16) applies again. Hence it remains to consider the case that

$$\rho = \mu = 1 \quad \text{and} \quad u_{kt}, u_{it} \quad \text{are not left regular,}$$

and, in view of (5.17),

$$k = i \quad \text{or} \quad k = q \quad \text{and} \quad t = j .$$

So, (5.19) reduces to

$$\text{(i)} \qquad \varphi(\hat{u}_{qj}\hat{u}_{ji}) = u_{qj}u_{ji} \quad \text{and} \quad \varphi(\hat{u}_{ij}\hat{u}_{ji}) = u_{ij}u_{ji} .$$

Since $c_i = c_{qi} = 1$, we have

$$\text{(ii)} \qquad \hat{u}_{qj}\hat{u}_{ji} = \alpha\,\hat{u}_{qi} \quad \text{and} \quad \hat{u}_{ij}\hat{u}_{ji} = \beta\,\hat{u}_{ii} \quad , \ \alpha, \beta \in K .$$

On the other hand, by our induction hypothesis, we know that

$$\text{(iii)} \quad \hat{u}_{qj}\hat{u}_{jq} = \delta\,\hat{u}_{qq} \quad \text{with} \quad \delta = 1 \ \text{or}\ 0\,,$$

according as $S = A_S$ or $S = A'_S$.

We claim that $\alpha = \beta = \delta$. To see this we observe that (5.17c) implies:

$$\text{either} \quad u^2_{ji} \mid u_{qi} \quad \text{and} \quad u_{iq} \mid u_{ij}$$

$$\text{or} \quad u_{qi} \mid u_{ji} \quad \text{and} \quad u^2_{ij} \mid u_{iq}$$

holds. Without restriction we can assume the first. From $u^2_{ji} \mid u_{qi}$ it follows that

$$u_{iq} \mid u^{\nu}_{jq} \quad \text{for} \quad \nu = 1 \ \text{or}\ 2$$

and from this and (5.17c) that $\nu = 1$. Thus we have

$$u_{ji}u_{iq} = u_{jq} \quad \text{and} \quad u_{iq}u_{qj} = u_{ij}$$

which by the preceeding discussion of (5.19) gives

$$\text{(iv)} \quad \hat{u}_{ji}\hat{u}_{iq} = \hat{u}_{jq} \quad \text{and} \quad \hat{u}_{iq}\hat{u}_{qj} = \hat{u}_{ij}\,.$$

Since u_{iq} is regular, the discussion of (5.19) also shows that

$$\hat{u}_{qi}\hat{u}_{iq} = \hat{u}_{qq} \quad \text{and} \quad \hat{u}_{iq}\hat{u}_{qi} = \hat{u}_{ii}\,.$$

Hence (ii), (iii) and (iv) imply

$$\delta = \alpha = \beta$$

This proves (i) and so proves part b) of (5.1) in case 1.

<u>Case 2</u>. Re_i is the only member of $Re_1, \ldots\ldots, Re_n$ which is not right regular.

We claim that in this case S is an indecomposable, weakly symmetric generalized uniserial algebra.

From [9,Theorem 2 and 2.3] we have

(5.20) Let $E = \sum_{c_j > 1} e_j$. Then ERE is an indecomposable, weakly symmetric generalized uniserial algebra.

Hence it is enough to show that

(5.21) $c_k = 1$ implies $k = i$.

Suppose not. Then, since R is indecomposable, there exists a $k \neq i$ with $c_k = 1$ such that

(i) $c_{ki} \neq 0$ or

(ii) $c_{kt} \neq 0$ for some t with $c_t > 1$

holds.

Choose u_{ji} which is not right regular. Then from (ii) it follows by the same argument as in the proof of (5.17b) that (i) holds. However, (i) and $c_k = 1$ imply as in the proof of (5.17d) that u_{kj} is not left regular. Hence Re_k is not regular, a contradiction which proves (5.21). Thus S has the property stated above.

As a consequence of the fact that S is an indecomposable generalized uniserial algebra and that Re_i is uniserial and $c_{ji} \leq 2$ for all j, the order of the composition factors of Re_i is, in suitable notation,

$$F_i = F_1, F_2, F_3, \dots, F_n, F_2, F_3, \dots, F_q, F_1 \quad , \quad 1 < q \le n.$$

In particular, we have

$$B_1 = \left\{ e_1, \hat{u}_{21} = x_{21} \ , \ \hat{u}_{31}, \dots\dots, \hat{u}_{q1}^2, \ \hat{u}_{11} \right\}$$

$$B_1^* = \left\{ e_1, \hat{u}_{1q} = y_{1q} \ , \ \hat{u}_{1q-1} \ , \dots, \hat{u}_{12}^2, \ \hat{u}_{11} \right\} \ .$$

Now, the discussion of case 1 shows that, since in the present case S is regular, the multiplication on $\hat{B}$ is the same as on B, except possibly for the products

$$\hat{u}_{1j}\hat{u}_{j1} \quad \text{with} \quad c_{j1} = 2 \ , j \neq 1 \ , \text{i.e.} \quad j = 2, \dots, q \ .$$

For these, as in case 1, we have

$$\hat{u}_{1j}\hat{u}_{j1} = \alpha_j \hat{u}_{11} \ , \ \alpha_j \in K \ .$$

Moreover, let $\alpha = \alpha_2$, then

$$\alpha_j = \alpha \quad \text{for} \quad j = 2, \dots, q \ ,$$

which easily follows from the fact that

$$\hat{u}_{j1} = \hat{u}_{jk}\hat{u}_{k1} \quad \text{for} \quad 1 < k < j \le n \qquad \text{and}$$

$$\hat{u}_{1k} = \hat{u}_{1j}\hat{u}_{jk} \quad \text{for} \quad 1 < k < j \le q$$

holds.

If char $K \neq 2$, we put

$$v_{2n} = \hat{u}_{2n} + \frac{1-\alpha}{2} \hat{u}_{2n}\hat{u}_{nn}$$

$$\text{(i)} \qquad v_{1q} = y_{1q} + (1-\alpha) y_{1q}\hat{u}_{qq}$$

$$v_{21} = x_{21}$$

$$v_{j+1\,j} = \hat{u}_{j+1\,j} \quad \text{for} \quad j = 2, \dots, n-1 \ .$$

Then, observing that S is a generalized uniserial algebra, that by (5.2), (5.6), (5.7) and $[8, F_6]$, we have

$$\hat{u}_{21}\hat{u}_{1j} = \hat{u}_{2j}^2 \quad \text{for } j = 2, \dots, q \ ,$$

$$\hat{u}_{21}\hat{u}_{1j} = \hat{u}_{2j}^3 \quad \text{for } j > q \ ,$$

and that the elements $v_{k\ell}$ generate the radical N of R, it is not hard to check that the map

$$v_{kj} \longrightarrow u_{kj}$$

induces an isomorphism between R and A_R.

The same argument applies for char $K = 2$ and $\alpha \neq 0$, if, instead of (i), we put

$$\begin{aligned} v_{2n} &= \alpha \hat{u}_{2n} \\ v_{1q} &= \alpha y_{1q} \\ v_{21} &= \alpha x_{21} \\ v_{j+1j} &= \hat{u}_{j+1j} \quad \text{for } j = 2,\dots,n-1 . \end{aligned}$$

Finally, if $\alpha = 0$, the definition of A'_R immediately implies $R \cong A'_R$ which completes the proof of the proposition.

Bibliography

[1] K. Bongartz, Algebras of finite representation type without cycles, to appear.

[2] C.W. Curtis and I. Reiner, Representation theory of finite groups and associative algebras, Interscience Publ. New York, 1962.

[3] P. Gabriel, Indecomposable representations II, Symposia Math. XI (1973), 81 - 104.

[4] P. Gabriel, Finite representation type is open, Lecture Notes in Math. No. 488, 132 - 155, Springer-Verlag, New York/Berlin, 1975.

[5] J.P. Jans, Indecomposable representations of algebras, Ann. of Math. 66 (1957), 418 - 429.

[6] G. Janusz, Indecomposable modules for finite groups, Ann. of Math. 89 (1969), 209 - 241.

[7] H. Kupisch, Beiträge zur Theorie nichthalbeinfacher Ringe mit Minimalbedingung, J. Reine Angew. Math. 201 (1959), 100 - 112.

[8] H. Kupisch, Symmetrische Algebren mit endlich vielen unzerlegbaren Darstellungen I,II, J. Reine Angew. Math. 219 (1965), 1 - 25; 245 (1970), 1 - 14.

[9] H. Kupisch, Quasi-Frobenius-algebras of finite representation type, Lecture Notes in Math. No. 488, 184 - 200, Springer-Verlag, New York/Berlin, 1975.

[10] H. Kupisch, Basisalgebren symmetischer Algebren und eine Vermutung von Gabriel, J. of Algebra 55 (1978), 58 - 73.

[11] H. Kupisch, Correction to : Basisalgebren symmetrischer Algebren, to appear.

[12] T. Nakayama, On Frobenius-algebras II, Ann. of Math. 42 (1941), 1 - 21.

[13] T. Nakayama and C. Nesbitt, Note on symmetric algebras, Ann. of Math. 39 (1938), 659 - 668.

[14] Ch. Riedtmann, Algebren, Darstellungsköcher, Überlagerungen und zurück, to appear.

[15] J. Waschbüsch, Über Bimoduln in Artinringen vom endlichen Typ, Com. Algebra 8 (1980), 105 - 151 .

SOME REMARKS ON LOEWY LENGTHS OF PROJECTIVE MODULES

Peter Landrock*

Let A be a finite dimensional algebra, b a block of A, $M \in b$ any indecoposable module (i.e. a b-module). We denote the socle of M by $s_1(M)$ and define as usual the socle series by

$$s_i(M)/s_{i-1}(M) = s_1(M/s_{i-1}(M))$$

Also, we let $r_1(M) = \mathrm{rad}\, M$ and

$$r_i(M) = \mathrm{rad}(r_{i-1}(M))$$

which defines the Loewy series. (In [1], the socle and Loewy series are called the upper and lower Loewy series). Also, we call $M/\mathrm{rad}\, M$ the head of M.

It is a well known and trivial fact that the Loewy series of M terminates in 0 in the k'th step if and only if the socle series terminates in M in the k'th step, and this number, $\ell(M)$, is called the Loewy length of M (A proof may be found in [1]). It should also be pointed out that the Loewy series of M is the dual of

Supported in part by The Danish Natural Science Research Council.

the socle series of the dual of M.

Assume in the following that A is quasi-Frobenius. Thus a b-module is projective if and only if it is injective. Moreover, if P is projective, P is indecomposable if and only if either the socle of P is simple of the head of P is simple. If S is a simple module, we denote the projective cover of S by P_S.

Our purpose here is simply to find relations, if any, between the Loewy lengths of the various indecomposable projective modules of b. Any such general information will obviously be helpful, if one wants to determine the Loewy- og socle series of the indecomposable projective modules of b.

In the following, we let M be any indecomposable module of b. Most proofs will be shorter than the statements.

Lemma 1. Let $N \subseteq M$. Then the socle series of N is obtained by intersecting N with the socle series of M.

Proof. Clearly, $s_i(M) \cap N \subseteq s_i(N)$. On the other hand if by induction $s_{i-1}(N) \subseteq s_{i-1}(M)$,

$$(s_i(N) + s_{i-1}(M)/s_{i-1}(M)$$

is semi-simple, i.e. $s_i(N) \subseteq s_i(M)$.

Lemma 2. (Stripping a factor of the socle series). M contains a simple module S in its i'th socle factor if and only if M has an indecomposable submodule N of Loewy length i with simple head S.

Proof. By Lemma 1, one way is trivial. Conversely, assume S occurs in $s_i(M)/s_{i-1}(M)$. Choose $N \subseteq M$ minimal such that $(N+s_{i-1}(M))/s_{i-1}(M) \simeq S$. Then $\ell(N) = i$ by Lemma 1. Now minimality of N assures that the head of N is just S.

Corollary 1. Let P be an indecomposable projective module of b, and let S be a simple factor of the $\ell(P)-1$'st socle factor of P. Then

$$\ell(P) \leq \ell(P_S)$$

Proof. Let M above equal P. Then N is an epimorphic image of P_S, and the kernel is proper, since N is not projective. Hence $\ell(P)-1 = \ell(N) < \ell(P_S)$.

Corollary 2. Let $S \in b$ be simple, and assume that P_S has larger Loewy length than all other indecomposable projective modules of b. Then only the simple module S occurs in $s_{\ell(P_S)-1}(P_S)$. In particular, $\mathrm{Ext}_A^1(S,S) \neq 0$

Proof. Follows immediately from Corollary 1.

Our next results will deal with the following

Definition 1. An indecomposable projective b-module P is said to be upper- (lower-) stable if

$$r_2(P) = s_{\ell(P)-2}(P) \quad (r_{\ell(P)-2}(P) = s_2(P))$$

P is said to be stable, if its Loewy and socle series coincide.

We observe that, in view of Lemma 1 & 2, an equivalent definition of upper-stable is that for any simple module $S \subseteq r_1(P)/r_2(P)$, P has a submodule M of Loewy length $\ell(P)-1$ with a head isomorphic to S, such that $(M+r_2(P))/r_2(P) = S$.

To continue we need the following which most likely has been observed in other contexts, too.

Lemma 3. Let $\mathcal{S}$ be a symmetric algebra, b a block of A. Let $R,S \in b$ be simple module. Then there exists a sequence of simple modules

$$(*) \qquad R = T_1, T_2, \ldots, T_n = S$$

such that $\mathrm{Ext}^1_{\mathcal{S}}(T_i, T_{i+1}) \neq 0$ for all i.

Proof. Let M be any module, $T'' \in r_i(M)/r_{i+1}(M)$ any simple module. Then there exists a simple module $T' \in r_{i-1}(M)/r_i(M)$ such that $\mathrm{Ext}^1_{\mathcal{S}}(T', T'') \neq 0$.

To prove the lemma, we choose indecomposable projectives

$$P_R = P_1, P_2, \ldots, P_m = P_S$$

in b such that P_i and P_{i+1} have a simple composition factor A_i in common (see [1]). Let B_i equal the socle of P_i, which is isomorphic to the head of P_i by assumption. By the remark above, we may find sequences

$$A_i = X_1, \ldots, X_s = B_i, \quad B_i = Y_1, \ldots, Y_t = A_{1+i}$$

such that $\mathrm{Ext}^1_{\mathcal{S}}(X_i, X_{i+1}) \neq 0$ for all i and $\mathrm{Ext}^1_{\mathcal{S}}(Y_j, Y_{j+1}) \neq 0$ for all j, and the lemma follows.

We remark that in particular, (*) defines an equivalence relation on the simple modules of $\mathcal{S}$.

From this lemma and Corollary 2 of Lemma 2 it immediately follows that if all indecomposable projective modules of b are upper-(lower-) stable and R,S are any two simple modules of b, then

$$\ell(P_R) \leq \ell(P_S) \leq \ell(P_R)$$

Thus we have

Theorem 1. Let S be a symmetric algebra, b a block of S. Assume that any indecomposable projective module of b is upper-(lower-) stable. Then all projective modules of b have the same Loewy length.

By weakening the conditions slightly we also get

Lemma 5. Let $S \in b$ be simple and suppose all indecomposable projective modules $P_1,\dots,P_n$ of b but possibly P_S are upper-(lower-) stable. Then

$$\ell(P_S) \geq \min_i \ell(P_i)$$

If furthermore $\mathrm{Ext}^1_S(S,S) = 0$,

$$\max_i \ell(P_i) \geq \ell(P_S)$$

holds as well.

Proof. We may assume that there exists a simple module $T \in b$ such that $\mathrm{Ext}^1_S(T,S) \neq 0$. Also, we may assume that $T \not\simeq S$ since otherwise S is the only simple module of b. Hence a proper homomorphic image of P_S has Loewy length $\ell(P_T)-1$, proving the first inequality. The second follows from Corollary 2 of Lemma 2.

Finally we give a sufficient condition that two indecomposable projective modules of a block in a symmetric algebra have the same Loewy length.

Definition 2. Let $S_1, S_2 \in b$ be simple modules, b a block of a symmetric algebra. Write $S_1 \sim S_2$ if there exist non-split extensions X, Y of T_1 by T_2, $\{T_1, T_2\} = \{S_1, S_2\}$ such that the following diagrams commute

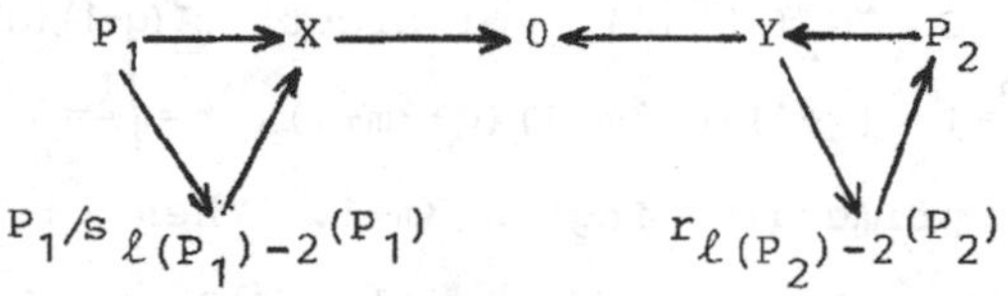

where P_i is the projective cover ($\sim$ injective hull) of T_i. In general we call simple modules R_1, R_2 of b strongly linked if either $R_1 \simeq R_2$ or there exists a sequence of simple modules

$$R_1 = S_1, S_2, \ldots, S_n = R_2$$

such that $S_i \sim S_{i+1}$ for all i.

Clearly $R_1 \equiv R_2$ if R_1 and R_2 are strongly linked defines an equivalence relation on the simple modules of b.

Lemma 6. Assume R_1 and R_2 are strongly linked. Then $\ell(P_{R_1}) = \ell(P_{R_2})$.

Proof: Clear from our previous discussions.

Thus this definition distribute the simple modules of b into classes with the property that all corresponding projective covers have the same Loewy length. It is natural to ask: How accurate is this sorting? Our examples below seem to indicate that it is difficult to improve it.

Example. Let $G = {}^2G_2(3^{2n+1})$, of order $q^3(q-1)(q^3+1)$, $q = 3^{2n+1}$. Then $q^3+1 = (q+1)(q-3m+1)(q+3m+1)$, $m=\sqrt{\frac{q}{3}}=3^n$.

Let p be any prime dividing $q-3m+1$. Then a Sylow p-subgroup of G is cyclic, and it easily follows from the character table of G (see [3]) that the principal p-block of G consists of exactly 6 simple modules I (the trivial module), M^-, X_1, X_2, X_3, X_4 with projective covers

$$
\begin{array}{cccccc}
M^- & & X_1 & X_2 & X_3 & X_4 \\
 & X_1 & X_2 & X_3 & X_4 & M^- \\
 & X_2 & X_3 & X_4 & . & . \\
I & X_3 & X_4 & M^- & . & . \\
M^-, \ I & X_4, & M^-. & X_1, & ., & ., \\
I & M^- & . & . & . & . \\
 & . & . & . & . & . \\
 & . & . & . & . & . \\
 & . & . & . & . & . \\
 & X_4 & M^- & X_1 & X_2 & X_3 \\
M^- & & X_1 & X_2 & X_3 & X_4
\end{array}
$$

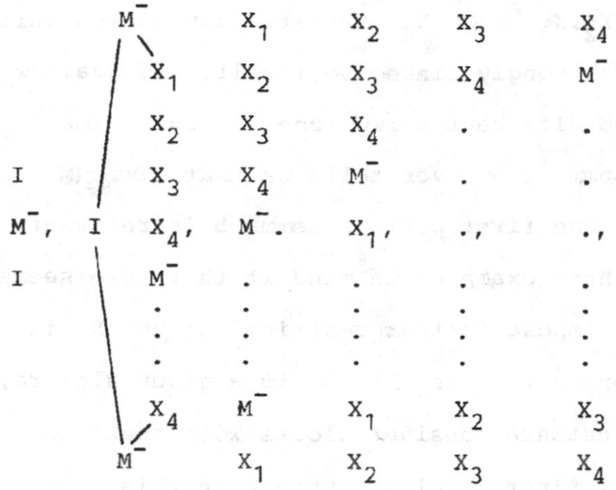

We observe that M^-, X_1, X_2, X_3 and X_4 are strongly linked while I is only strongly linked to itself. Correspondingly we have exactly two different Loewy lengths. Also Lemma 5 informs that $\ell(P_I) \leq \ell(P_{M^-}) \leq \ell(P_{X_i})$, since $Ext^1_{FG}(M^-, M^-) = 0$.

Let p be any prime dividing $q + 3m + 1$. Again, a Sylow p-subgroup is cyclic and the principal p-block contains 6 simple modules $I, M^+, X_1, .., X_4$. This time their projective covers have the form

$$
\begin{array}{ccccccc}
 & & & & & & M^+ \\
 & & & & & & M^+ \\
I & X_1 & X_2 & X_3 & X_4 & & M^+ \\
X_1 & X_2 & M^+ & X_4 & I & X_3 & . \\
X_2 & M^+ & . & . & . & X_4 & . \\
M^+, & X_3, & ., & ., & ., & I & . \\
X_3 & X_4 & . & . & . & X_1 & . \\
X_4 & I & X_1 & M^+ & X_3 & X_2 & . \\
I & X_1 & X_2 & X_3 & X_4 & & M^+ \\
 & & & & & & M^+
\end{array}
$$

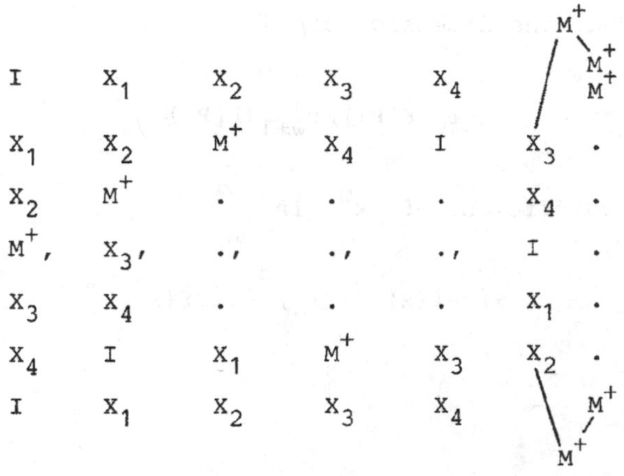

Here I, X_1, X_2, X_3 and X_4 are strongly linked while M^+ is only strongly linked to itself, and again we also get two different Loewy lengths. This time Corollary 2 of Lemma 2 however tells us that $\mathrm{Ext}^1_{FG}(M^+, M^+) \neq 0$, whence only the first part of Lemma 5 is relevant.

With these examples in mind it therefore seems necessary to impose further restrictions on S in order to get stronger results. If S is a group algebra, we might for instance consider blocks with specific defect groups. This first result to obtain in this line is obviously

Proposition 1. Let P be a p-group, F a field of characteristic p. Then $F[P]$ is stable.

Proof. This result really is due to S.A. Jennings [2], who determined the Loewy series of $F[P]$. He proved (Th. 3.7) that the dimension of

$$r_w(F[P])/r_{w+1}(F[P])$$

equals the coifficient of x^w in

$$(*) \qquad F(x) = f(x)^{d_1} f(x^2)^{d_2} \ldots f(x^m)^{d_m}$$

where $f(x) = 1 + x + .. + x^{p-1}$ and P has a certain central series

$$P = K_1 \geq K_2 \geq \dots \geq K_{m+1} = 1$$

where $K_\lambda/K_{\lambda+1}$ is elementary abelian of order p^{d_λ}.
But by (*),

$$F(x) = x^{\Sigma\lambda d_\lambda (p-1)} F(\tfrac{1}{x})$$

which implies that the coefficient of x^w equals that of $x^{\Sigma\lambda d_\lambda (p-1)-w}$, from which the proposition follows.

We immediately get

__Corollary__. Let G be p-closed, F a field of characteristic p. Then all indecomposable $F[G]$-modules are stable.

__Proof__. Let $G = P.K$, $P \in Syl_p(G)$, $P \trianglelefteq G$. Then Clifford Theory yields that Loewy series are preserved under restriciton from G to P.

Unfortunately this does not continue:

__Example__. Consider the symmetric group on four letters, Σ_4, and let F be a splitting field of characteristic 2. Then $F[G]$ has only one block, with Cartan matrix

$$\begin{Bmatrix} 4 & 2 \\ 2 & 3 \end{Bmatrix}$$

Let I and X be the simple modules, I the trivial, X of dimension 2. As Σ_4 has a unique normal subgroup of index 2, $\dim_F \mathrm{Ext}^1_{F[G]}(I,I) = 1$. Clearly $\mathrm{Ext}^1_{F[G]}(X,I) \neq 0$ as well, and it immediately follows that P_I has Loewy and socle series

$$\begin{matrix} & I & \\ I & & X \\ X & & I \\ & I & \end{matrix}$$

Hence the Loewy length of P_X is at most 4 by Corollary 2 of Lemma 2. However as X only occurs with multiplicity 3 in P_X this corollary in fact implies that $\ell(P_X) = 4$. It immediately follows that

$$\Omega X/X \simeq X \oplus \begin{matrix} I \\ I \end{matrix}$$

where ΩX is defined by $0 \to \Omega X \to P_X \to X \to 0$. Thus P_X is neither upper- nor lower-stable. As Σ_4 has a normal subgroup of index 2 which is 2-closed, there is no hope of generalizing the above corollary except of course to remark that the argument actually is that if $H \Delta G$ of index prime to p, then indecomposable projective modules of H are stable if and only if those of G are.

R E F E R E N C E S

1. Artin, E., Nesbitt, C., Thrall, R.M. Rings with minimum condition. University of Michigan, Ann. Arbor, 1944.

2. Jennings, S.A. The structure of the group ring of a p-group over a modular field. Trans. Amer. Math. Soc. 50 (1941), 175-185.

3. Ward, H.N. On Ree's series of simple groups. Trans. Amer. Math. Soc. 121 (1966), 62-89.

Matematisk Institut
Aarhus Universitet
8000 Århus C
DENMARK

REFLECTION FUNCTORS

Nikolaos Marmaridis

Let Λ be a basic 1-hereditary artin algebra and modΛ the category of finitely generated left Λ-modules. Our main result is the following theorem.

THEOREM A : Let S be a simple non-injective Λ-module. Let $X=\mathrm{TrD}S\coprod Q$, where P is the projective cover of S and Q is given by $\Lambda=P\coprod Q$. Let $\Gamma=\mathrm{End}_\Lambda(X)^{op}$ and consider the functor

$$F = \mathrm{Hom}_\Lambda(X,-) : \mathrm{mod}\Lambda \longrightarrow \mathrm{mod}\Gamma$$

Then $T=\mathrm{Ext}^1_\Lambda(X,S)$ is a simple Γ-module and F induces an equivalence between the full subcategory $\underline{C}$ of modΛ whose objects are the M in modΛ with $\mathrm{Hom}_\Lambda(M,S)=0$ and the full subcategory $\underline{D}$ of modΓ whose objects are the N in modΓ with $\mathrm{Hom}_\Gamma(T,N)=0$.

The functor F is called a reflection functor.

Theorem A is a generalization in the case of 1-hereditary artin algebras of theorem due to M.Auslander-M.I. Platzeck-I.Reiten ([3], Th.1.11). In [3] is considered an arbitrary basic algebra but our module S has to be a projective one. In the case of categories of representations of finite ordered sets these reflection functors are the functors, which are constructed in [7] and [8].

Also, in this paper we given all finite ordered sets of tame representation type which have one of the following form :

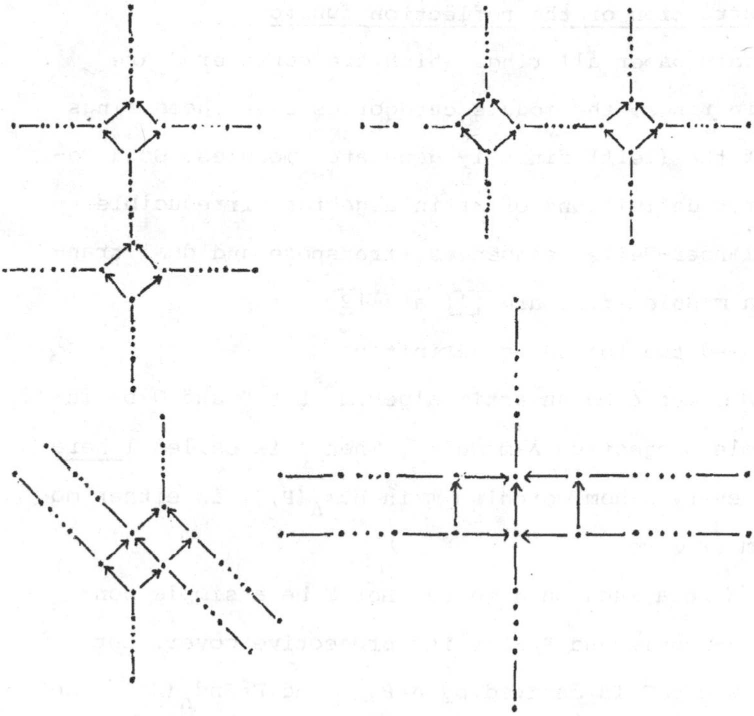

Notice that the above ordered sets consist of two commutative relations, where both cycles are of the form 

In most of the cases the representation type is computed by applying reflection functors.

It must be noticed that Sheila Brenner and M.C.R. Butler in [5] using other methods study independently, functors which coincide with the reflection functors in the case of 1-hereditary artin algebras.

I would like to thank Sheilla Brenner and M.C.R.Butler for the helpful discussions during ICRA II.

1. The construction of the reflection functors

In this paper all rings which are considered are basic artin rings, the module categories over these rings consist of the (left) finitely generated modules. Good references for definitions of artin algebras, irreducible maps, Auslander-Reiten sequences, transpose and dual transpose of a module et.c. are [1] and [2].

We need the following definition :

DEFINITION : Let Λ be an artin algebra. Let P and Q be indecomposable projective Λ-modules. Then Λ is called 1-hereditary if every Λ-homomorphism φ in $\mathrm{Hom}_\Lambda(P,Q)$ is either monomorphism or zero.

Let Λ be a such an algebra. Let S be a simple non-injective Λ-module and $P(S)=P$ its projective cover. Let $X=\mathrm{TrDS}\coprod Q$ where Q is defined by $\Lambda=P\coprod Q$ and $\Gamma=\mathrm{End}_\Lambda(X)^{op}$. Let $\underline{r}$ be the radical of Λ.

1.1. LEMMA : If $0 \longrightarrow K \xrightarrow{\alpha} Z \xrightarrow{f} \mathrm{TrDS} \longrightarrow 0$ is an exact sequence with (Z,f) a projective cover of TrDS then :

(i) $K \cong P(S) = P$

(ii) $T = \mathrm{Ext}^1_\Lambda(X,S)$ is a simple Γ-module.

(iii) P is not isomorphic to a direct summand of Z.

Proof. (i) Since S is a simple non-injective module we know from [1] Proposition 5.3 that $K/\underline{r}K \cong S$. So the projective cover of K is isomorphic to P. Considering the map $P \xrightarrow{p} K \xrightarrow{\alpha} Z \xrightarrow{\pi_{Z'}} Z'$, where Z' is an indecomposable direct summand of Z we see, that $\pi_{Z'} \circ \alpha \circ p \neq 0$, since $\pi_{Z'} \circ \alpha \neq 0$ for every Z'. Because Λ is an 1-hereditary algebra it follows that $\pi_{Z'} \circ \alpha \circ p$ is a monomorphism and then p is an isomorphism.

(ii) It follows by [1] Propositions 4.1 and 5.1.

(iii) Assume, that P is isomorphic to some indecomposable summand Z' of Z. Then, the non-zero map $\pi_{Z'} \circ \alpha : P \longrightarrow Z'$ is an isomorphism. So, there is a $\sigma : Z' \longrightarrow P$ with $\sigma \circ \pi_{Z'} \circ \alpha = 1_P$ and α is a splitable monomorphism, which gives a contradiction.

1.2. <u>LEMMA</u> : <u>Let</u> M <u>be in</u> modΛ.

$\operatorname{Hom}_\Lambda(M,S) = 0$ <u>iff</u> $\operatorname{Ext}^1_\Lambda(\operatorname{TrDS},M) = 0$

<u>Proof</u> "$\longrightarrow$". Assume, that there is a non-split sequence $0 \longrightarrow M \longrightarrow L \longrightarrow \operatorname{TrDS} \longrightarrow 0$. Considering the Auslander-Reiten sequence of $S : 0 \longrightarrow S \longrightarrow Y \longrightarrow \operatorname{TrDS} \longrightarrow 0$ we can construct the following commutative diagram

$$\begin{array}{ccccccccc} 0 & \longrightarrow & M & \longrightarrow & L & \longrightarrow & \operatorname{TrDS} & \longrightarrow & 0 \\ & & \downarrow{\scriptstyle\varphi} & & \downarrow & & \| & & \\ 0 & \longrightarrow & S & \longrightarrow & Y & \longrightarrow & \operatorname{TrDS} & \longrightarrow & 0 \end{array}$$

Since φ must be zero it follows that the lower sequence splits. Which is a contradiction.

"$\longleftarrow$" Assume, that there is a φ in $\operatorname{Hom}_\Lambda(M,S)$ not equal to zero. There is a $\psi : P \longrightarrow M$, with $\varphi \circ \psi = \pi$, where (P,π) is the projective cover of S. Consider the commutative diagram.

$$\begin{array}{ccccccccc} 0 & \longrightarrow & P & \xrightarrow{\alpha} & Z & \xrightarrow{f} & \operatorname{TrDS} & \longrightarrow & 0 \\ & & \downarrow{\scriptstyle\psi} & & \downarrow & & \| & & \\ 0 & \longrightarrow & M & \longrightarrow & Z \coprod_P M & \longrightarrow & \operatorname{TrDS} & \longrightarrow & 0 \end{array}$$

Assuming that the lower sequence splits, we get that there is an epimorphism from Z to S. This is a contradiction because P is not isomorphic to a direct summand of Z.

1.3. LEMMA : Let M be in modΛ. Then $\mathrm{Ext}^1_\Lambda(X,M) \overset{\Gamma}{\cong} \mathrm{Ext}^1_\Lambda(X,S)^r$ for some non-negative integer r.

Proof. It is enough to prove that the canonical epimorphism $\pi : M \longrightarrow M/\mathrm{rad}M$ induces an injection from $\mathrm{Ext}^1_\Lambda(\mathrm{TrDS},M)$ to $\mathrm{Ext}^1_\Lambda(\mathrm{TrDS},M/\mathrm{rad}M)$. Let $[E_1] : 0 \longrightarrow M \longrightarrow A \longrightarrow \mathrm{TrDS} \longrightarrow 0$ be a non-zero element of $\mathrm{Ext}^1_\Lambda(\mathrm{TrDS},M)$ and $[E_2]$: $0 \longrightarrow S \longrightarrow Y \longrightarrow \mathrm{TrDS} \longrightarrow 0$ the Auslander-Reiten sequence of S. It is not difficult to see, that we have the following commutative diagram, where $\varphi : M \longrightarrow S$ is not equal to zero.

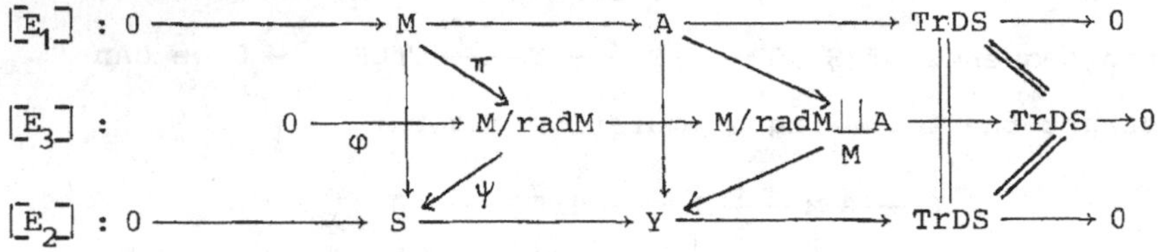

Since $[E_2]$ is induced from $[E_3]$ by ψ, it follows that if $[E_3]$ splits then $[E_2]$ is a splitting sequence. Which gives a contradiction.

1.4. LEMMA : Let N be an indecomposable module in modΓ. Then there is some M in modΛ and a nonnegative integer r such that there is an exact sequence

$$0 \longrightarrow T^r \longrightarrow N \longrightarrow \mathrm{Hom}_\Lambda(X,M) \longrightarrow 0$$

Proof. Having the Lemma's 1.1, 1.2 and 1.3 we follow the same proof as in [3], Lemma 1.10.

The proof of Theorem A. We repeat slightly modificated the proof of Theorem 1.11 in [3]. We know that the restriction of F to $\underline{\underline{C}}$ is full and faithfull [3], Proposition 1.4. By Lemma 1.1 we know that $T=\mathrm{Ext}^1_\Lambda(X,S)$ is a simple Γ-module. Let N be an object in $\underline{\underline{D}}$. There is some exact sequence $0 \longrightarrow T^r \longrightarrow N \longrightarrow \mathrm{Hom}_\Lambda(X,M) \longrightarrow 0$. Since $\mathrm{Hom}_\Gamma(T,N) = 0$ we have an isomorphism from N to $\mathrm{Hom}_\Lambda(X,M)$. We state without proof the dual of Theorem A.

THEOREM B. *Let Γ be a basic l-hereditary artin algebra with a simple non-projective module* T. *Let* E(T) *the injective hull of* T *and* I *the sum of one copy of each of the other indecomposable injective Γ-modules. Let* $Y = D(I) \amalg \mathrm{Tr}T$, *and* $\Lambda = \mathrm{End}_\Gamma(Y)$. *Then* $S = \mathrm{Tor}^r_1(Y,T)$ *is a simple* Λ-*module and the functor* $G = Y\otimes - : \mathrm{mod}\Gamma \longrightarrow \mathrm{mod}\Lambda$ *induces an equivalence of categories between the full subcategory of* modΓ *whose objects are the* N *in* modΓ *with* $\mathrm{Hom}_\Gamma(T,N) = 0$ *and the full subcategory* $\underline{\underline{C}}$ *of* modΛ *whose objects are the* M *in* modΛ *with* $\mathrm{Hom}_\Lambda(M,S) = 0$.

2. APPLICATIONS

Let I be a finite (partially) ordered set, k a (commutative) field and ${}_kI$ the category of the representations of I. (See [7] and [8]).

It is well known that ${}_kI$ is equivalent to the category mod $\mathrm{End}(P_1 \amalg \ldots \amalg P_n)^{op}$, where $P_1,\ldots,P_n$ are indecomposable projective objects. The ring $\Lambda = \mathrm{End}(P_1 \amalg \ldots \amalg P_n)^{op}$ is an indecomposable l-hereditary algebra. (See [4]). So, we may apply the theory which is developed in the first part of this paper.

For every indecomposable projective object P of ${}_kI$

the ring EndP is isomorphic to k. Let (P,Q) be a pair of non isomorphic projective with $H = \mathrm{Hom}(P,Q) \neq 0$. Then H is a simple EndQ-EndP bimodule. In case $H \neq 0$ we will write $P<Q$.

<u>DEFINITION</u> : Let P and Q be two indecomposable non isomorphic projective objects. If $P<Q$ and there is no other indecomposable projective object Z, ($Z \not\cong P$ and $Z \not\cong Q$), with $P<Z<Q$ then P and Q are called <u>neighbours</u>.

Let S be a simple non injective object of ${}_kI$, (P,p) its projective cover and (Z,f) the projective cover of TrDS.

2.1. <u>LEMMA</u> : <u>Every two indecomposable summands of Z are not isomorphic</u>.

Proof. Let $Z_1^{m_1} \amalg \ldots \amalg Z_n^{m_n}$ be a decomposition of Z in indecomposable direct summands, where $Z_i \not\cong Z_j$ if $i \neq j$ and $m_i \in \mathbb{N}$, for every $i = 1,\ldots,n$. Consider the exact sequence $[E_1] : 0 \longrightarrow P \xrightarrow{\alpha} Z \xrightarrow{\beta} \mathrm{TrDS} \longrightarrow 0$ and the Auslander-Reiten sequence of S $[E_2] : 0 \longrightarrow S \longrightarrow Z/\underline{r}P \longrightarrow \mathrm{TrDS} \longrightarrow 0$.

Since $\mathrm{Hom}(P,Z_i)$ is a simple $\mathrm{End}Z_i$ - EndP bimodule for every $i = 1,\ldots,n$ the morphism $\alpha : P \longrightarrow Z$ factorizes over $Z_1 \amalg \ldots \amalg Z_n$. So the following diagram is commutative :

$$\begin{array}{ccc} & (Z_1 \amalg \ldots \amalg Z_n)/\underline{r}P & \\ {}^{\mu}\nearrow & & \searrow^{\nu} \\ S & \xrightarrow[\beta]{} & Z/\underline{r}P \end{array}$$

Because β is an irreducible morphism and μ cannot be a split monomorphism, it follows that ν is a split epimorphism. So we have $l(Z_1 \amalg \ldots \amalg Z_n) \geq l(Z)$ and therefore $Z_1 \amalg \ldots \amalg Z_n = Z$.

In the next Proposition we will use the notations of Lemma 2.1.

2.2. PROPOSITION : The indecomposable projective object R of $_kI$ with P<R is a neighbour of P iff R is isomorphic to a direct summand of Z.

Proof. " $\longrightarrow$ " Since P<R, there is a non zero morphism $\varphi : P \longrightarrow R$. The induced morphism $\varphi : S \longrightarrow R/\underline{r}P$ is not a split monomorphism, so there is a morphism $\psi : Z/\underline{r}P \longrightarrow R/\underline{r}P$ such the folowing diagram is commutative :

$$\begin{array}{ccc} S & \xrightarrow{\alpha} & Z/\underline{r}P \\ {\scriptstyle\varphi}\downarrow & \swarrow{\scriptstyle\psi} & \\ R/\underline{r}P & & \end{array}$$

Then there is a non-zero morphism $x : Z \longrightarrow R$ and exists an indecomposable summand Z' of Z, such that $x_{|Z'} : Z' \longrightarrow R$ is not zero. Since $P \xrightarrow{\alpha} Z \xrightarrow{\pi_{Z'}} Z'$ is always not zero, we have P<Z'<R. Therefore Z' is isomorphic to R.

" $\longleftarrow$ " We have proved that every neihbour R of P with P<R appears as direct summand of Z. Let $Z_1, \ldots, Z_n$ be the non isomorphic neighbours of P with $P<Z_i$, $i = 1, \ldots, n$. Consider the induced commutative diagram :

$$\begin{array}{ccccc} & & (Z_1 \amalg \ldots \amalg Z_n)/\underline{r}P & & \\ & {\scriptstyle\mu}\nearrow & & \searrow{\scriptstyle\nu} & \\ S & & \longrightarrow & & Z/\underline{r}P \end{array}$$

The morphism β is an irreducible one and μ cannot be a split monomorphism. Therefore ν is a split epimorphism and so $l(Z) \leq l(Z_1 \amalg \ldots \amalg Z_n)$. This implies $Z \simeq Z_1 \amalg \ldots \amalg Z_n$.

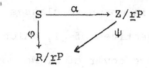

It is not difficult to see, that P together with $Z_1,\ldots,Z_n$ and the morphism $\alpha_i = \pi_{Z_i} \circ \alpha_i : P \longrightarrow Z_i$ is a direct system of objects of ${}_kI$.

2.3. <u>COROLLARY</u> : <u>The object</u> TrDS <u>is isomorphic to the direct limit of</u> (P, Z_i, α_i).

Let ${}_kI$ be the category of representations of an ordered set I. Let S be a simple non injective object and $F = \mathrm{Hom}(X,-) : {}_kI \longrightarrow \mathrm{modEnd}(X)^{op}$ the reflection functor as in Theorem A.

We construct a quiver J with relations as follows : The points i of J are the isomorphism classes $[P_i]$ i=1,..,n of the indecomposable projectives of ${}_kI$, which are not isomorphic to the projective cover of S. We consider an additional point, the isomorhism class $[P_{n+1}]$ of TrDS. There is an arrow from i to j $(i \neq j)$ iff $\mathrm{Hom}(P_j, P_i) \neq 0$ and there is no P_l $(l \neq i, l \neq j)$ with $\mathrm{Hom}(P_j, P_l) \neq 0$ and $\mathrm{Hom}(P_l, P_i) \neq 0$. It is easy to see, that the quiver J can have commutativity or zero relations depending of the construction of TrDS. The category ${}_kJ$ of representations of J is equivalent to $\mathrm{modEnd}(X)^{op}$.

Let V be an object of ${}_kI$. The object $FV = \mathrm{Hom}(X,V)$ of $\mathrm{End}(X)^{op}$ can be considered as an object of ${}_kJ$ as follows For every i in J, $(FV)_i \cong \mathrm{Hom}(P_i, V)$ and there is a k-linear morphism from $(FV)_i$ to $(FV)_j$ iff there is a non-zero morphism from P_j to P_i. It turns out by Corollary 2.3 that this construction of the reflection functor F, or dually of G as in Theorem B, coinside, up to naturale equivalence, with the functors which are constructed in [7] and [8].

Using the reflection functors we get all the ordered sets of tame representation type of the form (*) which are considered at the beginning of the paper.

In the list which follows the representation type of (13), (14), (15) and (16) is computed with a method which can be found in [6].

The list of the ordered sets of tame type

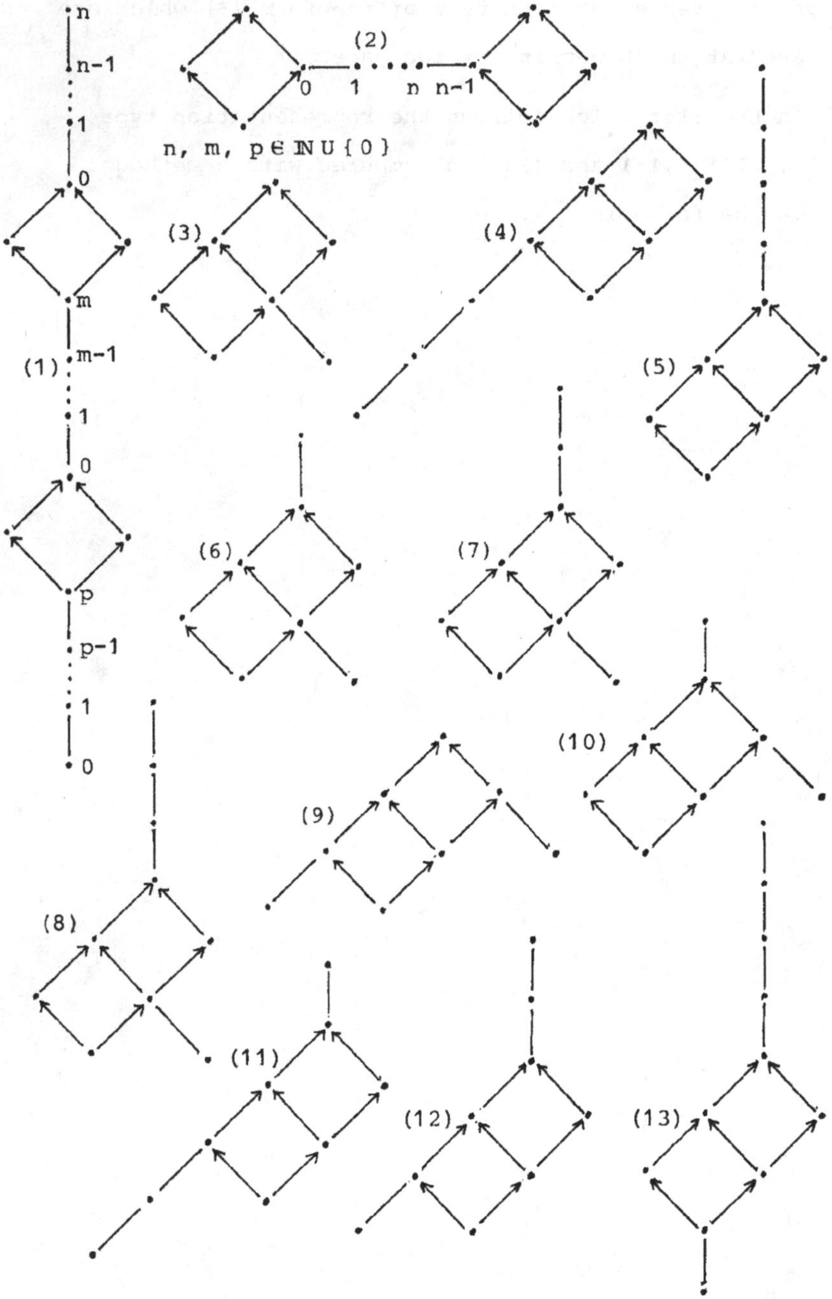

(continued)

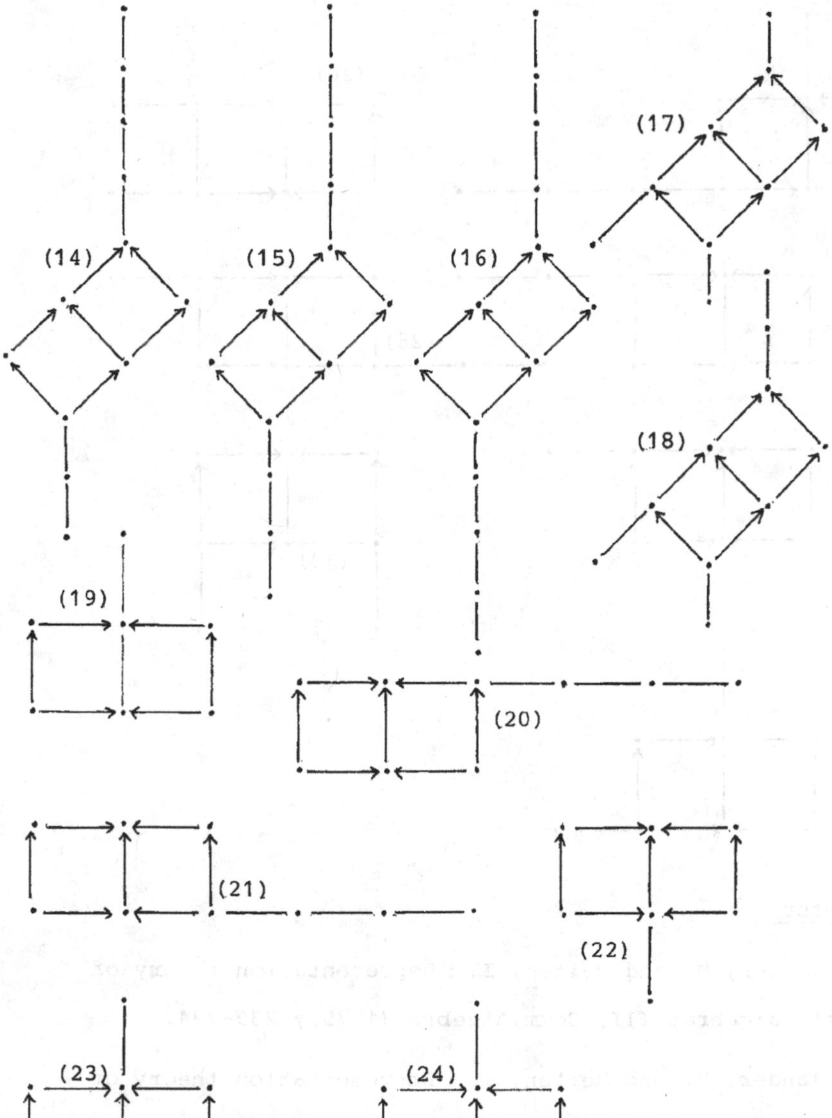

(continued)

(25)

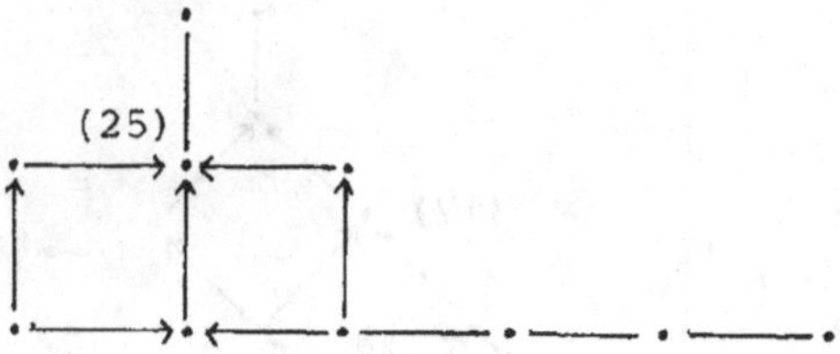

(26)

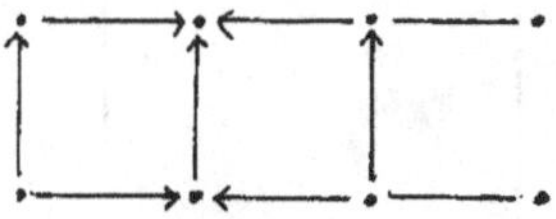

(27)

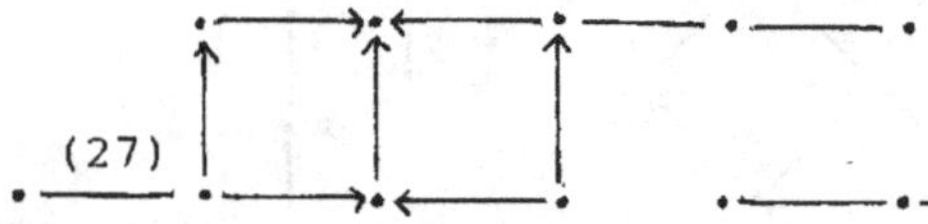

(28)

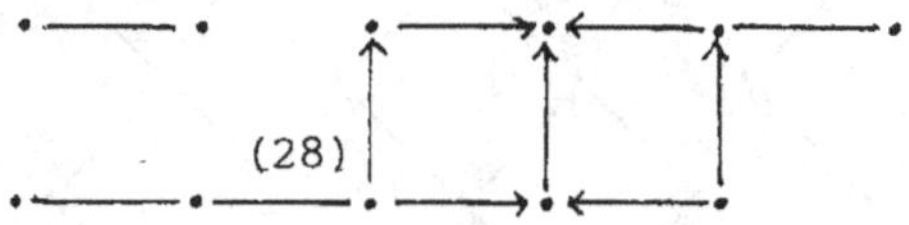

(29)

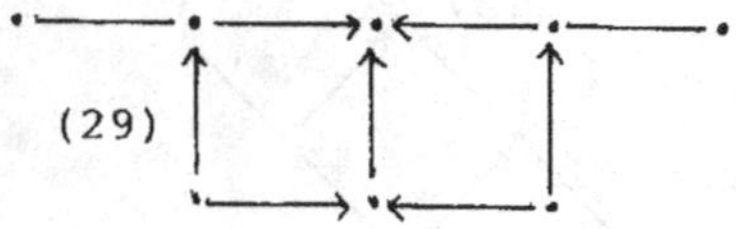

(30)

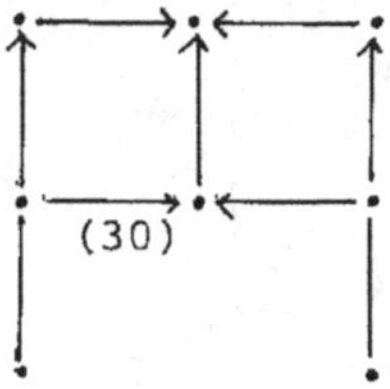

(31)

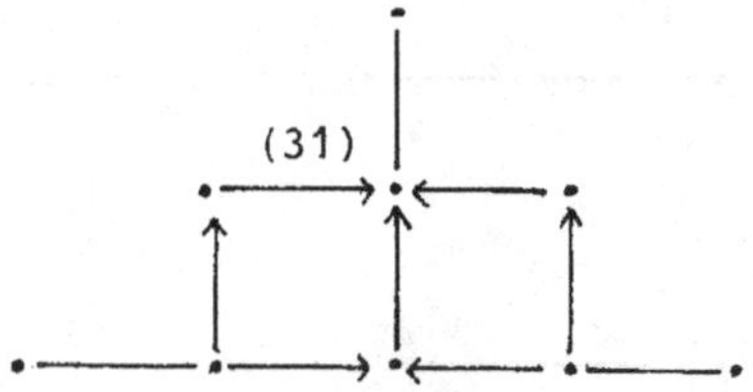

<u>Literature</u>

1. Auslander, M. and Reiten, I. : Representation theory of Artin algebras III, Comm.Algebra (1975), 239-294.

2. Auslander, M. and Reiten, I. : Representation theory of Artin algebras IV, Comm.Algebra (1977), 443-518.

3. Auslander, M ; Platzek, M.I., and Reiten, I : Coxeter functors without diagrams, Preprint 1977

4. Bautista, R : On algebras close to hereditary algebras, Communicaciones Internas No 35, 1978.

5. Brenner, S. and Butler M.C.R : On generalizations of the Bernstein-Gelfand-Ponomarev reflection functors, Preprint.

6. Loupias, M. Représentations indécomposables des ensembles ordonnés finis, These, 1975 (Tours).

7. Marmaridis, N : Darstellungen endlicher Ordnungen, Dissertation 1978 (Zürich).

8. Marmaridis, N. Représentations linéaires des ensembles ordonnés C.R.Acad.Sc.Paris t.288, 1979.

Nikolaos Marmaridis
University of Crete
Department of Mathematics
Iraklion - crete
Greece.

ALGEBRAS STABLY EQUIVALENT TO ℓ-HEREDITARY

Roberto Martínez-Villa

An artin algebra Λ is ℓ-hereditary, if given any pair of indecomposable projectives P,Q any non zero map $f: P \to Q$ is a monomorphism.

ℓ-hereditary algebras are factors of hereditary ones, and generalize quivers with commutativity conditions.

Auslander-Reiten have characterized the algebras stably equivalent to the hereditary [3].

We prove that ℓ-hereditary algebras have a similar characterization:

An artin algebra is stably equivalent to an ℓ-hereditary if and only if the following conditions hold:

i) If $f: P \to Q$ is not zero with P,Q indecomposable projective, then f is a monomorphism or $\operatorname{Im} f$ is simple.

ii) If $\operatorname{Im} f$ is simple non projective, then it is a factor of an injective.

To prove the sufficient condition we consider the following zero relations:

An algebra Λ has a "node" Q if Q is an indecomposable non simple projective such that for any pair of indecomposable projectives P_i, P_j and non zero maps $f: P_i \to Q$ $g: Q \to P_j$ $gf = 0$.

We prove the following theorem:

Any artin algebra with "nodes" is stably equivalent to an algebra without "nodes".

I thank Idun Reiten by the simplifications suggested in the proof of the first theorem.

1) Necessary conditions:

A simple, non projective, non injective module S, will be called a "node", if the almost split sequence for S, $0 \to S \to P \to trD(S) \to 0$ has P projective.

A simple projective, non injective will be called a "sink" and a simple injective non projective a "source".

We know from Auslander-Reiten [1], that "sinks" have almost split sequence: $0 \to S \to P \to trD(S) \to 0$ with P, projective, and "sources" have almost split sequence: $0 \to Dtr(S) \to I \to S \to 0$ with I, injective.

Moreover, we know that a non injective module A has almost split sequence: $0 \to A \to P \to trD(A) \to 0$ with P projective if and only if:

i) A is non injective simple

ii) A is not a composition factor of $rI/soc\ I$ for any indecomposable injective I.

We have dually:

A non projective module C has almost split sequence: $0 \to Dtr(C) \to I \to C \to 0$, with I injective if and only if:

i) C is a non projective simple

ii) C is not a composition factor of $rP/soc\ P$ for any indecomposable projective P.

Nodes can be seen as generalization of radical square zero conditions. This is a consequence of the following:

LEMMA 1. Let S be a simple module and Q its projective cover, then the following conditions are equivalent:

i) For any pair of indecomposable projectives P_i, P_j and non zero, non isomorphism maps: $f: P_i \to Q$, $g: Q \to P_j$ the composition $gf = 0$.

ii) S is not a composition factor of $rI/\mathrm{soc}\, I$ for any indecomposable injective I.

iii) S is not a composition factor of $rP/\mathrm{soc}\, P$ for any indecomposable projective P.

iv) For any pair of indecomposable projectives P_i^*, P_j^* Λ^{op}-modules and non zero maps: $P_i^* \xrightarrow{f} Q^* \xrightarrow{g} P_j^*$, $gf = 0$, $(Q^* = \mathrm{Hom}(Q,\Lambda))$.

v) S is either injective or S has almost split sequence: $0 \to S \to P \to \mathrm{trD}(S) \to 0$ with P projective.

vi) S is either projective or S has almost split sequence: $0 \to \mathrm{Dtr}(S) \to I \to S \to 0$ with I injective.

Proof: We know: v) $\Leftrightarrow$ ii), ii) $\Leftrightarrow$ vi)

i) $\Leftrightarrow$ iv) is obvious.

i) $\Leftrightarrow$ iii)

iii) Is equivalent to: $\mathrm{Hom}_\Lambda(Q, rP/\mathrm{soc}\, P) = 0$ for any indecomposable projective P and is also equivalent to:

Any non zero map $f: Q \to P$ is either isomorphism or $\mathrm{Im} f$ is simple.

Let $f: Q \to P_j$ be a non zero, non isomorphism map, $\mathrm{Im} f = H$ and let S be in the socle of H, P_i the projective cover of S. Then, there exists a map $g: P_i \to Q$ such that: $fg(P_i) = S$.

Hence; $fg \neq 0$ and i) implies g is an isomorphism.

iii) ⇒ i) Let $f: P_i \to Q$ $g: Q \to P_j$ be non zero, non isomorphisms.

Hence; $\mathrm{Im}g$ is simple and $\mathrm{Im}f \subset \mathrm{Ker}\ g = rQ$.

ii) ⇔ iv)

ii) Is equivalent to: $\mathrm{Hom}_\Lambda(rI/\mathrm{soc}\ I, I(S)) = 0$ for any indecomposable injective I and $I(S)$ is the injective envelope of S. So, it is equivalent to:
$0 = \mathrm{Hom}_{\Lambda^{op}}(D(I(S)), D(rI/\mathrm{soc}\ I)$. It is known:
$D(I(S)) = Q^*$, $D(rI/\mathrm{soc}\ I) = \dfrac{rD(I)}{\mathrm{soc}\ D(I)}$.

Hence; ii) is equivalent to:
$\mathrm{Hom}_{\Lambda^{op}}(Q^*, rP^*/\mathrm{soc}\ P^*) = 0$ for any indecomposable projective P^*.

By i) iii), already proved, this is equivalent to: $Q^*/rQ^* = D(S)$, has almost split sequence:
$0 \to D(S) \to P^* \to trD(S) \to 0$ with P^*, projective.

Applying the duality D,

ii) is equivalent to iv).

<u>COROLLARY</u>: Let Λ be an artin algebra satisfying the following conditions:

i) For any pair of indecomposable projectives P_i, P_j, any non zero map $f: P_i \to P_j$ is either monomorphism or $\mathrm{Im}f$ is simple.

ii) If $\mathrm{Im}f$ is a non projective simple, then $\mathrm{Im}f$ is a factor of an injective.

Then, every simple torsionless module is a "node" or a "sink".

Proof: Let S be a torsionless, non projective, simple and Q its projective cover.

Let P be an indecomposable projective, $f: Q \to P$ a non zero map.

Assume; f is a monomorphism.

By ii), S is a factor of an injective I. Let $\pi\ Q \to S$, $t: I \to S$ be non zero maps. Since Q is projective, there exists $g: Q \to Z$ such that $tq = \pi$, and I injective implies, there exists a map $h: P \to I$ such that $hf = g$. Hence; f has to be an isomorphism. By iii) this proves the corollary.

LEMMA 2. Let Λ be an artin algebra and assume; every non injective, torsionless, simple Λ-module, is either a "node" or a "sink".

If P is an indecomposable projective, I an indecomposable injective, and $f: I \to P$ a non zero, non isomorphism map, then Imf is semisimple.

Proof: Assume to the contrary that Imf has an indecomposable summand K, which is not simple, and take S in the socle of K.

S is torsionless and is not a composition factor of rI/soc I.

Let Λ and Γ be stably equivalent artin algebras, and let $F: \underline{\text{mod}}_\Lambda \to \underline{\text{mod}}$ be the stable equivalence. F induces an equivalence $H = \text{Dtr } F \text{ trD}$, $H: \overline{\text{mod}}_\Lambda \to \overline{\text{mod}}_\Gamma$.

The induced equivalence H takes "nodes" and "sources" in either a "node" or a "source".

<u>LEMMA 3</u>. Let $H: \overline{\mathrm{mod}}_\Lambda \to \overline{\mathrm{mod}}_\Gamma$ be the induced equivalence given above. An almost split sequence: $0 \to A \to B \to C \to 0$ has B projective if and only if the almost split sequence: $0 \to H(A) \to B' \to C \to 0$, has B' projective.

<u>Proof</u>: Follows from Auslander-Reiten [2].

Simple torsionless modules over ℓ-hereditary algebras are projectives.

We can prove that algebras stably equivalent to ℓ-hereditary satisfy the conditions of Lemma 1.

More generally:

<u>LEMMA 4</u>. Let Λ and Γ be stably equivalent artin algebras and assume that each torsionless, non injective, simple Γ-module is either a "node" or a "sink". Then each torsionless, non injective, simple Λ-module, is either a "node" or a "sink".

<u>Proof</u>: Let S_1 be a simple, torsionless, non injective Λ-module, $H: \overline{\mathrm{mod}}_\Lambda \to \overline{\mathrm{mod}}_\Gamma$ the induced equivalence and H^{-1} its inverse.

Auslander-Reiten [3], proved that H, gives a 1-1 correspondence between the non injective torsionless modules. Hence; $H(S_1)$ is torsionless.

Take any simple S in the socle of $H(S_1)$ and $j: S \to H(S_1)$ the inclusion map.

By hipothesis, S is a "node" or a "sink". The induced map $\bar{j}: S \to H(S_1)$ in $\overline{\text{mod}}_\Gamma$ is not zero. So, the map $H^{-1}(\bar{j}): H^{-1}(S) \to S_1$ is not zero.

From Lemma 2, $H^{-1}(S)$ is a "node" or a "sink" in any case simple.

Hence; $H^{-1}(\bar{j})$ is an isomorphism. Then $\bar{j}$ is an isomorphism.

This shows $H(S_1)$ is a node.

We can prove now the first main theorem:

<u>THEOREM</u>. Assume that Λ is stably equivalent to an ℓ-hereditary algebra Γ. Then the following conditions hold:

i) If $f: P \to Q$ is not zero, where P and Q are indecomposable projective, then f is a monomorphism or $\text{Im}f$ is semisimple.

ii) If $\text{Im}f$ is simple and not projective, it is a factor of an injective module.

<u>Proof</u>: Let $f: P \to Q$ be a non zero map, with P, Q indecomposable projectives.

We can assume P is not simple, and from Lemma 2 and 4 that P is not injective.

We assume also, f is not a monomorphism. Consider the map $f_1: P \to \text{Im}f = K$ and let S be a simple submodule of $\text{Ker}\, f_1$ and let $j: S \to P$ be the inclusion map. $H: \overline{mod}_\Lambda \to \overline{mod}_\Gamma$ is the induced equivalence. If $\overline{f}_1 = \overline{0}$ in $\overline{mod}_\Lambda$, we are done by Lemmas 2 and 4.

So we can assume $\overline{f}_1 \neq \overline{0}$.

Consider the maps: $H(S) \xrightarrow{H(j)} H(P) \xrightarrow{H(f_1)} H(K)$.

$H(S)$ is a node or a sink, since S is node or sink. Hence: $H(S)$ is a simple projective.

Auslander-Reiten [3], proved that H gives a 1-1 correspondence between the non injective, indecomposable projectives and between the non injective indecomposable torsionless.

Hence; $H(P)$ is projective, and $H(K)$ is torsionless.

Λ ℓ-hereditary implies, the map $H(f_1)$ is a monomorphism.

$j: S \to P$ a monomorphism implies $\overline{j} \neq \overline{0}$; so, $H(\overline{j}) \neq \overline{0}$, and $H(S)$ simple implies $H(\overline{j})$ is a monomorphism.

$H(\overline{f})\ H(\overline{j}) \neq 0$, since it is the composition of two monomorphisms.

This is a contradiction, because $f \cdot j = 0$. This finish the proof.

2. <u>Sufficient conditions</u>.

We will prove here that any artin algebra with "nodes" is stably equivalent to an artin algebra without "nodes".

Let Λ be an artin algebra with "nodes" the

indecomposable projectives; $Q_1, Q_2, \ldots, Q_n$ and $S_1, S_2, \ldots, S_n$ the corresponding simples.

$S = S_1 \oplus S_2 \oplus \ldots \oplus S_n$, $a = \tau_\Lambda(S)$ is the trace of S in Λ.

b is the left anhilator of a. $a \subset r$, where r is the radical of Λ. Then it is clear that:

a is a two sided ideal and $a^2 = 0$. The projective cover of a Λ/a-module X, is of the form P/aP with P the projective cover of X as Λ-module.

Λ/b is semisimple and a is a $\Lambda/b - \Lambda/a$ bimodule.

The following Lemma follows very easily by standard isomorphisms, we include it here for reference.

<u>LEMMA 2.1</u>. The following propositions hold:

i) Let C be a two sided ideal of the ring Λ, X a Λ/C-module, then there is a natural isomorphism: $C \otimes_{\Lambda/C} X \simeq C \otimes_\Lambda X$. If P is a projective Λ-module then: $C \otimes_{\Lambda/C} P/CP \simeq C \otimes_\Lambda P \simeq CP$.

ii) Let X be any Λ-module, where Λ is an artin ring and $a = \tau_\Lambda(X)$ the trace of X in Λ, then: $aP = \tau_P(X)$.

We need also the following:

<u>LEMMA 2.2</u>. Let S be the sum of the simple "nodes" of the artin algebra Λ, $a = \tau_\Lambda(S)$, then the following propositions hold:

i) For any Λ-module X, $aX \subset \tau_X(S)$.

ii) If T is simple $a \cdot T = 0$.

iii) If X has no semisimple summand then $aX = \tau_X(S)$.

Proof:

i) Let P be the projective cover of X.

$\pi: P \to X$ the natural map.

By Lemma 2.1, $aP = \tau_P(S)$.

Hence; $\pi(aP) = a\pi(P) = aX = \tau_X(S)$.

ii) If T is a non projective simple, Q its projective cover, and $\pi: Q \to T$ a non zero map, then:

$aQ \subset rQ$ implies, $\pi(aQ) = a\pi(Q) = aT = 0$.

If T is simple projective then: $aT = \tau_T(S) = 0$.

iii) Let X be a Λ-module with no semisimple summand.

By i), $\tau_X(S) = 0$ implies, $aX = \tau_X(S) = 0$.

Assume: $\tau_X(S) \neq 0$, and let T be a simple submodule of $\tau_X(S)$.

Since T is a "node", there exists an almost split sequence:

$$0 \to T \xrightarrow{\alpha} Q \to \mathrm{trD}(T) \to 0 \quad \text{with } Q \text{ projective.}$$

By the properties of the almost split sequences, the inclusion map extends to Q, i.e., there exists a map $f: Q \to X$ such that $f\alpha$ is the inclusion map $T \to X$.

But, $\alpha(T) \subset aQ = \tau_Q(S)$ implies:

$$f\alpha(T) \subset f(aQ) = af(Q) \subset aX$$

Hence; $T \subset aX$.

This shows: $aX = \tau_X(S)$.

<u>COROLLARY</u>: A submodule X/aP of a projective Λ/a-module P/aP, such that the map:

$$a \underset{\Lambda/a}{\otimes} X/aP \xrightarrow{1 \otimes j} a \underset{\Lambda/a}{\otimes} P/aP \text{ , with } j: X/aP \to P/aP$$

the inclusion map, is zero. Is of the form: $X/aP \simeq X'$, where X' is a submodule of P with $X' \cap aP = 0$.

<u>Proof</u>: We have the following commutative diagram:

$$\begin{array}{ccc} a \otimes X/aP & \xrightarrow{1 \otimes j} & a \otimes P/aP \\ \downarrow m_1 & & \downarrow m_2 \\ aX & \to & aP \end{array}$$

where $m_1(a \otimes \bar{x})$ ax and $m_2(a \otimes \bar{p}) = ap$.

$1 \otimes j = 0$ implies, $aX = 0$.

Let $X = \bigoplus_{1=1}^{t} S_i \oplus X'$ where each S_i is a node and no summand of X' is a node.

$$X/aP \simeq \frac{X'}{aP \cap X'} = \frac{X'}{\tau_{X'}(S)} = \frac{X'}{aX'} = X'.$$

We will prove now that Platzeck's [6] construction produces an algebra with no node.

<u>PROPOSITION 2.3</u>. The matrix ring $\hat{T} = \begin{pmatrix} \Lambda/a & 0 \\ a & \Lambda/b \end{pmatrix}$ has no node.

<u>Proof</u>: Let $\hat{T}$ be a simple, torsionless, non projective T-module.

The indecomposable projective Γ-modules are of the form: $\hat{P} = (P/aP,\ a \otimes P/aP,\ id)$ and $\hat{Q} = (0,Q,0)$, Q a simple Λ/b-module.

Hence; $\hat{T}$ has to be of the form: $\hat{T} = (T,0,0)$ with T a simple torsionless Λ/a-module.

If $\hat{P}$ is an indecomposable projective with $\hat{T} \subset \hat{P}$, then the inclusion map: $j\colon T \to P/aP$, induces a zero map:

$$1 \otimes j\colon a \otimes T \to a \otimes P/aP.$$

By Corollary of Lemma 2.2, T is a simple submodule of P with $aP \cap T = 0$.

Assume; $\hat{T}$ is a node of Γ, we will prove that T is a node of Λ, this will be a contradiction.

Let $\hat{Q} = (Q/aQ,\ a \otimes Q/aQ,\ id)$ be the projective cover of $\hat{T}$.

Then; Q and Q/aQ are projective covers of T as Λ and Λ/a-module, respectively.

Let P be an indecomposable projective non isomorphic to Q and $f\colon Q \to P$ a non zero map. f induces a map $\bar{f}\colon Q/aQ \to P/aP$. If $\bar{f} = 0$ then $f(Q) \subset aP$.

Hence; T is a node.

$\bar{f} \neq 0$ induces a non zero map:

$$f_1 = \bar{f},\ (f_1,f_2)\colon (Q/aQ,\ a \otimes Q/aQ,\ id) \to (P/aP,\ a \otimes P/aP, id)$$

Since $\hat{Q}$ is a node, (f_1,f_2) factors through $\hat{T}$.

So, $\bar{f}$ factors:

$$\begin{array}{ccc} Q/aQ & \xrightarrow{\bar{f}} & P/aP \\ & {}_{g}\searrow \quad \nearrow_{i} & \\ & T & \end{array}$$

and by Corollary of Lemma 2.3, $T \cap aP = 0$.

So, we have the diagram:

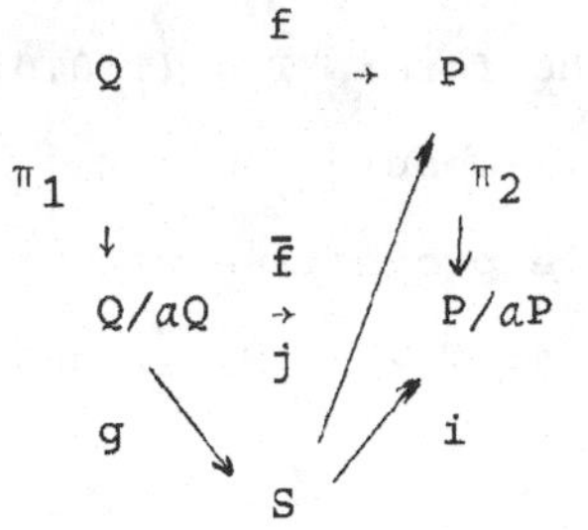

and $\pi_2 j = i$ $\quad \bar{f} = ig$

$\pi_2 f = \bar{f}\pi_1$

$\pi_2 f = \pi_2(jg\pi_1)$ implies, $\pi_2(f - jg\pi_1) = 0$.

If $f - jg\pi_1 \neq 0$, then there exists a non zero map $Q \to aP$, and T has to be a node.

$f - jg\pi_1 = 0$ implies f factors through T.

Hence; T is a node.

<u>LEMMA 2.4</u>. Let Λ be an artin algebra satisfying the following conditions:

i) Given any pair or indecomposable projectives P, Q any non zero map $f: P \to Q$ is either a monomorphism or $\mathrm{Im} f$ is simple.

ii) If $\mathrm{Im} f$ is simple and not projective it is a factor of an injective module.

Then, $\Gamma = \begin{pmatrix} \Lambda/a & 0 \\ a & \Lambda/b \end{pmatrix}$ is ℓ-hereditary.

<u>Proof</u>: Note that Λ/a is ℓ-hereditary:

If P/aP is an indecomposable projective Λ/a-module and $f: P/aP \to \Lambda/a$ a non zero map, then there exists a map $g: P \to \Lambda$ such that the induced map is f, and we have a commutative diagram:

$$\begin{array}{ccccccccc} 0 & \to & aP & \to & P & \to & P/aP & \to & 0 \\ & & \downarrow h & & \downarrow g & & \downarrow f & & \\ 0 & \to & aQ & \to & Q & \to & Q/aQ & \to & 0 \end{array}$$

$f \neq 0$ implies, g is a monomorphism, then there exists a map: $0 \to \text{Ker } f \to \text{Coker } h$. Since, $P = \tau_p(S)$ and Coker h is a summand of Q, Ker $f = 0$.

Since Λ/b is semisimple, it is enough to prove that given indecomposable projectives of the form:

$$\hat{P} = (P/aP,\ a \otimes P/aP,\ \text{id})\ \hat{Q} = (Q/aQ,\ a \otimes Q/aQ,\ \text{id})$$

any non zero map $f = (f_1, 1 \otimes f_1)$ is a monomorphism.

So we have to prove that:

$1 \otimes f_1: a \otimes P/aP \to a \otimes Q/aQ$ is a monomorphism.

Let $\hat{f} : P \to Q$ be the lifting of f_1.

We have a commutative diagram:

$$\begin{array}{ccc} a \otimes P/aP & \to & aP \\ \downarrow 1 \otimes f_1 & & \downarrow g \\ a \otimes Q/aQ & \to & aQ \end{array}$$

with the raw maps isomorphisms, and g the restriction of $\hat{f}$ to aP.

$\hat{f}$ has to be a monomorphism, because otherwise its image is a node. Hence; g is a monomorphism and $1 \otimes f_1$ is so.

Λ is any artin algebra with nodes a the trace of the simple nodes in Λ and $\Gamma = \begin{pmatrix} \Lambda/a & 0 \\ a & \Lambda/b \end{pmatrix}$ the ring constructed above.

We have as in Platzeck [6] the functor: $F\colon \text{mod}_\Lambda \to \text{mod}_\Gamma$ given by: $F(X) = (X/aX,\ aX,\ m)$ where m is multiplication, and defined on morphisms in the obvious way.

We will prove F induces an stable equivalence.

We need first some lemmas.

<u>LEMMA 2.5</u>. Let $\Gamma = \begin{pmatrix} \Lambda/a & 0 \\ a & \Lambda/b \end{pmatrix}$ be the algebra constructed above, F the functor described, X, Y Λ-modules, $(f_1,f_2)\colon (Y/aY,\ aY,m) \to (X/aX,\ aX,m)$ a Γ-morphism from $F(Y)$ into $F(X)$, and suppose there exists a map $f\colon Y \to X$ such that:

$$
*)\qquad \begin{array}{ccc}
Y & \xrightarrow{\ \pi\ } & Y/aY \\
\downarrow{\scriptstyle f} & & \downarrow{\scriptstyle f_1} \\
X & \xrightarrow{\ \pi\ } & X/aX
\end{array}
$$

commutes. With π the canonical surjections.

Then, the restriction of f to aY is f_2.

<u>Proof</u>: *) Induces a commutative diagram:

$$
\begin{array}{ccc}
a \underset{\Lambda}{\otimes} Y & \xrightarrow{\ 1 \otimes \pi\ } & a \otimes Y/aY \\
\downarrow{\scriptstyle 1 \otimes f} & & \downarrow{\scriptstyle 1 \otimes f_1} \\
a \underset{\Lambda}{\otimes} X & \xrightarrow{\ 1 \otimes \pi\ } & a \otimes X/aX
\end{array}
$$

We also have the commutative diagram:

$$\begin{array}{ccc} \mathcal{a} \otimes Y/\mathcal{a}Y & \xrightarrow{m} & \mathcal{a}Y \\ \downarrow {\scriptstyle 1 \otimes f_1} & & \downarrow {\scriptstyle f_2} \\ \mathcal{a} \otimes X/\mathcal{a}X & \xrightarrow{m} & \mathcal{a}X \end{array}$$

Putting the two diagrams together we have the commutative square:

$$\begin{array}{ccc} \mathcal{a} \otimes Y & \xrightarrow{m(1\otimes \pi)} & \mathcal{a}Y \\ \downarrow {\scriptstyle 1 \otimes f} & & \downarrow {\scriptstyle f_2} \\ \mathcal{a} \otimes X & \xrightarrow{m(1\otimes \pi)} & \mathcal{a}X \end{array}$$

Then, if $a \in \mathcal{a}$, $y \in Y$.

$f_2 m(1 \otimes \pi)(a \otimes y) = f_2 m\ (a \otimes \bar{y}) = f_2(ay)$

$= m(1 \otimes \pi)(1 \otimes f)(a \otimes y) = m(1 \otimes \pi)(a \otimes f(y)) =$

$= m(a \otimes f(\bar{y})) = af(y)$.

Therefore: $af(y) = f(ay) = f_2(ay)\ \forall\ a \in \mathcal{a},\ y \in Y$.

LEMMA 2.6. Let P be a projective Λ-module, X an a arbitrary Λ-module and $(f_1, f_2)\colon (P/\mathcal{a}P, \mathcal{a}P, m) = F(P) \to (X/\mathcal{a}X, \mathcal{a}X, m) = F(X)$ a Γ-morphism, then there exists $f\colon P \to X$ such that $F(f) = (f_1, f_2)$.

Since P is projective, there exists a map $f\colon P \to X$ such that:

$$\begin{array}{ccc} P & \xrightarrow{\pi} & P/aX \\ f\downarrow & & \downarrow f_1 \\ X & \xrightarrow{\pi} & X/aX \end{array}$$ commutes.

By Lemma 2.5, $F(f) = (f_1, f_2)$.

Now, we can prove F is full.

<u>PROPOSITION 2.7</u>. The functor $F\colon \text{mod}_\Lambda \to \text{mod}_\Gamma$ given above is full.

<u>Proof</u>: Let $f = (f_1, f_2)\colon F(X) = (X/aX,\ aX, m) \to F(Y) = (Y/aY,\ aY, m)$ be a Γ-map, and P the projective cover of X as Λ-modules, let K be the Kernel of the surjection $h\colon P \to X$. Then; $\operatorname{Ker} h \cap aP = \operatorname{Ker} h_2$, where h_2 is the restriction of h to aP.

We have the commutative exact diagram:

$$\begin{array}{ccccccccc} & & 0 & & 0 & & 0 & & \\ & & \downarrow & & \downarrow & & \downarrow & & \\ 0 & \to & aP \cap K & \xrightarrow{\beta} & K & \xrightarrow{q} & K/\tau_K(S) & \to & 0 \\ & & \gamma_2\downarrow & & \gamma\downarrow & & \gamma_1\downarrow & & \\ 0 & \to & aP & \xrightarrow{\alpha} & P & \xrightarrow{\pi} & P/aP & \to & 0 \\ & & h_2\downarrow & & h\downarrow & & h_1\downarrow & & \\ 0 & \to & aX & \xrightarrow{\alpha_1} & X & \xrightarrow{\pi_1} & X/aX & \to & 0 \\ & & \downarrow & & \downarrow & & \downarrow & & \\ & & 0 & & 0 & & 0 & & \end{array}$$

Since; P is projective, there exists a map: $g: P \to Y$ such that:

$$\begin{array}{ccc} P & \xrightarrow{\pi} & P/aP \\ g \downarrow & & \downarrow f_1h_1 \\ Y & \xrightarrow{\pi_2} & Y/aY \end{array} \quad \text{commutes.}$$

$F(h)F(f) = (h_1f_1, h_2f_2) = F(h \circ f)$.

By Lemma 2.4, the restriction of g to aP, $g_2 = g|aP = h_2f_2$

So, $g_2\gamma_2 = f_2h_2\gamma_2 = 0$.

We want to prove that the composition $g\gamma = 0$. Suppose; $g\gamma \neq 0$

$\pi_2 g\gamma = f_1h_1\pi\gamma = f_1(h_1\gamma_1)q = 0$

So, $\operatorname{Im} g\gamma \subset aY \subset \tau_Y(S)$.

Then; we have $g\gamma: K \to \bigoplus_{i=1}^{n} S_i \to 0 \quad \bigoplus_{i=1}^{n} S_1 \subset \tau_Y(S)$

Let; Q_i be the projective cover of S_i. There exists a map $t: Q_i \to K$ such that: $g\gamma t(Q_i) = S_i$ $\gamma: K \to P$.

Q_i a node implies, γt is either an split monomorphism or Imt is simple. If γt is an split monomorphism. Then, Imt is a non superfluous submodule of P. A contradiction, since P was choosen the projective cover of X.

Then, the image of the restriction of $g\gamma$ to $\tau_K(S)$ is the image of $g\gamma$, i.e., $\operatorname{Im} g\gamma = \operatorname{Im} g\gamma\beta$.

But; $g\gamma\beta = g\alpha\gamma_2 = \alpha_2 g_2 \gamma_2 = 0$.

A contradiction.

So, we proved $g\gamma = 0$.

Therefore; there exists $f: X \to Y$ such that $fh = g$.

$$\pi_2 fh = \pi_2 g = f_1 h_1 \pi = f_1 \pi_1 h$$

h epi implies, $\pi_2 f = f_1 \pi_1$.

We have proved F is full.

Let Λ be an artin algebra with nodes. $V(X,Y)$ denotes, the set of maps: $f: X \to Y$ such that: f factors:

$$\begin{array}{ccc} X & \xrightarrow{f} & Y \\ & {\scriptstyle g}\searrow \quad \nearrow{\scriptstyle h} & \\ & \oplus S_i & \end{array}$$

, where $\oplus S_i$ is a sum of simple nodes and for any splittable monomorphism $j: S_i \to \oplus S_i$ hj is not an splittable monomorphism.

It is easy to see (and will be a consequence of next lemma), $V(X,Y)$ is an abelian subgroup of $\mathrm{Hom}_\Lambda(X,Y)$.

We can form the category $\underset{\sim}{\mathrm{mod}}_\Lambda$, whose objects are the same as in mod_Λ, and maps:

$$\underset{\sim}{\mathrm{Hom}}(X,Y) = \frac{\mathrm{Hom}_\Lambda(X,Y)}{V(X,Y)}.$$

We will prove; the functor F induces a faithfull functor on $\underset{\sim}{\mathrm{mod}}_\Lambda$.

<u>LEMMA 2.8</u>. Let F be the functor defined above and X,Y be a pair of Λ-modules, then:

i) The Kernel of the abelian group homorphism $F: \operatorname{Hom}_\Lambda(X,Y) \to \operatorname{Hom}_\Lambda(F(S), F(Y))$ is $V(X,Y)$.

ii) $V(X,Y) \subset P(X,Y)$, where $P(X,Y)$ denotes the subgroup of $\operatorname{Hom}_\Lambda(X,Y)$ of all maps, that factors through a projective module.

Proof:

i) It follows from the commutative diagram:

$$\begin{array}{ccccccccc} 0 & \to & aX & \to & X & \to & X/aX & \to & 0 \\ & & f_2 \downarrow & & \downarrow f & & \downarrow f_1 & & \\ 0 & \to & aY & \to & Y & \to & Y/aY & \to & 0 \end{array}$$

$F(f) = 0$ if and only if f factors:

$$\begin{array}{ccc} X & \xrightarrow{f} & Y \\ \searrow & & \nwarrow \\ & X/aX \to aY & \end{array}$$

Assume; $f \neq 0$ and $F(f) = 0$.

$Y = Y' \oplus Y''$, where Y' has no summand isomorphic to a node, and Y'' is a sum of nodes.

By Lemma 2.2, $aY = aY' = \tau_{Y'}(S)$.

Hence; f factors:

$$\begin{array}{ccc} X & \xrightarrow{f} & Y \\ & \searrow \quad \nearrow j & \\ & aY' & \end{array}$$

and for any splittable monomorphism $i: S_i \to aY'$ ji is not an splittable monomorphism.

Assume now, $f: X \to Y$ factors:

$$\begin{array}{ccc} X & \xrightarrow{f} & Y \\ & {\scriptstyle g}\searrow \quad \nearrow{\scriptstyle h} & \\ & \oplus S_i & \end{array}$$

where $\oplus S_i$ is a sum of nodes, and for any splittable monomorphism $j: S_i \to \oplus S_i$ hj is not an splittable mono.

$Y = Y' \oplus Y''$, $X = X' \oplus X''$, where Y', X' have no summand isomorphic to a node, and X'', Y'' are sums of simples nodes. Let, $p: Y \to Y''$ be the projection map. If $ph \neq 0$, then there exist simples S_i, S_j and morphisms, $j: S_i \to \oplus S_i$, $q: Y'' \to S_j$ such that: $qphj: S_i \to S_j$, $qphj \neq 0$.

Hence; $qphj$ is an isomorphism and hj is an splittable monomorphism. So, we proved: $\mathrm{Im}h \subset Y'$.

Hence; $\mathrm{Im}f \subset \mathrm{Im}h \subset aY' = \tau_{Y'}(S) = aY$.

Let, g' be the restriction of g, $g|X'$, and suppose $g(aX') \neq 0$. As before, there exists simples S_k, S_ℓ and maps, $s: S_k \to X$, $t: \oplus S_i \to S$ such that $tg's: S_k \to S_\ell$ is a isomorphism, so $X' \xrightarrow{tg'} S_\ell$ is an splittable epi.

This proves, $g(aX') = g(aX) = 0$ and we have showed; $X \xrightarrow{f} Y$ factors:

$$\begin{array}{ccc} X & \xrightarrow{f} & Y \\ \downarrow & & \uparrow \\ X/aX & \rightarrow & aY \end{array}$$

ii) Let $f \in V(X,Y)$ $\begin{array}{ccc} X & \xrightarrow{f} & Y \\ & \searrow \quad \nearrow{\scriptstyle h} & \\ & \oplus S_i & \end{array}$ $h = (h_1, \ldots, h_k)$

where; $h_i: S_i \to Y$ are non splittable monos.

Each S_i is a node, so they have almost split sequence: $0 \to S_i \to P_i \to trD(S_i) \to 0$ with P_i projective, hence; each h_i factors through P_i.

This proves, h factors through a projective.

<u>PROPOSITION 2.9</u>. Let $mod_S\Gamma$ be the full subcategory of mod_Γ of all Γ-modules with no summand of the form: $(0,S_i,0)$ and S_i a Λ/b module, then; the functor:

$$F: mod_\Lambda \to mod_{S\ \Gamma} \text{ is dense.}$$

<u>Proof</u>: If (X_1,X_2,ϕ) is an object in $mod_S\Gamma$ then the map: $\phi: a \otimes X_1 \to X_2$ is an epimorphism.

Then; we have a commutative diagram:

$$\begin{array}{ccccc}
 & & 0 & & \\
 & & \downarrow & & \\
0 & \to & Ker\ \phi & & \\
\downarrow & id & \downarrow & & \\
a \otimes X_1 & \to & a \otimes X_1 & & \\
\| & \phi & \downarrow \phi & & \\
a \otimes X_1 & \to & X_2 & \to & 0 \\
 & & \downarrow & & \\
 & & 0 & &
\end{array}$$

We will prove first, that given a Γ-module of the form: $(X_1,\ a \otimes X_1,\ id)$, there exist a Λ-module X such that: $(X/aX,\ aX,\ m)$ is isomorphic to $(X_1,\ a \otimes X_1,\ id)$.

Let $Q/aQ \xrightarrow{f_1} P/aP \xrightarrow{h_1} X_1 \to 0$ be a minimal presentation of X_1.

Then, it is known: (see Fosum-Griffiths-Reiten [5])

$$(Q\mathfrak{a}Q, \mathfrak{a}\otimes Q/\mathfrak{a}Q, \mathrm{id}) \xrightarrow{(f_1, 1\otimes f_1)} (P/\mathfrak{a}P, \mathfrak{a}\otimes P/\mathfrak{a}P, \mathrm{id}) \xrightarrow{(h_1, 1\otimes h_1)} (X_1, \mathfrak{a}\otimes X_1, \mathrm{id}) \to 0$$

is a minimal projective presentation of $(X_1, \mathfrak{a} \otimes X_1, \mathrm{id})$.

We also have the commutative exact diagram:

$$\begin{array}{ccccccccc}
 & & & & & & 0 & & \\
 & & & & & & \downarrow & & \\
 & & & & & & K & & \\
 & & & & & & \downarrow & & \\
0 & \to & \mathfrak{a}Q & \to & Q & \to & Q/\mathfrak{a}Q & \to & 0 \\
 & & \downarrow f_2 & & \downarrow f & & \downarrow f_1 & & \\
0 & \to & \mathfrak{a}P & \to & P & \to & P/\mathfrak{a}P & \to & 0 \\
 & & \downarrow & & \downarrow h & & \downarrow h_1 & & \\
 & & \mathfrak{a}\otimes X_1 & \to & X & \to & X_1 & \to & 0 \\
 & & \downarrow & & \downarrow & & \downarrow & & \\
 & & 0 & & 0 & & 0 & &
\end{array}$$

where, X is the cokernel of f, and K the kernel of f_1.

We want to prove that the map:

$s: K \to \mathfrak{a} \otimes X_1$ obtained from the Snake lemma, is zero.

Suppose $s \neq 0$, $\mathfrak{a} \otimes X_1$ is semisimple and each indecomposable summand of $\mathfrak{a} \otimes X_1$ is a node.

Let $\mathrm{Im} s = \bigoplus_{i=1}^{n} S_i$ and $Q_i/\mathfrak{a}Q_i$ be the projective cover of S_i.

Then, there exists a map $t_i: Q_i/\mathfrak{a}Q_i \to K$ such that $st_i(Q_i/\mathfrak{a}Q_i) = S_i$.

Hence; we have:

$$\begin{array}{ccc} Q_i/aQ_i & \longrightarrow & S_i \\ \downarrow t_i & & \downarrow \\ K & \longrightarrow & \oplus\, S_i \\ & \searrow j & \\ & & Q/aQ \end{array}$$

jt_i lifts to:

$$\begin{array}{ccccc} Q_i & & \xrightarrow{\ \ell\ } & & Q \\ \downarrow p_i & & & & \downarrow p \\ Q_i/aQ_i & \rightarrow & K & \rightarrow & Q/aQ \end{array}$$

Since Q_i, is a node; ℓ is either an split monomorphism or $\mathrm{Im}\ell \subset aQ$.

The second statment is impossible, so ℓ is an split monomorphism.

Hence; $Q_i/aQ_i \to K \to Q/aK$ is an split monomorphism.

But this contradicts, the fact K is an superfluous submodule of Q/aQ.

We have proved: $0 \to a \otimes X_1 \overset{\alpha}{\to} X \overset{\beta}{\to} X_1 \to 0$ is exact.

Also; $h(aP) = ah(P) = aX \simeq a \otimes X_1$ and $X_1 \simeq X/aX$.

One verifies easily that under this identifications, $(X_1, a \otimes X_1, \mathrm{id})$ is isomorphic to $(X/aX, aX, m)$.

We have in general:

$$\begin{array}{ccc} a \otimes X/aX & \xrightarrow{m} & aX \\ \downarrow{\scriptstyle 1 \otimes f_1} & & \downarrow{\scriptstyle f_2} \\ a \otimes X_1 & \xrightarrow{\phi} & X_2 \end{array}$$

with f_1 an isomorphism and f_2 epi.

Consider the push out:

$$\begin{array}{ccccccccc} 0 & \to & aX & \to & X & \to & X/aX & \to & 0 \\ & & \downarrow{\scriptstyle f_2} & & \downarrow{\scriptstyle f} & & \| & & \\ 0 & \to & X_2 & \to & Y & \to & X/aX & \to & 0 \\ & & \downarrow & & \downarrow & & & & \\ & & 0 & & 0 & & & & \end{array}$$

$f(aX) = af(X) = aY$ and $Y/aY \simeq X/aX$.

It follows (X_1, X_2, ϕ) is isomorphic to $(Y/aY, aY, m)$, so we have proved F is dense.

We restate the previous propositions as a theorem.

__THEOREM 2.10.__ Let Λ be an artin algebra with simple nodes, $S_1, S_2, \ldots S_k$, $a = \tau_\Lambda(S_1 \oplus S_2, \ldots, \oplus S_k)$, the trace of $\bigoplus_{i=1}^{k} S_i$ in Λ, and b the left anhilator of a. Then:

a) The matrix ring: $\Gamma = \begin{pmatrix} \Lambda/a & 0 \\ a & \Lambda/b \end{pmatrix}$ has no "nodes".

b) Denote by $\underset{\sim}{\mathrm{mod}}_\Lambda$ the category with the same objects as mod_Λ and maps: $\underset{\sim}{\mathrm{Hom}}_\Lambda(X,Y) = \dfrac{\mathrm{Hom}_\Lambda(X,Y)}{V(X,Y)}$

where $V(X,Y)$ denotes the maps $f\colon X \to Y$ that factors through a sum of simple nodes:

$$\begin{array}{ccc} X & \xrightarrow{\quad f \quad} & Y \\ & {\scriptstyle g}\searrow \quad \nearrow{\scriptstyle h} & \\ & \oplus S_i & \end{array}$$

and for any splittable mono $S_i \underset{j}{\to} \oplus S_i \underset{h}{\to} Y$, hj is not an splittable mono.

Denote by $\underset{S}{\text{mod}}_\Gamma$, the full subcategory of mod_Γ of all modules with no summand of the form:

$(0, S_i, 0)$ and S_i a $\Lambda/\mathfrak{b}$-module. Then:

$F\colon \underset{\sim}{\text{mod}}_\Lambda \to \underset{S}{\text{mod}}_\Gamma$, given by

$F(X) = (X/\mathfrak{a}X, \mathfrak{a}X, m)$ $m\colon \mathfrak{a}\otimes X/\mathfrak{a}X \to X$

m multiplication, is an equivalence of categories.

c) F induces an stable equivalence:

$F\colon \underline{\text{mod}}_\Lambda \to \underline{\text{mod}}_\Gamma$.

d) If Λ is an artin algebra satisfying the conditions:

i) For any pair of indecomposable projectivs P_i, P_j any non zero map; $f\colon P_i \to P_j$ is either mono or $\operatorname{Im} f$ is simple.

ii) If $\operatorname{Im} f$ is non projective simple, then $\operatorname{Im} f$ is a factor of an injective.

Then; Γ is an ℓ-hereditary algebra.

<u>Proof</u>:

c) Follows from the fact:

$V(X,Y) \subset P(X,Y)$ and the objects of the form:

$(0, S_i, 0)$ are projective Γ modules.

3. Appendix

We sketch here some other relations, and show how to construct algebras with "nodes".

a) Algebras with nodes generalice the radical square zero condition

Say; the indecomposable projective Q is a node of Λ:

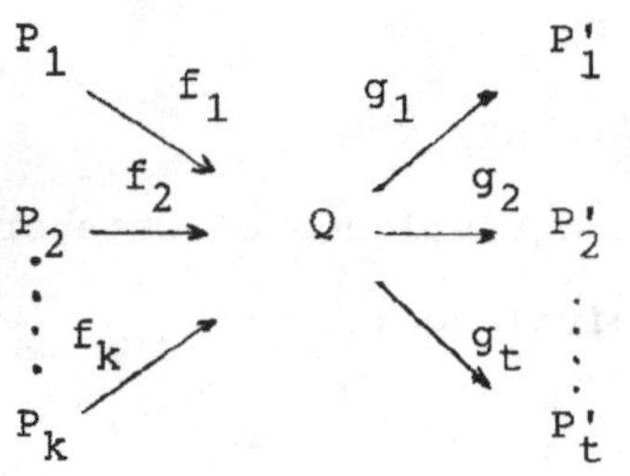

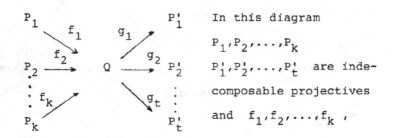

In this diagram $P_1, P_2, \ldots, P_k$ $P_1', P_2', \ldots, P_t'$ are indecomposable projectives and $f_1, f_2, \ldots, f_k$, $g_1, g_2, \ldots, g_t$ non zero, non isomorphism maps. They satisfy the condition $g_i\ f_j = 0$ for $1 \leq i \leq t$ $1 \leq j \leq k$.

We want to construct a new algebra without these zero relations.

We would like to have instead of Q, two indecomposable projectives Q_0, Q_1 such that:

$\mathrm{Hom}(Q_0, Q_1) = 0$ $\mathrm{Hom}(Q_1, Q_0) = 0$ and non zero non isomorphism maps:

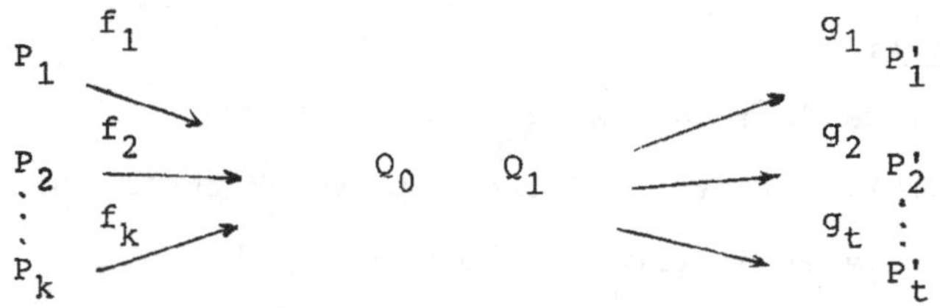

In other words, we want to separate the "node", in a "sink" and a "source".

Let Λ be an algebra with "nodes" the indecomposable projectives, $Q_1, Q_2, \ldots, Q_\ell$ $S_1, S_2, \ldots, S_\ell$, the corresponding tops, and $P_1, P_2, \ldots, P_r$ the indecomposable projectives that are non nodes.

We may assume Λ-basic.

Denote by $S = \bigoplus_{i=1}^{\ell} S_i$ and by Δ the endomorphism ring of $\Lambda \oplus S$, $\Delta = \mathrm{End}_\Lambda(\Lambda \oplus S)^{op}$.

In this new ring, we have added a new indecomposable projective for each "node".

Δ has matrix form:

$$\Delta = \begin{pmatrix} \mathrm{End}_\Lambda(\Lambda)^{op} & \mathrm{Hom}_\Lambda(\Lambda,S) \\ \mathrm{Hom}_\Lambda(S,\Lambda) & \mathrm{End}_\Lambda(S)^{op} \end{pmatrix}$$

In order to be, the new indecomposable projectives of Δ "sinks" and those corresponding to the nodes Q_i sources, we need to factor Δ module the ideal:

$$C = \begin{pmatrix} a & \mathrm{Hom}_\Lambda(\Lambda,S) \\ 0 & 0 \end{pmatrix}$$

where a us the ideal of $\mathrm{End}_\Lambda(\Lambda)^{op}$ generated by the maps that factor through some simple summand of S.

We get the ring:

$$\Delta/C = \begin{pmatrix} \mathrm{End}_\Lambda(\Lambda)^{op} & 0 \\ \mathrm{Hom}_\Lambda(S,\Lambda), & \mathrm{End}_\Lambda(S)^{op} \end{pmatrix}$$

This construction is very similar to the one given by: Auslander-Reiten [3].

We show it is equivalent to the one given in part 2.

We identify Λ and $\mathrm{End}_\Lambda(\Lambda)^{op}$ in the usual way,

By means of this isomorphism, we can see a as a two sided ideal of Λ.

It is easy to check, a is isomorphic to the trace of S in Λ, $\tau_\Lambda(S)$.

Since Λ is basic, S is isomorphic as $\mathrm{End}_\Lambda(S)^{op}$-module to $\mathrm{End}_\Lambda(S)^{op}$.

Hence; the canonical morphisms:

$$\psi : S \underset{\mathrm{End}_\Lambda(S)}{\otimes} \mathrm{Hom}(S,\Lambda) \to \tau_\Lambda(S)$$

$$s \otimes f \qquad \to \quad f(s)$$

and

$$\phi : \mathrm{Hom}_\Lambda(S,\Lambda) \to \mathrm{End}(S) \underset{\mathrm{End}(S)}{\otimes} \mathrm{Hom}_\Lambda(S,\Lambda) \; S \simeq \underset{\mathrm{End}(S)}{\otimes} \mathrm{Hom}(S,\Lambda)$$

$$f \qquad \to \quad 1 \otimes f$$

are isomorphism of Λ-Λ bimodules and Λ^{op}-modules, respectively.

b is as above, the left anhilator of $\tau_\Lambda(S)$.

S torsionless implies; left anhilator of a = left anhilator of S.

As Λ/r-module, S is isomorphic to $\Lambda/r\, u$, where u is a central idempotent and $\Lambda/r = \frac{\Lambda}{r}u \oplus \frac{\Lambda}{r}(1-u)$.

It is easy to see, $b/r \simeq \Lambda/r\ (1-u)$ as two sided ideal.

Hence; $\Lambda/b \simeq \Lambda/r/b/r \simeq \Lambda/r\, u \simeq u\, \Lambda/r\, u \simeq \mathrm{End}_{\Lambda/r}(S)^{op} \simeq \mathrm{End}_\Lambda(S)^{op}$.

Using the ring isomorphism: $\Lambda/b \overset{\simeq}{\to} \mathrm{End}_\Lambda(S)^{op}$.

We can see $\mathrm{Hom}_\Lambda(S,\Lambda)$ as Λ/b-module.

The isomorphism: ϕ: $\mathrm{Hom}_\Lambda(S,\Lambda) \to \tau_\Lambda(S)$ is a $\Lambda/b - \Lambda$ bimodule isomorphism.

Using these identifications, Λ/C is isomorphic to the ring Γ of part 2.

Denote the composite functor:

$(\Delta/C \underset{\Delta}{\otimes} -) \circ \mathrm{Hom}_\Lambda(\Lambda \oplus S, -)$ by $\widetilde{\mathrm{Hom}}(\Lambda \oplus S, -)$, $\widetilde{\mathrm{Hom}}(\Lambda \oplus S, -)$ is different from F defined in part 2, but they coincide in the full subcategory $\underset{S}{\mathrm{mod}}_\Lambda$ of mod_Λ of all modules with no summand isomorphic to a "node"

Moreover; let $U(X,Y)$ be the set of all map f: $X \to Y$ such that f factors

$$\begin{array}{ccc} X & \xrightarrow{\ f\ } & X \\ & {}_g\searrow \quad \nearrow_h & \\ & \oplus S_i & \end{array}$$

in such a way, that for any map p: $\oplus S_i \to S$ pg is not an splittable epimorphism.

$U(X,Y)$ gives rise to an additive subfunctor

$U(-,?)$ of $\mathrm{Hom}\,(-,?)$.

$U(X,Y) \subset I(X,Y)$, $I(X,Y)$ the set of maps that factors through an injective module and

$$\frac{\mathrm{Hom}_\Lambda(\Lambda \oplus S, -)}{U(\Lambda \oplus S, -)} \text{ is isomorphic to } \widetilde{\mathrm{Hom}}(\Lambda \oplus S\,, -).$$

b) Construction of algebas with "nodes".

Let Λ be a basic artin algebra with a simple projective S_0 and a simple injective S_1 and assume $\mathrm{End}_\Lambda(S_0) \simeq \mathrm{End}_\Lambda(S_1)$.

We can construct an stably equivalent algebra Γ identifying S_0, and S_1 and putting zero relations.

This can be done formally as follows:

Let P_1 be the projective cover of S_1.

Then $\Lambda = P_1 \oplus Q \oplus S_0$ where Q is a projective module with no summand isomorphic neither to P_1 nor to S_0.

It is easy to see that Λ has the following matrix form:

$$\Lambda \simeq \mathrm{End}_{\Lambda^{op}}(\Lambda^{op}) = \begin{pmatrix} \mathrm{End}_\Lambda(P_1)^{op} & 0 & 0 \\ \mathrm{Hom}_\Lambda(Q,P_1) & \mathrm{End}_\Lambda(Q)^{op} & 0 \\ \mathrm{Hom}_\Lambda(S_0,P_1) & \mathrm{Hom}_\Lambda(S_0,Q) & \mathrm{End}_\Lambda(S_0)^{op} \end{pmatrix}$$

$$\text{i.e.: } \Lambda \simeq \begin{pmatrix} D & 0 & 0 \\ M_{21} & R & 0 \\ M_{31} & M_{32} & D \end{pmatrix}, \ D \text{ a division ring}$$

${}_R M_{21}{}_D$, ${}_D M_{31}{}_D$, ${}_D M_{32}{}_R$ are bimodules and multiplication is given by the D - D map: $\psi: M_{32} \underset{R}{\otimes} M_{21} \to M_{31}$.

We construct Γ as follows:

$D \ltimes M_{21}$ is the trivial extension, and

$D \ltimes M_{31} \to D$ the canonical map.

By scalar extension, M_{21} , M_{32} are $D \ltimes M_{31}$-modules.

Let $\psi: M_{32} \underset{R}{\otimes} M_{21} \to D \ltimes M_{31}$ be the composite map: $M_{32} \underset{R}{\otimes} M_{21} \to M_{31} \to D \ltimes M_{31}$ and $0: M_{21} \underset{D}{\otimes} M_{32} \to R$ the zero map.

Construct the ring: $\Gamma = \begin{pmatrix} D \ltimes M_{31} & M_{32} \\ M_{21} & R \end{pmatrix}$

with multiplication given by: $0,\psi$.

Conversely; if Γ is a basic artin algebra with node Q, $\Gamma = Q \oplus P$ where P has no projective summand isomorphic to Q.

Γ has matrix form:

$$\Gamma = \begin{pmatrix} \mathrm{End}_\Gamma(Q)^{op} & \mathrm{Hom}_\Gamma(Q,P) \\ \mathrm{Hom}_\Gamma(P,Q) & \mathrm{End}_\Gamma(P)^{op} \end{pmatrix}$$

Let $T = \mathrm{End}_\Gamma(Q)^{op}$, $M = \mathrm{Hom}_\Gamma(Q,P)$

$N = \mathrm{Hom}_\Gamma(P,Q)$ $R = \mathrm{End}_\Gamma(P)^{op}$

$\psi_1: M \underset{R}{\otimes} N \to T$, $\psi_2: N \otimes M \to R$ are given by composition, i.e. $\psi_i(g \otimes f) = f \circ g$

$$\Gamma = \begin{pmatrix} T & M \\ N & R \end{pmatrix}$$

has the following properties:

a) T is local with radical r, and $r^2 = 0$.

b) $rM = Nr = 0$.

c) $\operatorname{Im} \psi_1 \subset \operatorname{rad} T$.

d) $\psi_2 = 0$.

Given an algebra $\Gamma = \begin{pmatrix} T & M \\ N & R \end{pmatrix}$ satisfying conditions a), b), c), d), the indecomposable projective $P_1 = \begin{pmatrix} T & 0 \\ N & 0 \end{pmatrix}$ is a "node" of Γ.

Taking $a = \tau_\Gamma(P_1/rP_1)$, the trace of P_1/rP_1 in Γ and b the left anhilator of a, the ring: $\begin{pmatrix} \Gamma/a & 0 \\ a & \Gamma/b \end{pmatrix}$ is isomorphic to:

$$\Lambda = \begin{pmatrix} T/r & 0 & 0 \\ N & R & 0 \\ r & M & T/r \end{pmatrix}$$

Hence; Γ is stably equivalent to Λ.

We call to the construction Γ, "The construction of a node".

There is a particular case for which "the construction of a node" takes a simpler form.

If $\Lambda = \Lambda_2 \times \Lambda_1$ where Λ_1, Λ_2 are algebras with matrix form: $\Lambda_1 = \begin{pmatrix} R & 0 \\ M & D \end{pmatrix}$ $\Lambda_2 = \begin{pmatrix} D & 0 \\ N & T \end{pmatrix}$

with D a division ring.

$$\Lambda = \Lambda_2 \times \Lambda_1 = \begin{pmatrix} D & 0 & 0 & 0 \\ N & T & 0 & 0 \\ 0 & 0 & R & 0 \\ 0 & 0 & M & D \end{pmatrix}$$ the ring

Γ constructed from Λ is isomorphic to:

$$\begin{pmatrix} R & 0 & 0 \\ M & D & 0 \\ 0 & N & T \end{pmatrix}$$ with multiplication given

by the obvious maps.

Now it is easy to construct examples.

Let Γ_1, Γ_2 be algebras with matrix form:

$$\Gamma_1 = \begin{pmatrix} K & 0 & 0 & 0 & 0 \\ 0 & K & 0 & 0 & 0 \\ K & K & K & 0 & 0 \\ K & 0 & K & K & 0 \\ 0 & K & K & 0 & K \end{pmatrix} \qquad \Gamma_2 = \begin{pmatrix} K & 0 & 0 & 0 \\ K & K & 0 & 0 \\ K & 0 & K & 0 \\ K & K & K & K \end{pmatrix}$$

where K is a field, multiplication given as ordinary matrices but defining in Γ_1

$a_{53} \cdot a_{31} = 0$, a_{43} $a_{32} = 0$.

The Gabriel graphs of Γ_1 and Γ_2 are:

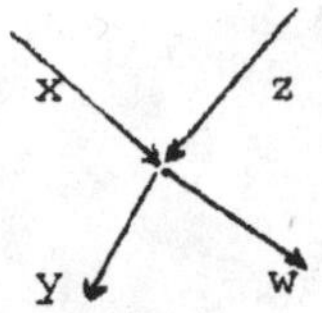

with relations: $xw = 0$, $zy = 0$ for Γ_1.

and 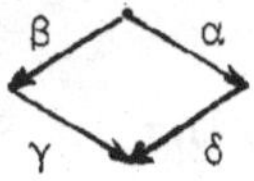

with $\beta\gamma = \alpha\delta$ for Γ_2.

$$\Gamma = \begin{pmatrix} K & 0 & 0 & 0 & 0 & 0 & 0 & 0 \\ 0 & K & 0 & 0 & 0 & 0 & 0 & 0 \\ K & K & K & 0 & 0 & 0 & 0 & 0 \\ K & 0 & K & K & 0 & 0 & 0 & 0 \\ 0 & K & K & 0 & K & 0 & 0 & 0 \\ 0 & 0 & 0 & 0 & K & K & 0 & 0 \\ 0 & 0 & 0 & 0 & K & 0 & K & 0 \\ 0 & 0 & 0 & 0 & K & K & K & K \end{pmatrix}$$

With multiplication given by:

$b_{53}\, b_{31} = 0$, $b_{43}\, b_{32} = 0$ $b_{65}\, b_{53} = 0$

$b_{75}\, b_{53} = 0$.

The graph of Γ is:

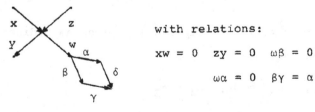

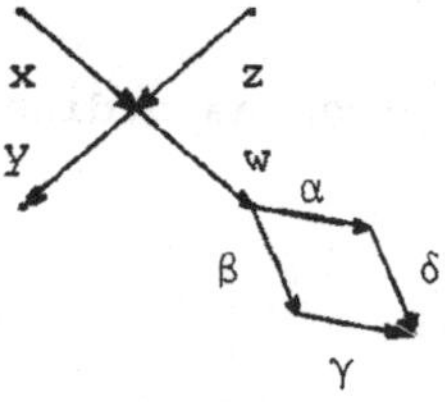

with relations:

$xw = 0$ $zy = 0$ $\omega\beta = 0$

$\omega\alpha = 0$ $\beta\gamma = \alpha\delta$.

Γ is stably equivalent to $\Gamma_1 \times \Gamma_2$.

Similar results where obtained for algebras stably equivalent to hereditary in an independent way by Bongartz-Riedtmann [4].

REFERENCES

[1] Auslander, M., Reiten, I.: Representation theory of Artin algebras III, Comm. Algebra 3, (3), 239-294 (1975).

[2] Auslander, M., Reiten, I.: Representation theory of Artin algebras V, Comm. Algebra 5, (5), 519-554 (1977).

[3] Auslander, M., Reiten, I.: Stable equivalence of Artin algebras. Proc. Conf. on Orders, Group Rings and related topics. Springer Lecture Notes 353, 8-71 (1973).

[4] Bongartz, K., Riedtmann Ch.: Algèbrers stablement héréditaires. C.R. Acad. Sc. Paris 288, 703-706 (1979).

[5] Fossum, R.M., Griffiths, P.A., Reiten, I.: Trivial extensions of abelian categories. Lecture Notes 456 Springer 1975.

[6] Platzeck, M.I.: Representation theory of algebras stably equivalent to an hereditary Artin algebra. Trans. Amer. Math. Soc. 238, 89-128 (1978).

INSTITUTO DE MATEMATICAS
U. N. A. M.
México 20, D.F.
MEXICO

HEREDITARY ALGEBRAS THAT ARE NOT PURE-HEREDITARY

Frank Okoh

Abstract

Let R be a hereditary finite-dimensional algebra of tame type. In this note it is shown that any pure-injective R-module contains an indecomposable direct summand. This result is used to prove that if R is not countable and the category of R-modules contains a full sub-category equivalent to the category of representations of $\tilde{A}_1$ then R is not pure-hereditary.

Let R be a finite-dimensional algebra, M an R-module. We recall that a submodule, N, of M is said to be a <u>pure</u> submodule if whenever $N \subset L \subset M$ with L/N finite-dimensional then N is a direct summand of L. An exact sequence, $0 \to A \overset{\alpha}{\to} B \to C \to 0$, is said to be <u>pure</u> if $\operatorname{Im}\alpha$ is a pure submodule of B. An R-module is said to be <u>pure-injective</u> (<u>pure-projective</u>) if it is injective (projective) with respect to pure exact sequences.

From now on R will stand for a finite-dimensional algebra of tame representation type over an algebraically-closed field, K.

<u>THEOREM 0. The indecomposable pure-injective R-modules are:</u>

(i) <u>indecomposable pre-injective modules</u>

(ii) <u>indecomposable regular torsion modules</u>

(iii) <u>indecomposable preprojective modules</u>

(iv) <u>rank one torsion-free divisible modules</u>

(v) Prüfer modules

(vi) p-adic modules.

Proof. See [5]

PROPOSITION 1. A pure-injective R-module M contains an indecomposable direct summand.

Proof. Suppose M does not contain a direct summand of the form (i), (iii), (iv), (v) or (vi), (see above). Then as in the proof of Theorem 0, we may suppose that for some R-module, N

$$M \oplus N = \prod_{t \in T} L_t$$

where L_t is a module over a complete discrete valuation ring, D_t and $T = \mathcal{P}_1(K)$, the projective line over K, see [6] for details. Each L_t is fully invariant, so $L_t = (M \cap L_t) \oplus (N \cap L_t)$. For some $t_0 \in T$, $M_{t_0} = M \cap L_{t_0} \neq 0$ since $M \neq 0$. Since M_{t_0} is a module over a complete discrete valuation ring, it must have an indecomposable direct summand. □

Every R-module has a pure-projective resolution and so the functors Pext^n, $n \geqq 1$ are defined. The usual long exact sequences can be obtained from short exact sequences. It is known that Pext^1 is a subfunctor of Ext^1, see [7]. In [2], it is stated, though not used, that Pext^n is a subfunctor of Ext^n, $n \geqq 1$. We shall show that this is not the case. For that we shall need representations of $\tilde{A}_1$, i.e. $\cdot \rightrightarrows \cdot$.

Let (a,b) be a fixed basis of K^2, $\tilde{K} = K \cup \{\infty\}$

$$b_\theta = \begin{cases} b - \theta a, & \text{if } \theta \in K \\ a, & \text{if } \theta = \infty . \end{cases}$$

A K-representation of $\tilde{A}_1$ gives rise to a K-bilinear map $K^2 \times V$ to W where V and W are K-vector spaces. Conversely a pair of K-vector spaces (V,W) together with a K-bilinear map $K^2 \times V$ to W gives rise to a K-representation of $\tilde{A}_1$. Such objects have been studied extensively under the name systems.

The results on systems that we shall be using can be found in [4]. The results there do not depend on the complex numbers. The indecomposable preprojective (preinjective) systems are those of type III^k, (I^k) $k = 1, 2, \ldots$. Let (V_k, W_k) be of type III^k. In [4] it is shown that

(1) $\prod_{k=1}^{\infty} (V_k, W_k) / \bigoplus_{k=1}^{\infty} (V_k, W_k)$ has no direct summand of the form (i), (ii), (iii) or (v).

Let V be a torsion-free θ-adic module i.e. V is a module over $\widehat{K[\zeta]}_{(\zeta-\theta)}$ - the completion of the polynomial ring $K[\zeta]$ at $\zeta - \theta$, $\theta \in K$. The case $\theta = \infty$ is handled as in the theory of complex variables. One makes (V, V) a system by setting $av = v$ and $bv = \zeta v$. We shall call such a system a θ-adic system.

Torsion-free rank one divisible systems and θ-adic systems have the property that for any nonzero element v in the range space the equation

(2) $b_\eta \, x_\eta = v$

has Card (K) linearly independent solutions.

If (V_k, W_k) is of type III^k then (2) has at most $k-1$ linearly independent solutions.

The category of systems is hereditary i.e. $Ext^2 = 0$. The next result shows that if K is not countable then $Pext^2 \neq 0$.

<u>Theorem 1</u>. <u>If K is not countable then the category of systems is not pure-hereditary.</u>

<u>Proof</u>. Let $(V,W) = \prod_{k=1}^{\infty} (V_k, W_k)$, (V_k, W_k) as above.

$$(V_0,W_0) = \bigoplus_{k=1}^{\infty} (V_k, W_k)$$

Since (V_0, W_0) is an ascending union of direct summands of (V,W) it is a pure subsystem of (V,W). So we have the pure exact sequence:

$$0 \longrightarrow (V_0,W_0) \longrightarrow (V,W) \longrightarrow (V,W)/(V_0,W_0) \longrightarrow 0$$

This yields the exact sequence:

$$Pext^1(- , (V_0,W_0)) \longrightarrow Pext^1(- , (V,W)) \longrightarrow$$
$$\longrightarrow Pext^1(- , (V,W)/(V_0,W_0)) \longrightarrow Pext^2(- ,(V_0,W_0))$$

where $-$ denotes an arbitrary system. Since (V,W) is a direct product of finite-dimensional systems it is pure-injective. Suppose $Pext^2(- , (V_0,W_0)) = 0$. Then $Pext^1 (- , (V,W)/(V_0,W_0)) = 0$. So by Theorem 0, Proposition 1 and (1), $(V,W)/(V_0, W_0)$ must have a θ-adic system or a torsion-free rank one divisible system as a direct summand. In $(V,W)/(V_0,W_0)$ we have $b_\theta(v+V_0) = b_\theta v + W_0$. Since Equation (2) has only $k-1$ linearly independent solutions in (V_k,W_k) it can have only countably many solutions in $(V,W)/(V_0,W_0)$. So if K is not countable, $\mathrm{Pext}^2((X,Y), (V_0,W_0)) \neq 0$ for some system, (X,Y) □

<u>REMARK</u> An immediate consequence of a theorem in [3] is that if K has cardinality $\aleph_r$ then any finite-dimensional K-algebra has $\mathrm{P\,gl\,dim} \leqq r + 1$. In particular, if K is countable, the category of systems is pure-hereditary i.e. pure subsystems of pure-projective systems are pure-projective. However an arbitrary subsystem of a pure-projective system is not necessarily pure-projective. The example of such a subsystem in [4] is independent of the cardinality of the field.

References

1. Fuchs, L.: Infinite abelian groups, Vol. I, Academic Press, New York and London, (1970).
2. Griffith, P.: On the decomposition of modules and generalized left uniserial rings, Math. Ann. 184 (1970) 300-308.
3. Jensen, C.U. and Simson D.: Purity and generalized chain conditions, Journal of Pure and Applied Algebra, 14 (1979) 297-305.
4. Okoh, F.: Direct sums and direct products of canonical pencils of matrices, Lin. Alg. and Appl. 25, (1979) 1-26.
5. ______: Indecomposable pure-injective modules over finite-dimensional algebras of tame type, to appear.
6. Ringel, C.M.: Infinite-dimensional representations of finite-dimensional hereditary algebras, Symposia Math. Ins. Nat. Alta. Mat. 23 (1979).
7. Walker, C.P.: Relative homological algebra and abelian groups, Illinois J. Math. 10, (1966) 186-209.

Frank Okoh
Department of Mathematics
York University
4700 Keele Street,
Downsview, Ontario
M3J 1P3
Canada

PROJECTIVE LATTICES OVER GROUP ORDERS AS AMALGAMATIONS OF IRREDUCIBLE LATTICES

Wilhelm Plesken

I. Introduction

Throughout this article let G be a finite group, K an algebraic number field, R a valuation ring with quotient field K and maximal ideal π, and $F = R/\pi$ the residue class field of R. We assume that both, K and k, are splitting fields for G, and we denote the characteristic of F by p.

In this note we want to study the group ring RG by a careful analysis of the embedding of RG into $\bigoplus_{s=1}^{h} e_s RG$, where $e_1, e_2, \ldots, e_h$ are the central primitive idempotents of KG. It turns out that the R-orders $e_s RG$ $(s=1,2,\ldots,h)$ can be characterized by few parameters in case certain of the Brauer decomposition numbers are equal to zero or one. Some methods to determine these parameters are sketched and demonstrated by an example of a group all of which decomposition numbers are equal to zero or one. The method is applied to blocks with cyclic defect groups. As a corollary one obtains congruence properties for the nonexceptional characters. A more detailed account of the results sketched here will appear elsewhere.

II. General Results

A particularly simple type of R-orders are graduated orders [Zas 75], which will frequently turn out to be epimorphic images of group rings over R.

<u>(II.1) Definition.</u> Let Λ be an R-order in the matrix algebra $K^{n\times n}$. Then Λ is called a graduated order, if Λ contains a set of n orthogonal idempotents ($\neq 0$).

Clearly, the maximal order $R^{n\times n}$ is a graduated order. To get more general examples, consider the following subset of $R^{n\times n}$:

$$\Lambda(n_1,\ldots,n_l;(m_{ij})):=\{(a_{ij}) \mid a_{ij}\in\pi^{m_{ij}}R^{n_i\times n_j}, 1\leq i,j\leq l\},$$

where $l,n_1,\ldots,n_l\in\mathbb{N}$ with $n_1+\ldots+n_l=n$ and $(m_{ij})\in\mathbb{Z}_{\geq 0}^{l\times l}$. $\Lambda(n_1,\ldots,n_l;(m_{ij}))$ is an R-order iff $m_{ii}=0$ and $m_{ij}+m_{jk}\geq m_{ik}$ for all i,j,k with $1\leq i,j,k\leq l$, and in this case it is already a graduated order.

<u>(II.2) Lemma.</u> [Zas 75] Let Λ be a graduated R-order in $K^{n\times n}$ with radical $\mathrm{Jac}(\Lambda)$. Then $\Lambda/\mathrm{Jac}(\Lambda)\simeq\bigoplus_{i=1}^{l}F^{n_i\times n_i}$ for suitable $l,n_1,\ldots,n_l\in\mathbb{N}$ with $n_1+\ldots+n_l=n$. Furthermore there exists a matrix $M=(m_{ij})\in\mathbb{Z}_{\geq 0}^{l\times l}$ satisfying

$$(*)\quad\begin{cases} m_{ii}=0 & \text{for } 1\leq i\leq l,\\ m_{ij}+m_{jk}\geq m_{ik} & \text{for } 1\leq i,j,k\leq l,\\ m_{ij}+m_{ji}>0 & \text{for } 1\leq i,j\leq l \text{ and } i\neq j\end{cases}$$

such that Λ is isomorphis to $\Lambda(n_1,\ldots,n_l;M)$.

In this context we want to call M an exponent matrix of Λ and whenever we talk about an exponent matrix, we assume that all three conditions of (*) are fulfilled. However there is more than one exponent matrix for a graduated order Λ. With a prescibed order of the components of $\Lambda/\mathrm{Jac}(\Lambda)$ the exponent matrices of Λ are in 1-1-correspondence with the isomorphism classes of irreducible Λ-lattices [Zas 75].Note that the submodules of an irreducible Λ-lattice form a distributive lattice . Indeed, graduated R-orders can be characterized as the intersections of the maximal orders containing an R-order in $K^{n\times n}$ the submodule lattice of one of its irreducible lattices is distributive (cf. [Ple 77] , Theorem 3.19).

The next theorem gives necessary and sufficient conditions, when an epimorphic image of the group ring RG is a graduated order. At the same time it gives an idea, how far such an epimorphic image is from being a direct summand of the group ring. Before we formulate the theorem, we have to fix some notation:

$A_1, A_2, \ldots, A_r$ (set of representatives of the) simple kG-modules;

$P_1, P_2, \ldots, P_r$ projective RG-covers of $A_1, A_2, \ldots, A_r$;

$V_1, V_2, \ldots, V_h$ (set of representatives of the) irreducible KG-modules ;

$e_1, e_2, \ldots, e_h$ central primitive idempotents of KG with $e_s V_s = V_s$ for $s = 1,2,\ldots,h$;

finally the decomposition number d_{si} is the multiplicity of A_i in $L_s/\pi L_s$, where L_s is some RG-lattice satisfying KL_s $(:= K\otimes_R L_s) \simeq V_s$ ($1\leq s\leq h$, $1\leq i\leq r$).

<u>(II.3) Theorem.</u> Let $1 \leq s \leq h$.

(i) $e_s RG$ is a graduated order if and only if the decomposition numbers d_{si} are equal to zero or one for all i with $1 \leq i \leq r$.

(ii) Let $\phi: e_s RG \to \Lambda(n_1,\ldots,n_l;M)$ be an isomorphism of R-orders with M satisfying (*) of (II.2). Then

$$\phi(e_s RG \cap RG) = \Lambda(n_1,\ldots,n_l;t^{(s)}J-M^{tr}) ,$$

where M^{tr} is the transposed matrix of M, $J = \begin{pmatrix} 1 & \cdots & 1 \\ \vdots & & \vdots \\ 1 & \cdots & 1 \end{pmatrix} \in \mathbb{Z}^{l\times l}$, and $t^{(s)} \in \mathbb{Z}_{\geq 0}$ is defined by $\pi^{t^{(s)}} = \frac{|G|}{\dim_K V_s} R$.(Clearly the n_i are the F-dimensions of those A_{j_i} for which the the decomposition numbers d_{sj_i} are equal to one.)

The proof of the nonobvious direction of (II.3)(i) proceeds by lifting idempotents. Note, since K is a splitting field, RG is a semi-perfect ring. The proof of (II.3)(ii) is more involved and uses Schur's relations and the fact that RG is a Gorenstein order (cf. [Rog 70] ,pg. 252). Since $e_s RG \subseteq RG$ is contained in $e_s RG$ the inequalities $0 < m_{ij} + m_{ji} \leq t^{(s)}$ are an immediate consequence of (II.3)(ii). This still leaves many possibilities for the exponent matrix M. To get further restrictions, we make a general assumption for the rest of this chapter.

<u>Assumption</u> for the rest of Chapter II:
All decomposition numbers d_{si} of G are equal to zero or one.

Clearly,one obtains restrictions for the exponent matrices from automorphisms and antiautomorphisms of RG. However, a more powerful tool to interrelate the exponent matrices of the different $e_s RG$ for $s = 1,2,\ldots,h$ can be developed by the study of the embedding of $\bigoplus_{s=1}^{h} e_s RG$ in RG. To simplify the considerations we rather look at the projective indecomposable RG-lattices and study these as amalgamations or subdirect products of irreducible RG-lattices. First we have to fix some more notation: Let

$$S_i = \{s \mid 1 \leq s \leq h \text{ , } d_{si} = 1\} \quad \text{for} \quad i = 1,2,\ldots,r,$$

$$Z_s = \{i \mid 1 \leq i \leq r \text{ , } d_{si} = 1\} \quad \text{for} \quad s = 1,2,\ldots,h,$$

and let $n_i = \dim_F A_i$ for $i = 1,2,\ldots,r$. Application of Brauer's reciprocity theorem and of Theorem (II.3) to our assumption that all decompostion numbers are equal to zero or one yields

(i) $KP_i = \bigoplus_{s\in S_i} e_sKP_i$ for each $i = 1,2,\dots,r$, where e_sKP_i is an irreducible KG-module for each $s \in S_i$;

(ii) $e_sRG \simeq \Lambda(n_i \mid i\in Z_s; M^{(s)})$ for each $s = 1,2,\dots,h$,where $M^{(s)} = (m_{ij}^{(s)}) \in \mathbb{Z}_{\geq 0}^{|Z_s|\times|Z_s|}$ is an exponent matrix. To make $M^{(s)}$ unique, we assume that the first collumn is equal to zero.

As an immediate consequence of (i) we get

$$\bigoplus_{s\in S_i} (P_i \cap e_sP_i) \subseteq P_i \subseteq \bigoplus_{s\in S_i} e_sP_i$$

for $i = 1,2,\dots,r$. The e_sP_i and $P_i \cap e_sP_i$ are irreducible RG- resp. e_sRG-lattices for $s \in S_i$. It follows that $\bigoplus_{s\in S_i} e_sP_i$ is the unique minimal lattice in KP_i which contains P_i and is completely decomposable. Similarly $\bigoplus_{s\in S_i} (P_i \cap e_sP_i)$ is the unique maximal completely decomposable sublattice of P_i . To investigate the embedding of P_i into $\bigoplus_{s\in S_i} e_sP_i$ we introduce two RG-modules:

$O(P_i) := (\bigoplus_{s\in S_i} e_sP_i)/P_i$ ("upper amalgamating factor"),

$U(P_i) := P_i/ \bigoplus_{s\in S_i} (P_i \cap e_sP_i)$ ("lower amalgamating factor")

for $i = 1,2,\dots,r$. Both, $O(P_i)$ and $U(P_i)$, are R-torsion modules, indeed $|G|O(P_i) = |G|U(P_i) = 0$ by Theorem (II.3). In some sense $O(P_i)$ and $U(P_i)$ are dual to each other. E. g. $O(P_i)$ has a simple socle and $U(P_i)$ has a simple head. Among the many properties one can prove about the $O(P_i)$ and $U(P_i)$ we only point out the following two.

<u>(II.4) Theorem.</u> Let $1 \leq i \leq r$.
(i) For each $s \in S_i$ there exists an epimorphism of $U(P_i)$ onto $e_sP_i/(P_i \cap e_sP_i)$ and a monomorphism of $e_sP_i/(P_i \cap e_sP_i)$ in $O(P_i)$. Furthermore $e_sP_i/(P_i \cap e_sP_i)$ and $(1-e_s)P_i/(P_i \cap (1-e_s)P_i)$ are isomorphic RG-modules.
(ii) For $j = 1,2,\ldots,r$ let o_{ij} resp. u_{ij} be the multiplicity of A_j in $O(P_i)$ resp. $U(P_i)$. Then

$$o_{ij} + u_{ij} = \sum_{s \in S_i} (t^{(s)} - m_{ij}^{(s)} - m_{ji}^{(s)}) ,$$

where $t^{(s)}$ is defined as in (II.3). Moreover $o_{ij} \leq (|S_i|-1)u_{ij}$ and $u_{ij} \leq (|S_i|-1)o_{ij}$. In case A_i and A_j are selfdual, equality holds: $o_{ij} = u_{ij}$.

The proof of (II.4)(i) makes use of the fact that P_i is subdirect product of e_sP_i and $(1-e_s)P_i$ with amalgamating factor $e_sP_i/(P_i \cap e_sP_i)$, i. e. P_i is the pullback of
$e_sP_i \rightarrow e_sP_i/(P_i \cap e_sP_i) \simeq (1-e_s)P_i/(P_i \cap (1-e_s)P_i) \leftarrow (1-e_s)P_i$.
For the proof of (II.4)(ii) one has to use (II.3)(ii). To apply Theorem (II.4) for finding the exponent matrices we introduce the "amalgamation matrix" which allows one to keep track of the composition factors of the amalgamating factors.

<u>(II.5) Definition.</u> For $i = 1,2,\ldots,r$ the <u>amalgamation matrix</u> $A(P_i)$ of the projective indecomposable RG-lattice P_i is defined as follows: The columns are indexed by those j, $1 \leq j \leq r$, for which A_j belongs to the block of P_i. The rows are indexed by the elements of S_i. The (s,j)-entry of $A(P_i)$ is the multiplicity of A_j in a composition series of $e_sP_i/(P_i \cap e_sP_i)$, which is given by
$t^{(s)} - m_{ij}^{(s)} - m_{ji}^{(s)}$.

By Theorem (II.4)(ii) the sum of the entries in the j-th column of $A(P_i)$ is equal to $o_{ij} + u_{ij}$, which again is equal to the sum of the entries of the i-th column of $A(P_j)$ ($1 \leq i,j \leq r$). The complete knowledge of the amalgamation matrices gives some information about the embedding of RG into $\bigoplus_{s=1}^{h} e_s RG$.

III. Examples and Applications

As a straightforward example we discuss the group ring RG of the simple group $G = PSL(2,7)$ of order 168, where R is the localization of $\mathbb{Z}[\frac{1+\sqrt{-7}}{2}]$ at one of the prime ideals containing 2. There are six irreducible KG-modules $V_1,...,V_6$ and four irreducible FG-modules $A_1,A_2,...,A_4$ up to isomorphism. The decomposition matrix is given by

degree		1	3	3	8
		A_1	A_2	A_3	A_4
1	V_1	1	0	0	0
3	V_2	0	1	0	0
3	V_3	0	0	1	0
6	V_4	0	1	1	0
7	V_5	1	1	1	0
8	V_6	0	0	0	1

.

One obtains immediately:

$e_1RG \simeq R$; $e_2RG \simeq e_3RG \simeq R^{3\times 3}$ $(\simeq \Lambda(3;(0)))$;

$e_4RG \simeq \Lambda(3,3;\begin{pmatrix} 0 & x \\ 0 & 0 \end{pmatrix})$ for a suitable natural number x;

$e_5RG \simeq \Lambda(1,3,3;\begin{pmatrix} 0 & a & b \\ 0 & 0 & c \\ 0 & d & 0 \end{pmatrix})$ for suitable $a,b,c,d \in \mathbb{Z}_{\geq 0}$;

and finally $e_6RG \simeq R^{8\times 8}$.

The amalgamation matrix $A(P_1)$ of P_1 is given by

$A(P_1)$	A_1	A_2	A_3
$e_1P_1/(P_1 \cap e_1P_1)$	3	0	0
$e_5P_1/(P_1 \cap e_5P_1)$	3	3-a	3-b

Application of Theorem (II.4) yields $3-a = 3-b = 0$ and hence $a = b = 0$. The amalgamation matrix of P_2 is given by

$A(P_2)$	A_1	A_2	A_3
$e_2P_2/(P_2 \cap e_2P_2)$	0	3	0
$e_4P_2/(P_2 \cap e_4P_2)$	0	2	2-x
$e_5P_2/(P_2 \cap e_5P_2)$	3-a	3	3-c-d

from which we conclude $2-x = 3-c-d$. Since the outer automorphisms of G interchange A_2 and A_3, one gets $c = d$. Hence one ends up with $x = c = d = 1$ and $a = b = 3$, and the group ring RG can be described in a somewhat lax way up to Morita equivalence by

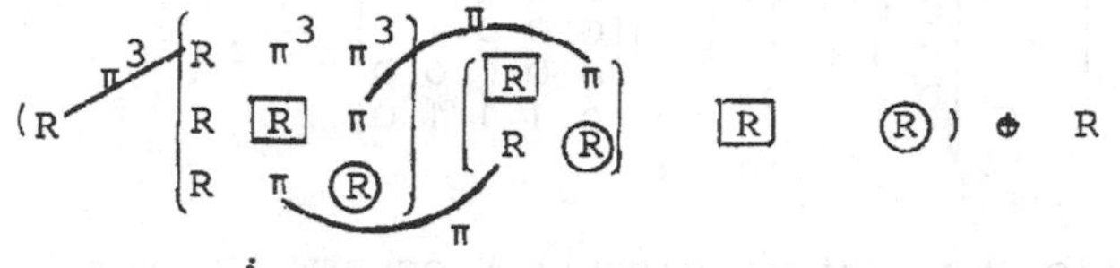

where $\overline{\pi^i}$ indicates a linear congruence modulo π^i, the encircled copies of R resp. those in squares are amalgamated by linear congruences modulo π^3 which in this case (but not in general) can be obtained from the character table of G.

We note that the lattice of submodules of the irreducible RG-lattices can easily be computed from the exponent matrices. E. g. in the last example the lattice of submodules of e_5P_1 is given by

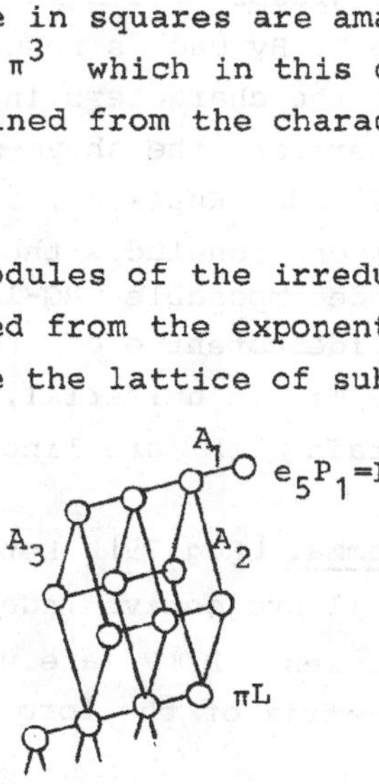

Note that the ten lattices which are not contained in $\pi L = \pi e_5 P_1$ form a set of representatives of the isomorphism classes of the nonzero sublattice of $e_5 P_1$ (cf. [Ple 77], Theorem 2.5).

In many applications of Theorem (II.4) one knows already some of the exponent matrices and wants to determine the others. As a typical example we choose $G = S_{10}$ (symmetric group on ten elements) and R as the localization of $\mathbb{Z}$ at the prime 5. By using the fact that the group ring of S_9 is contained in RG one can see that the exponent matrices of those $e_s RG$ for which Z_s contains only two indices are equal to $\begin{pmatrix} 0 & 1 \\ 0 & 0 \end{pmatrix}$. Then a somewhat longer, but essentially not more difficult computation as above shows that all occurring exponent matrices are take. rom the set

$$\left\{ (0), \begin{pmatrix} 0 & 1 \\ 0 & 0 \end{pmatrix}, \begin{pmatrix} 0 & 1 & 2 \\ 0 & 0 & 1 \\ 0 & 0 & 0 \end{pmatrix}, \begin{pmatrix} 0 & 1 & 1 & 2 \\ 0 & 0 & 1 & 1 \\ 0 & 1 & 0 & 1 \\ 0 & 0 & 0 & 0 \end{pmatrix}, \begin{pmatrix} 0 & 1 & 2 & 2 & 1 \\ 0 & 0 & 1 & 1 & 1 \\ 0 & 0 & 0 & 1 & 0 \\ 0 & 0 & 1 & 0 & 0 \\ 0 & 1 & 1 & 1 & 0 \end{pmatrix} \right\} .$$

As a final application we want to discuss blocks with cyclic defect groups. Let B be a block of RG with cyclic defect group D. By Dade's results [Dad 66] the decomposition numbers of the characters in the block B are equal to zero or one. Therefore the above method is applicable. From the results of Dade, Kupisch , Janusz, and Peacock (cf. [Pea 77]) one concludes the following: (**) For each projective indecomposable RG-lattice P in B and each central primitive idempotent e of KG with $eP \neq 0$, the factor module $eP/\pi eP$ is uniserial, i. e. the sublattices of eP which contain πeP are linearly ordered by inclusion.

<u>(III.1) Lemma.</u> [Rog 79] Let Λ be a graduated R-order such that for all projective indecomposable Λ-lattices P the factor modules $P/\pi P$ are uniserial, then Λ admits an exponent matrix of the form aH_1 for suitable

natural numbers a and l , where $H_l = \begin{pmatrix} 0 & 1 & \dots & 1 \\ \vdots & \ddots & \ddots & \vdots \\ \vdots & & \ddots & 1 \\ 0 & \dots & \dots & 0 \end{pmatrix} \in \mathbb{Z}^{l\times l}$.

Because of (**) Lemma (III.1) can be applied and all that remains to be determined is the parameter a in (III.1). This can be done by an application of Theorem (II.4) in case the irreducible Frobenius character belonging to the idempotent e is nonexceptional.

<u>(III.2) Theorem.</u> Let B be a block of RG with cyclic defect group D. If e is a central primitive idemponent of KG belonging to B such that the ordinary irreducible character associated with e is nonexceptional, then eRG is isomorphic to the graduated order $\Lambda(n_1,\dots,n_l;dH_l)$ for suitable natural numbers $l,n_1,\dots,n_l$, where H_l is defined in (III.1) and d by $\pi^d = |D|R$.

Brauer (cf. [Bra 41] ,Theorem 11) has proved a result as in (III.2) for all characters in a block with a defect group of prime order. Theorem (III.2) answers the question raised at the end of [Dad 66] for all nonexceptional characters. The proof of (III.2) uses the particularly simple form of the decomposition matrix of a block with cyclic defect group. This is also used for the proof of the following corollary concerning congruences for characters.

<u>(III.3) Corollary.</u> Under the assumption of (III.2) let γ be the Brauer tree of B. and let $\hat{\gamma}$ be the union of trees obtained from γ by deleting the exceptional vertex and the adjacent edges. If χ_1 and χ_2 are two irreducible Frobenius characters belonging to the same connected component of $\hat{\gamma}$, then

$$\frac{\chi_1(g)\,|g^G|}{\chi_1(1)} \equiv \frac{\chi_2(g)\,|g^G|}{\chi_2(1)} \quad (\text{ mod } |D|R\ )$$

for all $g \in G$, where g^G denotes the conjugacy class of g in G .

References

[Bra 41] R. Brauer: Investigations on Group Characters. Ann. of Math. 42, 4 (1941), 936-958.

[Dad 66] E. C. Dade: Blocks with cyclic defect groups. Ann. of Math. 84, 2 (1966), 20-48.

[Pea 77] R. M. Peacock: Ordinary character theory in a block with a cyclic defect group. J. of Algebra 44 (1977), 203-220.

[Ple 77] W. Plesken: On absolutely irreducible representations of orders. 241-262 in H. Zassenhaus (ed.): Number Theory and Algebra. Academic Press 1977 .

[Rog 70] K. W. Roggenkamp: Lattices over Orders II. Springer Lecture Notes in Mathematics 115 (1970).

[Rog 79] K. W. Roggenkamp: Private communication.(1979).

[Zas 75] H. Zassenhaus: Graduated Orders. Manuscript, California Institute of Technology, Pasadena 1975.

RWTH Aachen
Lehrstuhl D für Mathematik
Templergraben 64
51 Aachen, WEST GERMANY

REPRESENTATION-FINITE SELFINJECTIVE ALGEBRAS OF CLASS A_n

Christine Riedtmann (Basel and Zürich)

Throughout this article k denotes an algebraically closed field, Λ a finite-dimensional associative k-algebra with unit element $1_\Lambda = 1$ and $\operatorname{mod} \Lambda$ the category of unitary right Λ-modules with finite k-dimension. We assume Λ to be <u>connected</u> (i.e. any direct factor-algebra is either 0 or Λ) and <u>representation-finite</u>. The last condition means that there are finitely many non-isomorphic indecomposable modules $M_1,\dots,M_r \in \operatorname{mod} \Lambda$ such that each $N \in \operatorname{mod} \Lambda$ is isomorphic to $M_1^{m_1} \oplus \dots \oplus M_r^{m_r}$ for some well determined $m_1,\dots,m_r \in \mathbb{N}$. In fact we choose such modules $M_1,\dots,M_r$ once for all.

From paragraph 2 onwards we shall moreover assume that Λ is <u>self-injective</u> (i.e. is an injective right module over itself). Our purpose is to prove some general statements on representation-finite selfinjective algebras and to classify completely those of tree-class A_n ([6], 1.7 and § 2, Hauptsatz). The classification of the representation-finite selfinjective algebras of class D_n, E_6, E_7 or E_8 will appear in subsequent publications. Those of tree-class A_n are divided into two subclasses, the <u>wreath-like algebras</u> already considered in [5] and the so-called <u>Moebius algebras</u>. The starting impetus for our results was the observation that the methods used in [5] may be extended to Moebius algebras. Despite history however, we present a new approach aiming at a direct description of the Auslander algebra of Λ (and leading to some changes in notation). I thank P. Gabriel for helpful discussions on the presentation of my results.

§ 1 Stable algebras

1.1 The Auslander algebra $E_\Lambda = E$ of Λ is the k-algebra of endomorphisms of $M = M_1 \oplus \ldots \oplus M_r$. Its isomorphism class does not depend on the choice of $M_1, \ldots, M_r$; it is basic and has finite dimension over k . The canonical projections e_i of M onto $M_i \subset M$ supply a partition $1_E = e_1 + \ldots + e_r$ of 1_E into orthogonal primitive idempotents, and the direct summand $\mathrm{Hom}_\Lambda(M_j, M_i)$ of $E \cong \bigoplus_{i,j} \mathrm{Hom}_\Lambda(M_j, M_i)$ is identified with $e_i E e_j$. Moreover, the radical R_E of E is given by

$$R_E = R_\Lambda(M,M) \cong \bigoplus_{i,j} R_\Lambda(M_j, M_i) \cong \bigoplus_{i,j} e_i R_E e_j ,$$

where R_Λ is the radical of the category $\mathrm{mod}\,\Lambda$ ([6] , 3.1). Similarly we have

$$R_E^2 = R_\Lambda^2(M,M) \cong \bigoplus_{i,j} R_\Lambda^2(M_j, M_i) \cong \bigoplus_{i,j} e_i R_E^2 e_j ,$$

and

$$[e_i(R_E/R_E^2)e_j : k] = [e_i R_E e_j / e_i R_E^2 e_j : k] \leq 1$$

for all i,j ([6], 3.1). In other words, the Auslander-Reiten-quiver Γ_Λ of Λ ([6], 1.3) is identified with the ordinary quiver of the algebra E and looks as follows: The vertices are identified with $e_1, \ldots, e_r$ and two vertices e_i, e_j are connected by an arrow $e_j \to e_i$ iff $[e_i(R_E/R_E^2)e_j : k] = 1$. We shall say that a vertex e_i represents a module $N \in \mathrm{mod}\,\Lambda$ if $N \cong M_i$.

1.2 The quiver-algebra $k[\Gamma_\Lambda]$ of Γ_Λ consists in the formal linear combinations of paths: a path from e_i to e_j of length n is a sequence $(e_j|\alpha_n, \ldots, \alpha_1|e_i)$ of vertices and arrows of Γ_Λ such that $n \geq 0$, e_i = source(α_1), range(α_m) = source(α_{m+1}) if $1 \leq m < n$ and e_j = range(α_n) ; moreover we assume that $i=j$ if $n=0$. The multiplication of $k[\Gamma_\Lambda]$ is defined by

$$(e_m|\beta_p,\dots,\beta_1|e_l)(e_j|\alpha_n,\dots\alpha_1|e_i) = \begin{cases} 0 & \text{if } j\neq l \\ (e_m|\beta_p,\dots,\beta_1,\alpha_n,\dots,\alpha_1|e_i) & \text{if } j=l \end{cases}$$

In practice we often write $\alpha_n\dots\alpha_1$ instead of $(e_j|\alpha_n,\dots,\alpha_1|e_i)$, thus identifying an arrow with a path of length 1.

Let us now choose an element $f_{ij}\in(e_iR_Ee_j)\setminus(e_iR_E^2e_j)$ for each arrow $e_j \overset{\alpha}{\to} e_i$ of Γ_Λ . Our choice uniquely determines a homomorphism

$$\phi_\Lambda : k[\Gamma_\Lambda] \to E_\Lambda$$

such that $\phi_\Lambda(e_i||e_i) = e_i$ for any i and $\phi_\Lambda(e_i|\alpha|e_j) = f_{ij}$ for any arrow $e_j \overset{\alpha}{\to} e_i$. The homomorphism ϕ_Λ is <u>surjective</u> and its kernel I_Λ is contained in J_Λ^2 , where J_Λ denotes the (twosided) ideal of $k[\Gamma_\Lambda]$ generated by all "<u>arrows</u>" $(e_i|\alpha|e_j)$. If (x_s) is a family of generators of the ideal I_Λ , we say that E_Λ is defined by the quiver Γ_Λ and the <u>relations</u> $x_s=0$.

Our purpose is to describe E_Λ by quiver and relations in case Λ is representation-finite and selfinjective. In the present article we achieve the description for selfinjective algebras of tree-class A_n .

1.3 For each vertex e of Γ_Λ we denote by e^+ (resp. by e^-) the set formed by the ranges (resp. the sources) of the arrows starting (resp. ending) at e . We set further

$$P = \{e_i : M_i \text{ is not projective}\}$$

$$J = \{e_i : M_i \text{ is not injective}\} .$$

The <u>Auslander-Reiten translation</u> of Γ_Λ is a bijection $\tau : P \to J$ such that $(\tau e)^+=e^-$ for each $e\in P$ ([6], 1.1 and 1.2). It is uniquely determined by the requirement that $\tau e_i=e_j$ iff there is an Auslander-

Reiten sequence ([6], 3.4)

$$0 \to M_j \to N \to M_i \to 0$$

The <u>stable algebra</u> $\bar{E}_\Lambda = \bar{E}$ associated with Λ is the residue algebra $E/\sum Ee_iE$, where e_i ranges through the non-stable vertices of Γ_Λ (a vertex e is called <u>stable</u> iff $\tau^n e$ is defined for any $n \in \mathbb{Z}$). It may also be interpreted as being the algebra of endomorphisms of $M = M_1 \oplus \ldots \oplus M_r$ in the <u>stable category</u> <u>mod</u> Λ ([6] , 2.2). Its ordinary quiver is by construction the stable part ${}_s\Gamma_\Lambda$ of the Auslander-Reiten quiver Γ_Λ . As we know by ([6] , § 2, Hauptsatz) ${}_s\Gamma_\Lambda$ is a disjoint union of quivers of the form $\mathbb{Z}B/\Pi$, where $\mathbb{Z}B$ ranges through the infinite quivers of figure 1 and where Π denotes an admissible group of automorphisms ([6] , 1.5) . Diverging from the notations chosen in [6] we parametrize the vertices of $\mathbb{Z}B$ in such a way that $\tau(p,q)=(p-1,q)$.

The partition of the <u>representation-quiver</u> ${}_s\Gamma_\Lambda$ (i.e. of the quiver together with its Auslander-Reiten translation, [6], 1.1) into connected components is related to the unique decomposition $\bar{E} = B_1 \times B_2 \times \ldots \times B_t$ of $\bar{E}$ into "connected" direct factors B_i . We call such a factor a <u>stable</u> A-<u>block</u>, D-<u>block</u> or E-<u>block</u> of Λ according as its ordinary quiver is isomorphic to $\mathbb{Z}A_n/\Pi$, $\mathbb{Z}D_n/\Pi$ or $\mathbb{Z}E_n/\Pi$. The structure of A-blocks is described in the following proposition. The corresponding (more complicated) structure theorems for D- and E-blocks will appear in subsequent publications.

Let us recall what the admissible groups Π of automorphisms of $\mathbb{Z}A_n$ look like ([6], 4.3): if τ is the translation $(p,q)\mapsto(p-1,q)$ of $\mathbb{Z}A_n$, the cyclic group $\tau^{r\mathbb{Z}}$ is admissible for any $r\geq 1$, $r\in\mathbb{N}$. The corresponding residue quiver $A_{n,r} = \mathbb{Z}A_n/\tau^{r\mathbb{Z}}$ has nr vertices. Besides these $A_{n,r}$ we can get a new family of residue quivers $A_{n,\tilde{r}}, n\geq 2$, by

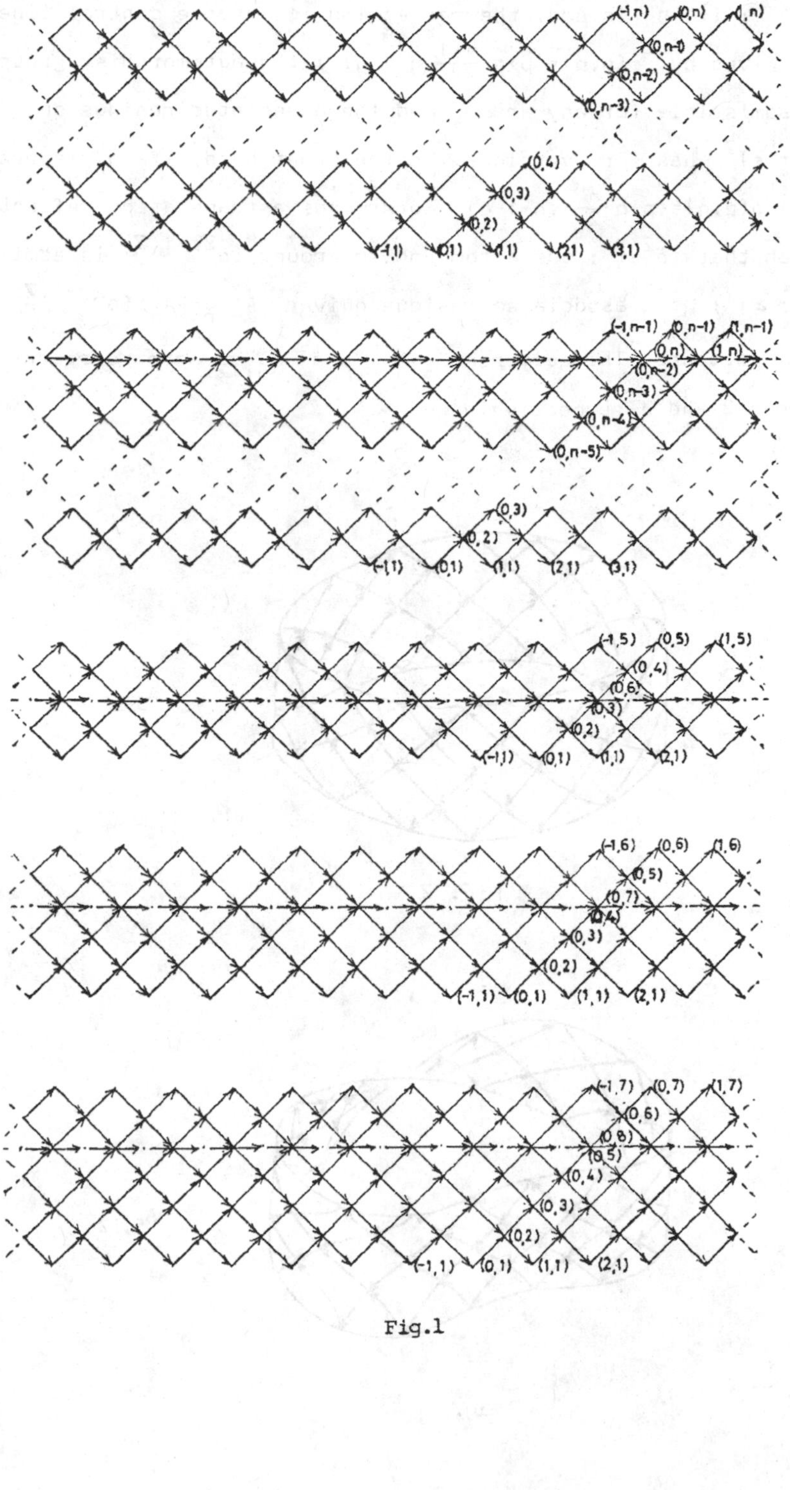

$\mathbb{Z}A_n$ $n \geq 1$

$\mathbb{Z}D_n$ $n \geq 4$

$\mathbb{Z}E_6$

$\mathbb{Z}E_7$

$\mathbb{Z}E_8$

Fig.1

twisting $\mathbb{Z}A_n$: if n is odd, the reflection ϕ at the central line of $\mathbb{Z}A_n$ is given by $\phi(p,q)=(p+q-\frac{n+1}{2},n+1-q)$; the automorphism group $(\tau^r\phi)^{\mathbb{Z}}$ is admissible for any $r \geq 1$ and the associated residue quiver $A_{n,\tilde{r}} = \mathbb{Z}A_n/(\tau^r\phi)^{\mathbb{Z}}$ has nr vertices. On the other hand, if n is even, the formula $\rho(p,q)=(p+q-\frac{n}{2}-1,n+1-q)$ determines a translation-reflection of $\mathbb{Z}A_n$ such that $\rho^2=\tau$; the automorphism group $(\rho^{2r+1})^{\mathbb{Z}}$ is admissible for $r \geq 1$; the associated residue quiver $A_{n,\tilde{r}}=\mathbb{Z}A_n/(\rho^{2r+1})^{\mathbb{Z}}$ has $n(r+\frac{1}{2})$ vertices. The quivers $A_{5,12}$ and $A_{5,\widetilde{12}}$ are represented in the figures 2 and 3 respectively.

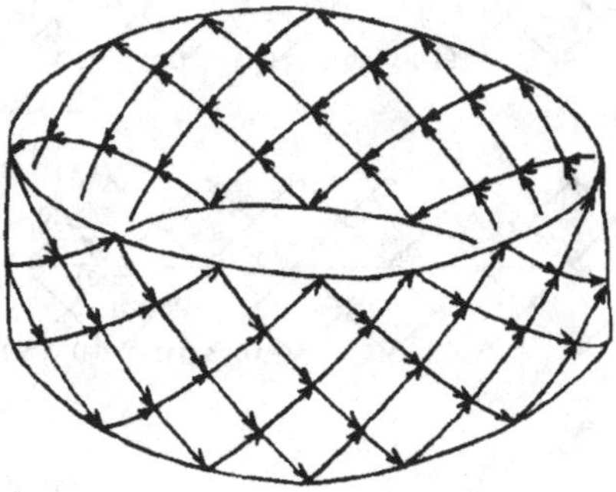

$A_{5,12}$

Fig. 2

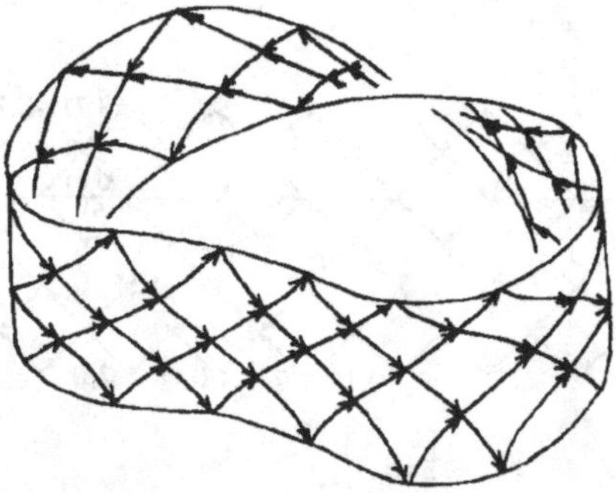

$A_{5,\widetilde{12}}$

Fig.3

1.4 Let Γ be a <u>finite</u> representation-quiver. For each vertex x such that τx is defined, we set $\xi_x = \sum_\alpha (x|\alpha,\sigma\alpha|\tau x) \in k[\Gamma]$, where α runs through all arrows with range x ([6],1.1). The elements ξ_x generate a twosided ideal of the quiver-algebra $k[\Gamma]$, which will be denoted by I_Γ. The residue algebra $k[\Gamma]/I_\Gamma$ is identified with $\bigoplus_{e,f} \mathrm{Hom}_{k(\Gamma)}(e,f)$, where e,f range through the vertices of Γ, and where $k(\Gamma)$ is the k-linear category assigned to Γ in [6], 2.1 (its objects are the vertices of Γ, its morphisms are generated by the arrows of Γ submitted to the relations $\sum_\alpha (x|\alpha,\sigma\alpha|\tau x)=0$).

Here we are interested in the cases $\Gamma = A_{n,r}$ and $\Gamma = A_{n,\tilde{r}}$. Then we write $I_{n,r}$ and $I_{n,\tilde{r}}$ instead of I_Γ.

<u>Proposition. Let Λ be representation-finite</u>. <u>A stable</u> A-<u>block of Λ with quiver $A_{n,r}$ is isomorphic to</u> $k[A_{n,r}]/I_{n,r}$.

<u>Proof.</u> If $\Gamma \tilde{\to} \mathbb{Z}A_n/\Pi$ is the connected component of ${}_s\Gamma_\Lambda$ related to the given A-block, it is sufficient to construct a fully faithful functor $H: k(\Gamma) \to \underline{\mathrm{mod}}\,\Lambda$ mapping a vertex $e_i \in \Gamma$ onto M_i (1.1). This will be done in the following sections.

1.5 <u>Definition: Let Γ be a connected component of the stable representation-quiver ${}_s\Gamma_\Lambda$, and let $\pi: \Delta \to \Gamma$ be a covering</u> ([6],1.6). <u>A k-linear functor</u> $F: k(\Delta) \to \underline{\mathrm{mod}}\,\Lambda$ <u>is called</u> well-behaved <u>iff it satisfies the following two conditions</u>: a) $Fe = M_i$ if $\pi e = e_i$ (1.1) ; b) $F\alpha: Ff \to Fe$ <u>is irreducible for any arrow</u> $f \overset{\alpha}{\to} e$ <u>of</u> Δ.

We simply write $F\alpha$ instead of $F\bar{\alpha}$, if $\bar{\alpha}$ is the morphism of $k(\Delta)$ associated with an arrow α of Δ. Proposition 2.2 of [6] claims that a well-behaved functor does exist if Δ is the universal covering of Γ. In this case the following statement is proved in [6], 2.3. In fact, the proof given there is valid for any covering Δ.

Proposition. For any two vertices e,f of Δ the map

$$\bigoplus_{Fh=Ff} \mathrm{Hom}_{k(\Delta)}(e,h) \to \underline{\mathrm{Hom}}_{\Lambda}(Fe,Ff) ,$$

induced by a well-behaved functor $F:k(\Delta)\to\underline{\mathrm{mod}}\,\Lambda$, is a bijection.

Corollary. A well-behaved functor $H:k(\Gamma)\to\underline{\mathrm{mod}}\,\Lambda$ is fully faithful.

In order to finish the proof of proposition 1.4 it remains for us to construct a well-behaved functor $F:k(\mathbb{Z}A_n)\to\underline{\mathrm{mod}}\,\Lambda$ such that $F(g\alpha)=F(\alpha)$ for any arrow α of $\mathbb{Z}A_n$ and any $g\in\Pi$. Indeed, such a functor F clearly factors through a functor $H:k(\mathbb{Z}A_n/\Pi)\to\underline{\mathrm{mod}}\,\Lambda$, to which we may apply the preceding corollary.

1.6 Construction of $F:k(\mathbb{Z}A_n)\to\underline{\mathrm{mod}}\,\Lambda$ in case $\Pi=\tau^{r\mathbb{Z}}$. We start from any well-behaved functor $F_0:k(\mathbb{Z}A_n)\to\underline{\mathrm{mod}}\,\Lambda$ and denote by Δ_p the "diagonal" $\{(p,q):1\le q\le n\}$ of $\mathbb{Z}A_n$. The functor F to be constructed will coincide with F_0 on β_1 (Fig. 4) and on all arrows connecting vertices of $\Delta_0\cup\Delta_1\cup\ldots\cup\Delta_{r-1}$. Furthermore we set $F\gamma_i = F_0(\tau^r\gamma_i)$.

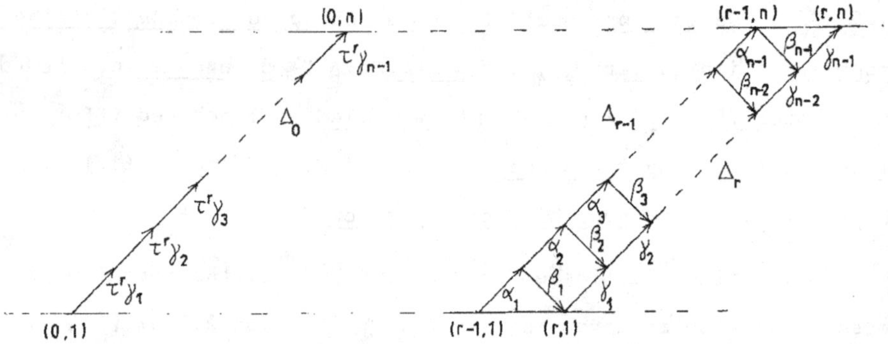

Fig. 4

It remains for us to construct $F\beta_p, p\geq 2$, in such a way that $(F\gamma_{n-1})(F\beta_{n-1}) = 0 = (F\gamma_1)(F\beta_1)+(F\beta_2)(F\alpha_2) = (F\gamma_2)(F\beta_2)+(F\beta_3)(F\alpha_3)=\ldots=(F\gamma_{n-2})(F\beta_{n-2})+$ $+(F\beta_{n-1})(F\alpha_{n-1})$. This will enable us to extend F to all of $\mathbb{Z}A_n$ by periodicity. The construction of $F\beta_p$ proceeds by induction on p : Suppose $F\beta_i$ is already constructed for $1<i<p<n$ in such a way that $(F\gamma_{i-1})(F\beta_{i-1})+(F\beta_i)(F\alpha_i) = 0$. In order to construct $F\beta_p$ consider the mesh

$$\begin{array}{ccccc} & & (r-1,p+1) & & \\ & \overset{\alpha_p}{\nearrow} & & \overset{\beta_p}{\searrow} & \\ (r-1,p) & & & & (r,p) \\ & \underset{\beta_{p-1}}{\searrow} & & \underset{\gamma_{p-1}}{\nearrow} & \\ & & (r,p-1) & & \end{array}$$

By [6], 3.4 a) there is an Auslander-Reiten sequence of the form

$$F(r-1,p) \xrightarrow{[F\alpha_p\ F\beta_{p-1}]^T} F(r-1,p+1)\oplus F(r,p-1) \xrightarrow{[\beta\ \gamma]} F(r,p)$$

in <u>mod</u> Λ . By [6], 3.1 and 3.5 we have

$\lambda\gamma - F\gamma_{p-1} \in \underline{R}^2_\Lambda(F(r,p-1),(r,p))$ for some non-zero $\lambda\in k$. We infer that

$$0 = \lambda[\gamma(F\beta_{p-1})+\beta(F\alpha_p)] = (F\gamma_{p-1})(F\beta_{p-1})+\gamma'(F\beta_{p-1})+\lambda\beta(F\alpha_p)$$

for some $\gamma'\in\underline{R}^2_\Lambda(F(r,p-1),F(r,p))$. Now $\gamma'(F\beta_{p-1})$ lies in $\underline{R}^3_\Lambda$ and can be written by the following lemma as $\gamma'(F\beta_{p-1})=\beta'(F\alpha_p)$, where $\beta'\in\underline{R}^2_\Lambda(F(r-1,p+1),F(r,p))$. We infer that $0 = (F\gamma_{p-1})(F\beta_{p-1})+$ $+(\beta'+\lambda\beta)(F\alpha_p)$, and we set $F\beta_p = \beta'+\lambda\beta$.

Proceeding in this way we construct by induction $F\beta_2, F\beta_3, \ldots, F\beta_{n-1}$, and all the requirements but $(F\gamma_{n-1})(F\beta_{n-1}) = 0$ are satisfied by construction. Now it follows from proposition 1.5 that $\underline{R}_\Lambda(F_0(r-1,n),F_0(r,n)) = 0$, since we have $\mathrm{Hom}_{k(\mathbb{Z}A_n)}((s,n),(r,n)) = 0$ for $s\neq r$ and $\mathrm{Hom}_{k(\mathbb{Z}A_n)}((r,n),(r,n)) \cong k$. As $(F\gamma_{n-1})(F\beta_{n-1})$ lies in $\underline{R}_\Lambda(F_0(r-1,n),F_0(r,n))$ it is automatically zero.

<u>Lemma</u>. <u>If</u> $p<n$ <u>each</u> $\mu\in\underline{R}^3_\Lambda(F_0(r-1,p),F_0(r,p))$ <u>can be written as</u> $\mu = \mu'(F_0\alpha_p)$, <u>where</u> $\mu'\in\underline{R}^2_\Lambda(F_0(r-1,p+1),F_0(r,p))$.

Proof. By the very definition of $\underline{R}^3_\Lambda$ and proposition 1.5 we have $\mu = \sum \lambda_w (F_0 w)$, where w ranges through paths of length ≥ 3 from $(r-1,p)$ to some (mr,p), $m \geq 1$. Such a path has either the form $w = u\alpha_p$ or the form $w = v\beta_{p-1}$, where the lengths of u, v are ≥ 2. In the second case v is decomposed according to figure 5. We infer that $F_0 w = (F_0 v_3)(F_0 v_2 v_1 \beta_{p-1}) = (-1)^{s+1}(F_0 v_3)(F_0 v_4) F_0 \alpha_p$. In any case $F_0 w$ may be written as $(F_0 u)(F_0 \alpha_p)$, where $F_0 u \in \underline{R}^2_\Lambda(F_0(r-1,p+1), F_0(r,p))$. OK.

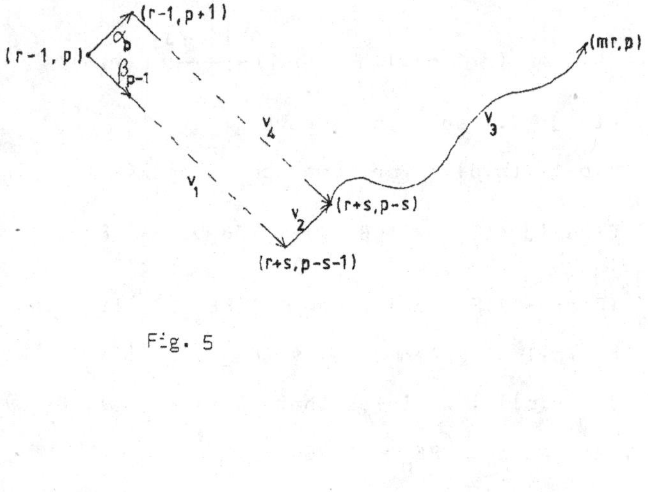

Fig. 5

1.7 Proposition. Let Λ be representation-finite. A stable A-block of Λ with quiver $A_{n,\tilde{r}}$ is isomorphic to $k[A_{n,\tilde{r}}]/I_{n,\tilde{r}}$ (1.4) either if $r \geq \left[\frac{n-1}{2}\right]$ or if $\operatorname{char} k \neq 2$.

Of course, $[x]$ denotes the integral part of a real number x. As we shall see in § 3, the case $r < \left[\frac{n-1}{2}\right]$ cannot occur if Λ is selfinjective. Therefore, we restrict our proof to the case $r \geq \left[\frac{n-1}{2}\right]$. The case $r < \left[\frac{n-1}{2}\right]$, $\operatorname{char} k \neq 2$, will be treated in a subsequent publication devoted to selfinjective algebras of class D_n. There we shall have to struggle with characteristic 2 even in the selfinjective case.

Proof in case $r \geq \left[\frac{n-1}{2}\right]$. By 1.3 we have $A_{n,\tilde{r}} = \mathbb{Z}A_n/(\rho^{2r+1})^{\mathbb{Z}}$ or $A_{n,\tilde{r}} = \mathbb{Z}A_n/(\tau^r\phi)^{\mathbb{Z}}$ according as n is even or odd. We examine in detail the first case, the second one being similar. Clearly, as in 1.5 and 1.6 we have to construct a well-behaved functor $F : k(\mathbb{Z}A_{2q}) \to \underline{\text{mod}}\,\Lambda$ such that $F(g\alpha) = F(\alpha)$ for each arrow α and each $g \in \Pi = (\tau^r\phi)^{\mathbb{Z}}$.

Let us start as in 1.6 from any well-behaved functor $F_0 : k(\mathbb{Z}A_{2q}) \to \underline{\text{mod}}\,\Lambda$. The functor F to be constructed will coincide with F_0 on β_1 and on the arrows lying between the diagonals Δ_{r-q} and $\rho^{2r+1}(\Delta_{r-q+1})$ (with the exception of α_1 in case $r=q-1$, the arrows lying on these diagonals are considered as lying also between). For the arrows lying on Δ_{r-q+1} we set $F\gamma_i = F_0(\rho^{2r+1}\gamma_i)$ (Fig. 6). The construction of $F\beta_p$, $p \geq 2$, proceeds by induction as in 1.6. The only trouble could arise in the preceding lemma, where $(r-1,p),(r,p)$ and (mr,p) have to be replaced by $(r-q,p),(r-q+1,p)$ and $g(r-q+1,p)$, $g \in \Pi$ respectively. As a matter of fact, there would be some trouble if $g(r-q+1,p)$ was below $(r-q,p)$ on the going down diagonal through $(r-q,p)$. But it is easily seen that this possibility is excluded by our assumption $r \geq p-1$.

Finally, F is extended to all of $\mathbb{Z}A_{2q}$ by using the "periodicity" $F\delta = F(\rho^{2r+1}\delta)$.

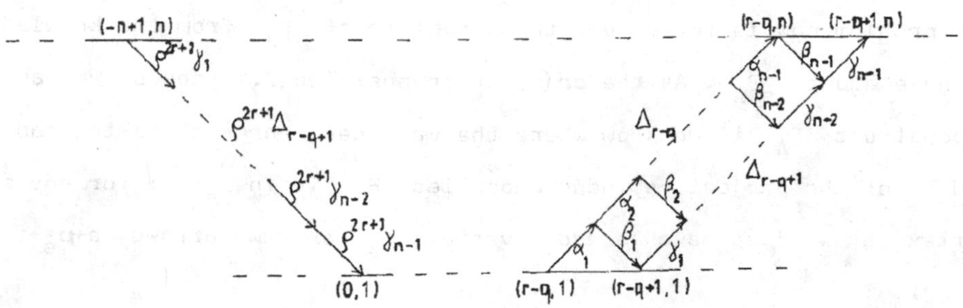

Fig. 6

§ 2 The Auslander-Reiten quiver of a representation-finite selfinjective algebra.

From now on we assume that Λ is representation-finite and selfinjective. The case $\Lambda=k$ being trivial we also suppose $1<[\Lambda:k]<\infty$.

2.1 Proposition. The stable part ${}_s\Gamma_\Lambda$ of the Auslander-Reiten quiver Γ_Λ of Λ is a connected representation-quiver.

Proof. As Λ is selfinjective, the only indecomposable Λ-modules represented by non-stable vertices of Γ_Λ are the projective (=injective) ones. Hence we get ${}_s\Gamma_\Lambda$ from Γ_Λ by deleting the vertices associated with projective Λ-modules. Now we know by [3], 3.1 that, if the vertex $e_i \in \Gamma_\Lambda$ represents a projective module $P \in \text{mod}\,\Lambda$, there is only one arrow $e_h \to e_i$ ending at e_i , and e_h represents the radical $\text{rad}\,P$ of P . The dual argument shows that there is also one and only one arrow $e_i \to e_j$ starting from e_i , and e_j represents $P/\text{soc}\,P$, where $\text{soc}\,P$ is the socle of P . We know furthermore by [3], 4.11 that $\tau e_j = e_h$. This implies that Γ_Λ cannot be split into disjoint τ-stable components by deleting vertices like e_i . OK .

2.2 We deduce from our proposition that ${}_s\Gamma_\Lambda$ is isomorphic to $\mathbb{Z}A_n/\Pi$, $\mathbb{Z}D_n/\Pi$, $\mathbb{Z}E_6/\Pi$, $\mathbb{Z}E_7/\Pi$ or $\mathbb{Z}E_8/\Pi$ for some admissible group Π . Our problem now is to recover the structure of Γ_Λ from the knowledge we have about ${}_s\Gamma_\Lambda$. As the proof of proposition 2.1 shows, we can reconstruct Γ_Λ if we know where the vertices representing the radicals $\text{rad}\,P$ of the projective indecomposables P lie in ${}_s\Gamma_\Lambda$: for any such vertex e we just have to add a vertex p_e and two arrows $e \to p_e \to \tau^{-1}e$ to ${}_s\Gamma_\Lambda$!

Notice in this connection that we may write $\text{rad}\,P$ as $\Omega(P/\text{rad}\,P)$, where $\Omega:\underline{\text{mod}}\,\Lambda \to \underline{\text{mod}}\,\Lambda$ is the loop-functor of Heller. This functor assigns to any $M \in \underline{\mathrm{mod}}\,\Lambda$ the kernel of its projective cover in $\mathrm{mod}\,\Lambda$. As Λ is selfinjective, Ω is a selfequivalence of the stable module category $\underline{\mathrm{mod}}\,\Lambda$; its quasi-inverse is the suspension-functor Σ assigning to any $M \in \underline{\mathrm{mod}}\,\Lambda$ the cokernel of its injective hull. We infer that, among the chosen representatives M_i (introduction), those isomorphic to $\mathrm{rad}\,P$ for some projective P form a set T satisfying the following two conditions: a) For any non-projective M_j there exists an $M_i \in T$ such that $\underline{\mathrm{Hom}}_\Lambda(M_j, M_i) \neq 0$; b) For any two M_i, M_j both belonging to T we have $\underline{\mathrm{Hom}}_\Lambda(M_i, M_j) = 0$ if $i \neq j$ and $\underline{\mathrm{Hom}}_\Lambda(M_i, M_i) \xrightarrow{\sim} k$. Indeed, Σ maps T onto a set S of representatives of the simple Λ-modules, and S satisfies the corresponding conditions both in $\mathrm{mod}\,\Lambda$ and in $\underline{\mathrm{mod}}\,\Lambda$:

2.3 <u>Definition. A configuration of a stable representation-quiver Γ is a set C of vertices satisfying the following two conditions:</u> a) <u>For any vertex</u> e <u>of</u> Γ <u>there exists a vertex</u> $f \in C$ <u>such that</u> $\mathrm{Hom}_{k(\Gamma)}(e,f) \neq 0$; b) <u>For any two</u> $e, f \in C$ <u>we have</u> $\mathrm{Hom}_{k(\Gamma)}(e,f) = 0$ if $e \neq f$ <u>and</u> $\mathrm{Hom}_{k(\Gamma)}(e,e) \xrightarrow{\sim} k$ (see 1.1 or [6], 2.1 for the definition of the category $k(\Gamma)$).

<u>Proposition. Consider a covering</u> $\pi : \Delta \to \Gamma$ <u>of the stable representation-quiver</u> Γ. <u>A set</u> C <u>of vertices of</u> Γ <u>is a configuration iff</u> $\pi^{-1}(C)$ <u>is a configuration in</u> Δ.

<u>Proof</u>. Use the obvious formula

$$\bigoplus_{\pi z = \pi y} \mathrm{Hom}_{k(\Delta)}(x,z) \xrightarrow{\sim} \mathrm{Hom}_{k(\Gamma)}(\pi x, \pi y) \; . \quad \mathrm{OK} \, .$$

2.4 <u>Proposition. The set</u> C_Λ <u>of vertices of</u> ${}_s\Gamma_\Lambda$ <u>representing the radicals of the projective modules is a configuration of</u> ${}_s\Gamma_\Lambda$.

Proof. This is easily seen in case ${}_s\Gamma_\Lambda \xrightarrow{\sim} A_{n,r}$ by using the fully faithful functor $H : k({}_s\Gamma_\Lambda) \to \underline{\text{mod}}\,\Lambda$ considered in 1.4 . Unfortunately, such a functor does not exist always. Hence we have to produce another proof: Let $\pi : \mathbb{Z}B \to {}_s\Gamma_\Lambda$ be the universal covering and $F : k(\mathbb{Z}B) \to \underline{\text{mod}}\,\Lambda$ a well-behaved functor. Using the formula

$$\bigoplus_{Fh=Ff} \text{Hom}_{k(\mathbb{Z}B)}(e,h) \xrightarrow{\sim} \underline{\text{Hom}}_\Lambda(Fe,Ff)$$

of proposition 1.5, we deduce from the conditions a) and b) of 2.2 that $F^{-1}(T)$ is a configuration in $\mathbb{Z}B$. Since we have $\pi^{-1}(C_\Lambda) = F^{-1}(T)$, C_Λ is a configuration of ${}_s\Gamma_\Lambda$ by proposition 2.3. OK.

2.5 Let C be a configuration of a stable representation-quiver Γ . We denote by Γ_C the representation-quiver obtained by adding to Γ a vertex $\mathbb{p}_e$ and arrows $e \to \mathbb{p}_e \to \tau^{-1}e$ for each $e \in C$. Furthermore we agree that the translation of Γ_C coincides with the translation of Γ on the common vertices and is not defined elsewhere. Our definition is motivated by the fact that $\Gamma_\Lambda \xrightarrow{\sim} ({}_s\Gamma_\Lambda)_{C_\Lambda}$ (see 2.2) .

Now consider the universal covering ${}_s\tilde{\Gamma}_\Lambda = \mathbb{Z}B$ of ${}_s\Gamma_\Lambda$, the fundamental group Π_Λ of ${}_s\Gamma_\Lambda$ and the canonical projection $\pi : {}_s\tilde{\Gamma}_\Lambda \to {}_s\Gamma_\Lambda$. We denote by $\tilde{C}_\Lambda$ the configuration $\pi^{-1}(C_\Lambda)$ of ${}_s\tilde{\Gamma}_\Lambda$, by $\tilde{\Gamma}_\Lambda$ the representation-quiver $({}_s\tilde{\Gamma}_\Lambda)_{\tilde{C}_\Lambda}$. There is a unique way to extend the action of Π_Λ to $\tilde{\Gamma}_\Lambda$ and the canonical projection to a morphism $\tilde{\Gamma}_\Lambda \to \Gamma_\Lambda$, which identifies Γ_Λ with $\tilde{\Gamma}_\Lambda/\Pi_\Lambda$ and will still be denoted by π . This "canonical projection" $\pi : \tilde{\Gamma}_\Lambda \to \Gamma_\Lambda$ is a <u>covering</u> ([6], 1.6) and is obviously the <u>universal</u> one. Moreover, if $e \in \tilde{C}_\Lambda$, $\pi\mathbb{p}_e$ and πe represent an indecomposable projective and its radical respectively.

<u>Definition. Let $\pi : \Delta \to \Gamma_\Lambda$ be a covering of the Auslander-Reiten quiver of Λ . A k-linear functor</u> $F : k(\Delta) \to \text{mod}\,\Lambda$ <u>is called well-behaved iff it satisfies the following two conditions:</u> a) $Fe = M_i$ if $\pi e = e_i$ (1.1) ;

b) $F\alpha : Ff \to Fe$ *is irreducible for any arrow* $f \overset{\alpha}{\to} e$ *of* Δ .

Theorem. a) *The universal covering* $\pi : \tilde{\Gamma}_\Lambda \to \Gamma_\Lambda$ *admits well-behaved functors.*

b) *For any covering* $\pi : \Delta \to \Gamma_\Lambda$, *any well-behaved functor* $F : k(\Delta) \to \operatorname{mod} \Lambda$ *and any two vertices* e, f *of* Δ *the induced map*

$$\bigoplus_{Fh=Ff} \operatorname{Hom}_{k(\Delta)}(e,h) \to \operatorname{Hom}_\Lambda(Fe,Ff)$$

is bijective.

Proof. "As" in [6] , 2.2 and 2.3 OK.

2.6 In order to complete our classification of the representation-finite selfinjective algebras we need a precise description of the possible configurations in each of the cases $\mathbb{Z}A_n$, $\mathbb{Z}D_n$, $\mathbb{Z}E_6$, $\mathbb{Z}E_7$, $\mathbb{Z}E_8$. We shall examine the cases $\mathbb{Z}D_n$, $\mathbb{Z}E_n$ in subsequent publications, restricting ourselves here to the case ${}_s\tilde{\Gamma}_\Lambda = \mathbb{Z}A_n$. By 1.4 the stable algebra $\bar{E}_\Lambda$ is then isomorphic either to $k[A_{n,r}]/I_{n,r}$ or to $k[A_{n,\tilde{r}}]/I_{n,\tilde{r}}$ for some n, r . The first case has already been investigated thoroughly in [5] from a different perspective.

Definition. *A Brauer relation of order* n *is an equivalence relation on the set* $\sqrt[n]{1} = \{\exp(2im\pi/n) : m \in \mathbb{Z}\} \subset \mathbb{C}$ *such that the convex hulls of distinct equivalence classes are disjoint.*

Let us set $\underline{e}_n m = \exp(2im\pi/n)$. We draw the convex hulls for one example in case $n=8$ (Fig. 7).

In the general case, if $\mathcal{B}$ is a Brauer relation of order n , we denote by $\bar{\beta}_{\mathcal{B}}$ the permutation of $\sqrt[n]{1}$ assigning to each point s its successor in the equivalence class of s endowed with the anti-clockwise orientation. In the preceding example we have for instance $\bar{\beta}_{\mathcal{B}}(\underline{e}_8 2) = \underline{e}_8 6$.

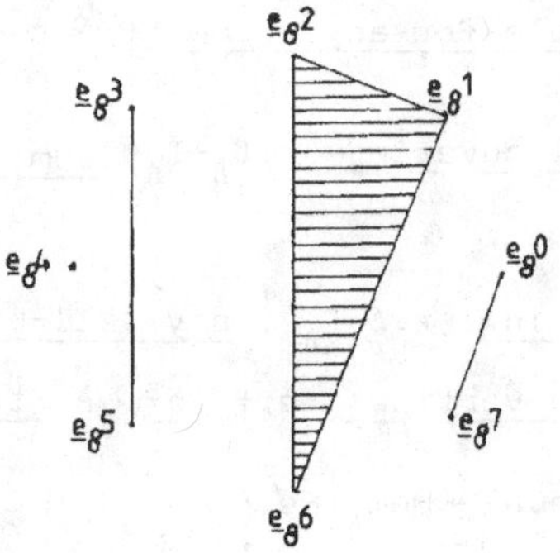

Fig. 7

Proposition. Let $\mathcal{B}$ be a Brauer relation of order n and denote by $C_{\mathcal{B}}$ the set of vertices (i,j) of $\mathbb{Z}A_n$ such that $\underline{e}_n(i+j)=\bar{\beta}_{\mathcal{B}}\underline{e}_n(i)$. The map $\mathcal{B}\mapsto C_{\mathcal{B}}$ is a bijection between the Brauer relations of order n and the configurations of $\mathbb{Z}A_n$.

For instance, if $\mathcal{B}$ is the Brauer relation of Fig. 7, the configuration $C_{\mathcal{B}}$ looks as in Fig. 8.

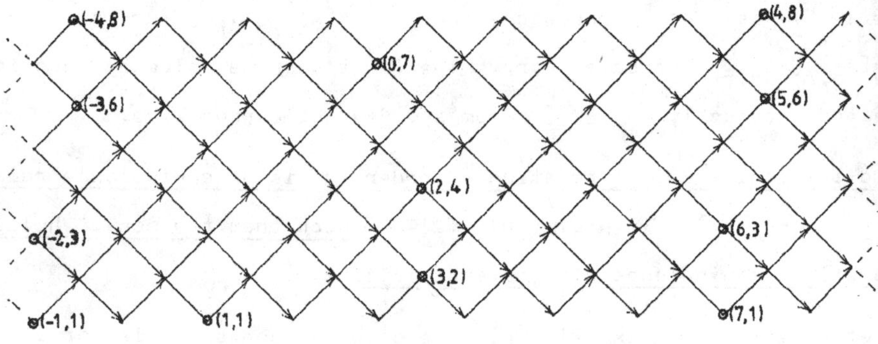

Fig. 8

The associated extended representation-quiver $(\mathbb{Z}A_n)_{C_{\mathcal{B}}}$ is given in Fig. 9.

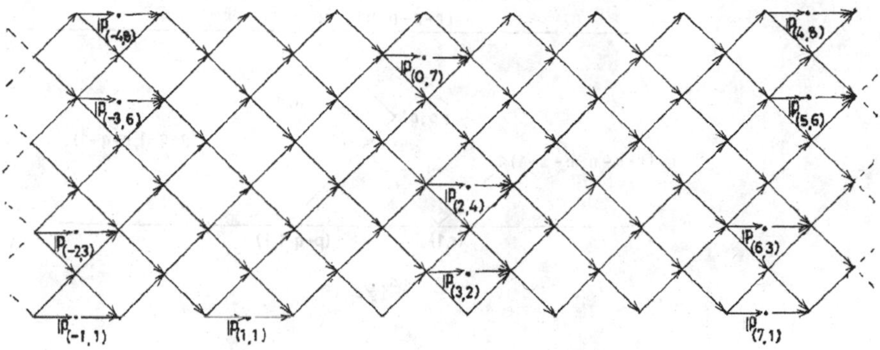

Fig. 9

The proof of the proposition is given in 2.7. It rests upon the lemma below, which we state without proof.

<u>Lemma</u>. <u>For any two vertices</u> $(p.q)$ <u>and</u> (r,s) <u>of</u> $\mathbb{Z}A_n$ <u>we have</u>

$$[\mathrm{Hom}_{k(\mathbb{Z}A_n)}((p,q),(r,s)):k] \leq 1 .$$

<u>The equality</u> $[\mathrm{Hom}_{k(\mathbb{Z}A_n)}((p,q),(r,s)):k] = 1$ <u>holds iff</u>

$$p \leq r < p+q \leq r+s \leq p+n .$$

Let us denote by Δ_p the "<u>going up diagonal</u>" $\{(p,q)\in\mathbb{Z}A_n : 1\leq q\leq n\}$, by ∇_r the "<u>going down diagonal</u>" $\{(p,q)\in\mathbb{Z}A_n : p+q=r\}$. The last inequalities mean that (r,s) lies within the set bounded by Δ_p, Δ_{p+q-1}, ∇_{p+q} and ∇_{p+n}. Such a quadrilateral (bounded by diagonals, with two opposite vertices on the upper and the lower border of $\mathbb{Z}A_n$) will simply be called a <u>rectangle</u> in the sequel. The preceding one "<u>starts</u>" at (p,q) and "<u>stops</u>" at $(p+q-1,n-q+1)$. Clearly, our inequalities may also be interpreted by saying that (p,q) lies in the rectangle stopping at (r,s) (Fig. 10).

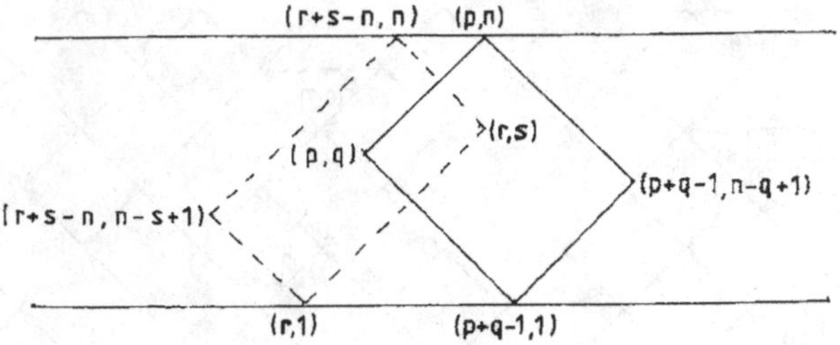

Fig. 10

2.7 <u>Proof of proposition 2.6</u>. Let C be a configuration of $\mathbb{Z}A_n$. Definition 2.3 means that each rectangle intersects C, and that a rectangle starting at a vertex $(p,q)\in C$ does not contain any other point of C. In particular, each going up diagonal Δ_p (=degenerated rectangle starting at (p,1)) contains a unique point $(p,q)\in C$. The induced map $p\mapsto p+q$ is denoted by $\beta_C=\beta$. Notice that by definition we have $p+1\leq\beta p\leq p+n$ for any $p\in\mathbb{Z}$. Furthermore, $\beta:\mathbb{Z}\to\mathbb{Z}$ is bijective: $\beta^{-1}r=p$ if $(p,r-p)$ is the unique point of C on ∇_r!

<u>First we claim that</u> $\underline{e}_n(\beta r)\neq\underline{e}_n(\beta p)$ <u>if</u> $r<p<r+n$. Proof by contradiction: Suppose $\underline{e}_n(\beta r)=\underline{e}_n(\beta p)$. Clearly, the inequalities $r<p<r+n$, $r<\beta r\leq r+n$ and $p<\beta p\leq p+n$ imply $\beta r-n<\beta p<\beta r+2n$. Therefore we have either $\beta p=\beta r$ or $\beta p=\beta r+n$. The former equality is impossible, since β is injective. Hence $\beta p=\beta r+n$ (see Fig. 11). The rectangle starting at $(r,\beta r-r)$ and that stopping at $(p,\beta p-p)$ contain no point of C but $(r,\beta r-r)$ and $(p,\beta p-p)$. We infer that the rectangle stopping at $(p,n-p+r)$ does not intersect C: contradiction!

By our claim $\underline{e}_n\beta(r),\underline{e}_n\beta(r+1),\dots,\underline{e}_n\beta(r+n-1)$ are n distinct points of $\sqrt[n]{1}$. The same holds for $\underline{e}_n\beta(r+1),\dots,\underline{e}_n\beta(r+n-1),\underline{e}_n\beta(r+n)$. Hence $\underline{e}_n\beta(r)=\underline{e}_n\beta(r+n)$. As $r<\beta r\leq r+n<\beta(r+n)\leq r+2n$ we conclude that $\beta(r+n)=\beta(r)+n$: <u>Each configuration</u> C <u>is periodic</u>: $\tau^n C=C$!

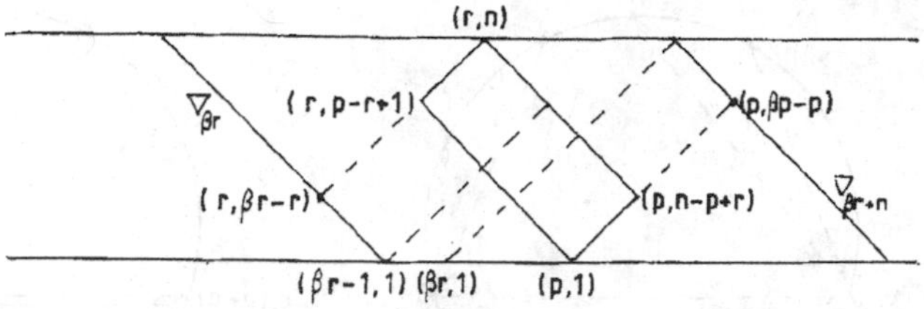

Fig. 11

Being periodic C is completely determined by its image $\bar{C}$ in $\mathbb{Z}A_n/\tau^{n\mathbb{Z}}$. The set of vertices of $\mathbb{Z}A_n/\tau^{n\mathbb{Z}}$ can be identified with $\sqrt[n]{1}\times\sqrt[n]{1}$ by assigning the pair $(\underline{e}_n i, \underline{e}_n(i+j))$ to the class $(\overline{i,j})\in\mathbb{Z}A_n/\tau^{n\mathbb{Z}}$. In the sequel we make such a pair $(\sigma,e)\in\sqrt[n]{1}\times\sqrt[n]{1}$ more concrete by identifying it with the naive vector (=oriented line segment) with origin σ and end e. Consequently, C is determined by the set $\bar{C}$ of all vectors $(\sigma,\bar{\beta}\sigma)$, $\sigma\in\sqrt[n]{1}$, where $\bar{\beta}=\bar{\beta}_C$ is the permutation of $\sqrt[n]{1}$ induced by β $(\bar{\beta}(\underline{e}_n i)=\underline{e}_n(\beta i))$.

For two integers p,q we set $[p,q[=\{t\in\mathbb{Z}:p\le t<q\}$, $[p,q]=[p,q+1[\ldots$. If (p,q) is a vertex of $\mathbb{Z}A_n$, the rectangle starting at (p,q) has as image in $\sqrt[n]{1}\times\sqrt[n]{1}$ the set of vectors (σ,e) such that $\sigma\in\underline{e}_n([p,p+q[)$ and $e\in\underline{e}_n([p+q,p+n])$. Similarly, the rectangle ending at (p,q) has as image the set of vectors (σ,e) such that $\sigma\in\underline{e}_n([p+q-n,p])$ and $e\in\underline{e}_n(]p,p+q])$. The union of both rectangles has as image the set of vectors "crossing" the vector $(\underline{e}_n(p),\underline{e}_n(p+q))$. "Crossing" is to be interpreted as follows: Two vectors meeting within the open unit disc do cross. Two disjoint vectors do not. What "crossing" means for vectors meeting on the unit circle is illustrated in Fig. 12, where the broken vectors cross $(\underline{e}_n p,\underline{e}_n(p+q))$, whereas the full ones do not. A vector of length 0 is represented by a loop.

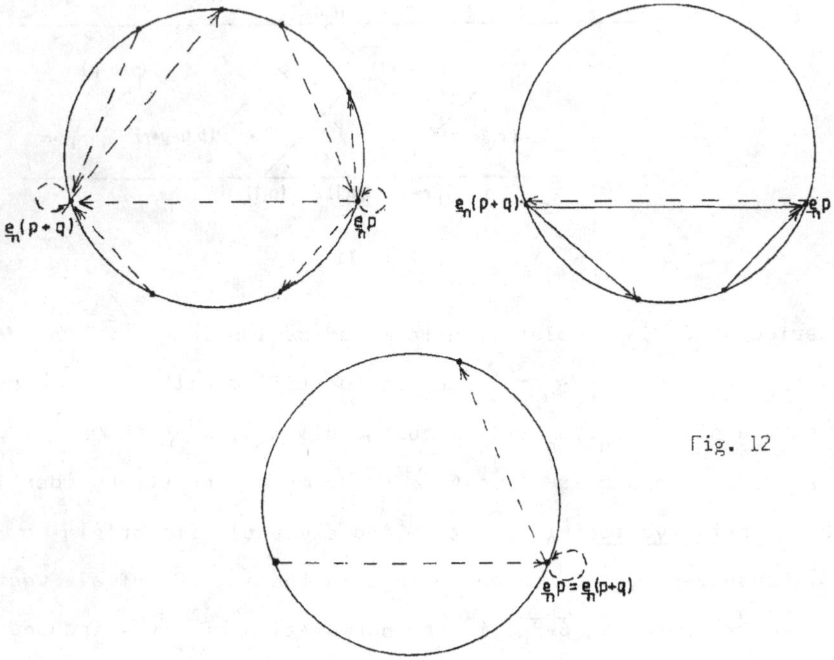

Fig. 12

We already noticed that the conditions imposed on configurations may be formulated in terms of rectangles. Now we know how to translate conditions on rectangles into conditions on vectors. In particular, no "crossing" is allowed in the vector-set $\vec{C}$ determining a configuration C . It follows easily that $\vec{C}$ consists in the anti-clockwise oriented edges of the convex hulls of the equivalence classes of some Brauer relation $\mathcal{B}$. Moreover, the permutation $\bar{\beta}_C$ induced by C coincides with $\bar{\beta}_{\mathcal{B}}$ (2.6). It is also easy to see by the same type of arguments that $C_{\mathcal{B}}$ is a configuration if $\mathcal{B}$ is a Brauer relation.

The vector-set $\vec{C}$ assigned to the configuration given as example in 2.6 is represented in Fig. 13 .

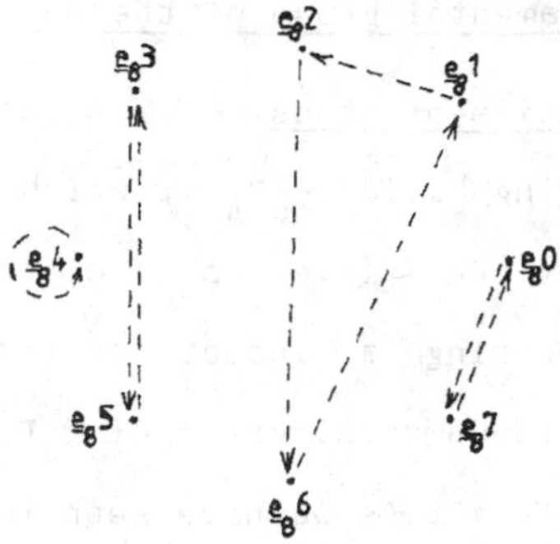

Fig. 13

2.8 <u>Remark.</u> <u>The number of simple Λ-modules</u> is equal to the cardinality of the configuration C_Λ . If Λ is of tree-class A_n , this number <u>depends only on the stable Auslander-Reiten quiver</u> ${}_s\Gamma_\Lambda$. <u>It is equal to</u> r <u>either if</u> ${}_s\Gamma_\Lambda \cong A_{n,r}$ or if ${}_s\Gamma_\Lambda \cong A_{n,\tilde{r}}$. The proof is trivial: call diagonal of ${}_s\Gamma_\Lambda \cong \mathbb{Z}A_n/\Pi_\Lambda$ the image of any diagonal of $\mathbb{Z}A_n$ (notice that there is no way of distinguishing going up and going down diagonals in $A_{n,\tilde{r}}$) . Let d be the number of diagonals in ${}_s\Gamma_\Lambda$. Each diagonal contains one point of C_Λ and each point of C_Λ is contained in 2 diagonals. So C_Λ has $\frac{1}{2}d$ points. Now we have $d=2r+1$ if n is even and ${}_s\Gamma_\Lambda \cong A_{n,\tilde{r}}$; so this case does not occur (compare with 3.2). In all the other cases we have $d = 2r$, and the cardinality of C_Λ equals r .

§ 3 The fundamental group of the Auslander-Reiten quiver

Let Λ be selfinjective of class A_n. As we know, this means the existence of a covering $\pi : \mathbb{Z}A_n \to {}_s\Gamma_\Lambda$, which is determined by ${}_s\Gamma_\Lambda$ up to an automorphism of $\mathbb{Z}A_n$ (i.e. up to a translation or a reflection-translation). The covering π induces an isomorphism $\mathbb{Z}A_n/\Pi \xrightarrow{\sim} {}_s\Gamma_\Lambda$, where $\Pi = \Pi_\Lambda$ is the fundamental group of ${}_s\Gamma_\Lambda$. This group does not depend on the choice of π. As we have seen in § 2, π determines also a configuration $C = C_\Lambda$ which is stable under Π and such that $(\mathbb{Z}A_n)_C/\Pi \xrightarrow{\sim} \Gamma_\Lambda$.

Now, the possible configurations C of $\mathbb{Z}A_n$ have been determined in § 2. In this paragraph we want to describe the automorphisms of the pairs $(\mathbb{Z}A_n, C)$ in order to get a list of possible fundamental groups.

3.1 An automorphism of $(\mathbb{Z}A_n, C)$ is either a translation or a reflection-translation of $\mathbb{Z}A_n$. The translations are easy to describe in terms of the associated Brauer relation $\mathcal{B}$ (2.6) : Assume that $C = C_{\mathcal{B}}$; let $\bar{\tau}$ be the cyclic permutation $\underline{e}_n i \mapsto \underline{e}_n(i-1)$ of $\sqrt[n]{1}$ and e the smallest divisor of n such that $\bar{\tau}^e(\mathcal{B}) = \mathcal{B}$. We call e the period of $\mathcal{B}$ or of C. The group of translations stabilizing C is clearly $\tau^{e\mathbb{Z}}$. In the example of Fig. 14 we have $n=6$, $e=2$.

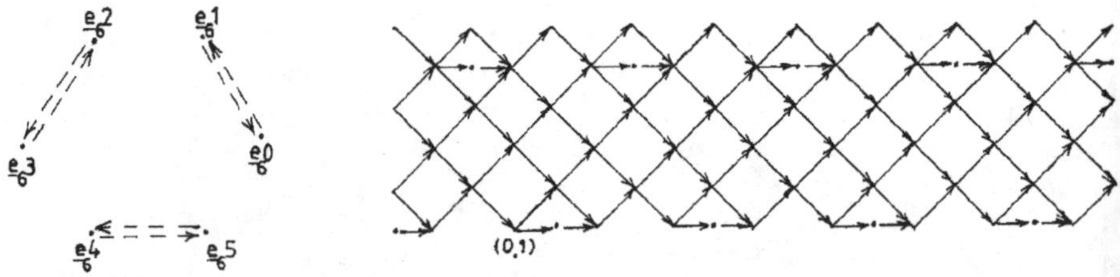

$\mathcal{B}$ $(\mathbb{Z}A_6)_{C_{\mathcal{B}}}$

Fig. 14

3.2 Proposition. Assume that there exists a reflection-translation of $\mathbb{Z}A_n$, $n>1$, which stabilizes the configuration C. Then n is odd, C intersects the central line $\{(p,\frac{n+1}{2}):p\in\mathbb{Z}\}$, and the stabilizer of C is

$$\mathrm{Aut}(\mathbb{Z}A_n,C) = \tau^{n\mathbb{Z}} \times \phi^{\mathbb{Z}/2\mathbb{Z}} \quad ,$$

where $\tau(p,q)=(p-1,q)$ and $\phi(p,q)=(p+q-\frac{n+1}{2},n+1-q)$ (1.3).

Proof. If n es even, a reflection-translation has the form ρ^{2r+1} with $\rho^2=\tau$ (1.3). If ρ^{2r+1} stabilizes C, so does $(\rho^{2r+1})^2=\tau^{2r+1}$. We infer that the period e of C divides $2r+1$ and n; hence we have $e=1$, and $C=\{(p,1):p\in\mathbb{Z}\}$ or $C=\{(p,n):p\in\mathbb{Z}\}$: contradiction!

If n is odd, let $\tau^r\phi$ be a reflection-translation stabilizing C. Composing $\tau^r\phi$ with τ^{se} if necessary, we are reduced to the case $0\leq r<e$. In this case we have $0\leq 2r<2e$, and $\tau^{2r}=(\tau^r\phi)^2$ stabilizes C. Since e is odd, this implies $r=o$ and $\phi(C)=C$. Now C has in $\mathbb{Z}A_n/\tau^{n\mathbb{Z}}$ an image $C/\tau^{n\mathbb{Z}}$ with odd cardinality n (C has one point on each going up diagonal!). Consequently, the involution ϕ has a fixed point in $C/\tau^{n\mathbb{Z}}$; this means that C intersects the central line. If $(p,\frac{n+1}{2})\in C$, the rectangles stopping or starting at $(p,\frac{n+1}{2})$ cover the interval $\{(p+x,\frac{n+1}{2}):-\frac{n-1}{2}\leq x\leq\frac{n-1}{2}\}$. Therefore $(p,\frac{n+1}{2})$ is the only point cf C in this interval, we have $e=n$, and the assertion becomes clear. OK.

Fig. 15 supplies us with an example of the situation doscribed in our proposition. If this situation occurs, we say that C is symmetric.

3.3 Proposition. For a given configuration C of $\mathbb{Z}A_n$ the admissible automorphism groups Π of $(\mathbb{Z}A_n)_C$ look as follows: If C is not symmetric, we have $\Pi=\tau^{se\mathbb{Z}}$, where $e|n$ is the period of C and $s\geq 1$. If C is symmetric, we have $e=n$ and $\Pi=\tau^{sn\mathbb{Z}}$ or

$\Pi = (\tau^{sn}\phi)^{\mathbb{Z}}$ with $s \geq 1$.

This follows immediately from 3.1, 3.2 and [6] , appendix 2 .

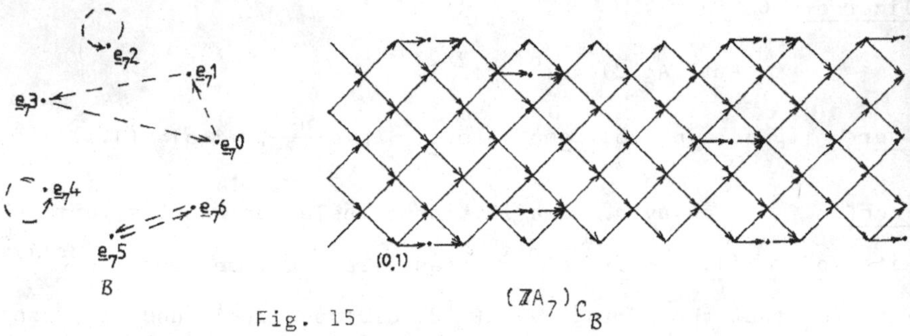

Fig. 15

3.4 In 2.7 we identified the class of a vertex (p,q) of $\mathbb{Z}A_n$ modulo $\tau^{n\mathbb{Z}}$ with the vector $(\underline{e}_n p, \underline{e}_n(p+q))$. In case $n=2m-1$, ϕ acts on the vectors through this identification by means of the formula $\phi(\sigma,e) = (\bar{\tau}^m e, \bar{\tau}^{m-1}\sigma)$ (for $\bar{\tau}$ see 3.1). We infer that the configuration C is symmetric iff the associated vector-set is stable under the transformation $(\sigma,e) \mapsto (\bar{\tau}^m e, \bar{\tau}^{m-1}\sigma)$.

This criterion seems artificial. In fact, we can find a more natural approach if we free ourselves from our "sin". Our original sin was to assign different rôles to the two borders of $\mathbb{Z}A_n$ in the definition of the map $\mathcal{B} \mapsto C_{\mathcal{B}}$ (2.6). In order to reestablish the symmetry, we want to relate the configuration C to a second Brauer relation, which is described in terms of the upper border of $\mathbb{Z}A_n$, $n \in \mathbb{N}$.

Let us first recall that we denote by $(p, \beta_C(p)-p)$ the unique point of C lying on the going up diagonal Δ_p ; its "height over the lower border" of $\mathbb{Z}A_n$ is $\beta_C(p)-p-1$. Similarly, the going down diagonal ∇_q (2.6) contains one point (x,y) of C ; we denote its "depth under the upper border" of $\mathbb{Z}A_n$ by $\alpha_C(q)-q-1$; hence we have by definition

$x+y=q$ and $n-y=\alpha_C(q)-q-1$, i.e. $(x,y)=(\alpha_C(q)-n-1,n+1-\alpha_C(q)+q)$. Of course, just as β_C , the map α_C is a permutation of $\mathbb{Z}$ satisfying $q<\alpha_C(q)\leq q+n$ and $\alpha_C(q+n)=\alpha_C(q)+n$; it induces a permutation $\bar{\alpha}_C$ of $\sqrt[n]{1}$ such that $\bar{\alpha}_C\underline{e}_n(i)=\underline{e}_n(\alpha_C i)$. We get a relation between α_C and β_C by turning to account the fact that a point of C may be written either as $(p,\beta_C(p)-p)$ or as $(\alpha_C(q)-n-1,n+1-\alpha_C(q)+q)$ (see Fig. 16). Equalizing these expressions yields $q=\beta_C(p)$ and

$$\alpha_C\beta_C p = p+n+1 .$$

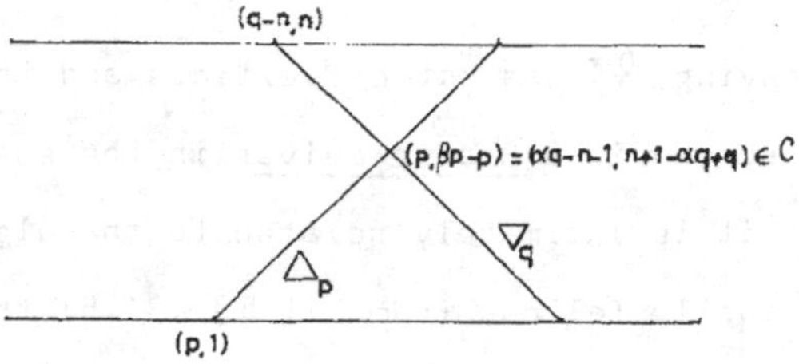

Fig. 16

The last equation implies $\bar{\alpha}_C\bar{\beta}_C=\bar{\tau}^{-1}$ and determines $\bar{\alpha}_C$ in terms of $\bar{\beta}_C$. We illustrate the permutations $\bar{\alpha}_C$ and $\bar{\beta}_C$ by drawing broken vectors $(\sigma,\bar{\beta}_C\sigma)$ and full vectors $(\sigma,\bar{\alpha}_C\sigma)$. The equation $\bar{\alpha}_C\bar{\beta}_C=\bar{\tau}^{-1}$ means that each broken vector (σ,e) is twinned with a full vector $(e,\bar{\tau}^{-1}\sigma)$ (see the figures 7,8,13 and 17). In this way we get two vector-sets, which give rise to two Brauer relations A_C and B_C ; their equivalence classes are the orbits of $\bar{\alpha}_C$ and $\bar{\beta}_C$ respectively (confer 2.7).

Let C be a configuration. The bijection $B\mapsto C_B$ of proposition 2.6 maps B_C onto C and A_C onto a configuration deduced from C by a reflection-translation. Hence there exists a reflection-translation stabilizing C iff $B_C=\bar{\tau}^m(A_C)$ for some m . If this holds we know by the beginning of the present section that $m=\frac{n+1}{2}$.

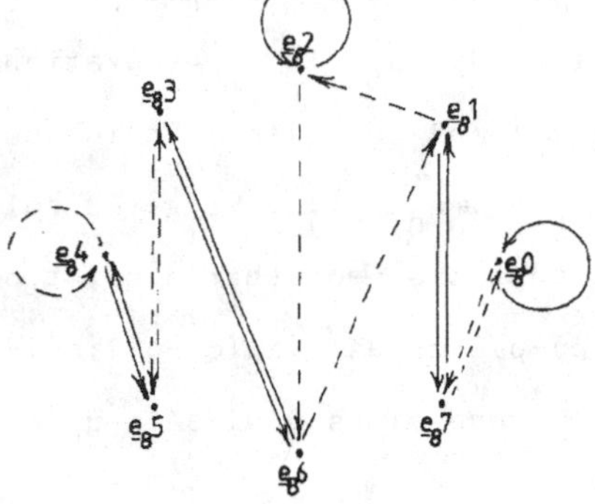

Fig. 17

The quiver Q_C having $\sqrt[n]{1}$ as set of vertices and the vectors $(\sigma, \bar{\beta}_C\sigma)$ and $(\sigma, \bar{\alpha}_C\sigma)$ as arrows is a <u>Brauer-quiver</u> in the sense of [5] , 1.4. We shall see that it is intimately related to the algebras Λ such that $\tilde{\Gamma}_\Lambda \stackrel{\sim}{\to} (\mathbb{Z}A_n)_C$. It easily follows from ([5] , 1.5) that it determines $(\mathbb{Z}A_n)_C$ up to an isomorphism. In particular, C is symmetric iff Q_C admits an automorphism exchanging the broken and the full arrows. Fig. 18 represents the Brauer-quiver corresponding to the configuration of Fig. 15.

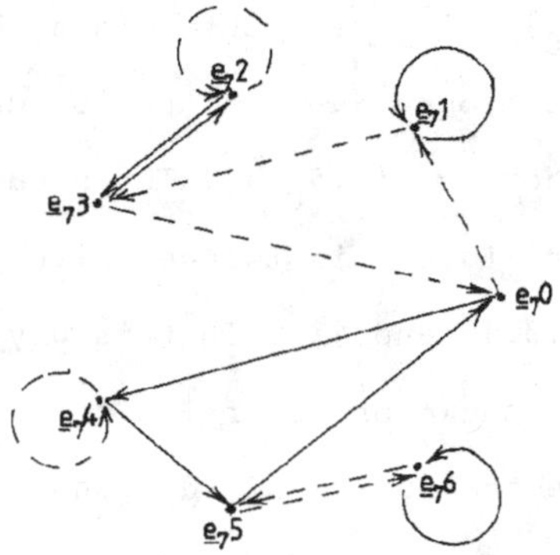

Fig. 18

3.5 <u>Proposition</u>. <u>Let</u> C <u>be a configuration of</u> $\mathbb{Z}A_n$ <u>with period</u> $e < n$ <u>and let</u> $\bar{\alpha}_C$ <u>and</u> $\bar{\beta}_C$ <u>be the permutations of</u> $\sqrt[n]{1}$ <u>associated with</u> C (3.4). <u>Then there is exactly one exceptional orbit, i.e. either an</u> $\bar{\alpha}_C$<u>-orbit or a</u> $\bar{\beta}_C$<u>-orbit which is stable under</u> $\bar{\tau}^e$ (see Fig. 19).

Proof. (Compare with [5], 1.10). Let T_C be the Brauer-tree associated with the Brauer-quiver Q_C ([5], 1.4): the vertices of T_C are the cycles (= $\bar{\alpha}_C$-orbits or $\bar{\beta}_C$-orbits) of Q_C; two cycles of Q_C are connected by an edge in T_C if they intersect (in one point); for each vertex v of T_C the permutations $\bar{\alpha}_C$ and $\bar{\beta}_C$ determine a cyclic ordering on the edges converging at v. As is well-known, an automorphism of a tree without fixed vertices has a fixed edge. We infer that an automorphism of T_C without fixed vertices exchanges the $\bar{\alpha}_C$-orbits and the $\bar{\beta}_C$-orbits. This is not true for $\bar{\tau}^e$. Therefore $\bar{\tau}^e$ has a fixed vertex v in T_C. As it is compatible with the cyclic ordering of the edges converging at v, v is the only fixed vertex. OK.

In the sequel we shall assume that the exceptional orbit is an $\bar{\alpha}_C$-orbit. This is no restriction, since the chosen covering $\pi: \mathbb{Z}A_n \to {}_s\Gamma_\Lambda$ may be composed with a reflection-translation of $\mathbb{Z}A_n$.

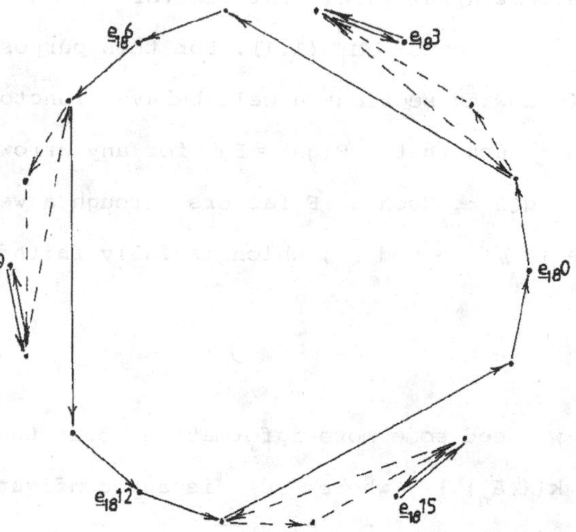

Fig. 19

§ 4 The structure of the Auslander algebra

Our aim in this paragraph is to describe the Auslander algebra E_Λ of the selfinjective basic algebra Λ of class A_n in terms of the configuration $\tilde{C}_\Lambda$ of $\mathbb{Z}A_n$ (2.5) and of the fundamental group Π_Λ of Γ_Λ ([6],1.7). Since the Λ-module $M = M_1 \oplus \ldots \oplus M_r$ (1.1) supplies an equivalence $X \mapsto \mathrm{Hom}_\Lambda(M,X)$ between mod Λ and the full subcategory proj E_Λ of mod E_Λ formed by the projective modules, this will imply that Λ itself is determined up to an isomorphism by $\tilde{C}_\Lambda$ and Π_Λ .

4.1 Theorem. The Auslander algebra E_Λ of a selfinjective algebra Λ of class A_n with Auslander-Reiten quiver Γ_Λ is isomorphic to $k[\Gamma_\Lambda]/I_{\Gamma_\Lambda}$ (1.4).

Since Γ_Λ is isomorphic to $(\mathbb{Z}A_n)_{\tilde{C}_\Lambda}/\Pi_\Lambda$, our theorem furnishes the wanted description of E_Λ in terms of $\tilde{C}_\Lambda$ and Π_Λ .

Proof. As we have $k[\Gamma_\Lambda]/I_{\Gamma_\Lambda} \tilde{\rightarrow} \bigoplus_{e,f} \mathrm{Hom}_{k(\Gamma_\Lambda)}(e,f)$ by 1.4, it is sufficient to construct a fully faithful functor $H : k(\Gamma_\Lambda) \rightarrow \mathrm{mod}\,\Lambda$ mapping a vertex e_i onto M_i (1.1). For this purpose we shall construct in the following sections a well-behaved functor $F : k((\mathbb{Z}A_n)_{\tilde{C}_\Lambda}) \longrightarrow \mathrm{mod}\,\Lambda$ such that $F(g\alpha) = F\alpha$ for any arrow α of $(\mathbb{Z}A_n)_{\tilde{C}_\Lambda}$ and any $g \in \Pi_\Lambda$. Such an F factors through a well-behaved functor $H : k((\mathbb{Z}A_n)_{\tilde{C}_\Lambda})/\Pi_\Lambda \rightarrow \mathrm{mod}\,\Lambda$, which is fully faithful by theorem 2.5.

4.2 First we need some more information about the morphisms of the category $k((\mathbb{Z}A_n)_C)$, where C is any configuration of $\mathbb{Z}A_n$. As usual we denote by $\bar{w} = \overline{\alpha_r} \ldots \overline{\alpha_1}$ the canonical image (= residue class) of a path $w = (f|\alpha_r,\ldots,\alpha_1|e)$ of $(\mathbb{Z}A_n)_C$ in

$k((\mathbb{Z}A_n)_C)$. We say that the path w is <u>stable</u> if e,f and each arrow α_i belong to $\mathbb{Z}A_n$. If this is so, we set $\tilde{w} = \tilde{\alpha}_r \ldots \tilde{\alpha}_2 \tilde{\alpha}_1$, where the morphism $\tilde{\alpha}_i$ of $k((\mathbb{Z}A_n)_C)$ is defined as follows: If $(i,j) \overset{\alpha}{\to} (i+1,j-1)$ is a "going down" arrow of $\mathbb{Z}A_n$, $\tilde{\alpha}$ equals $\bar{\alpha}$, whereas we set $\tilde{\alpha} = (-1)^j \bar{\alpha}$ if the arrow $(i,j) \overset{\alpha}{\to} (i,j+1)$ goes up.

Furthermore, two stable paths w,w' are called C-<u>neighbours</u> if they have the form $w=w_2\beta\alpha w_1$, $w'=w_2\delta\gamma w_1$ or the form $w=w_2\delta\gamma w_1$, $w'=w_2\beta\alpha w_1$, where $\alpha,\beta,\gamma,\delta$ are the four arrows of a mesh

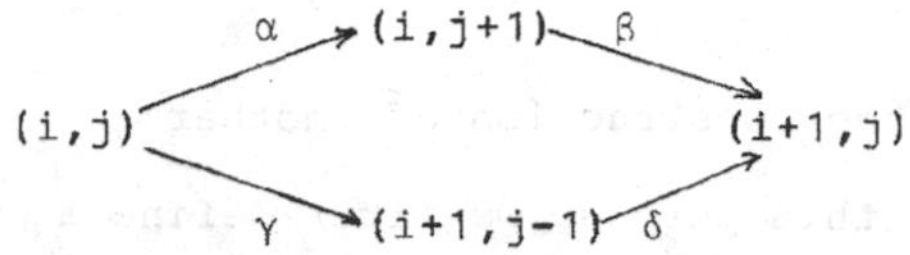

such that $1<j<n$ and $(i,j)\notin C$. Two stable paths w,w' are called C-<u>homotopic</u> if they may be linked together by a sequence $w = w_o, w_1, \ldots, w_m = w'$ of stable paths such that w_i and w_{i+1} are C-neighbours for any i . Finally, we say that the path w is C-<u>marginal</u> if it is C-homotopic to some $w_2\beta\alpha w_1$ or some $w_2\delta\gamma w_1$, where $\beta\alpha$ and $\delta\gamma$ are "<u>crenels</u>" of the following forms

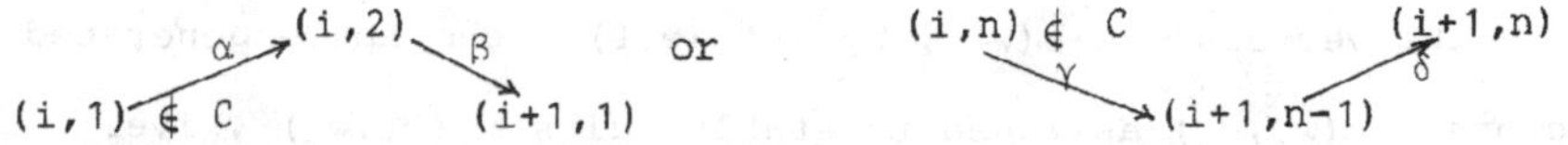

A stable path, which is not C-marginal, is called C-<u>essential</u>.

<u>Proposition</u>. a) <u>A stable path</u> w <u>is C-marginal iff</u> $\tilde{w} = 0$.

b) <u>Two</u> C-<u>essential paths</u> v <u>and</u> w <u>are</u> C-<u>homotopic iff</u> $\tilde{v} = \tilde{w}$.

c) <u>For any two stable vertices</u> e <u>and</u> f <u>we have</u>

$$\mathrm{Hom}_{k((\mathbb{Z}A_n)_C)}(e,f) = \oplus k\tilde{w},$$

<u>where</u> w <u>ranges through representatives of the</u> C-<u>homotopy classes of</u> C-<u>essential paths from</u> e <u>to</u> f .

Proof. Let e,f be two stable vertices, $W(e,f)$ the vector space freely generated by all paths of $(\mathbb{Z}A_n)_C$ from e to f, $S(e,f)$ the subspace generated by the stable paths. To any pair (w_1,w_2), where w_1 and w_2 are paths of the form $e \xrightarrow{w_1} \tau x$ and $x \xrightarrow{w_2} f$, we assign a vector $\rho(w_1,w_2) = \Sigma w_2\alpha(\sigma\alpha)w_1 \in W(e,f)$, where α runs through all arrows of $(\mathbb{Z}A_n)_C$ with range x (1.4 and [6],1.1). These pairs generate a subspace $R(e,f)$, and we have by definition

$$\mathrm{Hom}_{k((\mathbb{Z}A_n)_C)}(e,f) = W(e,f)/R(e,f) .$$

The proof of our proposition rests on the construction of another family of generators of $R(e,f)$. For this purpose we first define a projection ρ of $W(e,f)$ onto $S(e,f)$: Decompose each path $w \in W(e,f)$ into a composition $w_m\kappa_m\iota_m w_{m-1}\ldots w_1\kappa_1\iota_1 w_o$, where the w_i are stable, whereas the κ_s and ι_s are non-stable arrows of the form $e_s \xrightarrow{\iota_s} \mathbb{P}_{e_s} \xrightarrow{\kappa_s} \tau^{-1}e_s$ (2.5). Then set $\rho(w) = (-1)^m \Sigma w_m\alpha_m(\sigma\alpha_m)w_{m-1}\ldots$ $\ldots w_1\alpha_1(\sigma\alpha_1)w_o \in S(e,f)$ where, for each s, α_s runs through all stable arrows with range $\tau^{-1}e_s$. Denote by $R_1(e,f)$ the space generated by the vectors $w-\rho(w)$, by $R_2(e,f)$ the space generated by the vectors $\rho(w_1,w_2)$ assigned to <u>stable pairs</u> (w_1,w_2), i.e. to the pairs with stable components $e \xrightarrow{w_1} \tau x$ and $x \xrightarrow{w_2} f$ such that $\tau x \notin C$. Our claim is that $R(e,f) = R_1(e,f)+R_2(e,f)$. The proof is easy. We skip it.

As ρ is a projection onto $S(e,f)$, $W(e,f)/R_1(e,f)$ is identified with $S(e,f)$ and $W(e,f)/R(e,f)$ with $S(e,f)/R_2(e,f)$. Now, if (w_1,w_2) is stable, $\rho(w_1,w_2)$ has either the form $w+w'$, where w and w' are C-neighbours satisfying $\bar{w}+\bar{w} = 0$ and $\tilde{w} = \tilde{w}'$; or $\rho(w_1,w_2)$ is simply a stable path w factoring through a "crenel" and satisfying $\bar{w} = 0 = \tilde{w}$. The proof ensues trivially.OK.

4.3 Lemma. Let (i,j) be a vertex of $\mathbb{Z}A_n$ such that $\tau(i,j) = (i-1,j) \in C$. A stable path from (i,j) to a vertex (p,q) such that $p \leq i+j-1$ is either C-marginal or C-homotopic to the composition

$$(i,j) \to (p,i+j-p) \to (p,q)$$

of a "going down" and a "going up" path (see Fig.20).

Of course, the second alternative can only occur if $i \leq p$ and $i+j \leq p+q$.

Proof. By induction on $2p+q$. Our claim is clearly true if $p \leq i$ or if $p+q \leq i+j$. Therefore let us assume $p>i$ and $p+q>i+j$, and let $(i,j) \xrightarrow{w} (p,q)$ be a path which is not C-marginal. Then w factors either through $(p,q-1) \xrightarrow{\delta} (p,q)$ or through $(p-1,q+1) \xrightarrow{\beta} (p,q)$. In the first case we have $w = \delta w_1$ and we know by induction that w_1 is C-homotopic to the composition $(i,j) \xrightarrow{u_1} (p,i+j-p) \xrightarrow{v_1} (p,q-1)$. Hence w is C-homotopic to $(\delta v_1)u_1$, where u_1 goes down and δv_1 goes up.

In the second case we have $w = \beta w_2$, where w_2 is C-homotopic to the composition

$$(i,j) \xrightarrow{u_2} (p-1,i+j+1-p) \xrightarrow{v_2} (p-1,q) \xrightarrow{\alpha} (p-1,q+1) .$$

If $(p-1,q) \notin C$, $\beta\alpha$ is C-homotopic to $\delta\gamma$, where γ is the arrow from $(p-1,q)$ to $(p,q-1)$; hence w is C-homotopic to $\delta\gamma v_2 u_2$. Now we know by induction that $\gamma v_2 u_2$ is C-homotopic to $v_1 u_1$. As a consequence w is C-homotopic to $(\delta v_1)u_1$.

It remains to be shown that the hypothesis $(p-1,q) \in C$ leads to a contradiction in the second case. Indeed, since the rectangle starting at $(i-1,j) \in C$ contains no point of C but $(i-1,j)$, we must have $p-1+q \geq i+n$. Similarly, since C has no point but $(p-1,q)$ in the rectangle stopping at $(p-1,q)$, there is no projective (=non-stable) vertex of $(\mathbb{Z}A_n)_C$ within the hatched region (see Fig.20). We infer that $\alpha v_2 u_2$ is C-homotopic to the composition $(i,j) \to a \to b \to c \to (p+q-n,n) \to (p-1,q+1)$, which is C-marginal: contradiction. OK.

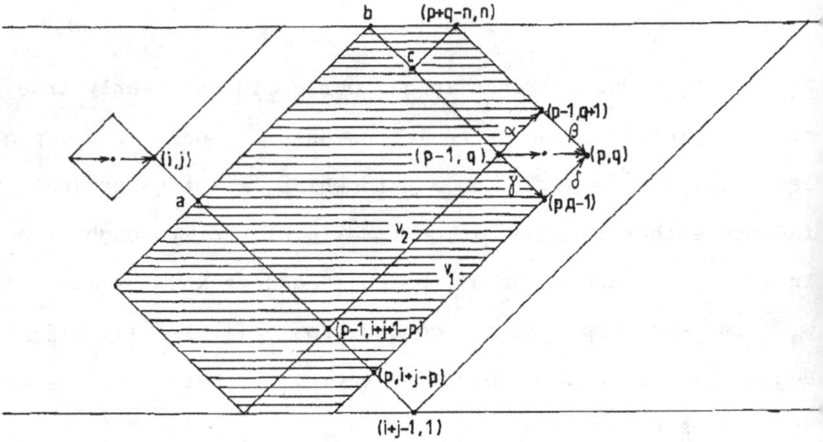

Fig. 20

4.4 Lemma. Every morphism $\mu : (x,y) \to (z,t)$ of $k((\mathbb{Z}A_n)_C)$ such that $y<n$ and $x+y \leq z$ factors through $\bar{\alpha} : (x,y) \to (x,y+1)$ (see Fig. 21). Every morphism $\mu : (x,n) \to (z,t)$ such that $x+n \leq z$ is zero.

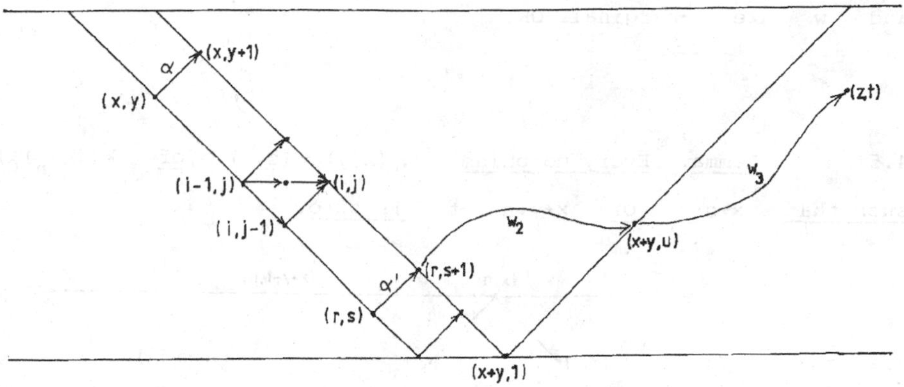

Fig. 21

<u>Proof</u>. First consider any $y \leq n$. It is enough to prove that each <u>stable</u> path $(x,y) \xrightarrow{w} (z,t)$ is either C-marginal or C-homotopic to a path factoring through α (if $y<n$). Clearly, w admits a decomposition

$$(x,y) \xrightarrow{w_1} (r,s) \xrightarrow{\alpha'} (r,s+1) \xrightarrow{w_2} (x+y,u) \xrightarrow{w_3} (z,t),$$

where $r+s = x+y$. Let $(i-1,j)$ with $i+j-1 = x+y$ be the point of C on the going down diagonal through (x,y). If $r \leq i-1$ and $y<n$, $\alpha' w_1$ is C-homotopic to the composition $(x,y) \overset{\alpha}{\to} (x,y+1) \xrightarrow{w_1'} (r,s+1)$, hence w is C-homotopic to $w_3 w_2 w_1' \alpha$. If $r \leq i-1$ and $y=n$, $\alpha' w_1$ and w are C-marginal. If $r>i-1$, the composition $(i,j-1) \to (r,s) \xrightarrow{\alpha'} (r,s+1)$ is C-homotopic to $(i,j-1) \to (i,j) \to (r,s+1)$. Therefore w is C-homotopic to a stable path factoring through the latter composition. This reduces the proof to the case where $(r,s) = (i,j-1)$. In this case, if w_2 is not C-marginal, it is C-homotopic to the composition $(i,j) \to (x+y,1) \to (x+y,u)$ by lemma 4.3. Consequently $w_2 \alpha'$ is C-marginal. In any case, the condition $r>i-1$ implies that $w_2 \alpha'$

and w are $\mathcal{C}$-marginal. OK.

4.5 Lemma. Every morphism $\mu : (x,y) \to (z,t)$ of $k((\mathbb{Z}A_n)_{\mathcal{C}})$ such that $x+n\leq z$ or $x+y+n\leq z+t$ is zero.

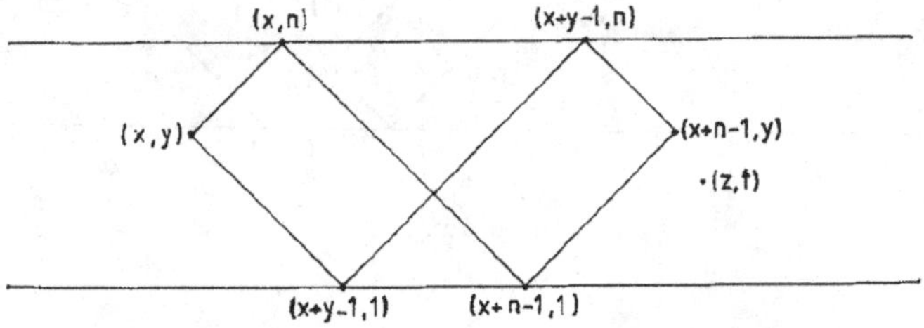

Fig. 22

The assumption of our lemma means that (z,t) lies strictly on the right of the rectangle stopping at $(x+n-1,y)$ (Fig.22).

Proof. If $x+n\leq z$, the first part of lemma 4.4 implies by induction on $n-y$ that each μ, is a composition $(x,y) \to (x,n) \overset{\nu}{\to} (z,t)$. By the second part of lemma 4.4 ν is zero. The symmetry of $\mathbb{Z}A_n$ with respect to the central line reduces the case $x+y+n\leq z+t$ to the preceding one. OK.

4.6 Proposition. If Λ is selfinjective of class A_n, there is a well-behaved functor $H : k(\Gamma_\Lambda) \to \mathrm{mod}\,\Lambda$.

As we noticed already, the proposition implies theorem 4.1. We prove it by constructing a well-behaved functor $F : k(\tilde{\Gamma}_\Lambda) \to \mathrm{mod}\,\Lambda$ such that $F(g\alpha) = F\alpha$ for any arrow α of $\tilde{\Gamma}_\Lambda \tilde{\to} (\mathbb{Z}A_n)_{\tilde{C}_\Lambda}$ and any element g of the fundamental group Π_Λ of Γ_Λ. The construction requires all the end of paragraph 4.

The construction of F in the "general case".

The general case is defined to be the case where Π_Λ has a generator of the form $\psi = \tau^m$, $m \geq n$, or $\psi = \tau^{sn}\varphi$, $s \geq 1$ (3.3). Set $C = \tilde{C}_\Lambda$. In the general case we start from any well-behaved functor $F_0 : k((\mathbb{Z}A_n)_C) \to \mathrm{mod}\Lambda$ and any going up diagonal $\Delta_r = \{(r,q) : 1 \leq q \leq n\}$ (compare with 1.6). We denote by $(r-1,t)$ the point of C on Δ_{r-1} (Fig.23). The functor F will coincide with F_0 on all arrows lying on the diagonals Δ_{r-1}, $\psi(\Delta_r)$ and between them, on β_1, on ι, and even on κ if $t=1$. Furthermore we set $F\gamma_i = F(\psi\gamma_i)$.

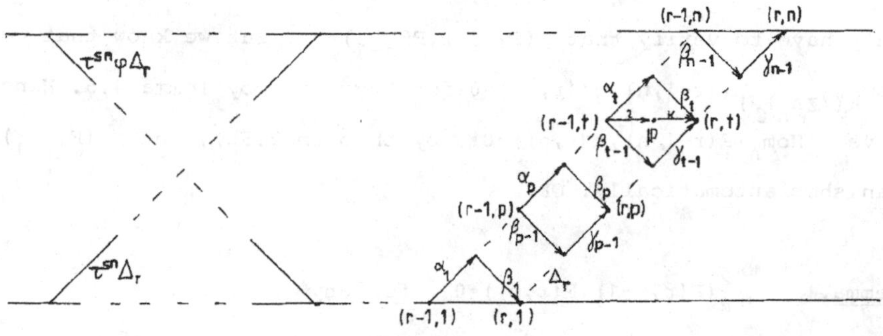

Fig. 23

The relations $(F\beta_1)(F\alpha_1)=0$ for $t>1$ and $(F\beta_1)(F\alpha_1)+(F\kappa)(F\iota)=0$ for $t=1$ are satisfied by construction. It remains for us to construct $F\kappa$ if $t>1$ and $F\beta_p$ for $p\geq 2$ in such a way that $(F\beta_p)(F\alpha_p)+(F\gamma_{p-1})(F\beta_{p-1})=0$ for $p\neq t$, $2\leq p\leq n-1$, $(F\gamma_{n-1})(F\beta_{n-1})=0$ for $t\neq n$, $(F\beta_t)(F\alpha_t)+(F\gamma_{t-1})(F\beta_{t-1})+(F\kappa)(F\iota)=0$ for $1<t<n$ and $(F\gamma_{n-1})(F\beta_{n-1})+(F\kappa)(F\iota)=0$ for $t=n$. This will enable us to extend F to all of $(\mathbb{Z}A_n)_C$ by using the "periodicity" $F(\psi\alpha)=F\alpha$.

The construction of $F\beta_p$ proceeds by induction on p: Suppose $F\beta_i$ is already constructed for $1<i<p<n$. First assume $p=t$: By [6], 3.4a) there is an Auslander-Reiten sequence of the form

$$F(r-1,t)\xrightarrow{[F\alpha_t\,F\iota\,F\beta_{t-1}]^T}F(r-1,t+1)\oplus F\,p\oplus F(r,t-1)\xrightarrow{[\beta\chi\gamma]}F(r,t)$$

in $\mathrm{mod}\,\Lambda$. By [6], 3.1 and 3.5 we have $\lambda\gamma-F\gamma_{t-1}\in\mathcal{R}^2_\Lambda(F(r,t-1),F(r,t))$ for some non-zero $\lambda\in k$. Since by the lemma below $\mathcal{R}^2_\Lambda(F(r,t-1),F(r,t))=0$ we infer that $F\gamma_{t-1}=\lambda\gamma$. We set $F\beta_t=\lambda\beta$ and $F\kappa=\lambda\chi$. In case $n>p\neq t$ we construct $F\beta_p$ in a similar way (confer 1.6).

It remains to be examined what happens at the upper "crenel". If $t=n$ we construct $F\kappa$ using the same argument as above. If $t<n$ we only have to verify that $(F\gamma_{n-1})(F\beta_{n-1})=0$. But we know that $\mathrm{Hom}_{k((\mathbb{Z}A_n)_C)}((r-1,n),\psi^j(r,n))=0$ for any j by lemma 4.5. Hence we have $\mathrm{Hom}_\Lambda(F(r-1,n),F(r,n))=0$ by theorem 2.5b), and $(F\gamma_{n-1})(F\beta_{n-1})$ vanishes automatically. OK.

<u>Lemma.</u> $\mathcal{R}^2_\Lambda(F(r,p-1),F(r,p))=0$ <u>for any</u> $1<p\leq n$.

Proof. Using theorem 2.5b) we only have to ascertain that $\mathrm{Hom}_{k((\mathbb{Z}A_n)_C)}((r,p-1),\psi^j(r,p))=0$ for any $j\neq 0$. This is true by lemma 4.5, since we are in the general case. OK.

4.7 The construction of F in the exceptional case. In this case the period e of the configuration $C=\tilde{C}_\Lambda$ is $<n$ and the fundamental group Π_Λ is generated by $\psi=\tau^{em}$ for some $m<f=\frac{n}{e}$. We keep the notations of 4.6 and start as there. A first deflection from 4.6 occurs in the inductive construction of β_p for $1<p=t<n$. In the present situation the relation $\lambda\gamma-F\gamma_{t-1}=\gamma'\in R^2_\Lambda(F(r,t-1),F(r,t))$ does not imply $F\gamma_{t-1}=\lambda\gamma$ any longer. Notwithstanding we get

$$0=\lambda[\gamma(F\beta_{t-1})+\chi(F\iota)+\beta(F\alpha_t)]=(F\gamma_{t-1})(F\beta_{t-1})+\gamma'(F\beta_{t-1})+\lambda\chi(F\iota)+\lambda\beta(F\alpha_t) .$$

Now $\gamma'(F\beta_{t-1})$ lies in R^3_Λ and may be written as $\gamma'(F\beta_{t-1})=\beta'(F\alpha_p)+\kappa'(F\iota)$ with β' and κ' both in R^2_Λ (use part a) of the lemma below). We infer that $0=(F\gamma_{t-1})(F\beta_{t-1})+(\kappa'+\lambda\chi)(F\iota)+(\beta'+\lambda\beta)(F\alpha_t)$, and we can set $F\kappa=\kappa'+\lambda\chi$, $F\beta_t=\beta'+\lambda\beta$.

In the case $p=t=n$ we construct $F\kappa$ in a similar way using part b) of the following lemma. The construction of $F\beta_p$ for $p\neq t$ and the verification of the equaltity $(F\gamma_{n-1})(F\beta_{n-1})=0$ for $n\neq t$ require more insight. In fact we can only adjust the argument given in 1.6 to the present situation for some particular choice of the diagonal Δ_r. This will be done in 4.8 and 4.9 below.

Lemma. a) If $1<t<n$ and $(r-1,t)\in C$ each $\mu\in R^3_\Lambda(F_0(r-1,t),F_0(r,t))$ can be written as $\mu=\kappa'(F_0\iota)+\beta'(F_0\alpha_t)$ with $\kappa'\in R^2_\Lambda(F_0(p),F_0(r,t))$ and $\beta'\in R^2_\Lambda(F_0(r-1,t+1),F_0(r,t))$.

b) If $(r-1,n)\in C$ each $\mu\in R^3_\Lambda(F_0(r-1,n),F_0(r,n))$ can be written as $\mu=\kappa'(F_0\iota)$ with $\kappa'\in R^2_\Lambda(F_0(p),F_0(r,n))$.

Proof. a) By the very definition of R^3_Λ and theorem 2.5b) we have $\mu=\Sigma\lambda_w w$, where w ranges through paths of length ≥ 3 from $(r-1,t)$ to some $\psi^j(r,t)$. Such a path has one of the following three forms $w=u_1\alpha_t$, $w=u_2\iota$ or $w=v\beta_{t-1}$, where the lengths of u_1, u_2, v are ≥ 2. In the third case v is decomposed according to Fig. 24: $v=v_3v_2v_1$. We infer that $F_0w=(F_0v_3)(F_0v_2v_1)(F_0\beta_{t-1})=$
$(-1)^s(F_0v_3)(F_0v_4\gamma_{t-1})(F_0\beta_{t-1})=(-1)^s(F_0v_3v_4)(F_0\gamma_{t-1}\beta_{t-1})$
$=-(-1)^s(F_0v_3v_4)(F_0\beta_t\alpha_t+F_0\kappa\iota)$
$=(-1)^{s+1}(F_0v_3v_4\beta_t)(F_0\alpha_t)+(-1)^{s+1}(F_0v_3v_4\kappa)(F_0\iota)$. In any case F_0w can be written as $\beta'(F_0\alpha_t)+\kappa'(F_0\iota)$, where $\beta'\in R^2_\Lambda(F_0(r-1,t+1),F_0(r,t))$ and $\kappa'\in R^2_\Lambda(F_0(p),F_0(r,t))$. The assertion b) is proved as in 1.6. OK.

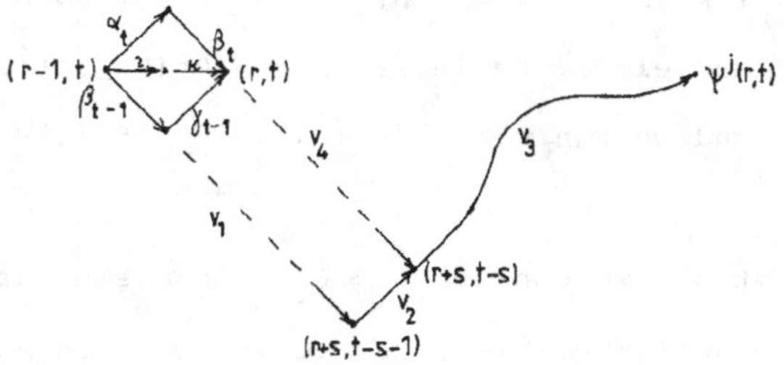

Fig. 24

4.8 Our aim is to find a diagonal Δ_r such that the two following conditions are satisfied: a) _If_ $1<p<n$ _and_ $(r-1,p)\in C$, _each morphism_ $\mu\in R^3_\Lambda(F_0(r-1,p),F_0(r,p))$ _can be written as_ $\mu=\beta'(F_0\alpha_p)$ _with_ $\beta'\in R^2_\Lambda(F_0(r-1,p+1),F_0(r,p))$; b) _If_ $(r-1,n)\notin C$, $\mathrm{Hom}_\Lambda(F_0(r-1,n),F_0(r,n))=0$. Clearly, for such a diagonal Δ_r the inductive construction of $F\beta_p$ goes through; so does the construction of $F\kappa$ if $t>1$ (compare with 1.6 and 4.6).

First suppose that for some vertex $(r,p)\notin C$, $1<p<n$, there is a morphism $\mu\in R^3_\Lambda(F_0(r-1,p),F_0(r,p))$ violating the condition a). By theorem 2.5b) this means that some morphism $\nu:(r-1,p)\to\psi^{-j}(r,p)$ of $k((\mathbb{Z}A_n)_C)$ is not factorized through $\bar{\alpha}_p:(r-1,p)\to(r-1,p+1)$. By lemma 4.4 (see also the proof of lemma 1.6) such an impediment can only occur in the situation of Fig. 25, where the path $(r-1,p)\to(r+je,p-je-1)\to(r+je,p)$ cannot be lifted over the projective vertex $\mathbb{P}$. This situation is characterized by the inequalities $r-1<u<r+je\leq p+r-2$. Using the permutation $\alpha=\alpha_C$ of $\mathbb{Z}$ introduced in 3.4 and setting $q=p+r-1$, we see that $u=\alpha q-n-1$ and that the preceding inequalities may be written in the form

(*) $r<\alpha q-n\leq r+je<q$.

We say that the vertex $(r,p)=(r,q+1-r)$ is C-critical, if the inequalities (*) hold for some $j=1,\ldots,f-1$.

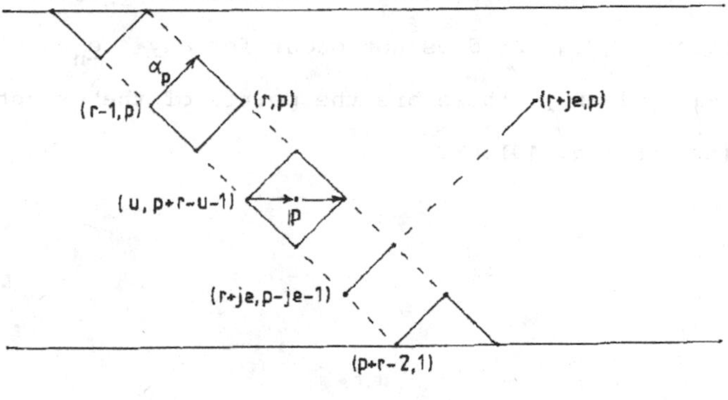

Fig. 25

Now suppose that for some vertex $(r-1,n)\notin C$ there is a non-zero morphism in $\mathrm{Hom}_\Lambda(F_0(r-1,n),F_0(r,n))$, i.e. condition b) is violated. By lemma 4.4 this can only occur in the situation of Fig. 25, if $(r-1,p)$ lies at the upper border and α_p vanishes. But then (r,n) is C-critical according to the preceding definition.

To sum up, we conclude that the proof given in 4.6 goes through in the exceptional case too, if the diagonal Δ_r does not contain any C-critical point. The existence of such a diagonal is granted by lemma 4.9. The assumption made there is permissible by 3.5.

4.9 Lemma. Suppose that the period e of C is $<n$ and that the exceptional orbit of the Brauer-quiver is an $\bar{\alpha}$ -orbit (3.5). Then Δ_r does not contain any C-critical vertex (4.8) iff $\underline{e}_n r$ belongs to the exceptional orbit (3.5).

Proof. The relation $q=p+r-1$ of 4.8 implies $q \leq r+n-1$. Therefore the inequalities $r<\alpha q-n \leq r+je<q$ of 4.8 mean that the relative position of $\underline{e}_n r$, $\underline{e}_n(\alpha q)$, $\underline{e}_n(r+je)$ and $\underline{e}_n q$ on the unit circle is that of Fig. 26 (where the limiting case $\underline{e}_n(\alpha q)=\underline{e}_n(r+je)$ is admitted). We are interested in the points $\underline{e}_n r \in \sqrt[n]{1}$ such that the position of Fig. 26 does not occur for any $\underline{e}_n q \in \sqrt[n]{1}$ and any $j=1,\ldots f-1$. Clearly, these are the points of the exceptional orbit (try on Fig. 19). OK.

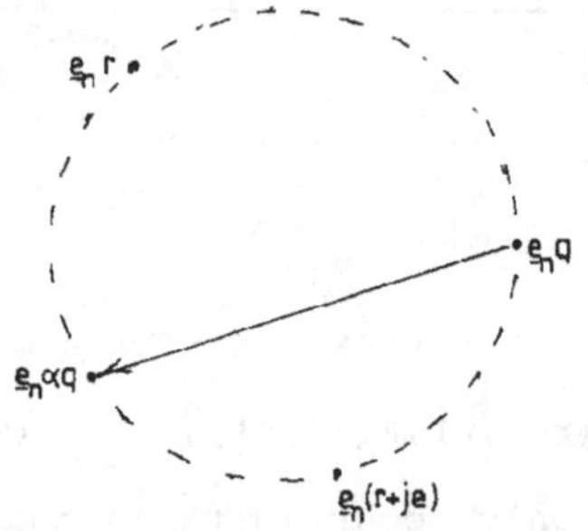

Fig. 26

§ 5. Configurations produce Auslander algebras

As we have seen in § 4, the Auslander algebra E_Λ of a selfinjective algebra of class A_n is isomorphic to $k[\Gamma_\Lambda]/I_{\Gamma_\Lambda}$ (1.4), where Γ_Λ is the Auslander-Reiten quiver of Λ. This quiver has the form $\Gamma_\Lambda \xrightarrow{\sim} (\mathbb{Z}A_n)_C/\Pi$ for some configuration $C=\tilde{C}_\Lambda$ and some admissible group $\Pi=\Pi_\Lambda$ of automorphisms of $\mathbb{Z}A_n$ stabilizing C. In this paragraph we choose once for all an arbitrary configuration C of $\mathbb{Z}A_n$ and an admissible automorphism group $\Pi \neq \{1\}$ stabilizing C. We set $\tilde{\Gamma}=(\mathbb{Z}A_n)_C$, $\Gamma=\Gamma_{C,\Pi}=\tilde{\Gamma}/\Pi$ and $E=E_{C,\Pi}=k[\Gamma]/I_\Gamma$ (1.4). Our aim is to prove the converse statement:

Theorem. $E_{C,\Pi}$ is the Auslander algebra of a selfinjective algebra of class A_n.

The proof of our theorem spreads over all of paragraph 5 and terminates our classification of selfinjective algebras of class A_n.

5.1 Auslander has shown ([1],§4) that a finite-dimensional, basic k-algebra F is the Auslander algebra (in the sense of 1.1) of some representation-finite, finite-dimensional algebra A iff F satisfies the following two conditions: a) The global homological dimension of F is ≤ 2, i.e. each $N \in \mathrm{mod}\, F$ has a projective resolution of the form $0 \to P_2 \to P_1 \to P_0 \to N \to 0$. b) In the minimal injective resolution $0 \to F \to J_0 \to J_1 \to ..$ of $F \in \mathrm{mod}\, F$, J_0 and J_1 are projective.

We shall prove that $E_{C,\Pi}$ satisfies these two conditions. This will imply that $E_{C,\Pi}$ is the Auslander algebra of a representation-finite basic algebra $\Lambda_{C,\Pi}$, which is uniquely determined up to an isomorphism. In order to ascertain that $\Lambda_{C,\Pi}$ is selfinjective, we shall need the following lemma, where a simple module is called <u>torsion-free</u>

if it is isomorphic to a submodule of some projective module.

__Lemma__. __Let__ E_A __be the Auslander algebra of a representation-finite, finite-dimensional algebra__ A. __Then__ A __is selfinjective iff the projective cover of any torsion-free simple module__ $S \in \text{mod } E_A$ __is injective__.

__Proof__. Let $N_1 \ldots, N_r \in \text{mod } A$ be representatives of the indecomposable A-modules, $N = N_1 \oplus \ldots \oplus N_r$ and $E_A = \text{Hom}_A(N,N)$. Since $A \in \text{mod } A$ is a direct summand of some N^s, $N \tilde{\to} \text{Hom}_A(A,N) \in \text{mod } E_A^{op}$ is a direct summand of $\text{Hom}_A(N^s,N) \tilde{\to} \text{Hom}_A(N,N)^s = E_A^s$. We infer that N is a projective left E_A-module and that the left adjoint $L: \text{mod } E_A \to \text{mod } A$, $Y \to Y \otimes_{E_A} N$ to the fully faithful functor $R: \text{mod } A \to \text{mod } E_A$, $X \to \text{Hom}_A(N,X)$ is exact. This is equivalent to saying that R transforms injectives into injectives.

Denote by I(S) the injective hull of a simple $S \in \text{mod } A$. Then $RI(S) \in \text{mod } E_A$ is injective and indecomposable. Its socle $\Sigma(S)$ is simple and torsion-free (RI(S) is projective, since R induces an equivalence between mod A and the full subcategory proj E_A of mod E_A formed by the projectives). Furthermore, since N is contained in some $\underset{j}{\oplus} I(S_j)$, each simple submodule of $RN = E_A \subset \underset{j}{\oplus} RI(S_j)$ is isomorphic to $\Sigma(S_j)$. Therefore, the torsion-free simple E_A-modules are those isomorphic to some $\Sigma(S)$.

Denote by P(S) the projective cover of a simple $S \in \text{mod } A$. The radical rad RP(S) of RP(S) consists in the morphisms $N \to P(S)$ without section (= right inverse). As these morphisms factor through rad P(S) , we have rad RP(S) $\tilde{\to}$ R(rad P(S)) . Now RP(S)/rad RP(S) is simple and RP(S)/R(rad P(S)) is "contained" in RS . We infer that RP(S) is the projective cover of $\Sigma(S) \subset RS$.

In conclusion, if A is selfinjective, each P(S) is injective. As a consequence , the projective cover RP(S) of $\Sigma(S)$ is injective. Conversely, if RP(S) is injective, its socle is a simple module $\Sigma(T)$. Hence RP(S) is isomorphic to RI(T) and P(S) is isomorphic to I(T) , since R is fully faithful.

5.2 Lemma. $E_{C,\Pi}$ has finite dimension over k .

Proof. The formula $\mathrm{Hom}_{k(\Gamma)}(\pi x, \pi y) \xrightarrow{\sim} \oplus \mathrm{Hom}_{k(\tilde{\Gamma})}(x,z)$, where $\pi:\tilde{\Gamma} \to \Gamma$ is the universal covering and z ranges through the vertices of $\tilde{\Gamma}$ such that $\pi z = \pi y$, implies $E_{C,\Pi} \xrightarrow{\sim} \oplus_{e,f} \mathrm{Hom}_{k(\Gamma)}(e,f) \xrightarrow{\sim} \oplus_{\tilde{e},z} \mathrm{Hom}_{k(\tilde{\Gamma})}(\tilde{e},z)$. Here e and f range through the finite set of vertices of Γ , $\tilde{e}$ is chosen in $\pi^{-1}(e)$ and z runs through all vertices of $\tilde{\Gamma}$. Since $\mathrm{Hom}_{k(\tilde{\Gamma})}(e,z)$ has finite dimension over k for any z, it is enough to show that $\mathrm{Hom}_{k(\tilde{\Gamma})}(\tilde{e},z)=0$ for all but finitely many z . This follows from 4.5.

5.3 To any $M \in \mathrm{mod}\, E$ we assign a (finite-dimensional) representation of $k(\Gamma)^{op}$, i.e. a k-linear contravariant functor $\bar{M}$ from $k(\Gamma)$ to mod k . For each object e of $k(\Gamma)$ (= vertex of Γ) we set $\bar{M}(e)=M\bar{e}$, where $\bar{e} \in E=k[\Gamma]/I_{\Gamma}$ is the residue class of the identical path $(e\|e) \in k[\Gamma]$ (1.2); for any $\mu \in \mathrm{Hom}_{k(\Gamma)}(e,f)=\bar{f}E\bar{e}$, we define $\bar{M}(\mu):M\bar{f} \to M\bar{e}$ as the multiplication -map $x \mapsto x\mu$. In this way we get an equivalence between mod E and the category mod $k(\Gamma)$ of all (finite-dimensional) representations of $k(\Gamma)^{op}$.

In the sequel we work with mod $k(\Gamma)$ rather than with mod E . By

5.1 our theorem will follow from the statements a),b),c),d) below. There k_e denotes the <u>simple representation</u> assigning the vector space k to the vertex e and 0 to the other vertices:

a) Each representation k_e admits a projective resolution of the form $0 \to P_2 \to P_1 \to P_0 \to k_e \to 0$.

b) In the minimal injective resolution $0 \to P \to I_0 \to I_1 \to \dots$ of an indecomposable projective representation P, I_0 and I_1 are projective.

c) A simple subrepresentation of an indecomposable projective representation is isomorphic to some k_p , where p is a projective (= non-stable) vertex.

d) If $p=\mathbb{p}_e$ and $q=\mathbb{q}_f$ where $f=\tau^n e$ (2.2), the projective cover of k_p is isomorphic to the injective hull of k_q.

5.4 The advantage of considering $\operatorname{mod} k(\Gamma)$ rather than $\operatorname{mod} E$ lies in the fact that the representations of $k(\Gamma)^{op}$ may be related to those of $k(\tilde{\Gamma})^{op}$: a (finite-dimensional) <u>representation</u> of $k(\tilde{\Gamma})^{op}$ is a k-linear contravariant functor X from $k(\tilde{\Gamma})$ to $\operatorname{mod} k$ such that $X(e')=0$ for all but finitely many vertices e' of $\tilde{\Gamma}$. We denote by $\operatorname{mod} k(\tilde{\Gamma})$ the category of these <u>contravariant</u> functors.

With any $X \in \operatorname{mod} k(\tilde{\Gamma})$ we associate a representation $\pi_* X \in \operatorname{mod} k(\Gamma)$ such that $(\pi_* X)(e) = \oplus X(e')$ where e' ranges through the vertices of $\tilde{\Gamma}$ such that $\pi e' = e$.

<u>Lemma</u>. <u>The functor</u> $\pi_*: \operatorname{mod} k(\tilde{\Gamma}) \to \operatorname{mod} k(\Gamma)$ <u>is exact, maps</u> $k_{e'}$ <u>onto</u> $k_{\pi e'}$ <u>and the socle</u> $\operatorname{soc} X$ <u>of any</u> $X \in \operatorname{mod} k(\tilde{\Gamma})$ <u>onto</u> $\operatorname{soc} \pi_* X$. <u>Furthermore</u> π_* <u>transforms the projective cover</u> $P_{e'}$

and the injective hull $I_{e'}$ of $k_{e'}$ into the projective cover $P_{\pi e'}$ and the injective hull $I_{\pi e'}$ of $k_{\pi e'}$ respectively.

Proof. The first two statements are clear. The statement on socles follows from their description: $(\mathrm{soc}\, X)(e')$ is formed by the $x \in X(e')$ which are annihilated by all arrows of $\tilde{\Gamma}$ stopping at e' ; the corresponding description works for $\mathrm{soc}\,\pi_* X$.
$P_{e'}$ and $P_{\pi e'}$ look as follows: $P_{e'}(f')=\mathrm{Hom}_{k(\tilde{\Gamma})}(f',e')$ and $P_{\pi e'}(f)=\mathrm{Hom}_{k(\Gamma)}(f,\pi e')$. The wanted formula $\pi_* P_{e'} \tilde{\to} P_{\pi e'}$ is equivalent to $\mathrm{Hom}_{k(\Gamma)}(f,\pi e') \tilde{\to} \oplus \mathrm{Hom}_{k(\tilde{\Gamma})}(f',e')$, where f' runs through all vertices of $\tilde{\Gamma}$ such that $\pi f'=f$. Similarly, $I_{e'}$ and $I_{\pi e'}$ are described by $I_{e'}(f')=\mathrm{Hom}_{k(\tilde{\Gamma})}(e',f')^T$ (= dual vector space) and $I_{\pi e'}(f)=\mathrm{Hom}_{k(\Gamma)}(\pi e',f)^T$. The wanted formula $\pi_* I_{e'} \to I_{\pi e'}$ is equivalent to $\mathrm{Hom}_{k(\Gamma)}(\pi e',f) \tilde{\to} \oplus \mathrm{Hom}_{k(\tilde{\Gamma})}(e',f')$, where f' is submitted to $\pi f'=f$. OK.

In the coming sections we shall prove that $\mathrm{mod}\, k(\tilde{\Gamma})$ satisfies the conditions a),b),c),d) of 5.3. This statement is independent of Π . The preceding lemma will then imply that the conditions a),b),c) and d) are also satisfied by $\mathrm{mod}\, k(\Gamma)$. In the sequel, whenever we shall speak of a representation without any further precision, we shall mean a representation of $k(\tilde{\Gamma})^{op}$.

5.5 First we need some more information on the C-homotopy classes of stable paths (4.2). Let $w=\alpha_\ell \ldots \alpha_2\alpha_1$ be a path of $\mathbb{Z}A_n$ from (i,j) to (p,q) $(\ell=2p+q-2i-j)$. The ranges and the domains of the arrows α_j are called vertices of w . For each integer r lying between $2i+j$ and $2p+q$, i.e. such that

$r \in [2i+j, 2p+q] = \{z \in \mathbb{Z} : 2i+j \leq z \leq 2p+q\}$, there is a unique vertex (x,y) of w satisfying $2x+y=r$ (notice that the vertices (x,y) of $\mathbb{Z}A_n$ such that $2x+y=r$ lie on a "vertical" line). The number y is called the <u>height of</u> w <u>at</u> r and is denoted by $w(r)$.

Let $P(e,f)$ be the set of all paths of $\mathbb{Z}A_n$ from $e=(i,j)$ to $f=(p,q)$. We define a partial order on $P(e,f)$ by saying that w is <u>higher</u> than v if $w(r) \geq v(r)$ for each $r \in [2i+j, 2p+q]$. Furthermore, we say that <u>a stable vertex</u> (s,t) (resp. <u>a projective vertex</u> $\mathbb{P}_{(s,t)}$) <u>lies between two paths</u> $v,w \in P(e,f)$, if $2s+t$ (resp. $2s+t+1$) lies between $2i+j$ and $2p+q$ and t between the minimum and the maximum of $v(2s+t)$ and $w(2s+t)$ (resp. of $v(2s+t+1)$ and $w(2s+t+1)$).

<u>Lemma</u>. <u>Let</u> $e=(i,j)$ <u>and</u> $f=(p,q)$ <u>be two vertices of</u> $\mathbb{Z}A_n$.

a) <u>Two paths</u> $v,w \in P(e,f)$ <u>are</u> C-<u>homotopic iff no projective vertex lies between</u> v <u>and</u> w.

b) <u>Each</u> C-<u>homotopy class</u> W <u>of</u> $P(e,f)$ <u>contains a highest path</u> h_W <u>and a lowest path</u> l_W. <u>Moreover</u> $W = \{e \in P(e,f) : l_W \leq w \leq h_W\}$.

c) <u>The paths of a</u> C-<u>homotopy class</u> W <u>are</u> C-<u>marginal</u> (4.2) <u>iff either</u> h_W <u>factors through an upper "crenel"</u> $C \not\ni (s,n) \xrightarrow{\gamma} (s+1,n-1) \xrightarrow{\delta} (s+1,n)$ <u>or</u> l_W <u>factors through a lower "crenel"</u> $C \not\ni (s,1) \xrightarrow{\alpha} (s,2) \xrightarrow{\beta} (s+1,1)$.

<u>Proof.</u> a) Assume for instance that $v(2s+t+1) < t < w(2s+t+1)$ for some projective vertex $\mathbb{P}_{(s,t)}$. It is easily seen that $u(2s+t+1) < t$ for any C-neighbour u of v, hence also for any path which is C-homotopic to v. Therefore v is not homotopic to w. Conversely, assume that no projective vertex lies between v and w. We prove by induction on $\|w-v\| = \frac{1}{2}\Sigma|w(r) - v(r)|$,

where r ranges through $[2i+j, 2p+q]$, that v and w are C-homotopic: This is clear if $\|w-v\|=0$. So assume that $w(r)>v(r)$ for some r (the case $w(r)<v(r)$ is similar). Choose an r satisfying $w(r)>v(r)$ and maximizing $w(r)$. Clearly, w has the form $w=w_2\beta\alpha w_1$, where α and β are the arrows of Fig. 27. Since no projective vertex lies between v and w, we have $(m,w(r)-1)\notin C$, and w is C-homotopic to $u=w_2\delta\gamma w_1$. Since $\|u-v\|=\|w-v\|-1$, u is C-homotopic to v by induction hypothesis.

b) If no projective vertex lies between v and w, no one lies between v and $\max(v,w)$. Hence v and $\max(v,w)$ are C-homotopic. Similarly, v and $\min(v,w)$ are C-homotopic. The rest follows easily.

Statement c) is trivial. OK.

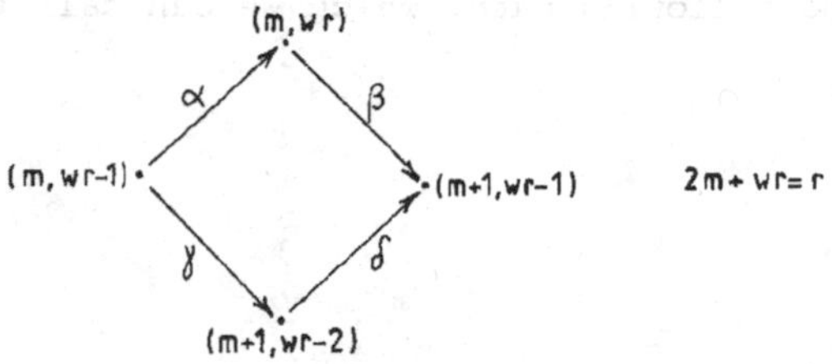

Fig. 27

In order to verify the conditions c) and d) of 5.3 for $\mathrm{mod}\ k(\tilde{\Gamma})$ we need a precise description of the highest and the lowest path in case $(p,q)\in C$ and $(i,j)=(p-n+1,q)$. This description is given in 5.6 below.

5.6 For any $r\in\mathbb{Z}$ let us denote by $c_r=(\alpha_C r-n-1, n+1-\alpha_C r+r)$

the only point of C on the going down diagonal ∇_r (3.4). Likewise, we denote by $\mathbb{P}_r$ the projective vertex of $(\mathbb{Z}A_n)_C$ lying between c_r and $\tau^{-1}c_r$. This notation enables us to transfer the action of the permutations $\alpha_C=\alpha$ and $\beta_C=\beta$ from $\mathbb{Z}$ to C and to the set of projective vertices of $(\mathbb{Z}A_n)_C$: just set $\alpha c_r=c_{\alpha r}$, $\beta c_r=c_{\beta r}$ and $\alpha\mathbb{P}_r=\mathbb{P}_{\alpha r}$, $\beta\mathbb{P}_r=\mathbb{P}_{\beta r}$.

The permutation α assigns to each $c_r\in C$ a string of configuration points: $c_r=(p_0,r-p_0)$, $\alpha c_r=(p_1,\alpha r-p_1)$, $\alpha^2c_r=(p_2,\alpha^2r-p_2),\ldots,$ $\alpha^{ar}c_r=(p_{ar},\alpha^{ar}r-p_{ar})=(p_0+n,r-p_0)$, where ar is the cardinality of the orbit of $\underline{e}_nr\in\sqrt[n]{1}$ under $\bar{\alpha}_C$ (see Fig.28). The formula $c_r=(\alpha r-n-1,n+1-\alpha r+r)$ yields $(p_0+1,n)=(\alpha r-n,n)$; in other words, the going up diagonal Δ_{p_C+1} through $\tau^{-1}c_r$ intersects the going down diagonal $\nabla_{\alpha r}$ through αc_r on the upper border. A similar statement holds if we replace r by some α^ir . Hence we get by composition the following path, which we can call the α-<u>path from</u> $\tau^{-1}c_r$ <u>to</u> $\tau^{-n}c_r$:

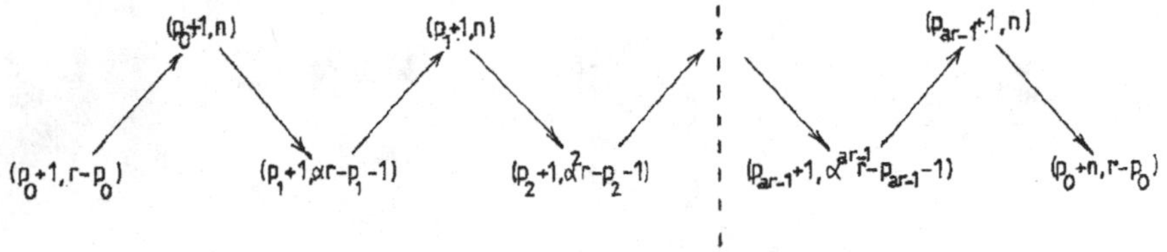

Like α the permutation β assigns to c_r a string of points of C : $c_r=(p_0,r-p_0)$, $\beta c_r=(q_1,\beta r-q_1)$, $\beta^2c_r=(q_2,\beta^2r-q_2)$,....., $\beta^{br}c_r=(q_{br},\beta^{br}r-q_{br})=(p_0+n,r-p_0)$, where br is the cardinality of the orbit of $\underline{e}_nr\in\sqrt[n]{1}$ under $\bar{\beta}_C$. The formulae $c_{\beta r}=(\alpha\beta r-n-1,n+1-\alpha\beta r+\beta r)$ and $\alpha\beta r=r+n+1$ of 3.4 imply

$\beta c_r = c_{\beta r} = (r, \beta r - r)$. Therefore, the going down diagonal ∇_{r+1} through $\tau^{-1}c_r$ intersects the going up diagonal Δ_{q_1} through βc_r at the vertex (r,1) of the lower border. Hence we get by composition the following path, which we call the <u>β-path from</u> $\tau^{-1}c_r$ <u>to</u> $\tau^{-n}c_r$:

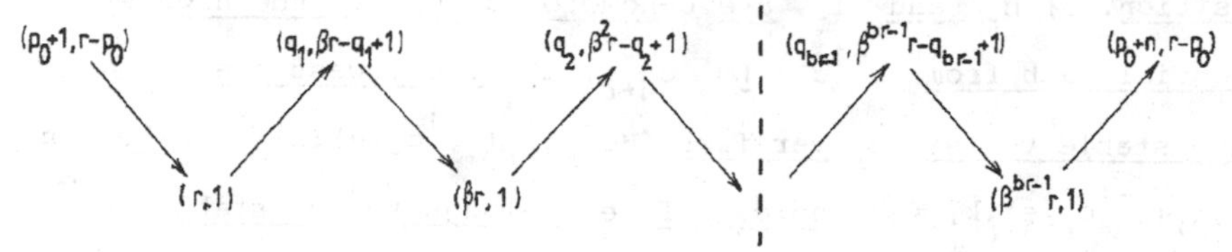

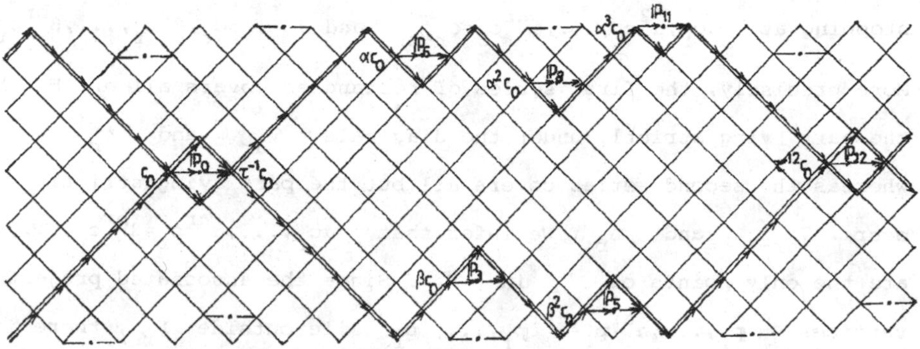

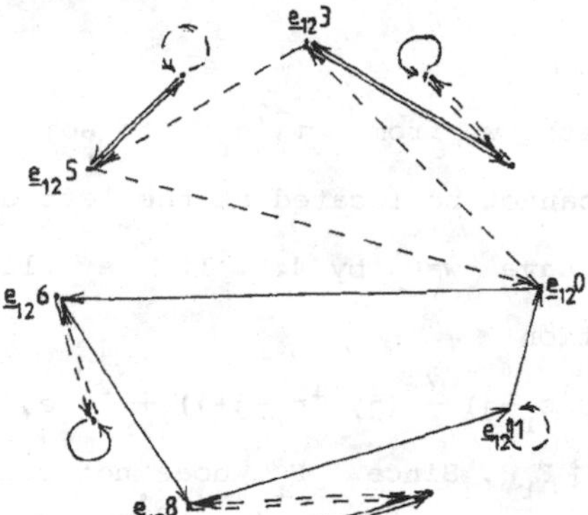

Fig. 28

5.7 If $c_r=(p,r-p)$ is a point of C, we denote by h_r the α-path from $\tau^{-1}c_r$ to $\tau^{-n}c_r=c_{r+n}$, by l_r the β-path. The stable C-fan F_r starting at $\tau^{-1}c_r$ (or stopping at c_{n+r}) is by definition the set of stable vertices lying between l_r and h_r (5.5).

Proposition. a) h_r and l_r are C-homotopic, h_r is the highest C-essential path from $\tau^{-1}c_r$ to c_{n+r}, l_r the lowest one.
b) Each stable vertex e verifies $[\mathrm{Hom}_{k(\tilde{\Gamma})}(\tau^{-1}c_r,e):k]\leq 1$. The equality $[\mathrm{Hom}_{k(\tilde{\Gamma})}(\tau^{-1}c_r,e):k]=1$ holds iff e belongs to the stable C-fan F_r starting at $\tau^{-1}c_r$. Similarly we have $[\mathrm{Hom}_{k(\tilde{\Gamma})}(e,c_{r+n}):k]\leq 1$, and the equality holds iff $e\in F_r$.

Proof. See Fig. 28. By construction F_r is covered by the rectangles stopping at $\alpha c_r,\alpha^2c_r,\ldots,\alpha^{ar}c_r=c_{r+n}$ and at $\beta c_r,\beta^2c_r,\ldots,\beta^{br}c_r=c_{r+n}$. More precisely, the first series of rectangles covers all of F_r but the part lying strictly under the diagonals $\nabla_{\alpha r}$ and Δ_{r-1}, whereas the second series covers all but the part lying strictly over $\nabla_{\alpha r-1}$ and Δ_r. We infer that $\alpha c_r,\ldots,\alpha^{ar}c_r=\beta^{br}c_r,\ldots,\beta c_r$ are the only points of C in F_r. Since the associated projective vertices $\alpha\mathbb{P}_r,\ldots,\alpha^{ar}\mathbb{P}_r=\beta^{br}\mathbb{P}_r,\ldots,\beta\mathbb{P}_r$ lie outside F_r, there is no projective vertex between l_r and h_r. By 5.5 l_r and h_r are C-homotopic.

Now consider a path w from $\tau^{-1}c_r$ to some stable vertex $e\notin F_r$. Obviously e cannot be located at the left of F_r. If it lies at the right, we have $\tilde{w}=0$ by 4.5. If e lies over $\nabla_{\alpha r}$ and Δ_r, w is a composition

$$\tau^{-1}c_r\xrightarrow{w_1}(j,\alpha^ic_r-j)\xrightarrow{\gamma}(j,\alpha^ic_r-j+1)\xrightarrow{w_2}e,$$ where w_1 lies in F_r and $(j,\alpha^ic_r-j)\notin F_r$. Since F_r does not contain any projective vertex, w_1 is C-homotopic to $\tau^{-1}c_r\xrightarrow{w_1'}(\alpha^ir-n,n)\xrightarrow{w_1''}(\alpha^ir-j,j)$

where w_1' is a "part" of h_r (see Fig. 29). The composition $\gamma w_1'' w_1'$ is C-marginal, and so is w. The "symmetric" argument shows that w is C-marginal if e is located under $\nabla_{\alpha r}$ and Δ_r. Our conclusion is that $\tilde{w}=0$ whenever $e \notin F_r$. The same type of considerations shows that $\tilde{v}=0$ for any path from $e \notin F_r$ to c_{r+n}.

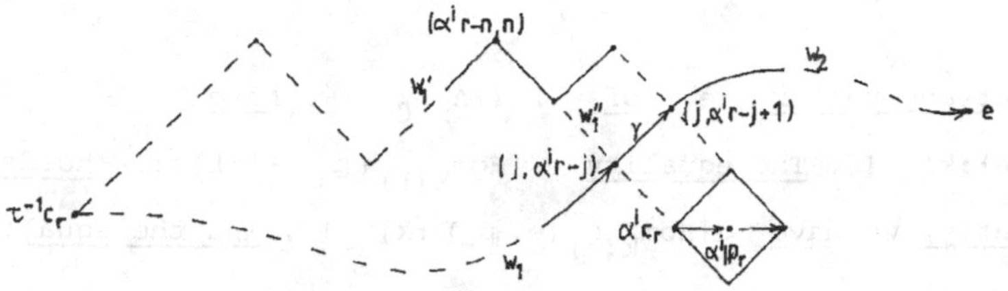

Fig. 29

Let us tackle the case of a vertex $e \in F_r$. The preceding lines imply that the C-essential paths from $\tau^{-1}c_r$ to e lies in F_r and they are C-homotopic (5.5a)). This yields $[\mathrm{Hom}_{k(\tilde{\Gamma})}(\tau^{-1}c_r,e):k] \leq 1$ and $[\mathrm{Hom}_{k(\tilde{\Gamma})}(e,c_{n+r}):k] \leq 1$ by symmetry. In order to show that the equalities hold, we choose within F_r a path u from $\tau^{-1}c_r$ to e and a path v from e to c_{r+n}; for u by way of example, we can start on h_r and end with the going down diagonal through e. The composed path vu is C-homotopic to h_r and l_r. But h_r is obviously the highest path from $\tau^{-1}c_r$ to c_{r+n} lying under $\alpha p_r, \alpha^2 p_r, \ldots, \alpha^{ar-1} p_r$; therefore it is the highest path in its

C-homotopy class. Similarly, 1_r is the lowest path. Using 5.5c) we conclude that $\tilde{h}_r=\tilde{v}\tilde{u}=\tilde{1}_r\neq 0$, and hence $\tilde{v}\neq 0$, $\tilde{u}\neq 0$, $\mathrm{Hom}_{k(\tilde{\Gamma})}(\tau^{-1}c_r,e)=k\tilde{u}$ and $\mathrm{Hom}_{k(\tilde{\Gamma})}(e,c_{r+n})=k\tilde{v}$ (4.2c)). OK.

5.8 The _complete_ C-_fan_ $\bar{F}_r$ starting at $\mathbb{P}_r$ and stopping at $\mathbb{P}_{r+n}$ is by definition the union of the stable C-fan F_r with the projective vertices $\mathbb{P}_r, \alpha\mathbb{P}_r, \alpha^2\mathbb{P}_r, \ldots, \alpha^{a_r}\mathbb{P}_r=\mathbb{P}_{r+n}, \beta\mathbb{P}_r, \beta^2\mathbb{P}_r, \ldots, \beta^{b_r}\mathbb{P}_r=\mathbb{P}_{r+n}$.

Corollary. For any vertex e _of_ $\tilde{\Gamma}=(\mathbb{Z}A_n)_C$ _we have_ $[\mathrm{Hom}_{k(\tilde{\Gamma})}(\mathbb{P}_r,e):k]\leq 1$. _The equality_ $[\mathrm{Hom}_{k(\tilde{\Gamma})}(\mathbb{P}_r,e):k]=1$ _holds iff_ $e\in\bar{F}_r$. _Similarly, we have_ $[\mathrm{Hom}_{k(\tilde{\Gamma})}(e,\mathbb{P}_r):k]\leq 1$, _and the equality holds iff_ $e\in\bar{F}_{r-n}$.

Proof. Clearly, if $e\neq\mathbb{P}_s$ we have $\mathrm{Hom}_{k(\tilde{\Gamma})}(e,c_s)\xrightarrow{\sim}\mathrm{Hom}_{k(\tilde{\Gamma})}(e,\mathbb{P}_s)$ and $\mathrm{Hom}_{k(\tilde{\Gamma})}(\tau^{-1}c_s,e)\xrightarrow{\sim}\mathrm{Hom}_{k(\tilde{\Gamma})}(\mathbb{P}_s,e)$. Therefore we can describe all morphisms of $k(\tilde{\Gamma})$ in terms of the morphisms between stable vertices. Our corollary follows immediately from this remark and from 5.7. OK.

5.9. Let p be a vertex of $\tilde{\Gamma}$. We recall that the projective cover $P_p\in\mathrm{mod}\,k(\tilde{\Gamma})$ of the simple representation k_p is described by $P_p(e)=\mathrm{Hom}_{k(\tilde{\Gamma})}(e,p)$. Similarly, the injective hull I_p of k_p is given by $I_p(e)=\mathrm{Hom}_{k(\tilde{\Gamma})}(p,e)^T$ (= dual vector space, 5.4). These are precisely the spaces determined in 5.8 for $p=\mathbb{P}_r$. _We infer that_ $\mathrm{mod}\,k(\tilde{\Gamma})$ _satisfies condition_ d) _of_ 5.3:

Proposition. $P_{\mathbb{P}_{r+n}}\xrightarrow{\sim}I_{\mathbb{P}_r}$ _for any_ $r\in\mathbb{Z}$.

Proof. By 5.8 we know that $[\mathrm{Hom}_{k(\tilde{\Gamma})}(\mathbb{P}_r,\mathbb{P}_{r+n}):k]=1$. Choose any

$0 \neq \varepsilon \in \mathrm{Hom}_{k(\tilde{\Gamma})}(\mathbb{P}_r, \mathbb{P}_{r+n})^T$ and assign to any $\nu \in P_{\mathbb{P}_{r+n}}(e) = \mathrm{Hom}_{k(\tilde{\Gamma})}(e, \mathbb{P}_{r+n})$ the linear from $\varepsilon\nu \in \mathrm{Hom}_{k(\tilde{\Gamma})}(\mathbb{P}_r, e)^T = I_{\mathbb{P}_r}(e)$ such that $(\varepsilon\nu)(\mu) = \varepsilon(\nu\mu)$. The map $\varepsilon \mapsto \varepsilon\nu$ gives rise to an isomorphism $P_{\mathbb{P}_{r+n}} \xrightarrow{\sim} I_{\mathbb{P}_r}$. OK.

5.10 Lemma. _Let w be a C-essential stable path from $f=(i,j)$ to e, where $j<n$. Denote by γ the arrow $(i-1,j+1) \to (i,j)$, by h_w the highest path in the C-homotopy class of w. Then $w\gamma$ is C-marginal iff h_w factors through $(i,j) \to (i,n)$ and none of the vertices $(i-1,j+1),(i-1,j+2),\ldots,(i-1,n)$ belong to C (see Fig.30)._

Proof. The condition is obviously sufficient. So suppose that $w\gamma$ is C-marginal. Denote by l_w the lowest path in the C-homotopy class of w. As w is C-essential, l_w does not factor through a lower crenel (5.5c); neither does $l_w\gamma = l_{w\gamma}$. Hence $h_{w\gamma}$ must be factorized through an upper crenel. Now $h_{w\gamma}$ has a decomposition

$(i-1,j+1) \to (i-1,p) \to (i,p-1) \xrightarrow{w'} e.$

Since no projective vertex lies between $w\gamma$ and $h_{w\gamma}$ (5.5a), no such vertex can lie between w and the composed path $(i,j) \xrightarrow{u} (i,p-1) \xrightarrow{w'} e$. We infer that w and $w'u$ are C-homotopic, hence that $w'u = h_w$. Since h_w is C-essential, $(i-1,n) \to (i,n-1) \to (i,n)$ is the only crenel through which $h_{w\gamma}$ can factor. This implies that $p=n$ and that w' is a composition $(i,n-1) \to (i,n) \xrightarrow{w''} e$. Finally, since no projective vertex lies between $h_w\gamma$ and $h_{w\gamma}$, the vertices $(i-1,j+1),\ldots,(i-1,n)$ do not belong to C. OK.

5.11 Lemma. _Let $w:(i,j) \to e$ be a stable C-essential path such that $w\gamma$ is C-marginal for each stable arrow γ stopping at_

(i,j) . *Then* $\tau(i,j)=(i-1,j)$ *is a point of* C *lying in the rectangle* R *starting at* $\tau^n e$.

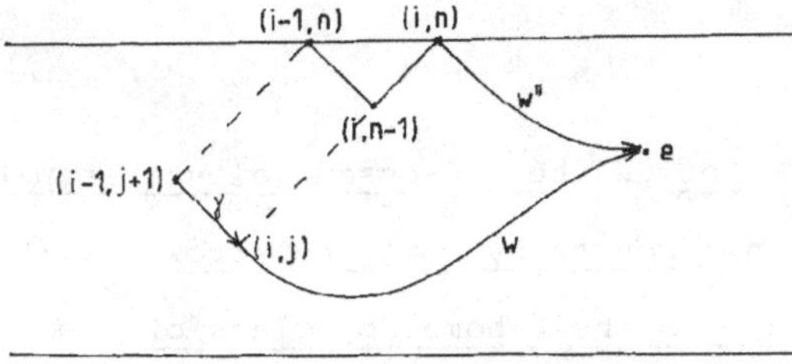

Fig. 30

Proof. Let us assume that $1<j<n$ for instance. By 5.10 the highest path h_w in the C-homotopy class of w factors through $(i,j) \to (i,n)$, and none of the vertices $(i-1,j+1),\ldots,(i-1,n)$ belong to C . The symmetric argument shows that the lowest path l_w C-homotopic to w factors through $(i,j) \to (i+j-1,1)$, and none of the vertices $(i,j-1),(i+1,j-2),\ldots,(i+j-2,1)$ belong to C . Hence the whole rectangle starting at (i,j) lies between l_w and h_w. Therefore it contains no projective vertex, and $\tau(i,j)=(i-1,j) \in C$ (see Fig. 31). Furthermore, our argument shows that e lies on or at the right of the diagonals through $(i+j-1,n-j+1)$. Using 4.5 we conclude that e belongs to the rectangle stopping at $\tau^{1-n}(i,j)$ or equivalently that $\tau(i,j) \in R$. OK.

Corollary. *Suppose* $c \in C$ *and let* $w:\tau^{-1}c \to e$ *be a stable* C- *essential path satisfying* $\tilde{w}\bar{\kappa}\bar{\iota}=0$ (Fig.32). *Then* c *lies in the rectangle* R *starting at* $\tau^n e$.

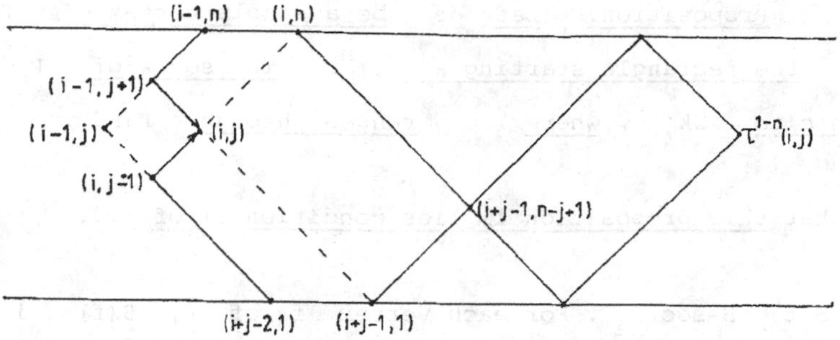

Fig. 31

Proof. First assume that c does not lie on a border of $\mathbb{Z}A_n$. We have $\tilde{w}\bar{\kappa}\bar{\iota}=-\tilde{w}\bar{\zeta}\bar{\varepsilon}-\tilde{w}\bar{\delta}\bar{\gamma}=\mp\tilde{w}\tilde{\zeta}\tilde{\varepsilon}\pm\tilde{w}\tilde{\delta}\tilde{\gamma}$. Since $\mathbb{P}_c$ lies between the paths $w\zeta\varepsilon$ and $w\delta\gamma$, they are not C-homotopic. Therefore the equality $\tilde{w}\bar{\kappa}\bar{\iota}=0$ implies $\tilde{w}\tilde{\zeta}\tilde{\varepsilon}=0=\tilde{w}\tilde{\delta}\tilde{\gamma}$ (4.2). By lemma 5.10 we know that $\tilde{w}\tilde{\delta}\tilde{\gamma}=0$ implies $\tilde{w}\tilde{\delta}=0$. A symmetric argument yields $\tilde{w}\tilde{\zeta}=0$. Hence $w\delta$ and $w\gamma$ are C-marginal, and our lemma implies $c\in R$. If c lies on a border of $\mathbb{Z}A_n$, the proof is quite similar. OK.

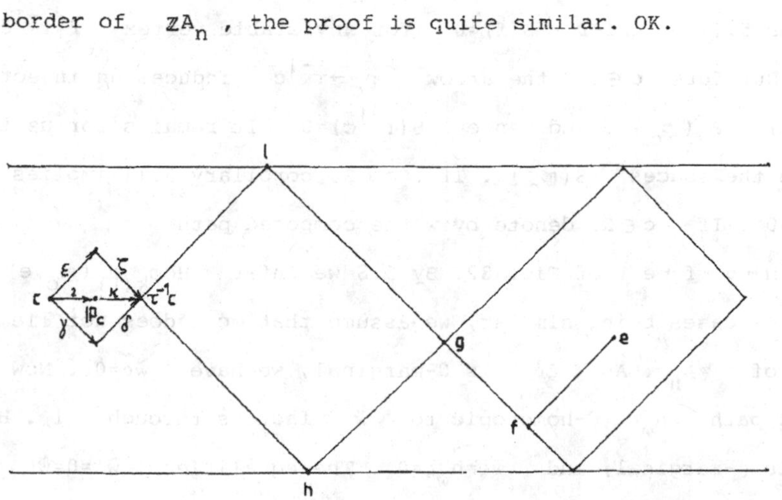

Fig. 32

5.12 Proposition. Let e be a stable vertex of $\tilde{\Gamma}=(\mathbb{Z}A_n)_C$ and R the rectangle starting at $\tau^n e$. The socle of P_e is isomorphic to $\oplus k_{\mathbb{P}_c}$, where c ranges through $R \cap C$.

Notice that this proposition implies condition c) of 5.3.

Proof. Set $S=\operatorname{soc} P_e$. For each vertex f of $\tilde{\Gamma}$, $S(f)$ is the space of all $\mu \in \operatorname{Hom}_{k(\tilde{\Gamma})}(f,e)=P_e(f)$ such that $\mu\tilde{\gamma}=0$ for every arrow γ stopping at f. If f is a stable vertex the morphism μ may be written as $\mu=\Sigma\lambda_w\tilde{w}$, where w runs through representatives of the C-homotopy classes of C-essential paths from f to e (4.2c)). If $w_1 \neq w_2$ are two such representatives, some projective vertex lies between them. Consequently, for any arrow γ stopping at f, $w_1\gamma$ is not C-homotopic to $w_2\gamma$, and the condition $\mu\tilde{\gamma}=0$ is equivalent to saying that $\tilde{w}\tilde{\gamma}=0$ for each w in the support of μ (i.e. such that $\lambda_w \neq 0$).

By lemma 5.11 we infer $S(f)=0$ for any stable vertex $f \neq \tau^{-1}c$, $c \in C$. But for $c \in C$ the arrow $\mathbb{P}_c \to \tau^{-1}c$ induces an injection $P_e(\tau^{-1}c) \to P_e(\mathbb{P}_c)$, and hence $S(\tau^{-1}c)=0$. It remains for us to examine the spaces $S(\mathbb{P}_c)$. If $c \notin R$, corollary 5.11 implies $S(\mathbb{P}_c)=0$. If $c \in R$, denote by w the composed path $\tau^{-1}c \xrightarrow{\xi} h \to g \to f \to e$ of Fig. 32. By 5.8 we infer $\operatorname{Hom}_{k(\tilde{\Gamma})}(\mathbb{P}_c,e) \xrightarrow{\sim} k\tilde{w}\bar{\kappa}$. The other cases being similar, we assume that c does not lie on a border of $\mathbb{Z}A_n$. As $\xi\delta$ is C-marginal, we have $\tilde{w}\tilde{\delta}=0$. Now the highest path h_w C-homotopic to w factors through ι. Hence $h_w\zeta$ is C-marginal, and $\tilde{w}\tilde{\zeta}=\tilde{h}_w\tilde{\zeta}=0$. The equalities $\tilde{w}\tilde{\delta}=0=\tilde{w}\tilde{\zeta}$ imply $\tilde{w}\bar{\kappa}\bar{\iota}=0$ and $S(\mathbb{P}_c)=k\tilde{w}\bar{\kappa}$. OK.

5.13. Each arrow $\alpha: e \to f$ induces a morphism $P_\alpha: P_e \to P_f$, which maps $\mu \in P_e(x) = \mathrm{Hom}_{k(\tilde{\Gamma})}(x,e)$ onto $\bar{\alpha}\mu \in P_f(x) = \mathrm{Hom}_{k(\tilde{\Gamma})}(x,f)$.

Proposition. a) <u>If</u> $c \in C$, <u>the simple representation</u> $k_{\mathbb{P}c}$ <u>has a projective resolution of the form</u> $0 \to P_c \xrightarrow{P_\iota} P_{\mathbb{P}c} \to k_{\mathbb{P}c} \to 0$, <u>where</u> ι <u>is the arrow of</u> $\tilde{\Gamma} = (\mathbb{Z}A_n)_C$ <u>from</u> c <u>to</u> $\mathbb{P}c$.

b) <u>If</u> s <u>is a stable vertex of</u> $\tilde{\Gamma}$, k_s <u>has a projective resolution of the form</u> $0 \to P_{\tau s} \xrightarrow{[P_{\sigma\alpha}]^T} \bigoplus_\alpha P_{d\alpha} \xrightarrow{[P_\alpha]} P_s \to k_s \to 0$, <u>where</u> α <u>runs through the arrows with range</u> s <u>and</u> $d\alpha$ <u>denotes the domain of</u> α.

<u>This proposition implies condition</u> a) <u>of</u> 5.3.

<u>Proof</u>. 1) Statement a) means that for each vertex $x \neq \mathbb{P}c$ the map $\mathrm{Hom}_{k(\tilde{\Gamma})}(x,c) \to \mathrm{Hom}_{k(\tilde{\Gamma})}(x,\mathbb{P}c)$, $\mu \mapsto \bar{\iota}\mu$ is bijective. This follows directly from the definition of the morphisms of $k(\tilde{\Gamma})$.

2) <u>Proof of statement b) if</u> $s=(p,q)$, $\tau s \notin C$ <u>and</u> $1<q<n$. It is clear that $\mathrm{Coker}[P_\alpha P_\beta] \xrightarrow{\sim} k_s$ and $\mathrm{Im}[P_{\sigma\alpha} P_{\sigma\beta}]^T \subset \mathrm{Ker}[P_\alpha P_\beta]$ (see Fig.33). <u>Let us show that</u> $\mathrm{Ker}[P_\alpha P_\beta] \subset \mathrm{Im}[P_{\sigma\alpha} P_{\sigma\beta}]^T$: Let $\lambda = \Sigma\lambda_v \tilde{v} \in P_r(x)$ and $\mu = \Sigma\mu_w \tilde{w} \in P_t(x)$ be such that $P_\alpha\lambda + P_\beta\mu = 0$, where λ_v, μ_w are non-zero scalars, and where v,w are representatives of distinct C-homotopy classes of stable paths. By 5.5a) we know that the C-homotopy classes of the composed paths αv are distinct. The corresponding statement holds for the compositions βw. The equality $P_\alpha\lambda + P_\beta\mu = 0$ therefore means that the paths v,w satisfying $\tilde{\alpha}\tilde{v} \neq 0 \neq \tilde{\beta}\tilde{w}$ are twinned in such a way that $\lambda_v - (-1)^q \mu_w = 0$ and $\tilde{\alpha}\tilde{v} = \tilde{\beta}\tilde{w}$ for any twinned pair (v,w) . So we are reduced to the following three cases: (i) $\lambda=0, \mu=\tilde{w}$ and $\tilde{\beta}\tilde{w}=0$; (ii) $\lambda=\tilde{v}, \mu=0$ and $\tilde{\alpha}\tilde{v}=0$; (iii) $\lambda=\tilde{v}$, $\mu=(-1)^q\tilde{w}$ and $\tilde{\alpha}\tilde{v}=\tilde{\beta}\tilde{w}$ (see 4.2). In case (i) the highest path h_w in the C-homotopy class of w factors through $(p+q-1-n,n) \xrightarrow{w_1} (p-1,q)$ and none of the vertices $(p+q-1-n,n),\ldots,(p-2,q+1),(p-1,q)$ belong to C (apply lemma 5.10 to the opposite representation-quiver $\tilde{\Gamma}^{op}$).

Hence h_w has the form $h_w=(\sigma\beta)w_1w_2$, and we get $\mu=\tilde{h}_w=P_{\sigma\beta}\tilde{w}_1\tilde{w}_2$ and $P_{\sigma\alpha}\tilde{w}_1\tilde{w}_2=\pm(\widetilde{\sigma\alpha})\tilde{w}_1\tilde{w}_2=0$ (see Fig. 33). In case (ii) the proof is similar. In case (iii) no projective vertex lies between βw and αv . Since x lies at the left of the going down diagonal through τs , v admits a decomposition $v=v_2\gamma v_1$ (see Fig. 33) and is C-homotopic to $(\sigma\alpha)v'$ with $v'=v_3v_1$. Similarly w is C-homotopic to some $(\sigma\beta)w'$. Since $\alpha(\sigma\alpha)v'$ and $\beta(\sigma\beta)w'$ are C-homotopic, so are v' and w' (5.5a) . We infer that $\lambda=\tilde{v}=(\widetilde{\sigma\alpha})\tilde{v}'=(-1)^q(\overline{\sigma\alpha})\tilde{v}'=(-1)^qP_{\sigma\alpha}\tilde{v}'$ (4.2) and that $\mu=(-1)^q\tilde{w}=(-1)^q(\widetilde{\sigma\beta})\tilde{w}'=(-1)^q(\overline{\sigma\beta})\tilde{w}'=(-1)^q(\overline{\sigma\beta})\tilde{v}'=(-1)^qP_{\sigma\beta}\tilde{v}'$. This finishes the proof of the equality $\mathrm{Ker}[P_\alpha(x)P_\beta(x)]=\mathrm{Im}[P_{\sigma\alpha}(x)P_{\sigma 3}(x)]^T$ if x is stable. If $x=\mathbb{P}_c$ is projective, we replace x by $\tau^{-1}c$.

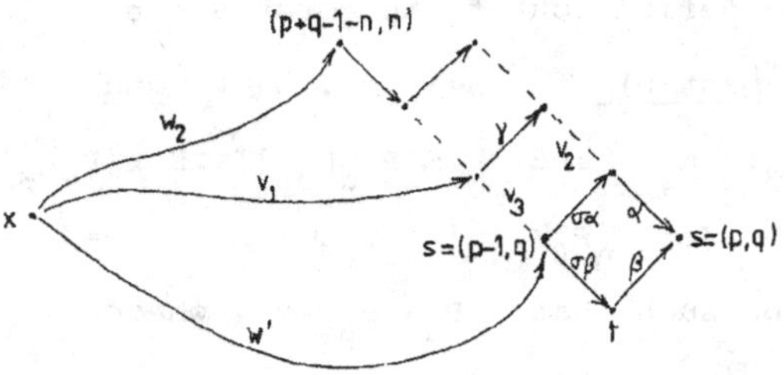

Fig. 33

In order to complete the proof of b) in the considered case, it remains to be shown that $[P_{\sigma\alpha}(x)P_{\sigma\beta}(x)]$ is injective. Otherwise τs would belong to the support of the socle of the projective representation $\mathrm{Hom}_{k(\tilde{\Gamma})}(x,?)$ of $k(\tilde{\Gamma})$; but that support contains only projective vertices (apply 5.12 to the opposite representation-quiver $\tilde{\Gamma}^{op}$).

3) In the cases $s=(p,1)$, $\tau s\notin C$ and $s=(p,n)$, $\tau s\notin C$, the proof of statement b) is similar to that of section 2).

4) <u>Proof of</u> b) <u>if</u> $s=(p,q)$, $\tau s \in C$, $1<q<n$. Again we easily reduce the proof to showing that $\mathrm{Ker}[P_\alpha(x) P_\kappa(x) P_\beta(x)] = \mathrm{Im}[P_{\sigma\alpha}(x) P_{\sigma\kappa}(x) P_{\sigma\beta}(x)]^T$ for each stable x (see Fig. 34). Let $\lambda \in P_r(x)$, $\mu \in P_t(x)$ and $(\overline{\sigma\kappa})\nu \in P_{\mathbb{P}}(x)$ be such that $P_\alpha\lambda + P_\beta\mu + P_\kappa(\overline{\sigma\kappa})\nu = 0$. We have $P_\kappa(\overline{\sigma\kappa})\nu = \bar\kappa(\overline{\sigma\kappa})\nu = -\bar\alpha(\overline{\sigma\alpha})\nu - \bar\beta(\overline{\sigma\beta})\nu$, hence $0 = P_\alpha\lambda + P_\beta\mu + P_\kappa(\overline{\sigma\kappa})\nu = P_\alpha(\lambda - (\overline{\sigma\alpha})\nu) + P_\beta(\mu - (\overline{\sigma\beta})\nu)$. We infer that $\lambda = (\overline{\sigma\alpha})\nu = P_{\sigma\alpha}\nu$ and $\mu = (\overline{\sigma\beta})\nu = P_{\sigma\beta}\nu$, since $[P_\alpha(x) P_\beta(x)]^T : P_r(x) \oplus P_t(x) \to P_s(x)$ is an isomorphism in the considered case. This is so because λ, μ, ν may be described in terms of stable paths as in point 2); moreover, non-homotopic C-essential paths w from x to t (resp. v from x to r) yield non-homotopic C-essentials paths βw (resp. αv) (use 4.2, 5.5 and lemma 5.10 applied to the opposite representation-quiver $\tilde\Gamma^{op}$); finally βw and αv are not C-homotopic (5.5).

5) In the cases $s=(p,1)$, $\tau s \in C$ and $s=(p,n)$, $\tau s \in C$, the proof is similar to that of section 4).

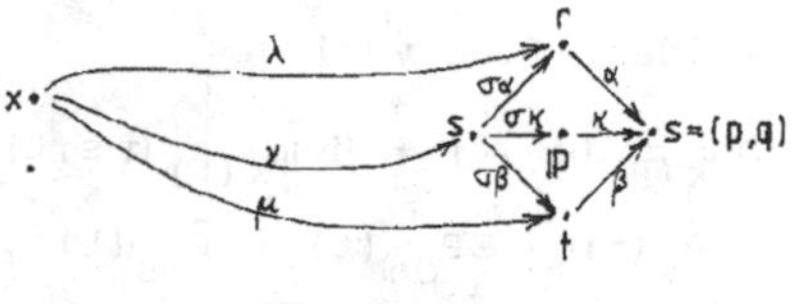

Fig. 34

5.14 <u>Proposition</u>. <u>Let</u> t <u>be a vertex of</u> $\tilde\Gamma = (\mathbb{Z}A_n)_C$ <u>and</u> $0 \to P_t \to I_0 \to I_1 \ldots$ <u>the minimal injective resolution of the projective representation</u> P_t . <u>Then</u> I_0 <u>and</u> I_1 <u>are projective.</u>

<u>This proposition implies condition</u> b) <u>of</u> 5.3.

Proof. If $t = \mathbb{p}_c$ is a projective vertex, we know by 5.9 that P_t is identified with the injective hull of $k_{\mathbb{P}_\tau n_c}$. In this case we have $I_0 = P_t$ and $I_1 = 0$. So we are reduced to the case of a stable vertex t. Then we know by 5.12 that the socle of P_t is a direct sum of some $k_{\mathbb{P}_c}$. Therefore I_0 is the direct sum of the corresponding injective hulls and is projective by 5.9.

It remains for us to show that I_1 is projective. Since I_1 has the same socle as I_0/P_t, it is enough to show that I_0/P_t contains only simple representations of the form $k_{\mathbb{P}_c}$, or equivalently that $\mathrm{Hom}(k_s, I_0/P_t) = 0$ for each stable vertex s. Since the exact sequence $0 \to P_t \to I_0 \to I_0/P_t \to 0$ induces a bijection $\mathrm{Hom}(k_s, I_0/P_t) \tilde{\to} \mathrm{Ext}^1(k_s, P_t)$, we finally have to show that $\mathrm{Ext}^1(k_s, P_t) = 0$ for any stable vertices s and t. To do this, we compute $\mathrm{Ext}^1(k_s, P_t)$ by means of the projective resolution of k_s given in proposition 5.13. So we are reduced to prove the exactness of the sequence

$$\mathrm{Hom}(P_s, P_t) \to \bigoplus_\alpha \mathrm{Hom}(P_{d\alpha}, P_t) \to \mathrm{Hom}(P_{\tau s}, P_t)$$

which is induced by the projective resolution of 5.13b). By the Yoneda lemma this sequence is identified with

$$\mathrm{Hom}_{k(\tilde{\Gamma})}(s,t) \to \bigoplus_\alpha \mathrm{Hom}_{k(\tilde{\Gamma})}(d\alpha, t) \to \mathrm{Hom}_{k(\tilde{\Gamma})}(\tau s, t)$$

or equivalently with $\bar{P}_s(t) \to \bigoplus_\alpha \bar{P}_{d\alpha}(t) \to \bar{P}_{\tau s}(t)$, where $\bar{P}_r$ denotes the projective representation $\mathrm{Hom}_{k(\tilde{\Gamma})}(r, ?)$ of $k(\tilde{\Gamma})$. Therefore, the exactness of the last sequence follows from Proposition 5.13b) applied to the opposite representation-quiver $\tilde{\Gamma}^{op}$.

5.15 Now we are ready for the conclusion. If C is a configuration of $\mathbb{Z}A_n$ and Π an admissible group of automorphism of $\mathbb{Z}A_n$ stabilizing C, we call (C,Π) a <u>configuration pair of class</u> A_n. In 2.5 we assigned a configuration pair $(\tilde{C}_\Lambda, \Pi_\Lambda)$ of class A_n to each selfinjective basic algebra Λ of tree-class A_n. In the present paragraph we associated the Auslander algebra $E_{C,\Pi}$ of some selfinjective, representation-finite and basic $\Lambda_{C,\Pi}$ with any configuration pair (C,Π) of class A_n. In order to establish a bijection between the types (= isomorphism classes) of selfinjective basic algebras of tree-class A_n and the types of configuration pairs of class A_n, we still have to ascertain that the Auslander-Reiten quiver of $\Lambda_{C,\Pi}$ is $\Gamma_{C,\Pi}=(\mathbb{Z}A_n)_C/\Pi$. Since $E_{C,\Pi}$ is defined by the quiver $\Gamma_{C,\Pi}$ and homogeneous relations of degree 2, $\Gamma_{C,\Pi}$ is the ordinary quiver of $E_{C,\Pi}$. Hence we have only to verify that the Auslander-Reiten translation of $\Lambda_{C,\Pi}$ (1.2) is defined precisely on $\mathbb{Z}A_n$ and coincides with the translation $\tau:(p,q)\mapsto(p-1,q)$. This follows immediately from 5.4 (which "lifts the problem to $(\mathbb{Z}A_n)''_C$) and from 5.13 (see [4], where the Auslander-Reiten sequences are interpreted by means of minimal projective resolutions; notice that the resolutions given in 5.13 are clearly minimal).

§ 6 Description of selfinjective algebras of class A_n by quivers and relations.

6.1 Our purpose in the last section of the present article is to describe a selfinjective basic algebra Λ of class A_n and its indecomposable modules in terms of the Auslander-Reiten quiver $\Gamma = \Gamma_\Lambda$. We set $\Pi = \Pi_\Lambda$, $C = \tilde{C}_\Lambda$ and $\tilde{\Gamma} = \tilde{\Gamma}_\Lambda \xrightarrow{\sim} (\mathbb{Z}A_n)_C$. By π we denote both the universal covering $\tilde{\Gamma} \to \Gamma$ and the induced functor $k(\tilde{\Gamma}) \to k(\Gamma)$.

The course to be adopted is clear. Λ is identified with $\mathrm{End}_\Lambda(\oplus M_p) \xrightarrow{\sim} \bigoplus_{p,q} \mathrm{Hom}_\Lambda(M_p, M_q)$, where M_p and M_q run through the projective modules among the representatives M_i chosen in the introduction. Accordingly, each well-behaved functor $H : k(\Gamma) \to \mathrm{mod}\,\Lambda$ (2.5) yields an isomorphism $\bigoplus_{p,q} \mathrm{Hom}_{k(\Gamma)}(e_p, e_q) \xrightarrow{\sim} \Lambda$, where e_p and e_q run through the projective (= non-stable) vertices of Γ. Similarly, since each module M_i is identified with $\mathrm{Hom}_\Lambda(\Lambda, M_i) \xrightarrow{\sim} \bigoplus_p \mathrm{Hom}_\Lambda(M_p, M_i)$, H gives rise to an isomorphism $\bigoplus_p \mathrm{Hom}_{k(\Gamma)}(e_p, e_i) \xrightarrow{\sim} M_i$. So in principle we only have to reap the fruits of the foregoing work.

6.2 Let $\alpha = \alpha_C$ and $\beta = \beta_C$ be the permutations of $\mathbb{Z}$ associated with C (2.7 and 3.4). We define an action of the fundamental group Π on $\mathbb{Z}$ by setting $c_{gr} = gc_r$ for every $g \in \Pi$ and every $r \in \mathbb{Z}$ (recall that c_r is the point of C on the going down diagonal V_r through $(r-1,1)$; see 5.6). Since g stabilizes C, we have $g\alpha = \alpha g$ and $g\beta = \beta g$ if g is a translation. On the contrary, if g is a translation-reflection, α and β are exchanged: $g\alpha = \beta g$ and $g\beta = \alpha g$. In any case g stabilizes the set $\{\alpha, \beta\}$.

Now we assign to each integer r two arrows $r \xrightarrow{\alpha_r} \alpha r$ and $r \xrightarrow{\beta_r} \beta r$. In this way we construct an <u>infinite quiver</u> $\tilde{Q}_C$ which has the integers as vertices. Since Π stabilizes $\{\alpha, \beta\}$, the action of Π on $\mathbb{Z}$ can be extended uniquely to $\tilde{Q}_C$. We denote by $Q_{C,\Pi}$ the <u>residue quiver</u>

$\tilde{Q}_C/\Pi$, by $\bar{\alpha}_r$ and $\bar{\beta}_r$ the residue classes of α_r and β_r modulo Π . Notice that if $\Pi = \tau^n\mathbb{Z}$, $Q_{C,\Pi}$ is identified with the Brauer-quiver Q_C defined in 3.4.

<u>Theorem.</u> Λ <u>is isomorphic to the algebra defined by the quiver</u> $Q_{C,\Pi}$ <u>and the relations</u>

$\bar{\beta}_{\alpha r}\bar{\alpha}_r = 0$, $\bar{\alpha}_{\beta r}\bar{\beta}_r = 0$ <u>and</u> $\bar{\alpha}_{\alpha^{ar-1}r}\cdots\bar{\alpha}_{\alpha r}\bar{\alpha}_r = \bar{\beta}_{\beta^{br-1}r}\cdots\bar{\beta}_{\beta r}\bar{\beta}_r$, <u>where</u> r <u>ranges through</u> $\mathbb{Z}$ <u>and</u> ar, br <u>are defined by</u> $\alpha^{ar}r = r+n = \beta^{br}r$.

Of course, the given set of relations depends only on the Π-orbit of r . Hence Λ is defined by a finite set of relations. Notice that in case ar=1 the last relation expresses $\bar{\alpha}_r$ in terms of the other arrows. Similarly, $\bar{\beta}_r$ is expressed in terms of the other arrows if br=1. As a consequence, the ordinary quiver of the algebra Λ is obtained from $Q_{C,\Pi}$ by deleting the residue classes of the arrows of the form $r \xrightarrow{\alpha_r} r+n$ and $r \xrightarrow{\beta_r} r+n$.

If $\Pi = \tau^{m\mathbb{Z}}$ is a translation group, Λ is a <u>wreath-like algebra</u> in the sense of [5] (the sign difference in the last relation is trifling). If $\Pi = (\tau^{sn}\phi)^{\mathbb{Z}}$ (3.2), we call Λ a <u>Moebius algebra</u>. Moebius algebras have also been studied by J. Waschbüsch (see [7] and [8]). If the Brauer-quiver has the form given in Fig. 35 (n=2p+1) and if $\Pi=(\tau^{sn}\phi)^{\mathbb{Z}}$, the ordinary quiver of Λ is that of Fig. 36. The relations of our theorem are then equivalent to the following ones:

$\bar{\alpha}_{tn+n-1}\cdots\bar{\alpha}_{tn+p+1}\alpha_{tn} = \bar{\beta}_{tn+p}\cdots\bar{\beta}_{tn+1}\bar{\beta}_{tn}$ for $0 \le t < s$,

$\bar{\alpha}_{tn}\bar{\beta}_{tn-p-1} = 0 = \bar{\beta}_{tn}\bar{\alpha}_{tn-1}$ for $1 \le t < s$, $\bar{\alpha}_0\bar{\alpha}_{sn-1} = 0 = \bar{\beta}_0\bar{\beta}_{sn-p-1}$, plus relations requiring that the compositions of any $p+2$ arrows are zero. We denote by $\Lambda_{p,s}$ the algebra defined by the quiver of Fig. 36 and the preceding relations. The propositions 1.4 and 3.2 imply that <u>each Moebius algebra is stably equivalent to some</u> $\Lambda_{p,s}$.

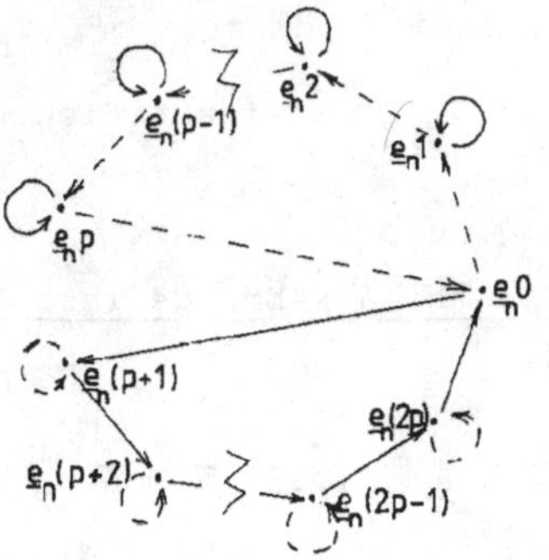
$e_n(p-1)$
$e_n 2$
$e_n 1$
$e_n p$
$e_n 0$
$e_n(p+1)$
$e_n(2p)$
$e_n(p+2)$
$e_n(2p-1)$

Fig. 35

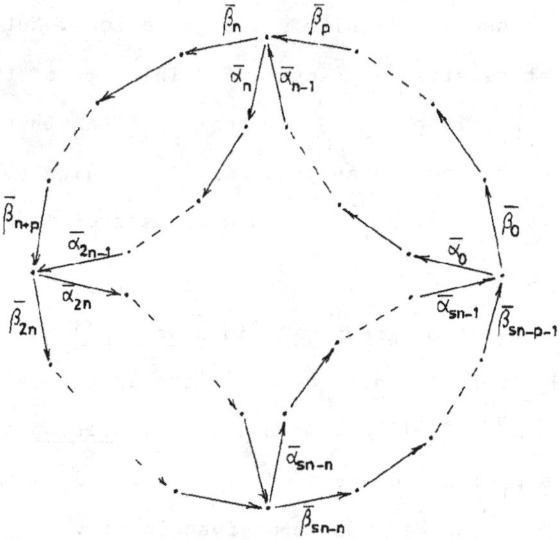
$\bar\beta_n$
$\bar\beta_p$
$\bar\alpha_n$
$\bar\alpha_{n-1}$
$\bar\beta_{n+p}$
$\bar\alpha_{2n-1}$
$\bar\alpha_{2n}$
$\bar\beta_{2n}$
$\bar\beta_0$
$\bar\alpha_0$
$\bar\alpha_{sn-1}$
$\bar\beta_{sn-p-1}$
$\bar\alpha_{sn-n}$
$\bar\beta_{sn-n}$

Fig. 36

6.3 Proof of theorem 6.2 in case $\Pi = \tau^{m\mathbb{Z}}$. First of all we identify $\mathbb{Z}/\Pi$ with the set of projective vertices of Γ by assigning to $\bar{r} = r \bmod \Pi$ the vertex $\pi\mathbb{p}_r$ (5.6). Then we define a homomorphism $\phi: k[Q_{C,\Pi}] \to \Lambda = \bigoplus_{\bar{r},\bar{s}} \mathrm{Hom}_{k(\Gamma)}(\bar{r},\bar{s})$ by setting $\phi(\bar{r}||\bar{r}) = 1\!\!1_{\bar{r}}$, $\phi(\overline{\alpha r}|\bar{\alpha}_r|\bar{r}) = -\pi(\iota_{\alpha r}\tilde{u}\kappa_r)$ and $\phi(\overline{\beta r}|\bar{\beta}_r|\bar{r}) = -\pi(\iota_{\beta r}\tilde{v}\kappa_r)$ (4.2). Here u and v denote any C-essential stable paths from $\tau^{-1}c_r$ to $c_{\alpha r}$ and to $c_{\beta r}$, and $\iota u\kappa, \iota v\kappa$ are the compositions
$\mathbb{p}_r \xrightarrow{\kappa_r} \tau^{-1}c_r \xrightarrow{u} c_{\alpha r} \xrightarrow{\iota_{\alpha r}} \mathbb{p}_{\alpha r}$ and $\mathbb{p}_r \xrightarrow{\kappa_r} \tau^{-1}c_r \xrightarrow{v} c_{\beta r} \xrightarrow{\iota_{\beta r}} \mathbb{p}_{\beta r}$.
It follows directly from Fig. 28 and proposition 5.7 that ϕ vanishes on $\bar{\beta}_{\alpha r}\bar{\alpha}_r$ and $\bar{\alpha}_{\beta r}\bar{\beta}_r$. Moreover $\phi(\bar{\alpha}_{\alpha^{ar-1}r}\dots\bar{\alpha}_{\alpha r}\bar{\alpha}_r) = -\pi(\iota_{r+n}\tilde{h}_r\kappa_r) = -\pi(\iota_{r+n}\tilde{l}_r\kappa_r) = \phi(\bar{\beta}_{\beta^{br-1}r}\dots\bar{\beta}_{\beta r}\bar{\beta}_r)$ (h_r and l_r denote the highest and the lowest C-essential paths from $\tau^{-1}c_r$ to c_{r+n}). Hence ϕ is factorized through the algebra $\bar{\Lambda}$ which is defined by the quiver $Q_{C,\Pi}$ and the relations of the theorem. It remains to be shown that the induced homomorphism $\bar{\phi}: \bar{\Lambda} \to \Lambda$ is bijective.

Let $\varepsilon_{\bar{r}} \in \bar{\Lambda}$ be the residue class of the identical path $(\bar{r}||\bar{r})$ in $Q_{C,\Pi}$. It follows directly from the relations that the k-vector space $\bar{\Lambda}\varepsilon_{\bar{r}}$ is generated by the residue classes of

$(\bar{r}||\bar{r}), (\overline{\alpha r}|\bar{\alpha}_r|\bar{r}), (\overline{\alpha^2 r}|\bar{\alpha}_{\alpha r},\bar{\alpha}_r|\bar{r}), \dots, (\overline{\alpha^{ar-1}r}|\bar{\alpha}_{\alpha^{ar-2}r},\dots,\bar{\alpha}_r|\bar{r})$,
$(\overline{\beta r}|\bar{\beta}_r|\bar{r}), \dots, (\overline{\beta^{br-1}r}|\bar{\beta}_{\beta^{br-2}r},\dots,\bar{\beta}_r|\bar{r})$ and $(\overline{r+n}|\bar{\alpha}_{\alpha^{ar-1}r},\dots,\bar{\alpha}_r|\bar{r}) \sim (\overline{r+n}|\bar{\beta}_{\beta^{br-1}r},\dots,\bar{\beta}_r|\bar{r})$. Hence we have $[\bar{\Lambda}\varepsilon_{\bar{r}}:k] \le ar+br$. By comparison we have $\Lambda 1\!\!1_{\bar{r}} = \bigoplus_{\bar{s}} \mathrm{Hom}_{k(\Gamma)}(\bar{r},\bar{s}) \xrightarrow{\sim} \bigoplus_{s} \mathrm{Hom}_{k(\tilde{\Gamma})}(\mathbb{p}_r,\mathbb{p}_s)$ and therefore $[\Lambda 1\!\!1_{\bar{r}}:k] = ar+br$ (5.8). Finally it is clear from 5.8 that the algebra Λ is generated by the elements $1\!\!1_{\bar{r}}, \phi(\overline{\alpha r}|\bar{\alpha}_r|\bar{r})$ and $\phi(\overline{\beta r}|\bar{\beta}_r|\bar{r})$. We infer that ϕ is surjective, and it is injective by dimension reasons. OK.

6.4 Proof of theorem 6.2 in case $\Pi = (\tau^{sn}\phi)^{\mathbb{Z}}$, $n=2p+1$.

The proof in this case is quite similar to that given in 6.3. The only trouble comes from the definition of $\phi(\overline{\alpha r}|\overline{\alpha}_r|\bar{r})$ and $\phi(\overline{\beta r}|\overline{\beta}_r|\bar{r})$. The reason is that $\pi(\bar{\imath}_{\alpha r}\tilde{u}\bar{\kappa}_r)$ and $\pi(\bar{\imath}_{\beta r}\tilde{v}\bar{\kappa}_r)$ now depend on the choice of r within the class $\bar{r}$, since we may have $\widetilde{gw} = -g\tilde{w}$ for some path w and some translation-reflection $g \in \Pi$. We get round the difficulty by replacing $w \longmapsto \tilde{w}$ by an ad hoc map $w \longmapsto w^{\varepsilon}$. Here ε denotes a Π-fixed <u>signature</u> of $\mathbb{Z}A_n$, which attaches a number $\varepsilon(\alpha) \in \{-1,1\}$ to each arrow α of $\mathbb{Z}A_n$. We require that one or three among the values $\varepsilon(\alpha)$, $\varepsilon(\beta)$, $\varepsilon(\gamma)$ and $\varepsilon(\delta)$ of ε on the arrows of a mesh

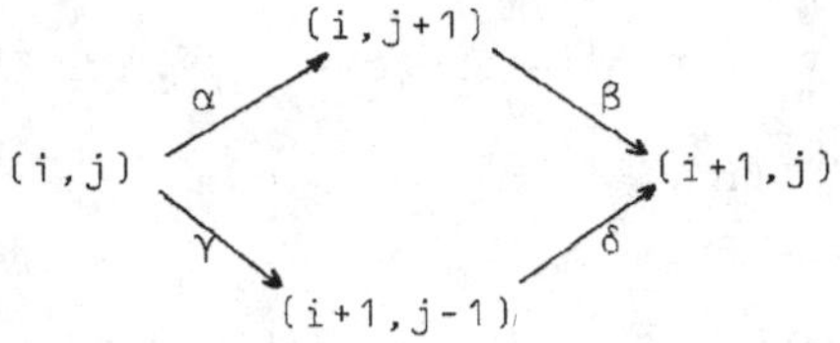

equal -1 . Moreover, ε should satisfy the equation $\varepsilon(\alpha) = \varepsilon(g\alpha)$ for each arrow α and each $g \in \Pi$. The existence of such signatures is illustrated by Fig. 37 in case n=5, s=1, where the arrows α such that $\varepsilon(\alpha) = -1$ are marked with a minus sign.

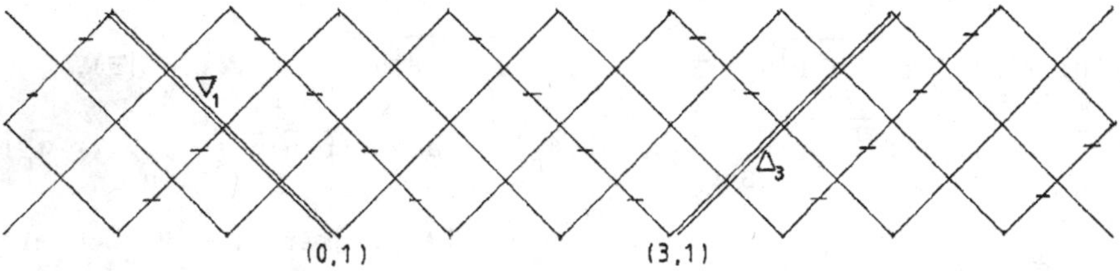

Fig. 37

If ε is a Π-fixed signature we set

$w^{\varepsilon} = \varepsilon(\alpha_t)\ldots\varepsilon(\alpha_1)\bar{\alpha}_t\ldots\bar{\alpha}_1 \in \mathrm{Hom}_{k(\tilde{\Gamma})}(i,j)$ for any stable path $w = (j|\alpha_t,\ldots,\alpha_1|i)$ of $\tilde{\Gamma}$. By construction we have $w^{\varepsilon} = \pm\tilde{w}$, and the equality $w^{\varepsilon} = v^{\varepsilon}$ holds whenever w is C-homotopic to v . We infer that the statements proved about the morphisms $\tilde{w}$ of $k(\tilde{\Gamma})$ remain valid if we replace $\tilde{w}$ by w^{ε}. In particular, the proof given in 6.2 goes through in the present case if we set $\phi(\overline{\alpha r}|\overline{\alpha_r}|\bar{r}) = -\pi(\bar{\iota}_{\alpha r}u^{\varepsilon}\bar{\kappa}_r)$ and $\phi(\overline{\beta r}|\overline{\beta_r}|\bar{r}) = -\pi(\bar{\iota}_{\beta r}v^{\varepsilon}\bar{\kappa}_r)$, where u and v denote C-essential stable paths from $\tau^{-1}c_r$ to $c_{\alpha r}$ and $c_{\beta r}$. OK.

6.5 Finally, we come to the point of describing the <u>indecomposable modules</u> over Λ. To any $N \in \mathrm{mod}\,\Lambda$ we assign a (finite-dimensional) <u>representation</u> $\bar{N}$ of $Q_{C,\Pi}^{op}$: For each vertex $\bar{r} \tilde{\to} \pi p_r$ of $Q_{C,\Pi}$ (see 6.3) we set $\bar{N}(\bar{r}) = N\mathbb{1}_{\bar{r}}$, where $\mathbb{1}_{\bar{r}} \in \mathrm{Hom}_{k(\Gamma)}(\bar{r},\bar{r}) \subset \Lambda$; for any arrow $\mu:\bar{r}\to\bar{s}$ of $Q_{C,\Pi}$, we define $\bar{N}(\mu):\bar{N}(\bar{s})\to\bar{N}(\bar{r})$ as the multiplication map $x \mapsto x\bar{\mu}$, where $\bar{\mu}$ is the canonical image of μ in $\mathrm{Hom}_{k(\Gamma)}(\bar{r},\bar{s}) \subset \Lambda$. The representations $\bar{N}$ are subjected to the relations defining Λ (theorem 6.2). We express this fact by saying that they are <u>bound</u>. In this way we get an equivalence between $\mathrm{mod}\,\Lambda$ and the <u>category</u> $\mathrm{mod}\,Q_{C,\Pi}$ <u>of all bound representations of</u> $Q_{C,\Pi}^{op}$.

The advantage of considering $\mathrm{mod}\,Q_{C,\Pi}$ rather than $\mathrm{mod}\,\Lambda$ lies in the fact that the representations of $Q_{C,\Pi}^{op}$ may be related to those of $\tilde{Q}_C^{op}$ (6.2) : a representation X of $\tilde{Q}_C^{op}$ is called <u>finite-dimensional</u> if $\sum[X(e):k] < \infty$, where e ranges through all vertices of $\tilde{Q}_C^{op}$. It is called <u>bound</u> if it satisfies the relations

$X(\alpha_r)X(\alpha_{\alpha r})\dots X(\alpha_{\alpha^{ar-1}r}) = X(\beta_r)X(\beta_{\beta r})\dots X(\beta_{\beta^{br-1}r})$, $X(\alpha_r)X(\beta_{\alpha r}) = 0$ and $X(\beta_r)X(\alpha_{\beta r})$ for all $r \in \mathbb{Z}$. The category of all finite-dimensional bound representations of $\tilde{Q}_C^{op}$ will be denoted by $\mathrm{mod}\,\tilde{Q}_C$.

With any $X \in \mathrm{mod}\,\tilde{Q}_C$ we associate a representation $\pi_* X \in \mathrm{mod}\,Q_{C,\Pi}$ such that $(\pi_* X)(\bar{r}) = \oplus X(r')$, where r' ranges through the set of all integers satisfying $\bar{r}' = \bar{r} (= \pi p_r)$. In fact we are interested in the case where $X = X_e = \mathrm{Hom}_{k(\tilde{\Gamma})}(p_?, e)$. Here e denotes any vertex of $\tilde{\Gamma}$; we set $X_e(r) = \mathrm{Hom}_{k(\tilde{\Gamma})}(p_r, e)$ for any $r \in \mathbb{Z}$; furthermore, the maps

$X_e(\alpha_r): X_e(\alpha r) \to X_e(r)$ and $X_e(\beta_r): X_e(\beta r) \to X_e(r)$ are defined as $\mu \mapsto -\mu \bar{1}_{\alpha r} u^{\varepsilon} \bar{\kappa}_r$ and $\mu \mapsto -\mu \bar{1}_{\beta r} v^{\varepsilon} \bar{\kappa}_r$, where u and v are C-essential stable paths from $\tau^{-1} c_r$ to $c_{\alpha r}$ and to $c_{\beta r}$, and where ε is a suitable signature (see 6.4; $u^{\varepsilon} = \tilde{u}$ if Π is a translation group). If $X = X_e$, $\pi_* X$ is the representation $X_{\pi e} \in \operatorname{mod} Q_{C,\Pi}$ such that $X_{\pi e}(\bar{r}) = \oplus \operatorname{Hom}_{k(\tilde{\Gamma})}(\mathbb{p}_{r'}, e) = \operatorname{Hom}_{k(\Gamma)}(\pi \mathbb{p}_r, \pi e)$, where r' runs through all integers satisfying $\bar{r}' = \bar{r}$. If $\pi e = e_i$ is the vertex of Γ associated with the indecomposable Λ-module M_i, $\pi_* X_e = X_{e_i}$ is just the representation in $\operatorname{mod} Q_{C,\Pi}$ corresponding to M_i (see 6.1). Hence <u>the classification of the indecomposable Λ-modules is reduced to the description of the representations</u> $X_e \in \operatorname{mod} \tilde{Q}_C$.

6.6 An <u>admissible walk</u> starting at s and stopping at t is by definition a subquiver W of $\tilde{Q}_C$ having the form illustrated in Fig. 38.

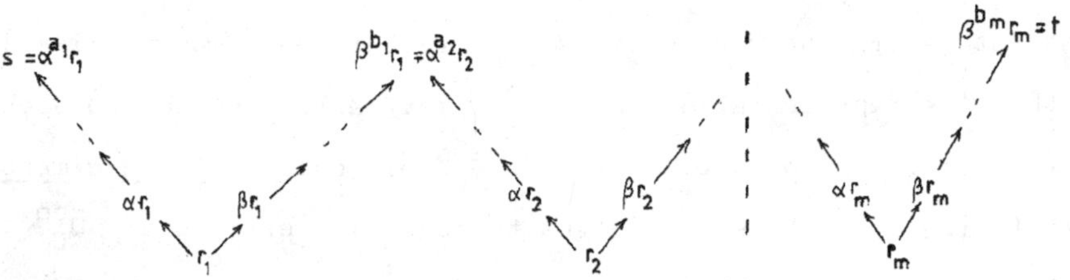

Fig. 38

The arrows tending to the upper left are of type α_r, those tending to the upper right are of type β_r. Moreover, we assume that $0 \le a_1 < ar_1$, $0 < b_1 < br_1$, $0 < a_2 < ar_2, \ldots$, $0 < a_m < ar_m$, $0 \le b_m < br_m$ and $m \ge 1$. We illustrate two admissible walks in Fig. 39 and Fig. 40. There we identify an integer r with the associated projective vertex $\mathbb{p}_r$ of $\tilde{\Gamma}$. In Fig. 39 we have $m = 3$, $s = r_1 = 3$, $r_2 = 4$, $t = r_3 = 6$,

$a_1 = 0$, $b_1 = a_3 = 1$, $b_2 = a_2 = 2$, $b_3 = 0$. In Fig. 40 we have $m = 2$, $s = r_1 = -3$, $r_2 = 0$, $a_1 = 0$, $b_1 = a_2 = b_2 = 2$.

With W we associate a bound representation $kW \in \operatorname{mod} \tilde{Q}_C$ by setting $kW(e) = k$ if e is a vertex of W and $kW(e) = 0$ otherwise. Accordingly, $kW(\gamma)$ is the identity, if γ is an arrow of W, and is zero otherwise. The corresponding representation $\pi_* kW$ in $\operatorname{mod} Q_{C,\Pi}$ is represented in the figures 39 and 40 for two possible values of Π : $\Pi = \tau^{12\mathbb{Z}}$ and $\Pi = \tau^{4\mathbb{Z}}$.

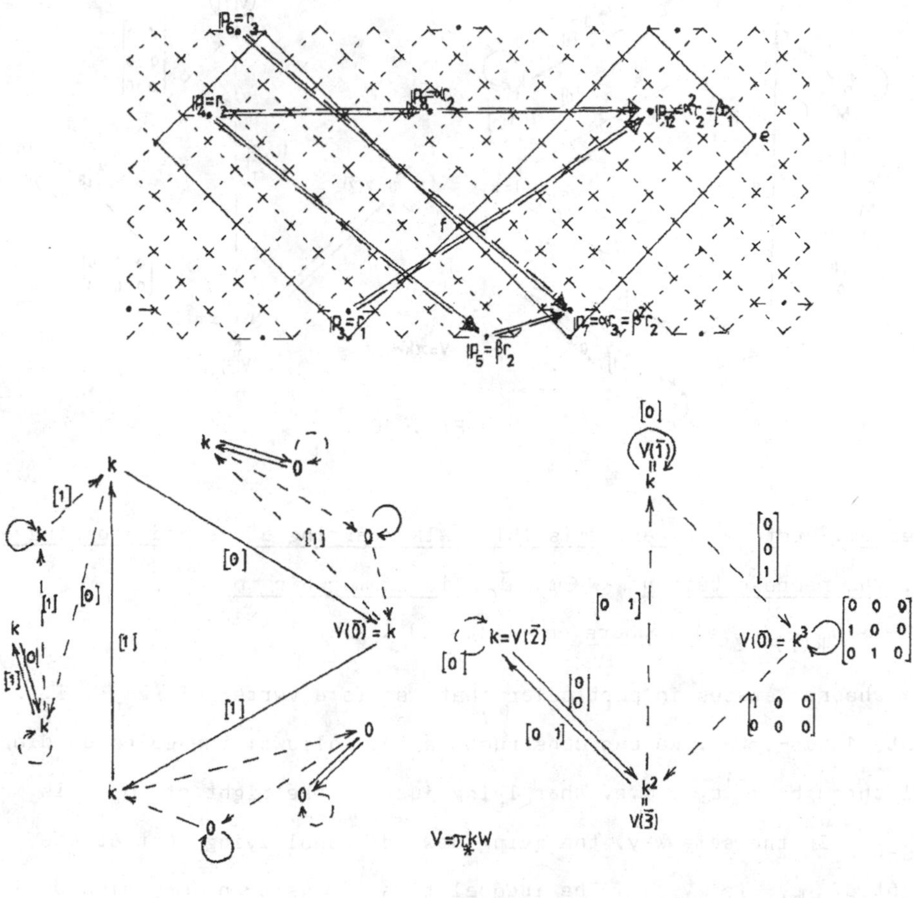

Fig. 39

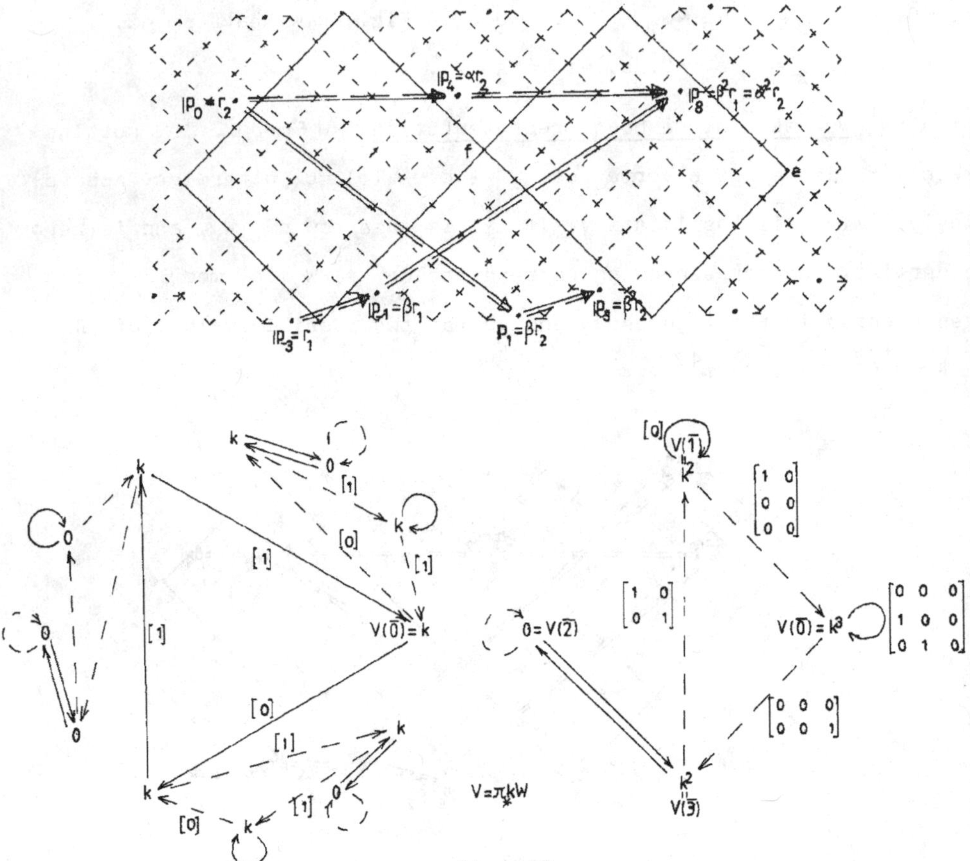

Fig. 40

<u>Theorem.</u> <u>Let</u> W <u>be an admissible walk starting at</u> s <u>and stopping at</u> t. <u>The representation</u> $kW \in \mathrm{mod}\,\tilde{Q}_C$ <u>is isomorphic to</u> $X_e = \mathrm{Hom}_{k(\tilde{\Gamma})}(\mathbb{P}_?, e)$, <u>where</u> $e = (t, \alpha s - t)$.

Our theorem states in particular that e is a vertex of $\mathbb{Z}A_n$, i.e. that $1 \leq \alpha s - t \leq n$. We can construct e as follows: the going up diagonal through $\tau^{-1}c_s$, i.e. that lying just at the right of $\mathbb{P}_s$, is $\Delta_{\alpha s-n}$. In the same way, the going down diagonal lying just at the right of $\mathbb{P}_t$ is ∇_{t+1}. The inequalities $1 \leq \alpha s - t \leq n$ may also be interpreted by saying $\Delta_{\alpha s-n}$ and ∇_{t+1} intersect at

$f = (\alpha s-n, n+t+1-\alpha s)$. The vertex e is the stopping vertex of the rectangle starting at f (see Fig. 39 and Fig. 40).

The theorem also shows that kW is completely determined by s and t, and so is clearly W . Since the indecomposable Λ-modules correspond to the representations $X_{\pi e} = \mathrm{Hom}_{k(\Gamma)}(\mathbb{P}_?, \pi e) \tilde{\to} \pi_* X_e$, we infer the

<u>Corollary.</u> <u>The map</u> $W \to \pi_*(kW)$ <u>yields a bijection between</u> Π-<u>congruence classes of admissible walks and non-projective indecomposable</u> Λ-<u>modules</u>.

The corollary ignores the projective bound representations, which have already been described in 5.8.

6.7 <u>Proof of theorem</u> 6.6. First, it is obvious that kW is a non-projective indecomposable bound representation in $\mathrm{mod}\,\tilde{Q}_C$. Second, each non-projective indecomposable bound representation in $\mathrm{mod}\,Q_{C,\Pi}$ is isomorphic to X_e for some stable vertex e of Γ (6.1). Taking $\Pi = (\tau^{mn})^{\mathbb{Z}}$ and passing to the limit as m tends to infinity, we infer the corresponding statement for $\tilde{Q}_C$. So kW is isomorphic to X_e for some $e = (p,q)$. The point is to determine p and q . In order to do so, we shall assume that $a_1 = 0$ and $b_m > 0$. This case occurs in Fig. 40. The proof in the other possible cases would be quite similar.

Since $kW \tilde{\to} \mathrm{Hom}_{k(\tilde{\Gamma})}(\mathbb{P}_?, e)$, we have $0 = (kW)(\beta^{-1}s) - \mathrm{Hom}_{k(\tilde{\Gamma})}(\mathbb{P}_{\beta^{-1}s}, e)$ and $0 = (kW)(\alpha s) = \mathrm{Hom}_{k(\tilde{\Gamma})}(\mathbb{P}_{\alpha s}, e)$ (in order to see that W does not reach $\beta^{-1}s$ and αs project W onto the Brauer-quiver $Q_C = \tilde{Q}_C/\tau^{n\mathbb{Z}}$). Now e lies within the C-fan starting at $\tau^{-1}c_s$ (5.7). The condition $\mathrm{Hom}_{k(\tilde{\Gamma})}(\mathbb{P}_{\alpha s}, e) = 0$ means that e satisfies $p+q \le \alpha s$ (i.e. e lies on $\nabla_{\alpha s}$ or at the left of it; have a look at Fig. 28). Similarly, the relation $\mathrm{Hom}_{k(\tilde{\Gamma})}(\mathbb{P}_{\beta^{-1}s}, e) = 0$ is equivalent to $\mathrm{Hom}_{k(\tilde{\Gamma})}(c, \mathbb{P}_{s'}) = 0$, where $s' = \beta^{bs-1}s$ (5.6). This implies $p+q \ge \alpha s$ (see Fig. 28).

On the other hand we have $0 = (kW)(\alpha^{-1}t) = \mathrm{Hom}_{k(\tilde{\Gamma})}(\mathbb{P}_{\alpha^{-1}t}, e)$ and $0 = (kW)(\beta t) = \mathrm{Hom}_{k(\tilde{\Gamma})}(\mathbb{P}_{\beta t}, e)$. Since e lies within the C-fan starting at $\tau^{-1}c_t$, the second relation implies $q \leq t$ (i.e. e lies on or at the left of the going up diagonal through $c_{\beta t}$; see Fig. 28). Similarly, $\mathrm{Hom}_{k(\tilde{\Gamma})}(\mathbb{P}_{\alpha^{-1}t}, e) = 0$ is equivalent to $\mathrm{Hom}_{k(\tilde{\Gamma})}(e, \mathbb{P}_{t'}) = 0$, where $t' = \alpha^{at-1}t$. This implies $q \geq t$ (see Fig. 28). OK.

Bibliography

[1] AUSLANDER, M.: Representation Theory of Artin algebras II, Comm. Algebra (1974), 269-310

[2] AUSLANDER, M. and REITEN, I.: Representation Theory of Artin algebras III, Comm. Algebra (1975), 239-294

[3] AUSLANDER, M. and REITEN, I.: Representation Theory of Artin algebras IV, Comm. Algebra (1977), 443-518

[4] AUSLANDER, M. and REITEN, I.: Representation Theory of Artin algebras V, Comm. Algebra (1977), 519-554

[5] GABRIEL, P. and RIEDTMANN, Chr.: Group representations without groups, Comm. Math. Helvetici (1979), 1-48

[6] RIEDTMANN, Chr.: Algebren, Darstellungsköcher, Ueberlagerungen und zurück, Comm. Math. Helvetici (1980), to appear

[7] GABRIEL, P.: Christine Riedtmann and the selfinjective algebras of finite representation type, in Ring Theory, Proceedings of the 1978 Antwerp Conference, 453-458, Marcel Dekker, New York 1979

[8] SCHERZLER, E. and WASCHBUESCH, J.: A class of selfinjective algebras of finite representation type, Proceedings of the second International Conference on Representation of Algebras, Ottawa 1979

REPRESENTATION THEORY OF BLOCKS OF DEFECT 1

K.W.Roggenkamp (Stuttgart)

Introduction and statement of the results.

Representation theory was developed at the end of the last century by Frobenius, Burnside and Schur to obtain information on the finite group G by studying its representations in $GL(n,\mathbb{C})$. In this spirit Brauer originated modular reprentation theory as a device to obtain even more specific properties of G by studying its representations over a field k with $\operatorname{char}(k) \mid |G|$. The success of Brauer's work lay in the interrelation between ordinary and modular representation theory. The link between characteristic zero and characteristic p representation was formed by integral representation theory; but integral representation theory was in this theory only a tool. However, I think it should be a vital part of representation theory in the sense of its founders and maybe also in the sense of Brauer. One reason is obvious: By studying global integral representations, one studies simultaneously ordinary representations and modular representations <u>for all p.</u> One probably will object, that global integral representation theory is too difficult. I must more or less agree to this objection. However, local integral representation theory still gives information for ordinary representation theory and modular representation theory at one prime p. And it seems to me that local integral representation theory - if studied more energetically - should give additional information on ordinary representation theory, modular representation theory and hence also on the group G . Moreover, I think that the proofs by local integral representation theory are less complicated than the ones using modular representation theory. In the following pages I shall try to substantiate this on one of the most developed subjects in modular representation theory: the study of blocks of defect one. Here - at least in my opinion - the integral proofs are less complicated and give more insight into the structure. In order to state the results, I introduce the following

<u>Notation:</u> p will be a fixed rational prime and R with field of quotients K a finite unramified extension of $\hat{\mathbb{Z}}_p$, the p-adic complete integers. Then $pR = \operatorname{rad} R$ and we put $\mathfrak{k} = R/pR$. G is assumed to be a finite group with $p \mid |G|$.

<u>Theorem I:</u> Let B be a block of RG with defect group D satisfying:

α) D is normal in G,

β) D is abelian,

γ) every subgroup of D is G-invariant.

Then:

(i) If $S_1, \ldots, S_t$ are the simple B-modules, then $\operatorname{End}_B(S_i) = \mathfrak{k}_o$ is the same for all $1 \le i \le t$.

(ii) The Schur-indices of all simple modules in $K \otimes_R B$ are bounded by $d = |\mathfrak{k}_o : \mathfrak{k}|$.

(iii) If e is a primitive central idempotent in $K \otimes_R B$, then Be is a hereditary order.

<u>Note</u> that if B contains one absolutely simple module (in particular if B is the principal block), then all simple modules in B are absolutely simple, and moreover, the Schur indices of all simple $K \otimes_R B$-modules are one.

<u>Remark:</u> W. Plesken has told me that $G = SL(2,8)$ at $p = 3$ does not satisfy (iii) of the theorem; hence the condition that D is normal seems to be necessary.

<u>Theorem II*:</u> Let B be a block of RG of defect one.

---

*I have learnt at this very same conference, that H. Jacobinski has also obtained these results.

Then:

(i) If $S_1,\dots,S_t$ are the simple B-modules, then $\mathrm{End}_B(S_i) = \mathfrak{f}_o$ is the same for all $1 \le i \le t$.

(ii) The Schur-indices of all simple $K \otimes_R B$-modules are bounded by $d = |\mathfrak{f}_o : \mathfrak{k}|$.

(iii) If ε is a primitive central idempotent of $K \otimes_R B$, then $B\varepsilon$ is a hereditary order.

(iv) There exists a hereditary R-order Γ in $K \otimes_R B = \prod_{j=1}^{\tau} (D_j)_{n_j}$ such that $B \subset \Gamma$ and $\mathrm{rad}\, B = \mathrm{rad}\, \Gamma$.

(v) If P is an indecomposable projective B-lattice, then $K \otimes_R P = U \oplus U'$ where U, U' are simple $K \otimes_R B$-modules. The Brauer-tree of B , T_K has as vertices the simple $K \otimes_R B$-modules U_j, $1 \le j \le \tau$, and U_j and $U_{j'}$ are joined by an edge if there exists P as above with $K \otimes_R P \simeq U_j \oplus U_{j'}$. T_K then is a tree.

(vi) If $\Gamma = \prod_{j=1}^{\tau} \Gamma_j$ is the decomposition of Γ into hereditary orders in simple algebras, then the Γ_j are in bijection with the vertices of T_K . A vertex j has order ν_j - i.e. ν_j edges meet at j - if and only if Γ_j has ν_j non-isomorphic indecomposable lattices.

The structure of hereditary orders is well understood [H],[Br]. Let D be a finite dimensional skewfield over K and Ω its unique maximal order, then - up to Morita-equivalence - a hereditary R-order Γ_o in $(D)_n$ has, up to conjugation, the form

$$\Gamma_o = \begin{pmatrix} \Omega & \Pi\Omega & \cdots & \Pi\Omega \\ \vdots & \ddots & \ddots & \vdots \\ \vdots & & \ddots & \Pi\Omega \\ \Omega & \cdots & \cdots & \Omega \end{pmatrix}_n , \text{ where } \Pi\Omega = \mathrm{rad}\,\Omega .$$

Moreover,

$$\mathrm{rad}\,\Gamma_o = \begin{pmatrix} \Pi\Omega & \cdots & \cdots & \Pi\Omega \\ \Omega & \ddots & & \vdots \\ \vdots & \ddots & \ddots & \vdots \\ \Omega & \cdots & \Omega & \Pi\Omega \end{pmatrix} .$$

In view of Theorem II, the structure of B is obtained as pullback

$$\begin{array}{ccc} \Gamma & \longrightarrow & \Gamma/\mathrm{rad}\,\Gamma \\ \vdots & & \uparrow \\ B & \dashrightarrow & B/\mathrm{rad}\,\Gamma \end{array} .$$

By (iv) we know Γ, provided we know the skewfields D_i. The Brauertree - moreover- tells us that the embedding

$$B/\mathrm{rad}\,\Gamma \to \Gamma/\mathrm{rad}\,\Gamma$$

is diagonal. Hence it is possible to describe B.

<u>Remark:</u> From (iii) we easily obtain the following generalization of a result of Brauer [B]. If U is a simple $K \otimes_R B$-module, and if the vertex of U in T_K has order ν, then there are exactly ν non-isomorphic B-lattices spanning U; $M_1, \ldots, M_\nu$ and the lattice of submodules of M_1 is linearly ordered

$$M_1 \supset M_2 \supset \ldots \supset M_\nu \supset \mathrm{rad}\,M_1 \supset \mathrm{rad}\,M_2 \ldots .$$

In a special situation, we may find the skewfields:

<u>Theorem III:</u> Assume that B contains an absolutely simple module. Then the Brauertree T_K coincides with the ordinary Brauertree - though K need not be a splitting field. Then the skewfields D_j, with the exception of the exceptional vertex j_o, coincide with K, and D_{j_o} is a totally ramified extension of K with ramification index the multiplicity of the exceptional vertex j_o.

<u>Example I:</u> The principal block B of the Mathieugroup M_{11} for p=11 has Brauertree o – o – o(–o, –o) – o 2, where 2 is the multiplicity of the exceptional vertex [A]. Hence B has the following structure

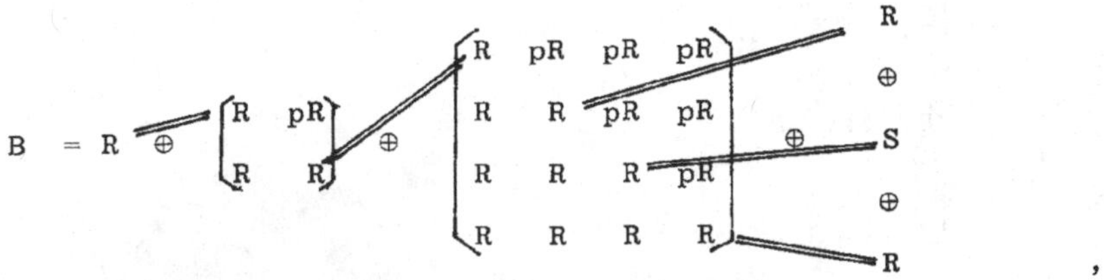

,

with $R = \hat{\mathbb{Z}}_{11}$, $pR = 11R$, S is a quadratic totally ramified extension of R, more precisely $S = \mathrm{Fix}_{C_5}(R[\sqrt[11]{1}])$, with C_5 the cyclic group of order 5. The bindings indicate congruences modulo 11; for example $R = S$ is the pullback of

$$\begin{array}{ccc} R & \longrightarrow & \mathbb{Z}/11\mathbb{Z} \\ \uparrow & & \uparrow \\ R = S & \dashrightarrow & S \end{array} .$$

Example II: Let G_1 be the Frobeniusgroup of order 21, and let G be the pullback

$$\begin{array}{ccccccccc} 1 & \to & C_7 & \to & G_1 & \to & C_3 & \to & 1 \\ & & \| & & \uparrow & & \uparrow & & \\ 1 & \to & C_7 & \to & G & \to & C_9 & \to & 1 \end{array} ,$$

C_i the cyclic group of order i.

If $R = \hat{\mathbb{Z}}_7$ then $RG = B_o \sqcap B_1 \sqcap B_2$, where B_o is the principal block and B_i are algebraically conjugate blocks, $i=1,2$. Then

$$K \otimes_R B_1 = L \sqcap D ,$$

where $L = K(\theta)$, θ a primitive 9th root of unity and D is the unique skewfield over $F = \mathrm{Fix}_{C_3}(K[\sqrt[7]{1}])$, $|F : K| = 2$, of index 3.

Let S be the ring of integers in F and Ω the maximal order in D, then $S/\mathrm{rad}\, S \simeq \Omega/\mathrm{rad}\, \Omega = \mathbb{F}_{27}$ and B_1 is the pullback

$$\begin{array}{ccc} S & \longrightarrow & \mathbb{F}_{27} \\ \uparrow & & \uparrow \\ B_1 = \; S = \Omega & \dashrightarrow & \Omega \end{array} .$$

Let B be as in Theorem II, then for the structure of the B-lattices we have

Theorem IV *: If e is the number of simple B-modules - i.e. the number of edges in T_K - then there are $2e$ non-isomorphic indecomposable

---

*Part of this result has been obtained by Reiner [R] and Pu [Pu].

non-projective B-lattices and e indecomposable projective ones. Let M_o be any non-projective indecomposable B-lattice. Then all indecomposable B-lattices occur in a projective resolution of M_o, which is constructed from T_K in the usual fashion. The projective cover P of an indecomposable B-lattice M is indecomposable, and

$$\mathbb{E}:\ 0 \to N \to P \to M \to 0$$

is the Auslander-Reiten sequence of M. [RS]

Because of IV, the Auslander-Reiten quiver of B, can be constructed easily from T_K. Recall that the vertices of the Auslander-Reiten quiver Q are the indecomposable B-lattices and arrows are irreducible maps. Let $\tilde{Q}$ be the dual of T_K and choose an orientation in $\tilde{Q}$ according to a projective resolution. Break each arrow into two arrows and insert a new middle point. The resulting quiver Q is the Auslander-Reiten quiver of B.

Example The Mathieugroup M_{11} :

T_K: 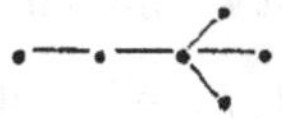

$\tilde{Q}$: 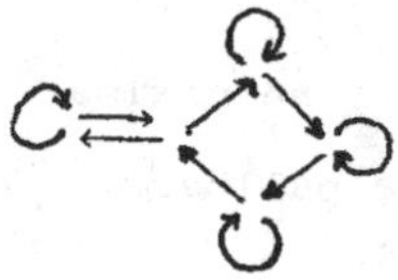

Q: 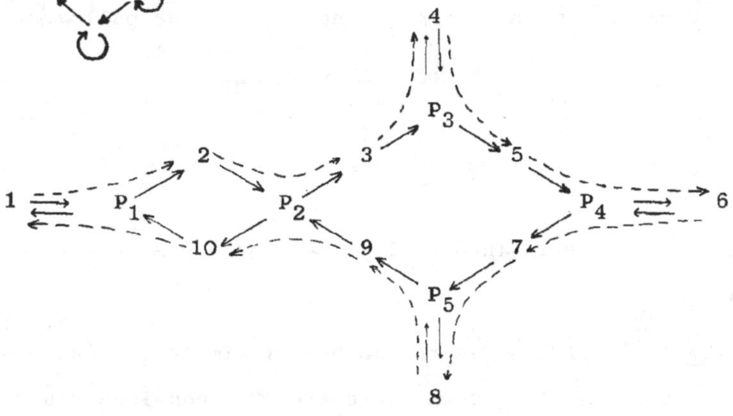

Here the solid arrows indicate irreducible maps and the dotted arrows Auslander-Reiten sequences.

I would like to thank W.Plesken and K.Erdmann for some valuable conversations.

§ 1 Normal defect groups

Proof of Theorem I (iii): Let B and D be as in Theorem I, and let η be the block idempotent of B.

(1) Let ϵ be a primitive idempotent in KD then ϵ is G-invariant.

Proof: Since D is a p-group and K is unramified, ϵ is also a primitive idempotent of $\mathbb{Q}D$; however, the idempotents in $\mathbb{Q}D$ are integral linear combinations of idempotents of the form $\epsilon_U = \frac{1}{|U|} \sum_{u \in U} u$, where U is a subgroup of D. Since by assumption all subgroups of D are G-invariant, the statement of (1) follows.

A slight generalization of [M1, 3.3] gives

(2) Let θ be the augmentation ideal of RD, then we have the exact sequence of B-lattices:

$$0 \to \theta \uparrow^G \eta \to B \to \widetilde{B} \to 0 ,$$

where $\widetilde{B}$ is a separable order and $\mathrm{rad}\, B \simeq \theta \cdot^G \eta + pB$.

Proof: $\widetilde{B}$ is a direct factor of $R\,G/D$ and it has defect zero. Hence $\widetilde{B}$ is a separable order [Ro]. The statement about the radical follows from reduction modulo p.

(3) Let ϵ be a primitive idempotent of KD, then $RD\epsilon$ is a maximal order.

Proof: Since D is commutative,

$$\mathbb{Q}D = \Pi L_i ,$$

where L_i are p^i-th cyclotomic extensions of $\mathbb{Q}$, $L_i = \mathbb{Q}(\theta_i)$. Hence we are given homomorphisms

$$\varphi_i : \mathbb{Q}D \to \mathbb{Q}(\theta_i) ,$$

$$\varphi_{i|_D} : D \to \mathbb{Q}(\theta_i)^* , \text{ the groups of units.}$$

The image of $\varphi_{i|_D}$ must contain a primitive p^i-th root of unity $\widetilde{\theta}_i$, and since there are exactly p^i p^i-th roots of unity in L_i, we conclude that $D/\mathrm{Ker}\,\varphi_i|_D \simeq C_{p^i} = \langle c_i \rangle$ is cyclic of order p^i, and so we may assume w.l.o.g.

$$\varphi_i : \; c_i \;\mapsto\; \theta_i \quad .$$

This then induces the homomorphism

$$\mathbb{Z}D/\mathfrak{h}\!\uparrow^D \;\simeq\; \mathbb{Z}C_{p^i} \;\to\; \mathbb{Z}[\theta_i] \quad ,$$

where $H = \mathrm{Ker}\,\varphi_i|_D$ and it remains to show that $\mathrm{rad}\,\hat{\mathbb{Z}}_p[\theta_i] \simeq \hat{\mathbb{Z}}_p[\theta_i]$; but this is obvious since $\mathrm{rad}\,\hat{\mathbb{Z}}_p[\theta_i] = (1-\theta_i)\,\hat{\mathbb{Z}}_p[\theta_i]$. Now R is unramified and the result follows.

(4) Let ε be a primitive idempotent in KD. Then

(i) $(\mathrm{rad}\,RD\,\varepsilon)\!\uparrow^G\eta \;=\; [\mathrm{rad}(RG\,\varepsilon)]\eta$

(ii) $[\mathrm{rad}(RG\,\varepsilon)]\,\eta \;\simeq\; (RG\,\varepsilon)\eta$.

<u>Proof:</u> Since $KD\eta \subset KG\eta \subset KG$, $\varepsilon\eta$ is an idempotent in $KG\eta$ and in KG. However, by hypothesis, ε is G-invariant and so <u>ε is a central idempotent in KG</u> and thus $\varepsilon\eta$ is also central in $KG\eta$. We have the exact sequence

$$(*) \qquad 0 \to \mathrm{rad}\,RD\,\varepsilon \to RD\,\varepsilon \to S \to 0 \quad ,$$

where $S = \mathfrak{k}_D$ is the trivial simple RD-module, RD being local. Since RG is a projective RD-module and B is a direct summand of RG, $B \otimes_{RD} -$ is an exact functor and so we obtain from $(*)$ the exact sequence

$$0 \to B \otimes_{RD} \mathrm{rad}\,RD\,\varepsilon \to B\varepsilon \to B \otimes_{RD} S \to 0 \quad .$$

Note that indeed $B\varepsilon \simeq B \otimes_{RD} RD\,\varepsilon$, under the canonical isomorphism $RG \simeq RG \otimes_{RD} RD$.

But $B \otimes_{RD} S$ is a $\widetilde{B}$-module and so $B \otimes_{RD} S$ is semi-simple; i.e. $B \otimes_{RD} \mathrm{rad}\,RD\,\varepsilon \supset \mathrm{rad}\,B\varepsilon$. But D is normal in G and ε is a central idempotent; thus $B \otimes_{RD} \mathrm{rad}\,RD\,\varepsilon$ is nilpotent modulo $p\,RG$, hence

$$\operatorname{rad} B\varepsilon \;=\; B \otimes_{RD} \operatorname{rad} RD\varepsilon \;\simeq\; B \otimes_{RD} RD\varepsilon \quad \text{by (3) and so}$$

$$B\varepsilon \;\simeq\; \operatorname{rad} B\varepsilon \;.$$

Proof of Theorem I (iii): Let e be a central primitive idempotent in $K \otimes_R B$. Then either $e\widetilde{B} \neq 0$, in which case Be is even separable by (2). Hence we may assume that e is a primitive idempotent in $K \otimes_R \theta\uparrow^G$. Then there exists a central primitive idempotent ε in KD with $\varepsilon e = e$. By (4) $\operatorname{rad}(B\varepsilon) \simeq B\varepsilon$ and so $B\varepsilon$ is hereditary [AG]. But then Be is a direct factor of $B\varepsilon$, and thus also hereditary.

Theorem I (i) and (ii) will follow directly from (iii), however, we shall postpone the proofs.

In the next proposition the structure of B is described more explicitely, in particular for later applications to blocks of defect one.

According to the exact sequence in (2) we write $\eta = \eta_1 + \eta_2$ with η_i orthogonal idempotents in $K \otimes_R B$ such that $\eta_2 \widetilde{B} = \widetilde{B}$ and $\eta_1\, \theta\uparrow^G = \theta\uparrow^G\eta$.

Proposition 1: Let e be a central primitive idempotent in $K \otimes_R \theta$ and put $\varepsilon = e\eta$. Then $B(\varepsilon + \eta_2)$ is a Bäckström order; i.e. $B\varepsilon \amalg B\eta_2$ is a hereditary order Γ and $\operatorname{rad}\Gamma = \operatorname{rad}(B(\varepsilon + \eta_2))$.

Proof: Since $\theta\uparrow^G\eta \subset \operatorname{rad} B$, we have the following commutative diagram with exact rows and columns:

$$\begin{array}{ccccccccccc}
 & & & & & & 0 & & 0 & & \\
 & & & & & & \uparrow & & \uparrow & & \\
 & & & & & & \widetilde{B}/\operatorname{rad}\widetilde{B} & \simeq & \widetilde{B}/\operatorname{rad}\widetilde{B} & & \\
 & & & & & & \uparrow & & \uparrow & & \\
\mathfrak{E}: & 0 & \longrightarrow & \theta\uparrow^G\eta & \longrightarrow & & B & \longrightarrow & \widetilde{B} & \longrightarrow & 0 \\
\downarrow & & & \| & & & \uparrow & & \uparrow & & \\
p\mathfrak{E}: & 0 & \longrightarrow & \theta\uparrow^G\eta & \longrightarrow & & \operatorname{rad} B & \longrightarrow & p\widetilde{B} & \longrightarrow & 0 \\
 & & & & & & \uparrow & & \uparrow & & \\
 & & & & & & 0 & & 0 & &
\end{array}$$

Let $\alpha: \theta\uparrow^{G}_{\eta} \to \theta\uparrow^{G}_{\varepsilon}$ be the natural projection, then we obtain the commutative diagram $\mathfrak{E}\alpha \to p\,\mathfrak{E}\alpha$:

$$\begin{array}{ccccccccccc}
 & & & & & & 0 & & 0 & & \\
 & & & & & & \uparrow & & \uparrow & & \\
 & & & & & & \widetilde{B}/p\widetilde{B} & = & \widetilde{B}/p\widetilde{B} & & \\
 & & & & & & \uparrow & & \uparrow & & \\
\mathfrak{E}\alpha: & & 0 & \longrightarrow & \theta\uparrow^{G}_{\varepsilon} & \longrightarrow & B(\varepsilon+\eta_2) & \to & \widetilde{B} & \to & 0 \\
\downarrow & & & & \| & & \uparrow & & \uparrow & & \\
p\,\mathfrak{E}\alpha: & & 0 & \longrightarrow & \theta\uparrow\varepsilon & \longrightarrow & \mathrm{rad}(B(\varepsilon+\eta_2)) & \to & p\widetilde{B} & \to & 0 \\
 & & & & & & \uparrow & & \uparrow & & \\
 & & & & & & 0 & & 0 & &
\end{array}$$

<u>Claim:</u> $p\,\mathfrak{E}\alpha$ is split exact.

<u>Proof of the claim:</u> $RG\varepsilon$ is hereditary with radical $\theta\uparrow^{G}_{\varepsilon}$. We have the homomorphism

$$\begin{array}{cccccc}
\rho: & RG(\varepsilon+\eta_2) & \to & RG\varepsilon & \to & 0 \\
 & \| & & \| & & \\
\rho: & B(\varepsilon+\eta_2) & \to & B\varepsilon & \to & 0
\end{array}$$

and the composite

$$\rho|_{\theta\uparrow^{G}_{\varepsilon}}: \quad \theta\uparrow^{G}_{\varepsilon} \to \mathrm{rad}(B(\varepsilon+\eta_2)) \to \mathrm{rad}\,B\varepsilon = \theta\uparrow^{G}_{\varepsilon}$$

is the identity, and so $p\,\mathfrak{E}\alpha$ is split exact. <u>This proves the claim.</u>

Thus $\mathrm{rad}\,B(\varepsilon+\eta_2) = \theta\uparrow^{G}_{\varepsilon} \oplus p\widetilde{B}$. Since both $RG\varepsilon$ with radical $\theta\uparrow^{G}_{\varepsilon}$ and $\widetilde{B}$ with radical $p\widetilde{B}$ are hereditary orders, the result is established.

We point out one consequence, in which the block B itself is a Bäckström order:

<u>Corollary 1:</u> Let B be a block with a normal defect group D of order p . Then B is a Bäckström order.

<u>Proof:</u> In this case the only possible idempotent e of $K \otimes_R \theta$ is the identity on $K \otimes_R \theta$ and so $\varepsilon = \eta_1$, and $\varepsilon+\eta_2 = \eta$. Hence the above proposition applies to $B = B(\varepsilon+\eta_2)$.

Structure of blocks of defect one with normal defect group.

Let B be a block with defect group D normal in G, of order p, and with block idempotent η. We assume that $\widetilde{B}$ has t non-isomorphic indecomposable projective lattices.

Let $\eta = \eta_1 + \eta_2$ be the orthogonal decomposition of η in KG such that $B\eta_2 = \widetilde{B}$. Then up to Morita equivalence -

$$K \otimes_R B = (D)_t \amalg K_1 \amalg \ldots \amalg K_t ,$$

and

$B \simeq$

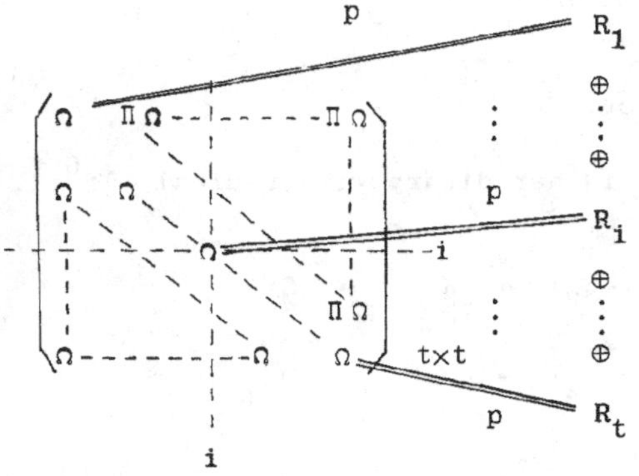

where Ω is the valuation ring in D, and K_i with p-adic rings of integers R_i are unramified extensions of K.

The obvious indecomposable lattices are the one-dimensional ones $M_i^+ = R_i$, the indecomposable projective ones P_i, and M_i^-, such that there are exact sequences

$$0 \to M_{i+1}^- \to P_i \to M_i^+ \to 0 ,$$

$1 \le i \le t$, and $i+1$ is to be taken modulo t.

Proof: We can classify B only up to Morita equivalence, so we assume B is basic, and so $\widetilde{B}/p\widetilde{B} = \prod_{i=1}^{t} \bar{\ell}_i$, where $\bar{\ell}_i$ are extension fields of ℓ. Since $p\widetilde{B} = \operatorname{rad}\widetilde{B}$, we must have

(*) $$\widetilde{B} = \oplus R_i , \quad 1 \le i \le t ,$$

where R_i are unramified extensions of R, i.e. $pR_i = \operatorname{rad} R_i$.

Claim: $B\eta_2$ has t non-isomorphic projective lattices.

Proof: For this it is enough to show that $\eta\theta\uparrow^G/\mathrm{rad}\,\eta\theta\uparrow^G$ has t non-isomorphic simple modules, $B\eta_2$ being hereditary. But $B\eta_2$ is a cyclic B-module and so it has at most t non-isomorphic lattices. On the other hand we have an epimorphism

$$B \to B\eta_2 \simeq \eta\theta\uparrow^G$$

and so if $\mathrm{rad}\,P_1 \simeq \mathrm{rad}\,P_2$, their projective covers are isomorphic, and so B has a projective with multiplicity >1, a contradiction. This proves the claim.

We now observe that $KB\eta_2$ is a simple K-algebra. In fact, if not, B would decompose two-sidedly; but this is impossible.

§ 2 Blocks of defect one.

In addition to the above notation, we assume that D is cyclic of order p with normalizer N in G. Let b be the Brauer correspondent of B.

From the structure of b in § 1 one easily derives:

(1) b has exactly $2t$ non-projective indecomposable lattices $M_1^+,\dots,M_t^+,M_1^-,\dots,M_t^-$ and t indecomposable projective ones. Moreover, all indecomposable b-lattices occur in a minimal projective resolution of M_1^+. $K\otimes_R b$ has $t+1$ simple modules U_o, $U_1,\dots,U_t$ such that $K\otimes_R P_i \simeq U_o \oplus U_i$, $K\otimes_R M_i^+ \simeq U_i$, $K\otimes_R M_i^- \simeq U_o$, $1\leq i\leq t$.

We denote by $\mathbb{F}$ and $\mathbb{G}$ the Green correspondence linking the modules in B with the modules in b and conversely resp.

(2) Let P_o be an indecomposable B-lattice, then $K\otimes_R P_o \simeq U\oplus V$, where U and V are non-isomorphic simple $K\otimes_R B$-modules.

<u>Proof:</u> Given any non-split exact sequence of B-lattices

$$0 \to M' \to M \to M'' \to 0$$

with M'' indecomposable. Then M is projective. In fact,

$$0 \to \mathbb{F}(M')\oplus P' \to \mathbb{F}(M)\oplus P \to \mathbb{F}(M'')\oplus P'' \to 0$$

is an exact sequence of b-lattices, P',P and P'' are projective b-lattices. The exact sequence

$$0 \to \mathbb{F}(M') \to \mathbb{F}(M)\oplus \widetilde{P} \to \mathbb{F}(M'') \to 0$$

is not split and $\mathbb{F}(M')$, $\mathbb{F}(M'')$ are non-projective indecomposable b-lattices, so by the structure of b, $\mathbb{F}(M) = 0$, but then - applying $\mathbb{G}$ we find that M must be a projective B-lattice.

Assume now that $K\otimes_R P_o = V_1\oplus V_2\oplus V_3$ where V_1 and V_2 are non-zero simple, and let $\varphi_i\colon K\otimes_R P_o \to V_i$ be the projections $i=1,2,3$. Let $M = \mathrm{Im}(\varphi_1+\varphi_2|_{P_o})$. Then we have an exact sequence

$$O \to M' \to M \to \operatorname{Im}\varphi_1|_{P_o} \to O \quad ,$$

which can not split, M being a local B-module. Hence by the above observation M must be projective, i.e. $V_3 = O$.

It remains to show that U_1 and U_2 are not isomorphic. We have the exact sequence

$$\mathscr{E}: \quad O \to N \to P_o \to M \to O$$

with $K \otimes_R N \simeq V_1$, $K \otimes_R M \simeq V_2$. We now restrict this sequence to b, and obtain the exact sequence of b-lattices

$$O \to L_1 \oplus P' \to P_o \to L_2 \oplus P'' \to O \quad ,$$

L_i indecomposable b-lattices, i=1,2.
From the structure of b (1) it follows that $K \otimes_R L_1 = U_o$ and $K \otimes_R L_2 = U_i$ for some i. But if $K \otimes_R N \simeq K \otimes_R M$, this can not happen (cf.(1)).

(3) If P is an indecomposable B-lattice, then there are exactly two B-lattices M^+ and M^- of which P is the projective cover.

<u>Proof:</u> By (2) $K \otimes_R P \simeq U \oplus V$ with U and V simple non-isomorphic, and so there are at least two non-isomorphic B-lattices M^+ and M^- onto which P maps. Assume now that M is another B-lattice onto which P maps. Then M is indecomposable and we have an epimorhism

$$\varphi: \; P \to M \quad ;$$

we may assume $K \otimes_R M \simeq K \otimes_R M^+$. We then also have the epimorphism

$$\varphi^+: \; P \to M^+ \quad .$$

But then $\operatorname{Ker}\varphi \simeq (K \otimes_R \operatorname{Ker}\varphi) \cap P = (K \otimes_R \operatorname{Ker}\varphi^+) \cap P \simeq \operatorname{Ker}\varphi^+$, and by Schanuel's lemma $M \simeq M^+$.

(4) Let $P_1,\ldots,P_t$ be the indecomposable projective B-lattices; then

$KP_i = U_i \oplus V_i$, where U_i and V_i are non-isomorphic simple $K \otimes_R B$-modules, $1 \leq i \leq t$. There are $2t$ non-projective non-isomorphic indecomposable B-lattices, $M_1^+, \ldots, M_t^+$, $M_1^-, \ldots, M_t^-$ with $K \otimes_R M_i^+ \simeq U_i$ and $K \otimes_R M_i^- \simeq V_i$. Moreover, in the projective resolution

$$0 \to M_1^+ \to Q_{2t} \searrow M_t^- \nearrow Q_{2t-1} \searrow M_t^+ \cdots Q_3 \searrow M_2^+ \nearrow Q_2 \searrow M_1^- \nearrow Q_1 \to M_1^+ \to 0$$

each indecomposable projective B-lattice P_i, $1 \leq i \leq t$, occurs exactly twice; in particular every indecomposable B-lattice M is local.

<u>Proof:</u> Since Green-correspondence commutes with taking syzygies, we know that the projective resolution of an indecomposable non-projective B-lattice M_1^+ has periodicity $2t$, and all indecomposable non-projective B-lattices occur. Since every indecomposable projective B-lattice is the projective cover of two non-isomorphic non-projective B-lattices, a counting argument gives the desired result.

(5) B is a Bäckström order with graph $\dot{\bigcup}_t A_3$ [RR §2].

<u>Proof:</u> Let Λ_o be the Bäckström order of B with associated hereditary order Γ [RR §2]. Then Λ_o has non-isomorphic indecomposable projective lattices $P_i' \simeq \Lambda_o \otimes_\Lambda P_i$, $1 \leq i \leq t$, and so $\Gamma \otimes_{\Lambda_o} P_i' = Q_i^1 \oplus Q_i^2$ with $Q_i^1 \not\simeq Q_i^2$, $1 \leq i \leq t$.

<u>Claim:</u> $Q_i^k \not\simeq Q_j^l$ for $(i,k) \neq (j,l)$.

<u>Proof:</u> Since $\Lambda_o \supset B$, every Λ_o-lattice is also a B-lattice, and so Λ_o has at most $2t$ non-isomorphic non-projective indecomposable lattices. If $Q_i^k \simeq Q_j^l$ for $(i,k) \neq (j,l)$ - note: $i \neq j$ - then the graph of Λ_o contains a subgraph of the form

a.) 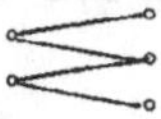 or b.) 

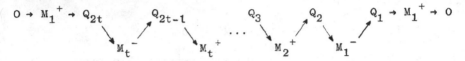

a.) has 10 indecomposable non-simple representations, but $P_i' \oplus P_j'$ gives rise to at most 6 indecomposable lattices, so a.) can not occur.

b.) has infinite representation type, and so this can not occur either.

<u>This proves the claim.</u>

Thus the graph of Λ_o is the disjoint union of t graphs of type A_3 and Λ_o has exactly 2 t non-isomorphic indecomposable non-projective lattices. Since $B \subset \Lambda_o$ and both orders have the same number of indecomposable lattices, they coincide.

(6) Note, (5) implies that for every central primitive idempotent e of $K \otimes_R B$, Be is a hereditary order.

(7) <u>Definitions:</u> (i) The Brauertree T_K of B is a graph which has as vertices the simple $K \otimes_R$ B-module U_j, $1 \le j \le \tau$, and U_j is joined by an edge to $U_j{}'$ if there exists an indecomposable projective B-lattice P_i with $K \otimes_R P_i \simeq U_i \oplus U_j'$. Because of the injectivity of the Cartan-map, T_K is indeed a tree..

(ii) A vertex j in T_K has <u>order ν_i</u> if there are ν_j edges meeting in j.

(iii) U_j is said to have <u>class number</u> γ_j, if there are exactly γ_j non-isomorphic indecomposable B-lattices which span U_j.

(8) For a vertex j the following are equivalent:

(i) j has order ν,

(ii) U_j has class number ν,

(iii) Be_j, where e_j is the central primitive idempotent corresponding to U_j, is a hereditary R-order with ν non-isomorphic indecomposable lattices.

Proof: j has order ν if and only if there are non-isomorphic indecomposable B-lattices $P_1,\ldots,P_\nu$ with $K \otimes_R P_i \simeq U_j \oplus U_i$. Since $P_i e_j \not\simeq P_k e_j$ for $i \neq k$, $B e_j$ is a hereditary order with ν lattices, and conversely.

Let us pause to see which parts of the theorems we have proved: Theorem II, (iii) is (6) ; Theorem II, (iv) is (5) ; Theorem II, (v) and (vi) are (2), (7), (8). Moreover, Theorem IV follows from (4). The remaining proofs will be given in the next section.

§ 3 Miscellaneous consequences.

Again we assume that R is the ring of integers in an unramified p-adic number field K and $\mathfrak{k} = R/pR$.

Proposition 2: Let B be a block of RG such that for every irreducible B-lattice M, $\mathrm{End}_B(M)$ is the maximal R-order in $\mathrm{End}_{K\otimes_R B}(K\otimes_R M)$. Let $S_1,\dots,S_t$ be the different simple B-modules. Then $\mathrm{End}_{RG}(S_i) = \mathfrak{f}_o$ is the same for all $1\le i\le t$ and $|\mathfrak{f}_o:\mathfrak{k}|$ bounds the Schur indices of all simple KB-modules.

Remark: The above hypotheses are obviously satisfied if for every central primitive idempotent ϵ of $K\otimes_R B$, $B\epsilon$ is a hereditary order. However, these are not the only cases.

Proof of Proposition 2: We first have to fix some notation:

$$K\otimes_R B = \prod_{j=1}^{\tau} (D_j)_{n_j} ,$$

L_j = centre (D_j), ν_j = Schur index of U_j, where U_j is a simple $(D_j)_{n_j}$-module, Ω_j with radical $\omega_j\Omega_j$ the maximal R-order in D_j and Δ_j with radical $\delta_j\Delta_j$ the maximal R-order in L_j. Moreover, we denote by ϵ_j the central primitive idempotents in $K\otimes_R B$, $1\le j\le\tau$. P_i, $1\le i\le t$ denotes the indecomposable projective B-modules. For the sake of simplicity we assume that B is basic. (Moreover, $\Gamma_j = Be_j$ are the hereditary orders, $1\le j\le\tau$.)

Claim: $P_i\epsilon_j = M_{ij}$ is an indecomposable projective $B\epsilon_j = \Gamma_j$-lattice and

$$\mathrm{End}_B(P_i/\mathrm{rad}\,P_i) \simeq \mathrm{End}_{\Gamma_j}(M_{ij}/\mathrm{rad}_{\Gamma_j}(M_{ij})) \simeq \Omega_j/\mathrm{rad}\,\Omega_j .$$

Proof of the claim: Since P_i is indecomposable, $P_i\epsilon_j = M_{ij}$ is an indecomposable projective Γ_j-lattice. Moreover, $P_i/\mathrm{rad}\,BP_i \simeq M_{ij}/\mathrm{rad}\Gamma_j M_{ij}$, since $\mathrm{rad}\,B\epsilon_j = \mathrm{rad}\,\Gamma_j$. To see this, we observe that $\mathrm{rad}\,B\epsilon_j$ is a

Γ_j-ideal which is nilpotent modulo π, and so $\mathrm{rad}\,\Gamma_j \supset \mathrm{rad}\, B\epsilon_j$. On the other hand if $P_{i'} \not\simeq P_i$ then $P_i\epsilon_j \not\simeq P_{i'}\epsilon_j$ by a projective cover argument and so $\Gamma_j/\mathrm{rad}\,\Gamma_j \simeq B\epsilon_j/\mathrm{rad}\,B\epsilon_j$; thus $\mathrm{rad}\,B\epsilon_j = \mathrm{rad}\,\Gamma_j$. So $P_i/\mathrm{rad}\,P_i \underset{B}{\simeq} M_{ij}/\mathrm{rad}_{\Gamma_j}(M_{ij})$; whence $\mathrm{End}_B(P_i/\mathrm{rad}\,P_i) \simeq \mathrm{End}_{\Gamma_j}(M_{ij}/\mathrm{rad}_{\Gamma_i}(M_{ij}))$. Moreover, each simple Γ_j-module has endomorphism ring $\Omega_j/\mathrm{rad}\,\Omega_j$. <u>This proves the claim.</u>

Since B is a block; i.e. it is a two-sidedly indecomposable RG-module,

(i) for every pair ϵ_j, $\epsilon_{j'}$ there exists a chain of projective B-lattices $P_{i_1},\dots,P_{i_\sigma}$ and of idempotents $\epsilon_{j_1},\dots,\epsilon_{j_{\sigma-1}}$ such that $\epsilon_j P_{i_1} \neq 0$, $\epsilon_{j_1} P_{i_1} \neq 0$ and $\epsilon_{j_1} P_{i_2} \neq 0 \dots \epsilon_{j_{\sigma-1}} P_{i_\sigma} \neq 0$, $\epsilon_{j'} P_{i_\sigma} \neq 0$,

(ii) for every pair P_i, $P_{i'}$ there exists similarly a chain of idempotents and projective lattices linking P_i to $P_{i'}$.

Thus we have shown:

(*) $\Omega_j/\mathrm{rad}\,\Omega_j \simeq \mathfrak{f}_o$ is the same for every $1 \leq j \leq \tau$,

$\mathrm{End}_B(P_i/\mathrm{rad}\,P_i) \simeq \mathfrak{f}_o$ is the same for every $1 \leq j \leq \tau$.

If now ν is the Schur index of a simple module U_j, then

$$\nu = |\Omega_j/\mathrm{rad}\,\Omega_j : \Delta_j/\delta_j\Delta_j| \leq |\mathfrak{f}_o : \mathfrak{k}| .$$

<u>This completes the proof of Proposition 2.</u>

<u>Corollary 2:</u> Under the hypotheses of Proposition 2 assume that B contains an absolutely simple module S_o - e.g. if B is the principal block - then all simple modules in B are absolutely simple, and in particular, all Schur indices are one.

Let now B be a block of RG of defect one. Then for every simple B-module S, $\mathrm{End}_B(S) = \mathfrak{f}_o$ is a finite extension of degree f over $\mathfrak{k}$. We now choose R_o to be an unramified extension of R with $R_o/pR_o = \mathfrak{f}_o$. In addition we keep the notation of the proof of Prop.2.

Proposition 3: (i) $R_o \otimes_R P_i \simeq \bigoplus_{k=1}^{f} P_{i_k}$, and $P_{i_k} \not\simeq P_{i_l}$ for $k \neq l$.

(ii) If $R_o \otimes_R B = \prod_{\sigma=1}^{s} B_\sigma$, then the union of the Brauertrees T_σ of B_σ forms an f-fold covering of the Brauertree T_K of B.

(iii) If $f = 1$, then the Brauertree of B coincides with the ordinary Brauertree of B over a splitting field. Moreover, Ω_j, $1 \leq j \leq \tau$ is commutative, and $\Omega_j = R$ except possibly for one $j = j_o$, corresponding to the exceptional "character"; i.e. corresponding to a simple KB-module U_{j_o} which splits into m non-isomorphic simple modules over a splitting field. Ω_{j_o} is then an extension of R of ramification index m and residue class degree 1. The vertex j_o will be called the exceptional vertex and m its multiplicity.

Proof: Since $\mathfrak{k}_o \otimes_{\mathfrak{k}} \mathfrak{k}_o \simeq \mathfrak{k}_o^{(f)}$, we conclude by the method of lifting idempotents, that $R_o \otimes_R P_i$ decomposes into f non-isomorphic indecomposable R_oB-lattices; whence the statement of (i), (ii) follows now immediately.

(iii): If $f = 1$, then $R/\mathrm{rad}\, R = \mathfrak{k}$ is a splitting field for $B/\pi B$, and since the ordinary Brauertree is defined in terms of modular splitting fields, we conclude that the Brauertree T_K coincides with the ordinary Brauertree. Moreover, from the proof of Proposition 2 it follows that all Ω_i must be totally ramified, and so there can not be any skewfields involved.

Moreover, the ordinary Brauertree can also be defined over a splitting field of characteristic zero. Thus we conclude that except for $j = j_o$ the Ω_j can not split by any groundring extension; i.e. Ω_j is also unramified; i.e. $\Omega_j = R$ for $j \neq j_o$. As for j_o, the multplicity of the exceptional vertex indicates the number of algebraic conjugates into which U_{j_o} breaks up in a splitting field. Thus Ω_{j_o}

is totally ramified of ramification index m .
This completes the proof of Proposition 3 .

Auslander-Reiten Sequences. Let B be a block of RG of defect one. Then there exists an irreducible B-lattice M_o such that M_o/pM_o is simple [M, 0.2]. Thus $\mathrm{Ext}^1_B(M_o,\Omega(M_o)) \simeq R_o/pR_o$, where $\Omega(-)$ denotes the first syzygy and $R_o = \mathrm{End}_B(M_o)$. Since R_o is unramified, $\mathrm{Ext}^1_B(M_o,\Omega(M_o'))$ is simple as R_o-module, and so the Auslander-Reiten sequence of M is its projective cover sequence. Now Auslander-Reiten sequences behave properly with respect to taking syzygies. Since every indecomposable B-lattice occurs in a projective resolution of M_o ,the statement of Theorem IV follows.

REFERENCES

[A] Alperin, J.L.: Resolution for finite groups. Finite groups. Proceedings of the Taniguchi Symposium: Division Mathematics, No.1 (1976)

[AG] Auslander, M., O.Goldman: Maximal Orders. Trans.Am.Math.Soc. 97 (1960), 1-24.

[B] Brauer, R.: Investigations on group characters. Ann. of Math. 42 (1941), 936-958

[B1] Brumer, A.: Structure of hereditary orders. Bull.Am.Math.Soc. 69 (1963), 721-729

[G] Green, J.A.: A transfer theorem for modular representations. J.Algebra 1 (1964), 73-84

[G1] Green, J.A.: Vorlesungen über modulare Darstellungstheorie endlicher Gruppen. Vorlesungen aus dem Math.Institut Giessen, Heft 2 (1974)

[H] Harada, M.: Structure of hereditary orders over local rings. J.of Mathematics, Osaka City Univ. 14 (1963), 1-22

[M1] Michler, G.O.: Green correspondence between blocks with cyclic defect groups I. J.Algebra

[M] Michler, G.O.: Green correspondence between blocks with cyclic defect groups II. Representations of Algebras, Springer Lecture Notes in Math. 488 (1975), 210-235

[Pu] Pu, M.L.: Integral representations of non-abelian groups of order pq. Mich.Math.J. 12 (1965), 231-246

[R] Reiner, I.: Relations between integral and modular representations. Mich.Math.J. 13 (1966), 357-372

[Ro] Roggenkamp, K.W.: Lattices over orders II. Springer Lecture Notes in Mathematics 142 (1970)

[R1] Roggenkamp,K.W.: Representation theory of finite groups. Séminaire de Mathématiques Supérieures,Été 1979 Les Presses de l'Université de Montréal to appear

[RR] Roggenkamp,K.W., C.M.Ringel: Diagrammatic methods in the representation theory of orders. J.of Algebra, 60(1979), 11-42.

[RS] Roggenkamp,K.W., J.Schmidt: Almost split sequences for integral group rings and orders. Com.in Algebra 4 (10),(1976), 893-917

K.W.Roggenkamp
Mathematisches Institut B
Universität Stuttgart
Paffenwaldring 57
D-7000 Stuttgart-80

A CLASS OF SELF-INJECTIVE ALGEBRAS OF FINITE REPRESENTATION TYPE

Eberhard Scherzler and Josef Waschbüsch

Recently P.Gabriel and Ch.Riedtmann (9) have described the structure of the finite dimensional self-injective algebras over algebraically closed fields, which are stably equivalent to self-injective Nakayama-algebras (= generalized uniserial algebras). In particular they show there that these algebras are two-serial in the sense that for each indecomposable projective module P, the subfactor module rad P/soc P is a direct sum of at most two uniserial modules. Conversely each two-serial symmetric algebra of finite representation type is stably equivalent to a symmetric Nakayama-algebra (7), (9), (12), but the corresponding statement is no longer true for self-injective algebras.

The purpose of this article is to determine the Morita-equivalence classes and the stable equivalence classes of two-serial self-injective algebras over algebraically closed fields which are of finite representation type. We describe full invariants for these classes.

The results

Throughout this paper let K be an algebraically closed field; all algebras should be K-algebras of finite vector space dimension. We recall that a Brauer-tree is a finite connected tree endowed with cyclic orderings on each set of edges converging at some common vertex.

The invariants for the Morita-equivalence classes of two-serial symmetric algebras of finite representation type are known to be a Brauer-tree G, one exceptional vertex S of G and the multiplicity $f \in \mathbb{N}$ of S. (For details see (7), (10), (11) or (9).)

For the self-injective algebras we need an additional invariant $n \in \mathbb{N}$ which gives us the orbit lengths of the Nakayama permutation. Also we have to extent f to a map from the vertices and the edges of G to the nonnegative integers in order to describe the algebras which are not stably equivalent to self-injective Nakayama-algebras.

Thus we define: A triple $T = (G,f,n)$, where G is a Brauer-tree, f a map from G to the nonnegative integers and n a natural number, is said to be a <u>two-serial</u> <u>invariant</u> <u>system</u> (for short: 2si-system) if the following conditions are satisfied:

1) $f(S) \neq 1$ and $f(\beta) \neq 1$ for at most one vertex S and one edge β of G.
2) If $f(S) \geq 1$ for all vertices S, then $f(\beta) = 1$ for all edges β, and f(S) and n are relatively prime for all vertices S.
3) If $f(S_o) = 0$ for a vertex S_o, then $f(\beta_o) = 0$ for an edge β_o having S_o as one of its endpoints, and n is even, $n = 2m$.
4) If G consists of just one vertex S, then $f(S) = 1 = n$.

Further we call two 2si-systems T and $T' = (G',f',n')$ <u>equivalent</u>, if $n = n'$ and if there exists an isomorphism $g : G \xrightarrow{\sim} G'$ of Brauer-trees such that $f = f'g$. Let $|G|$ be the number of edges in G and $|f| := \prod_{\text{vertices } S} f(S)$ (which equals the value of f at the exceptional vertex if one exists).

To each connected (= two-sided indecomposable) two-serial self-injective algebra A of finite representation type we attach a 2si-system T(A). T(A) describes how A is built up out of "elementary" algebras A_e^h and $B_{m,k}$, which are defined below (see (1.1)). Each vertex S of G corresponds to an elementary algebra e_sAe_s ($e_s^2 = e_s \in A$) whose type is given by n,f(S) and the position of S in G. Two vertices S and S' are neighbours if and only if $e_s \cdot e_{s'} \neq 0$.

Reversely we construct to each 2si-system T a two-serial self-injective algebra A(T) of finite representation type. A(T) is given by a quiver Q_T and relations I_T, whose forms are uniquely determined by T. Thus A(T) is described by a K-base and its multiplication table.

For the composition of the above mappings we get (2.13): A(T(A)) is Morita-equivalent to A and T(A(T)) is equivalent to T. This gives us

<u>Theorem 1</u>:

The maps $A \longmapsto T(A)$ and $T \longmapsto A(T)$ induce a bijection between the Morita-equivalence classes of connected two-serial self-injective algebras of finite representation type and the equivalence classes of 2si-systems.

We point out that theorem 1 classifies the two-serial self-injective algebras of finite representation type which split over their ground field K without any further restriction on K.

The "Brauer-quiver-algebras" (i.e. the self-injective algebras which are stably equivalent to self-injective Nakayama-algebras classified by P.Gabriel and Ch.Riedtmann in (9)) are exactly those algebras for which the corresponding 2si-system $T = (G,f,n)$ satisfies $|f| \geq 1$. Using this result we show, that the remaining algebras (i.e. the algebras A with $T(A) = (G,f,n)$ satisfying $|f| = 0$) are stably equivalent to the algebras $B_{m,k}$ in (1.1), whose indecomposable modules are easy to describe (see (1.2)):

<u>Theorem 2</u>:

Let $T = (G,f,n)$ be a 2si-system and A the algebra A(T).

1) (Gabriel, Riedtmann) If $|f| \geq 1$, then A is stably equivalent to the Nakayama-algebra A_e^h with $e = n \cdot |G|$ and $h = |f| \cdot |G|$.
2) If $|f| = 0$, then A is stably equivalent to the algebra $B_{m,k}$ with $m = n/2$ and $k = |G| + 1$.

Thus the category of left A-modules is uniquely determined up to stable equivalence by the three integers n, |f| and |G|.

P.Gabriel and Ch.Riedtmann (9) show conversely that each connected algebra which is stably equivalent to an algebra A_e^h with $h \geq 2$ is Morita-equivalent to an algebra A(T) for some 2si-system $T = (G,f,n)$ with $|f| \geq 1$. Using their methods we proof the corresponding

Theorem 3:

Each connected algebra which is stably equivalent to an algebra $B_{m,k}$, is Morita-equivalent to an algebra A(T) for some 2si-system $T = (G,f,n)$ with $|f| = 0$. Moreover the algebras A_e^h and $B_{m,k}$ represent the distinct stable equivalence classes of connected two-serial self-injective algebras of finite representation type.

Theorem 2 and Theorem 3 are also valid for arbitrary ground fields K if we replace "stable equivalence" by "K-linear stable equivalence".

In the first section we fix our terminology and describe the algebras A_e^h and $B_{m,k}$, in the second section we prove Theorem 1; Theorem 2 and Theorem 3 are proved in section 3.

1. Basic definitions

We adapt most of the basic definitions and most of the terminology in (9): A quiver Q consists in vertices and in arrows connecting these vertices together. An arrow $\alpha = i \longrightarrow j$ starts at the vertex i and ends in the vertex j. A sequence $w = (\alpha_s)_{s=1,2,\ldots n}$ of arrows is called a (directed) path of length n from i to j if α_1 starts at i, α_n ends in j and the starting point of α_{s+1} equals the end point of α_s for $s = 1,2,\ldots,n-1$; the latter vertices are called the inner points of w. For a path w from i to j and a path v from j to k we define a product vw as the whole path from i to k following first w and then v. The vertices i are considered as paths of length zero from i to i. Thus we have $w = (\alpha_s)_{s=1,2,\dots,n} = \alpha_n\alpha_{n-1}\dots\alpha_2\alpha_1$; $jw = wi = w$ for any path w from i to j and $i^2 = i\cdot i = i$. Two paths w and w' from i to j are <u>parallel</u>; a path from i to i is a <u>closed</u> path.

The free K-vector space K(Q) over the set of all paths in Q is an algebra, called the <u>quiver algebra</u> of Q over K, with the product induced by the product of paths, where $w\cdot w' = 0$ for paths w and w' which are not composable. An ideal I of K(Q) is called an <u>ideal of</u> Q, if the algebra K(Q)/I is finite dimensional over K. In this case we call (Q,I) a <u>bounden quiver</u> and K(Q,I) := K(Q)/I a <u>bounden quiver algebra</u>.

Let $\mathrm{mod}_K(Q)$ be the category of finite dimensional K-representations of Q; a representation V of Q consists in vector spaces V_i for each vertex i and in linear maps $V_\alpha : V_i \longrightarrow V_j$ for each arrow $\alpha = i \to j$ of Q; we write $V = (V_i, V_\alpha; i,\alpha \in Q)$. Each path w from i to j defines a linear map V_w: $V_i \longrightarrow W_j$ in an obvious way. Thus we get maps V_x: $V_i \longrightarrow V_j$ for each $x = jxi \in K(Q)$. If (Q,I) is a bounden quiver we denote by $\mathrm{mod}_K(Q,I)$ the full subcategory of $\mathrm{mod}_K(Q)$ given by all representations V of Q with the property $V_x = 0$ for all $x \in \bigcup_{i,j} jIi$. Thus $\mathrm{mod}_K(Q,I)$ is equivalent to the category of all finitely generated left K(Q,I)-modules.

Certain bounden quivers arising in this paper have additional structure given by a cyclic ordering of the arrows. We call such a bounden quiver S = (Q,I) a <u>track</u> if the arrows of Q form a closed path in their cyclic ordering. If Q has exactly n arrows we indicate their cyclic ordering by enumerating the arrows of Q by the elements of $\mathbb{Z}/n\mathbb{Z}$ (which bear a canonical cyclic ordering). The track S is said to be <u>nondegenerate</u> if $\alpha_{t+1}\alpha_t \notin I$ for all $t \in \mathbb{Z}/n\mathbb{Z}$. A vertex i of Q is called an r-fold point of S if r different arrows start at i.

Examples of tracks are the bounden quivers Z_e^h, $e,h \in \mathbb{N}$, in (<u>9</u>): $Z_e^h = (Z_e, I_e^h)$ where Z_e is the quiver in figure 1 with the natural ordering of the arrows and where I_e^h is the ideal generated by all paths of length h + 1.

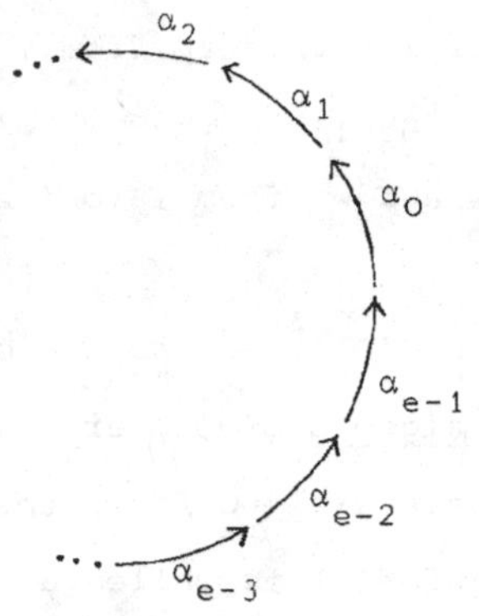

Figure 1

A further series of tracks $S_{m,k}$, $m,k \in \mathbb{N}$, $k \geq 2$, is given by quivers $Q_{m,k}$ in figure 2 with the indicated cyclic ordering of the 2mk arrows α_j, $j \in \mathbb{Z}/2mk\mathbb{Z}$ and by the ideals $I_{m,k}$ generated by all

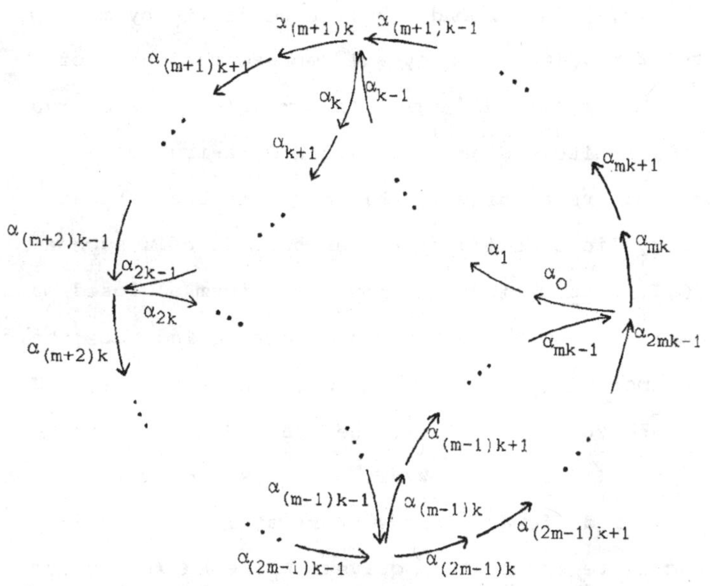

Figure 2

elements of the following forms:

(i) paths of length $k+1$

(ii) products of arrows $\alpha_{(i+m)k}\alpha_{ik-1}$ which are composable but not consecutive arrows in the cyclic ordering

(iii) differences of parallel paths of length k.

The corresponding bounden quiver algebras

(1.1) $A_e^h := K(Z_e, I_e^h)$; $e,h \in \mathbb{N}$

$B_{m,k} := K(Q_{m,k}, I_{m,k})$, $m,k \in \mathbb{N}$, $k \geq 2$

are connected generalized uniserial respectively two-serial self-injective algebras. A_e^h has e projective uniserial modules of length $h+1$. The indecomposable projective modules of $B_{m,k}$ correspond to the subquivers

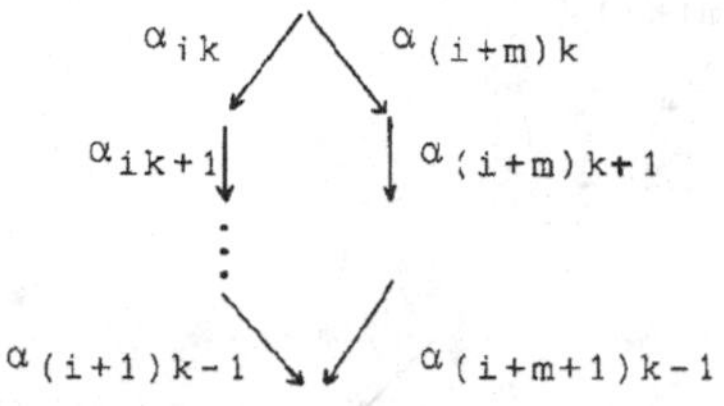

and

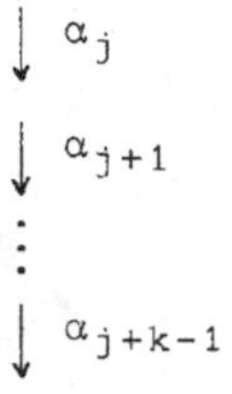

of $Q_{m,k}$ for $i = 0,1,\dots,m-1$, $1 \leq j \leq 2mk-1$, $j \not\equiv 0$ modulo k (put K at each vertex and id_K at each arrow!).

With the methods of P.W. Donovan and M.R.Freislich (6) (see also (2.6) below) one obtains a list $\mathcal{U}_{m,k}$ of the indecomposable non projective $B_{m,k}$-modules (here we use the notation $\alpha_j = a_j \rightarrow a_{j+1}$):

(1.2) a) $L_{j,s}$, $0 \leq j \leq 2mk-1$, $0 \leq s \leq k-1$, given by the subquivers

$a_j \rightarrow a_{j+1} \rightarrow \dots \rightarrow a_{j+s}$ (caution! $L_{ik,0} = L_{(i+m)k,0}$)

b) $V_{i,r,s}$, $0 \leq i \leq m-1$, $1 \leq r,s \leq k-1$, given by the subquivers

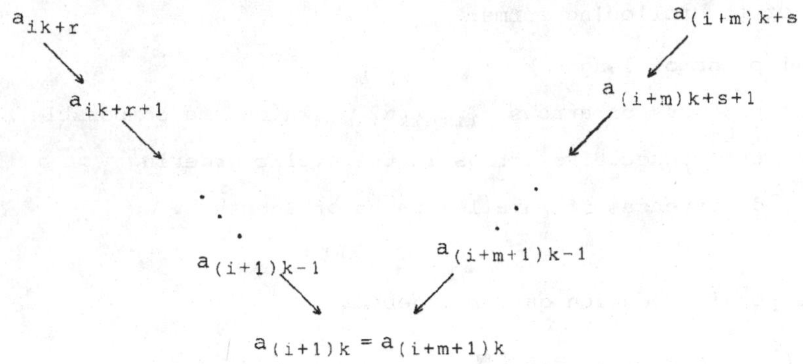

c) $\Lambda_{i,r,s}$, $0 \leq i \leq m-1$, $1 \leq r,s \leq k-1$, given by the quivers

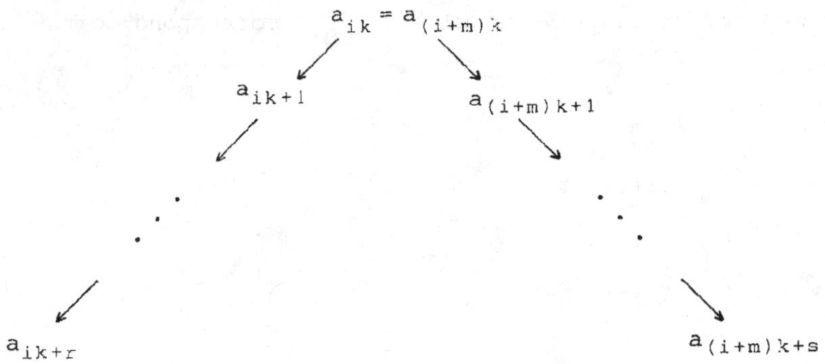

In a bounden quiver (Q,I) every closed path w which does not contain an arrow twice furnishes a track $S_w = (Q_w, I_w)$ where Q_w is the underlying quiver of w, I_w is the ideal $I \cap K(Q_w)$ of Q_w and where the cyclic ordering of the arrows of Q_w is given by w. Also if $S = (Q,I)$ is a track with cyclic ordering on the arrows given by $\alpha_r = a_r \longrightarrow a_{r+1}$, $r \in \mathbb{Z}/\mathbb{Z}n$, the lengths of the paths $w \in K(Q) \setminus I$ are bounded. Thus each arrow α_r can be extended to a maximal path $w(\alpha_r, S) = \alpha_{r+h_r} \cdots \alpha_{r+1}\alpha_r$ lying not in I; we also denote this path by $w(a_r, S)$. If i is a t-fold point of S there are t different paths of this kind starting at i. For $S = Z_e^h$ we have $w(\alpha_r, S) = \alpha_{r+h} \dots \alpha_{r+1}\alpha_r$; for $S = S_{m,k}$ the two maximal paths starting at the double point $a_{ik} = a_{(i+m)k}$ are $w(\alpha_{ik}, S)$ and $w(\alpha_{(i+m)k}, S)$ which are parallel.

2. The proof of Theorem 1

We now look at the (ordinary) quiver of a connected two-serial self-injective algebra A of finite representation type to prove Theorem 1. Thus we may assume that A is basic. Let N be the radical of A and $1 = e_1 + e_2 + \dots e_q$ a decomposition of the unit element of A in a sum of orthogonal primitive idempotents $e_i \in A$. With π we denote the Nakayama-permutation of A which is defined by soc $Ae_i \cong Ae_{\pi i}/Ne_{\pi i}$, $i = 1, 2, \dots, q$.

The quiver $\bar{Q}_A$ of A has $1, 2, \dots, q$ as vertices and the arrows of $\bar{Q}_A$ are in one-one-correspondence with the elements α_{ji} of a Cartan-base of N/N^2 (i.e. $\alpha_{ji} \in e_j(N/N^2)e_i$) where we consider α_{ji} as an arrow from i to j. To get a "higher symmetry" (see Lemma (2.2) below) we add arrows $\beta_i : i \to \pi i$ for each i with Ae_i uniserial and extend in this way $\bar{Q}_A$ to a quiver Q_A. If we choose representatives $a_{ji} \in N$ of the α_{ji}'s and for every β_i an element $0 \neq b_i \in$ soc Ae_i, we can extend the map $i \mapsto e_i$, $\alpha_{ij} \mapsto a_{ij}$, $\beta_i \mapsto b_i$ from Q_A into A to an algebra-epimorphism from $K(Q_A)$ onto A. Thus we have

(2.1) A is isomorphic to a bounden quiver algebra $K(Q_A, I_A)$ for a properly chosen ideal I_A of Q_A.

We remark, that the ideal I_A in (2.1) depends on the choice of $a_{ij} \in N$ representing a Cartan-base of N/N^2 and the elements b_i above. One of the problems will be to show that we can choose these elements in such a way that I_A has a "canonical" form.

From now on we assume that A is not Nakayama. The first lemma describes the neighbourhoods of the vertices in Q_A. The assertion is independent of the special choice of the elements a_{ij}, b_i above.

Lemma 2.2: 1) The Nakayama permutation π can be extended to an automorphism of Q_A.

2) Each arrow $\alpha = i \longrightarrow j$ of Q_A can be extended uniquely to a path $w = \alpha_r \dots \alpha_2\alpha_1$, $\alpha_1 = \alpha$, from i to πi and to a path $v = \alpha'_s \dots \alpha'_2\alpha'_1$, $\alpha'_s = \alpha$, from $\pi^{-1}j$ to j which are maximal in the set of all paths not belonging to I_A.

3) Each vertex i of Q_A is the endpoint of exactly two arrows α_1, α'_1 and the starting point of exactly two arrows α_2, α'_2. After a suitable renumbering we have

a) $\alpha'_2\alpha_1$, $\alpha_2\alpha'_1 \in I_A$

b) $\alpha_2\alpha_1 \notin I_A$ and

$\alpha'_2\alpha'_1 \in I_A$ if and only if Ae_i is uniserial.

Proof: 1) To each arrow $\alpha_{ji} = i \longrightarrow j$ there corresponds an arrow $\alpha_{\pi j,\pi i} = \pi i \longrightarrow \pi j$ given by the Nakayama-automorphism of A.

2) Let $Ne_i/\mathrm{soc}\, Ae_i$ be the direct sum of uniserial modules U_i and V_i ($V_i = 0$ if Ae_i is uniserial) and let α_{ji} be an element of U_i/NU_i. If the k-th factor $N^{k-1}U_i/N^kU_i$ of the composition series $U_i \supset NU_i \supset \dots \supset N^rU_i = 0$ of U_i is isomorphic to Ae_{i_k}/Ne_{i_k}, $k = 1,2,3,\dots,r$ we have a path $w = \alpha_{\pi i,i_r}\alpha_{i_r i_{r-1}} \dots \alpha_{i_3 i_2}\alpha_{i_2 i_1}$, $\alpha_{i_2 i_1} = \alpha_{ji}$, which is not an element of I_A; because U_i is uniserial, w is the unique maximal path as required. The assertion for v follows by the self-duality of A.

3) By assumption on A and K we have $\dim_k Ne_i/N^2e_i$ equals 2 or 1 (the latter holds if and only if Ae_i is uniserial). Thus by the construction of Q_A there are exactly two arrows starting at i and again by the self-duality of A two arrows ending in i.

Thus the neighbourhood of i looks like

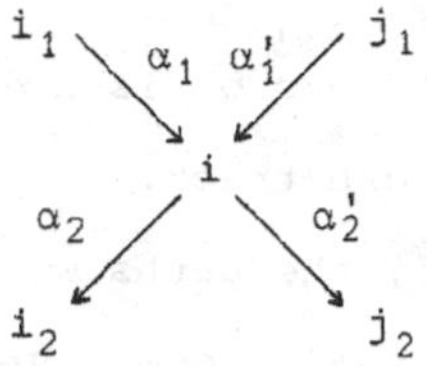

(we may have $i_1 = i_2$ or $j_1 = j_2$ or $i_1 = i_2 = i \ldots$)

If both $\alpha_2\alpha_1$ and $\alpha_2'\alpha_1$ are elements of I_A, we know by 2) that the length of Ae_i equals 2. But then A is Nakayama (see for example (4), Lemma 4.3). Thus we may assume that $\alpha_1\alpha_2 \notin I_A$ and this implies that $\alpha_2'\alpha_1 \in I_A$ by the definition of two-serial. The same arguments now show that $\alpha_2'\alpha_1' \notin I_A$ if Ae_i is not uniserial and $\alpha_2'\alpha_1' = \beta_i\beta_{\pi^{-1}i} \in I_A$ if Ae_i is uniserial.

Corollary 2.3: Each arrow $\alpha \in \bar{Q}_A \subset Q_A$ determines a unique closed path $w_\alpha = \alpha_{t-1}\alpha_{t-2} \cdots \alpha_1\alpha_0$, $\alpha_0 = \alpha$, with the properties

a) $\alpha_i \neq \alpha_j$ für $i \neq j$, $i,j = 0,1,\ldots,t-1$

b) $\alpha_{i+1}\alpha_i \notin I_A$ for all $i \in Z/tZ$.

Thus α determines a unique nondegenerate track $S_\alpha = S_{w_\alpha}$. If $S = (Q_S, I_S)$ is a nondegenerate track with $Q_S \subseteq Q_A$, $I_S = I_A \cap K(Q_S)$ we have $S = S_\alpha$ for each arrow $\alpha \in Q_S$.

Proof: The assertion follows immediately from lemma (2.2) using the fact that the Nakayama-permutation π has finite order.

For $\beta = \beta_i$ let S_β be the (degenerate) track given by the closed path $w_\beta = \beta_{\pi^{n-1}i} \cdots \beta_{\pi i}\beta_i$ where $n = \min(t\colon \pi^t i = i)$. With G_A we denote the set of all tracks $S_\alpha, \alpha \in Q_A$ (by definition we have $S_\alpha = S_{\alpha'}$ if α' is an arrow in S_α). In the following we refer to the elements of G_A as to the tracks of Q_A.

The first proposition tells us how Q_A is built up out of its tracks:

Proposition 2.4: 1) The automorphism π of Q_A maps each track S onto itself.

2) Each vertex of Q_A belongs to exactly two tracks or is a double point of one track. Each arrow belongs to exactly one track.

3) If i belongs to two tracks S and S', $S \neq S'$, the paths w(i,S) and w(i,S') have no common inner points. The set of all common vertices of two tracks is an orbit under the Nakayama-permutation π.

4) There exists a natural number $n = n_A$ such that two tracks meet either in n vertices or not.

5) If $S_1, S_2, \dots, S_t$ are all tracks meeting a given track S nontrivially - say in the vertices $i_{\ell p} \in S \cap S_p$ $\ell = 1,2,\dots,n$, $p = 1,2,\dots,t$ - then after a suitable renumbering of the S_i and of the $i_{\ell p}$ the sequence

$$i_{11}, i_{12}, \dots, i_{1t}, i_{21}, i_{22}, \dots, i_{nt}$$

is the natural ordering of these vertices given by the cyclic ordering of the arrows of S starting at these vertices.

Proof: 1) and 2) follow immediately from lemma (2.2); 3) holds by the "Jans-condition" (see (8), Prop. 9.2).

4) The track S_1 may meet the track S_2 in n vertices $i_0, i_1, \dots, i_{n-1}$ which are numbered according 2) such that $i_p = \pi^p i_0$, $p = 0,1,\dots,n-1$. If the path $w(i_0, S_1)$ meets a track S_3 in r vertices, then each of the n paths $w(i_p, S_1)$ meets S_3 in r vertices in the "same positions". Thus the intersection of S_1 and S_3 consists of at least n vertices. By symmetry S_1 meets any track S with $S_1 \cap S \neq \emptyset$ in exactly n vertices. If S and S' are any two tracks of Q_A, there is a sequence $S = S_1, S_2, \dots, S_r = S'$ such that $S_i \cap S_{i+1} \neq \emptyset$ for $i = 1,2,\dots,r-1$, because A (and so Q_A) is connected. Therefore n is an invariant of A.

5) Because of 1) at least one vertex of S_p is an inner point of $w(i_{\ell 1}, S)$ for $p = 1,2,\dots,t$. Thus the assertion follows by symmetry.

In order to get more information about the tracks of Q_A we consider now the indecomposable A-modules which are classified by P.W. Donovan and M.R. Freislich (6). We describe these modules in our terminology:

If i and j are two vertices of a track $S = (Q_S, I_S)$ we denote by $w(i,j,S)$ a path $w = \alpha_r\alpha_{r-1}\dots\alpha_2\alpha_1$ from i to j such that α_i and α_{i+1} are consecutive arrows in the cyclic ordering of the arrows of Q_S. $w(i,j,S)$ may run more than one time trough Q_S. A sequence

$$F = w(i_1,j_1,S_1), w(i_2,j_2,S_2),\dots,w(i_r,j_r,S_r)$$

of such paths in Q_A (i.e. $S \in G_A$ for $k = 1,2,\dots,r$) is called a walk in Q_A if the following conditions are satisfied:

(2.5) (i) Either $i_k = i_{k+1}$ or $j_k = j_{k+1}$ alternating with $k = 1,2,\dots,r-1$

(ii) $w(i_k,j_k,S_k)$ is a proper subpath of $w(i_k,S_k)$ for all k and has length ≥ 1 for $k = 2,3,\dots,r$.

(iii) $$w(i_k,j_k,S_k) \cap w(i_{k+1},j_{k+1},S_{k+1}) = \begin{cases} i_k & \text{if } i_k = i_{k+1} \\ j_k & \text{if } j_k = j_{k+1} \end{cases}$$

If we denote by ℓ_k the lenth of $w(i_k,j_k,S_k)$ then $\ell(F) := 1+\ell_1+\ell_2+\dots+\ell_r$ is called the length of F. Two walks F and F' are composable if they fit together in the sense of (2.5) and we write FF' for the composed walk given by the elements of F followed by the elements of F'. A walk F is called primitive if $F^m := FF\dots F$ (m times) is a walk for all $m \in \mathbb{N}$ but $F \neq F'^d$ for any walk F' and any natural number $d \geq 2$.

By the methods of P.W. Donovan and M.R. Freislich (6) one shows that to each walk F in Q_A there corresponds an indecomposable A-module M(F) of length $\ell(F)$ and to each primitive walk F there corresponds a family of indecomposable modules $M(F,\varphi)$ where φ runs through the set of irreducible automorphisms of K-vector spaces. $M(F,\varphi)$ has length $\ell(F)\cdot\text{rank}\,\varphi$. Each indecomposable A-module M is isomorphic to one of the modules M(F), $M(F,\varphi)$ above.

Because Q_A is a finite quiver we thus get the

Proposition 2.6: For a connected two-serial self-injective algebra A the following assertions are equivalent

(i) A is of finite representation type.

(ii) The lengths of the walks in Q_A are bounded.

(iii) There exists no primitive walk in Q_A.

A walk F of length $n \geq 3$ obviously has two extreme vertices i_1 and i_n and two extreme arrows α_1 and α_{n-1}. Fixing one direction we say that F starts with α_1 at i_1 and ends with α_{n-1} in i_n.

Lemma 2.7: Let $S_1, S_2, \ldots, S_r$ be a sequence of tracks such that $S_p \neq S_{p+1}$ and $S_p \cap S_{p+1} \neq \emptyset$ for $p = 1,2,\ldots,r-1$. Then for each vertex $i \in S_1 \cap S_2$ and for each $\alpha = i \longrightarrow j \in S_2$ or $\alpha = j \longrightarrow i \in S_2$ there exists a walk F of length $\ell(F) \geq r-1$ starting with α at i and ending in a vertex of S_r.

Proof: We denote by $w^{-1}(k,S)$ the path $w(\pi^{-1}k,S)$ and prove the case where $\alpha = i \longrightarrow j$ (in case $\alpha = j \longrightarrow i$ the proof is similar). The path $w(i,S_2)$ has a vertex $j_1 \in S_3$ as an inner point; the path $w^{-1}(j_1,S_3)$ has a vertex $i_2 \in S_4$ as an inner point ... continuing in this way we end in a vertex $i_\ell \in S_r$ (responsable $j_\ell \in S_r$) and the sequence

$$F = w(i,j_1,S_2), w(i_2,j_1,S_3), \ldots, w(i_\ell,j_{\ell-1},S_{r-1})$$

$$(\text{responsable } F = \ldots\ldots\ldots\ldots, w(i_\ell,j_\ell,S_{r-1}))$$

is the desired walk.

We now endow G_A with a Brauer-graph structure considering the elements of G_A as vertices:

1) Link two tracks by an edge if they meet nontrivially.
2) Order the edges $S - S_1, S - S_2, \ldots, S - S_t$ ending in some track S according to Proposition 2.4.,5.

From Lemma 2.7 we get immediately.

Proposition 2.8: G_A is a Brauer-tree.

Proof: If G_A had a cycle, there would be walks of arbitrary large lengths in Q_A by Lemma 2.7.

Another consequence of Lemma 2.7 is

Lemma 2.9: Let $t = t(S)$ be the number of tracks meeting a given track S nontrivially. Then S either is isomorphic to a track Z_{nt}^{at} where $a \in \mathbb{N}$, $(a,n) = 1$ or S is of the form $S_{n/2,t+1}$.

Here $n = n_A$ as in Proposition 2.4,4. And "$S = (Q_S, I_S)$ is of the form $S_{m,k}$", means that S is isomorphic to $S_{m,k}$ as a track except that eventually the Ideal I_S differs from $I_{m,k}$. Instead of $w - w'$ for parallel paths w, w' of length k the corresponding generators of I_S may have the form $cw - w'$ with coefficients $c \in K$. In Proposition 2.12 we will settle this question.

Proof of Lemma 2.9: If S has no double points, it is isomorphic to Z_{nt}^{h} for some $h \in \mathbb{N}$ because $\pi | S$ must be an automorphism of S of the form $\alpha_i \to \alpha_{i+h}$, $i \in \mathbb{Z}/nt\mathbb{Z}$. If S' is a track meeting S in the vertices $\pi^p i_0$, $p = 0,1,2,\ldots,n-1$ we know by Proposition 2.4 1) and 3) that $h = a \cdot t$ for some $a \in \mathbb{N}$ with $(a,n) = 1$. Let now $i_0, i_1, \ldots, i_{m-1}$ $(m \geq 1)$ be the set of all double points of S. Then S has some more vertices; this is clear for $m = 1$ and for $m \geq 2$ we otherwise would have walks of arbitrary large lengths in Q_A: Start with any arrow $\alpha_1 = i_{p_1} \to i_{p_2}$ and find recursively arrows $\alpha_{2k} = i_{p_{2k+1}} \to i_{p_{2k}}$ and $\alpha_{2k+1} = i_{p_{2k+1}} \to i_{p_{2k+2}}$. Then $\alpha_1, \alpha_2, \ldots, \alpha_\ell$ is a walk in Q_A for each $\ell \geq 1$.
Furthermore the paths $w(i_p, S)$ have no inner points which are double points of S. Otherwise we can construct walks of arbitrary large lengths which are of the form

$$w(i_{p_1}, i_{p_2}, S), w(i_{p_3}, i_{p_2}, S), \ldots, w(i_{p_\ell}, i_{p_{\ell+1}}, S).$$

Therefore the double points of S are just one orbit under π and they can be numbered in such a way that $i_p = \pi^p i_0, p = 0,1,\ldots,m-1$. This numbering is in accordance with their natural sequence given by the cyclic ordering of the arrows of S. In each vertex i_p there start two maximal paths $w(i_p,S)$ and $w'(i_p,S)$ which all have the same length $t+1$ because they are conjugate under π. Thus S has exactly $2m(t+1)$ arrows. By Proposition 2.4, 4) we know that $n = 2m$ and it is clear that S is of the form $S_{n/2,t+1}$. We point out that in case S is of the form $S_{m,k}$ one edge S-S' is singled out by the property that $S \cap S'$ consists in those vertices which are the end points of arrows starting at the double points of S.

We call a track S <u>exceptional</u>, if S is isomorphic to Z_{nt}^{at} with $a > 1$ or if S is of the form $S_{m,k}$. In the latter case we also call the edge $S - S'$ (singled out as above) <u>exceptional</u>. Next we will show that G_Λ has at most one exceptional track. To prove this we need the

<u>Lemma 2.10</u>: Let S be an exceptional track and S' another track meeting S nontrivially. Then for each vertex $i \in S \cap S'$ and for each arrow $\alpha = i \longrightarrow j \in S$ or $\alpha = j \longrightarrow i \in S$ there exists a walk F in Q_Λ (indeed $F \subset S$) starting with α at i and ending in a vertex $i' \in S \cap S'$.

<u>Proof</u>: First let $\alpha = i \longrightarrow j$. If S is isomorphic to Z_{nt}^{at}, $a \geq 2$, the path $w(i,S)$ has at least one inner point $i' \in S \cap S'$. Thus $F = w(i,i',S)$ is the desired walk. If S is of the form S_{mk} the path $w(i,S)$ has an inner point i_0 which is a double point of S and there is a second vertex $i' \in S \cap S'$ such that i_0 is an inner point of $w(i',S)$. Then $w(i,i_0,S)$, $w(i',i_0,S)$ is the desired walk. For $\alpha = j \longrightarrow i$ the proof is similar.

<u>Proposition 2.11</u>: Up to at most one track S_0 each track S is isomorphic to Z_{nt}^{t}, $t = t(S)$, while S_0 is isomorphic to Z_{nt}^{at} for a natural

number with $(a,n) = 1$ or S_0 if of the form $S_{n/2,t+1}$.

Proof: Assume that there are two exceptional tracks S and S'. Because A is connected there is a sequence $S = S_1, S_2, \ldots, S_r = S'$ of tracks such that $S_p \neq S_{p+1}$, $S_p \cap S_{p+1} \neq \emptyset$ for $p = 1,2,\ldots,r-1$. But then Lemma 2.7 and Lemma 2.10 give us walks of arbitrary large lengths in Q_A.

By Proposition 2.4, Lemma 2.9 and Proposition 2.11 Q_A is uniquely determined by the Brauer-tree G_A, the natural number n_A, the form of the tracks and, in case one track S_0 is of the form $S_{m,k}$, by an exceptional edge $S_0 - S'$. Now we show that these dates already determine the algebra A. To this end we show that the ideal I_A in (2.1) can be chosen in a "canonical form" in the sense of

Proposition 2.12: For properly chosen elements a_{ij} and b_i in (2.1) we have

1) $I_{S_0} = I_{m,k}$ if S_0 is a track of the form $S_{m,k}$.
2) The ideal I_A in (2.1) is generated by all elements of the following forms
   a) $w(i,S) + w(i,S')$ for vertices $i \in S \cap S'$, $S \neq S' \in G_A$
   b) $\alpha'\alpha$ for arrows $\alpha \in S$, $\alpha' \in S'$, $S \neq S' \in G_A$
   c) $u \in I_S$ for any track $S \in G_A$.

Proof: 1) This easily can be arranged by replacing some of the a_{ij} which correspond to arrows starting at double points of S by scalar-multiples.
2) In any case I_A is generated by elements

a') $c_i w(i,S) + w(i,S')$ for vertices $i \in S \cap S'$, $S \neq S' \in G_A$, $c_i \in K$

and elements of the form b) and c) above. Using the fact that G_A is a tree we can well order the vertices of G_A in such a way that each track S meets at most one track $S' < S$ nontrivially. Also we can choose the smallest track in this well-ordering to be the exceptional track if one exists. The assertion now follows with the arguments of P. Gabriel and Ch. Riedtmann in the proof of (9), 3.13.

Corollary 2.13: 1) Each connected two-serial self injective algebra A of finite representation type determines a 2si-system $T(A) = (G_A, f_A, n_A)$.

2) Each 2si-system T determines a connected two-serial self-injective algebra A(T) of finite representation type.

3) For the maps $A \mapsto T(A)$ and $T \mapsto A(T)$ we have

a) A(T(A)) is Morita-equivalent to A

b) T(A(T)) is equivalent to T.

Proof: 1) Let A^o be the basic algebra of A and put $G_A := G_{A^o}$, $n_A := n_{A^o}$. Define $f_A : G_A \longrightarrow \{0,1,2,\dots\}$ by $f_A(S) = a$ if S is a track which is isomorphic to $Z^{at}_{n_A t}$, $t = t(S)$, $f(S_o) = 0$ if S_o is of the form $S_{n/2,t+1}$ and $f(\beta) = 1$ for all nonexceptional edges β of G_A, $f(\beta_o) = 0$ for the exceptional edge (if one exists). Obviously $T(A) = (G_A, f_A, n_A)$ is a 2si-system.

2) Reversely a 2si-system $T = (G,f,n)$ determines a quiver Q_T by looking at the vertices of G as subquivers of Q_T, whose type is given by n,f and the number of neighbours in G (Prop.2.11). Also G describes how these subquivers fit together and overlap (Prop.2.4). By Proposition (2.12) T determines an ideal I_T of Q_T such that the algebra $A(T) := K(Q_T, I_T)$ is a connected two-serial self-injective algebra of finite representation type. — The latter holds because the lengths of walks in Q_T are bounded. Trivially we have that T(A(T)) is equivalent to T. That also A(T(A)) is Morita-equivalent to A follows from (2.1) and Proposition 2.12.

3. The proofs of Theorem 2 and Theorem 3.

In order to prove Theorem 2 and Theorem 3 we need some more terminology and some general remarks concerning the stable equivalence for which we refer to (1) or to (9).

If $\mathcal{C}$ is an abelian category, we denote by $\overline{\mathcal{C}}$ the stable category of $\mathcal{C}$ which has the same objects as $\mathcal{C}$ but with sets of morphisms $\overline{\mathcal{C}}(X,Y) := \mathcal{C}(X,Y)/P_{X,Y}$, where $P_{X,Y}$ is the subgroup of all morphisms $f: X \to Y$ which factor through a projective object. Two categories $\mathcal{C}$ and $\mathcal{C}'$ are said to be stably equivalent if the stable categories $\overline{\mathcal{C}}$ and $\overline{\mathcal{C}}'$ are equivalent. Two algebras A and A' are called stably equivalent if the categories Mod A and Mod A' of all left A-modules and of all A'-modules respectively are stably equivalent. If A is of finite representation type, A is stably equivalent to an algebra A' if and only if the categories mod A and mod A' of all finitely generated left A-modules and left A'-modules respectively are stably equivalent. The stable category of mod A is denoted by $\overline{\text{mod}}$ A and a left A-module M considered as an object of $\overline{\text{mod}}$ A is denoted by $\overline{M}$. Also we write $\overline{g}$ for the morphism $g + P_{M,M'} \in \overline{\text{Hom}}(\overline{M},\overline{M}') = \text{Hom}(M,M')/P_{M,M'}$.

$\overline{\text{mod}}$ A inherits from mod A the property that each object is a finite direct sum of indecomposable objects and the map $M \mapsto \overline{M}$ induces a bijection between the types (= isomorphism classes) of indecomposable objects of $\overline{\text{mod}}$ A and the types of indecomposable nonprojective A-modules. – From here on we use the word "module" as an abbreviation for "finitely generated left module".

Another concept we need in this section is that of the Auslander-Reiten-quiver $\Gamma(\mathcal{C})$ of an additive category $\mathcal{C}$. $\Gamma(\mathcal{C})$ has the types [M] of indecomposable objects as vertices and there is an arrow $[X] \to [Y]$ if and only if there exists an irreducible morphism $f: X \to Y$ in $\mathcal{C}$. A morphism $f: X \to Y$ is called irreducible, if it is neither a split monomorphism nor a split epimorphism and if for each factorization $X \xrightarrow{f_1} Z \xrightarrow{f_2} Y$ of f either f_1 is a split monomorphism or f_2 is a split epimorphism. For an algebra A we call $\Gamma_A := \Gamma_{\text{mod } A}$ the <u>Auslander-Reiten-quiver</u> of A and $\bar{\Gamma}_A := \Gamma_{\underline{\text{mod}}\, A}$ the <u>stable Auslander-Reiten-quiver</u> of A. $\bar{\Gamma}_A$ is the full subquiver of Γ_A corresponding to the types of indecomposable nonprojective A-modules.

We now come to the proof of Theorem 2. Part 1) of this Theorem is proved by (9), Theorem 1: To see this, let $T = (G,f,n)$, $|f| \neq 0$, be an 2si-system and let S_0 be a vertex of G with $|f| = f(S_0)$. To G there corresponds a Brauer-quiver Q (9), 1.4. Let Q_1 be an $|f|$-fold covering of Q with exceptional orbit S_0 (9), 1.11. Then Q_1 has exactly $h = |f||G|$ vertices and $|G|$ is a period of Q_1 (9), 1.8. From the construction in (9), 1.9 it follows that mod A(T) is equivalent to $\text{mod}_K^e(\tilde{Q}_1, \tilde{I}_1)$, $e = n|G|$, the category of e-periodic representations of the universal covering $\tilde{Q}_1$ of Q_1 satisfying the ideal $\tilde{I}_1$ (9) 1,6,1.8. Thus by (9), Theorem 1 A(T) is stably equivalent to A_e^h with $e = n|G|$ and $h = |f||G|$.

To prove part 2) of Theorem 2 we show first that in the case $T = (G,f,n)$ with $|f| = 0$, mod A(T) is a subcategory of mod $A(\hat{T})$ for a certain 2si-system $\hat{T} = (\hat{G},\hat{f},n)$ with $|\hat{f}| = 1$. Then we can use the result of P. Gabriel and Ch. Riedtmann above to prove our assertion.

In order to do this we need some general facts about bounden quivers with groups of automorphisms. Thus let (Q,I) be a bounden quiver and Ω a group of automorphisms of (Q,I), which means that each element $\gamma \in \Omega$ is an automorphism of the quiver Q mapping the ideal I onto itself by $w = \alpha_r \dots \alpha_2\alpha_1 \longrightarrow \gamma w = \gamma\alpha_r \dots \gamma\alpha_2\gamma\alpha_1$ and linear extension. Ω determines a quotient quiver Q/Ω and an ideal I/Ω of Q/Ω given by

(3.1) 1) The vertices of Q/Ω are the orbits $\langle i \rangle$ of the vertices i of Q under Ω.

2) The arrows of Q/Ω are the orbits $\langle\alpha\rangle$ of the arrows $\alpha \in Q$

under Ω, where $<\alpha> = <i> \longrightarrow <j>$ for $\alpha = i \longrightarrow j$.

3) The map $i \longmapsto <i>$, $\alpha \longmapsto <\alpha>$ is a quiver-epimorphism from Q onto Q/Ω and maps I onto an ideal I/Ω of Q/Ω.

Any automorphism γ of (Q,I) induces an automorphism of $\mathrm{mod}_K(Q,I)$ given by

(3.2) 1) $V = (V_i, V_\alpha; i,\alpha \in Q) \longrightarrow V^\gamma = (V_i^\gamma, V_\alpha^\gamma; i,\alpha \in Q)$ where $V^\gamma = V_{\gamma^{-1}i}$ and $V_\alpha^\gamma = V_{\gamma^{-1}\alpha}$ for all $i,\alpha \in Q$.

2) $g = (g_i)_{i\in Q} \longrightarrow g^\gamma = (g_i^\gamma)_{i\in Q}$ where $g_i^\gamma = g_{\gamma^{-1}i}$ for all vertices i. of Q.

We denote by $\mathrm{mod}_K^\Omega(Q,I)$ the subcategory of $\mathrm{mod}_K(Q,I)$ given by all objects V and all morphisms g with the properties $V^\gamma = V$, $g^\gamma = g$ for all $\gamma \in \Omega$.

If Ω is generated by just one element γ we denote $\mathrm{mod}_K^\Omega(Q,I)$ also by $\mathrm{mod}_K^\gamma(Q,I)$ and $(Q/\Omega, I/\Omega)$ by $(Q/\gamma, I/\gamma)$.

<u>Proposition 3.3</u>: The categories $\mathrm{mod}_K(Q/\Omega, I/\Omega)$ and $\mathrm{mod}_K^\Omega(Q,I)$ are equivalent.

<u>Proof</u>. Define a functor $\mathcal{E}$: mod $(Q/\Omega, I/\Omega) \longrightarrow \mathrm{mod}_K(Q,I)$ by $\mathcal{E}(V_{<i>}, V_{<\alpha>}; <i>,<\alpha> \in Q/\Omega) = (W_i, W_\alpha; i,\alpha \in Q)$ where $W_i = V_{<i>}$ and $W_\alpha = W_{<\alpha>}$ for all $i,\alpha \in Q$. As well $\mathcal{E}(g_{<i>})_{<i>\in Q/\Omega} = (h_i)_{i\in Q}$ where $h_i = g_{<i>}$ for all vertices i of Q. Then $\mathcal{E}$ is a faithful functor whose image is the category $\mathrm{mod}_K^\Omega(Q,I)$.

Now let $T = (G,f,n)$ be a 2si-system with $f(S_0) = 0$ for the vertex S_0 of G and $f(\beta_0) = 0$ for the edge β_0 of G. To T we construct another 2si-system $\hat{T}$ as follows: Let G' be a further Brauer-tree which is isomorphic to G under an isomorphism $S \longmapsto S'$, $\beta \longmapsto \beta'$ and let $\hat{G} = G\_\_G'$ be the Brauer tree given by the disjoint union of G and G' and by adding an edge between S_0 and S_0' which in the cyclic ordering on the set of edges ending in S_0 (respectively S_0') preceeds β_0 (respectively β_0'). Put

(3.4) $\hat{T} = (\hat{G}, 1, n)$ (here 1 denotes the constant map $1: \hat{G} \to \{0,1,2,\dots\}$ sending all on 1).

The reason for constructing $\hat{T}$ is the

Proposition 3.5: The bounden quiver $(Q_{\hat{T}}, I_{\hat{T}})$ has an involution $\hat{\mu}$ such that the quotient $(Q_{\hat{T}}/\hat{\mu}, I_{\hat{T}}/\hat{\mu})$ is isomorphic to (Q_T, I_T). Especially $\mathrm{mod}_K(Q_T, I_T)$ is equivalent to $\mathrm{mod}_K^{\hat{\mu}}(Q_{\hat{T}}, I_{\hat{T}})$.

Proof. Let $i_0, i_1, \dots, i_{n-1}$ be the vertices of $Q_{\hat{T}}$ which are the elements in the intersection of the tracks S_0 and S_0' of $Q_{\hat{T}}$ and let $i_p = \pi^p i_0$, $p = 0,1,\dots,n-1$, where π is the Nakayama permutation of $A(\hat{T})$. We now define automorphisms $\hat{\sigma}$ and $\hat{\rho}$ of $(Q_{\hat{T}}, I_{\hat{T}})$ by prescribing their action on the tracks:

$$\hat{\sigma} i_p = i_p, \quad \hat{\sigma} S = S', \quad \hat{\sigma} S' = S$$

$$\hat{\rho} i_p = i_{p+1}, \quad \hat{\rho} S = S, \quad \hat{\rho} S' = S'.$$

for all $p \in \mathbb{Z}/n\mathbb{Z}$, $S \in G$ and $S' \in G'$.

The automorphisms $\hat{\sigma}$ and $\hat{\rho}^m$, $m = n/2$, are commuting involutions of $(Q_{\hat{T}}, I_{\hat{T}})$. Therefore $\hat{\mu} := \hat{\sigma}\hat{\rho}^m$ is an involution of $(Q_{\hat{T}}, I_{\hat{T}})$. The images $\overline{S_0}$ and $\overline{S_0'}$ of S_0 and S_0' under the epimorphism $Q_{\hat{T}} \longrightarrow Q_{\hat{T}}/\hat{\mu}$ are equal and are a track of the form $S_{m,t+1}$, $t =$ number of edges in G ending in S_0. The vertices i_p are mapped onto the m double points of $\overline{S_0}$. The remaining tracks S and S' of $Q_{\hat{T}}$ are mapped isomorphically onto tracks $\overline{S} = \overline{S'}$ of $Q_{\hat{T}}/\hat{\mu}$. Thus $Q_{\hat{T}}/\hat{\mu}$ is isomorphic to Q_T by an isomorphism sending $\overline{S} \subseteq Q_{\hat{T}}/\hat{\mu}$ onto $S \subseteq Q_T$ and mapping $I_{\hat{T}}/\hat{\mu}$ onto I_T. The rest now follows from Proposition 3.3.

Proposition 3.5 can be used to determine the stable Auslander-Reiten-quiver $\bar{\Gamma}_{A(T)}$ of A(T). We first describe that of A($\hat{T}$):

We know that the algebra A($\hat{T}$) is stably equivalent to $A^{2k-1}_{n(2k-1)}$, where k is the number of vertices of G (=|G|+1). Since an equivalence induces bijections between the indecomposable objects and between the irreducible morphisms the stable Auslander-Reiten-quivers $\bar{\Gamma}_{A(\hat{T})}$ and $\bar{\Gamma}_{A^{2k-1}_{n(\ell k-1)}}$ are isomorphic. The latter is well known. We describe it in accordance with (9) 2.3 (see also (9) Lemma 2.6):

Let $\vec{u}$ and $\vec{v}$ be the vectors $\vec{u} = (-1,0)$, $\vec{v} = (-\frac{1}{2},\frac{1}{2})$ of $\mathbb{R}^2$ (in the standard coordinate system). Let $\mathfrak{Z}$ be the set of all points $(r,s) := r\vec{u} + s\vec{v} \in \mathbb{R}^2$ for $r,s \in \mathbb{Z}$. $\mathfrak{Z}$ becomes an (infinite) quiver if we connect the points of $\mathfrak{Z}$ by arrows $(r,s) \to (r,s-1)$ and $(r,s) \to (r-1,s+1)$ for all $r,s \in \mathbb{Z}$. The full subquiver of $\mathfrak{Z}$ given by all vertices (r,s) with $-(k-1) \leq s \leq k-1$ is denoted by $\mathfrak{A}_{2k-1}$. Each vertex $(r,s) \in \mathfrak{A}_{2k-1}$ determines two rectangles R(r,s) and R'(r,s) contained in $\mathbb{R}^2$ (see figure 3). R(r,s) is given by the four points (r,s), (r,-(k-1), (r-(k-1)+s,-s) and (r-(k-1)+s,k-1) whereas R'(r,s) is given by the four points (r+ k-1+s,-s), (r+ k-1+s,-(k-1), (r,k-1) and (r,s).

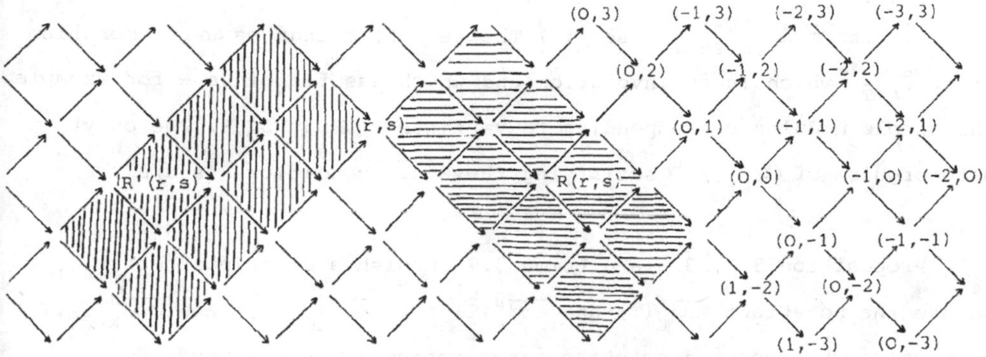

Figure 3 : $\mathfrak{A}_7$

$\mathfrak{A}_{2k-1}$ has two "essential" automorphisms: The reflection σ defined by $\sigma(r,s) = (r+s,-s)$ and the translation τ defined by $\tau(r,s) = (r-1,S)$. Let ρ be the automorphism $\rho = \tau^{2k-1}$: $\rho(r,s) = (r-(2k-1),s)$.

From (9) 2.3, Lemma 2.6 we get the

Proposition 3.6: There exists a surjective map $\bar{M}$ from $\mathfrak{A}_{2k-1}$ onto $\bar{\Gamma}_{A(\hat{T})}$ with the properties

1) $\bar{M}(r,s) = \bar{M}\rho^n(r,s)$ for all $(r,s) \in \mathfrak{A}_{2k-1}$
2) $\bar{M}$ induces an isomorphism from $\mathfrak{A}_{2k-1}/\rho^n$ onto $\bar{\Gamma}_{A(\hat{T})}$.

The automorphisms $\hat{\sigma}$, $\hat{\beta}$ of $(Q\ ,I\ )$ and the automorphisms σ, ρ of $\mathfrak{A}_{2k-1}$ induce permutations of the set $\mathfrak{M}_{A(\hat{T})}$ of types of indecomposable non-projective $A(\hat{T})$-modules.

Lemma 3.7: The permutations $\bar{M}(r,s) \longmapsto \bar{M}\sigma(r,s)$ and $\bar{M}(r,s) \longmapsto \bar{M}\rho(r,s)$ of $\mathfrak{M}_{A(\hat{T})}$ coincide with the permutations $[\bar{X}] \longmapsto [\bar{X}^{\hat{\sigma}}]$ responsable $[\bar{X}] \longmapsto [\bar{X}^{\hat{\beta}}]$. Thus we have $\bar{M}(r,s) = [\bar{X}]$ if and only if $\bar{M}\sigma\rho^m(r,s) = [\bar{X}^{\hat{\mu}}]$, where $m = \frac{n}{2}$.

Proof. For ρ the assertion follows from the connection between $\overline{\text{mod}}\ A(\hat{T})$ and $\overline{\text{mod}}\ A^{2k-1}_{n(2k-1)}$ (see (9) Theorem 1). $\hat{\sigma}$ induces an automorphism σ' of $\bar{\Gamma}_{A(\hat{T})}$ which is an involution and which has fixpoints - for example the simple modules corresponding to vertices $i_p \in S_o \cap S_o'$ -. The only automorphism of $\mathfrak{A}_{2k-1}/\rho^n$ satisfying these two conditions is σ/ρ^n.

Proposition 3.5, 3.6 and Lemma 3.7 furnish a description of $\bar{\Gamma}_{A(T)}$. We use the notations $\overline{\text{mod}}_K(\hat{T})$ and $\overline{\text{mod}}_K^{\hat{\mu}}(\hat{T})$ for $\overline{\text{mod}}_K(Q_{\hat{T}},I_{\hat{T}})$ and $\overline{\text{mod}}_K^{\hat{\mu}}(Q_{\hat{T}},I_{\hat{T}})$ respectively. Also we denote the automorphism $\sigma\rho^m$, $m = \frac{n}{2}$, by μ

Proposition 3.8: 1) Each indecomposable object $\bar{W}$ of $\overline{\text{mod}}_K^{\hat{\mu}}(\hat{T})$ has a decomposition $\bar{W} = \bar{X} \oplus \bar{X}^{\hat{\mu}}$ into indecomposables in $\overline{\text{mod}}_K(\hat{T})$.

2) Each irreducible morphism $\bar{f}$ between indecomposable objects of $\overline{\text{mod}}_K^{\hat{\mu}}(\hat{T})$ has a decomposition $\bar{f} = \bar{g} \oplus \bar{g}^{\hat{\mu}}$ into irreducibles in $\overline{\text{mod}}_K(\hat{T})$.

3) The map $\bar{M}$ in Proposition 3.6 induces an isomorphism from $\mathfrak{A}_{2k-1}/\mu$ onto $\bar{\Gamma}_{A(T)}$.

<u>Proof</u>: 1) If $\bar{W} = \bigoplus_{i=1}^{t} \bar{X}_i$ is a decomposition of $\bar{W}$ into indecomposables in $\overline{\text{mod}}_K(\hat{T})$, then $\bar{W} = \bar{W}^{\hat{\mu}} = \bigoplus_{i=1}^{t} \bar{X}_i^{\hat{\mu}}$. By the description of the morphisms in $\overline{\text{mod}}_K(\hat{T})$ in terms of the qiver $\mathfrak{A}_{2k-1}$ (see (9) 2.3, 2.4) and by Lemma 3.7 we know that $\bar{X}_1$ and $\bar{X}_1^{\hat{\mu}}$ have no common composition factors. Therefore the projections from $\bar{W}$ onto $\bar{X}_1$ and $\bar{X}_1^{\hat{\mu}}$ are orthogonal and $\bar{X}_1 \oplus \bar{X}_1^{\hat{\mu}}$ is an object of $\overline{\text{mod}}_K^{\hat{\mu}}(\hat{T})$. Thus $\bar{W}$ coinsides with $\bar{X}_1 \oplus \bar{X}_1^{\hat{\mu}}$.
2) follows now from 1) and from the fact, that for two indecomposable objects $\bar{X}$ and $\bar{Y}$ the inequality $\overline{\text{Hom}}(\bar{X},\bar{Y}) \neq 0$ implies $\overline{\text{Hom}}(\bar{X},\bar{Y}^{\hat{\mu}}) = 0$.
3) By lemma 3.7 we can choose representatives $\bar{X}_{r,s}$ of $\bar{M}(r,s)$ and irreducible maps $\bar{g}_\alpha : \bar{X}_{r,s} \longrightarrow \bar{X}_{r',s'}$ for each $\alpha = (r,s) \longrightarrow (r',s') \in \mathfrak{A}_{2k-1}$ such that $\bar{X}_{\mu(r,s)} = \bar{X}^{\hat{\mu}}_{r,s}$ and $\bar{g}_{\mu\alpha} = \bar{g}^{\hat{\mu}}_\alpha$. Therefore, using 1) and 2), the map $\bar{V}$: $\mathfrak{A}_{2k-1} \longrightarrow \bar{\Gamma}_{A(T)}$ determined by $\bar{V}_{(r,s)} = [\bar{X}_{r,s} \oplus \bar{X}^{\hat{\mu}}_{r,s}]$ induces the desired isomorphism $\mathfrak{A}_{2k-1}/\mu \longrightarrow \bar{\Gamma}_{A(T)}$.

From Proposition 3.8 we also get a description of the morphisms in the category $\overline{\text{mod}}\ A(T)$

<u>Corollary 3.9</u>: Let $\bar{V}$ be the map in the proof of Proposition 3.8. Then we have

1) $\overline{\text{Hom}}(\bar{V}_{(r,s)}, \bar{V}_{(r',s')}) \neq 0$ if and only if $(r',s') \in R(r,s)$ (equiv.: $(r,s) \in R'(r',s')$) for all $(r,s),(r',s') \in \mathfrak{A}_{2k-1}$.
2) $\text{Dim}_K \overline{\text{Hom}}(\bar{X},\bar{Y}) \leq 1$ for all indecomposable objects $\bar{X},\bar{Y}$ of $\overline{\text{mod}}\ A(T)$ and each morphism between indecomposable modules is a product of irreducible ones.

<u>Proof</u>: Both assertions follow from Proposition 3.8 and the stable

equivalence between $A(\hat{T})$ and $A^{2k-1}_{n(2k-1)}$ from the corresponding assertions for $A^{2k-1}_{n(2k-1)}$ in (9) 2.3, 2.4, using the fact that $2k-1 < n(2k-1)$ ($n \geq 2$!).

To prove Theorem 2,2) we have to compare $\bar{\Gamma}_{A(T)}$ and $\bar{\Gamma}_{B_{m,k}}$ where $m = n/2$, $k = |G|+1$. The 2si-system $T(B_{m,k})$ is the system $T_{m,k} = (G^k, f^k, 2m)$, where G^k is the Brauer-tree with vertices $0,1,2,\dots,k-1$ and just one edge $(0,p)$ between 0 and p, $p = 1,2,\dots,k-1$, in their natural ordering and where $f^k(0) = 0 = f^k(0,1)$. Thus the algebra $A(\hat{T}_{m,k})$ also is stably equivalent to $A^{2k-1}_{n(2k-1)}$ and by Proposition (3.8) the quivers $\bar{\Gamma}_{A(T)}$ and $\bar{\Gamma}_{B_{m,k}}$ are both isomorphic to $\mathfrak{A}_{2k-1}/\mu$.

Let us denote by $\mathfrak{U}_{A(T)}$ respectively $\mathfrak{U}_{m,k}$ a set of representatives for the types of indecomposable nonprojective $A(T)$ - respectively $B_{m,k}$ - modules, and let $\mathcal{C}(A(T))$ and $\mathcal{C}(B_{m,k})$ the corresponding skeletal categories of $\overline{\text{mod}}\ A(T)$ and $\overline{\text{mod}}\ B_{m,k}$ respectively. By Corollary 3.9 and by the isomorphism of $\bar{\Gamma}_{A(T)}$ and $\bar{\Gamma}_{B_{m,k}}$ we have a bijection $\varphi\colon \mathfrak{U}_{A(T)} \to \mathfrak{U}_{m,k}$ such that $\overline{\text{Hom}}(\bar{X},\bar{Y}) \underset{K}{\cong} \overline{\text{Hom}}(\varphi\bar{X},\varphi\bar{Y})$ for all $\bar{X},\bar{Y} \in \mathfrak{U}_{A(T)}$. If we put at the vertices $[\bar{X}]$ (resp. $\varphi[\bar{X}]$) of $\bar{\Gamma}_{A(T)}$ (resp. $\bar{\Gamma}_{B_{m,k}}$) the representatives $\bar{X} \in \mathfrak{U}_{A(T)}$ (resp. $\varphi(\bar{X}) \in \mathfrak{A}_{m,k}$) then we can choose irreducible morphisms between the indecomposable objects in such a way, that all squares corresponding to meshes

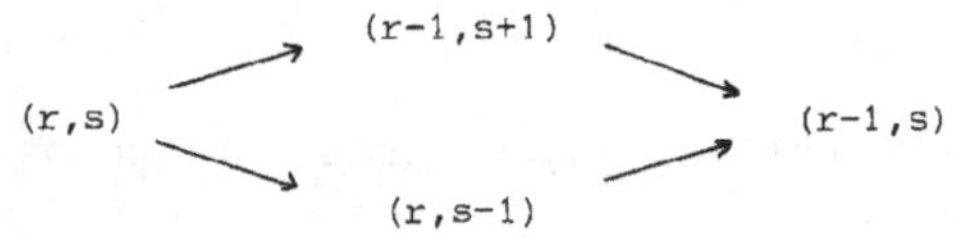

of $\mathfrak{A}_{2k-1}$ commute both in $\bar{\Gamma}_{A(T)}$ and in $\bar{\Gamma}_{B_{m,k}}$. The irreducible morphisms furnish K-bases of the K-spaces $\overline{\text{Hom}}(\bar{X},\bar{Y})$ and $\overline{\text{Hom}}(\varphi\bar{X},\varphi\bar{Y})$ respectively for all $\bar{X},\bar{Y} \in \mathfrak{U}_{A(T)}$. Thus we can construct isomorphisms

$$\varphi_{\bar{X},\bar{Y}} : \overline{\text{Hom}}(\bar{X},\bar{Y}) \xrightarrow[K]{\sim} \overline{\text{Hom}}(\varphi\bar{X},\varphi\bar{Y})$$

in such a way that $\varphi_{\bar{X},\bar{X}}(id_{\bar{X}}) = id_{\varphi\bar{X}}$ and $\varphi_{\bar{Y}\bar{Z}}(\bar{g})\varphi_{\bar{X}\bar{Y}}(\bar{f}) = \varphi_{\bar{X},\bar{Z}}(\bar{g}\bar{f})$ for all morphisms $f : \bar{X} \to \bar{Y}$, $g : \bar{Y} \to \bar{Z}$ between indecomposables. Extending φ to finite direct sums we get an isomorphism of categories $\mathcal{C}(A(T)) \cong \mathcal{C}(B_{m,k})$. Thus the stable categories $\overline{\text{mod}}\ A(T)$ and $\overline{\text{mod}}\ B_{m,k}$ are equivalent, and Theorem 2 is proved.

To prove Theorem 3 let $L : \overline{\text{mod}}\ A \longrightarrow \overline{\text{mod}}\ B_{m,k}$ be an equivalence of categories. With the assertions of Corollary 3.9 which hold for $B_{m,k} = A(T_{m,k})$ we can apply the whole proof of (9) Theorem 2 (see (9), section 3) to our case and get the desired result. But it suffices to show, that for each indecomposable projective A-module P the subfactor module rad P/soc P is a direct sum of at most two indecomposable modules P_α, P_β whose tops and socles are simple ((9) Propos. 3.2 and Lemma 3.4 as well as their duals). Namely this implies that P_α for example is a homomorphic image of P'/P'_α or P'/P'_β for a suitable indecomposable projective A-module P'. Thus also the second upper Loewy-factor NP_α/N^2P_α of P_α ($N = \text{rad}\ A$) is simple and by induction on the Loewy-length of P_α we conclude that P_α is uniserial. Because the same arguments work for P_β we have shown that A is two-serial. A is also self-injective by the results of I. Reiten (13). To finish the proof one has to observe that the algebras A_e^h and $B_{m,k}$ all are pairwise non stably equivalent, because their stable Auslander-Reiten-quivers are pairwise nonisomorphic.

Literature

(1) M.Auslander and I.Reiten, Stable equivalence of artin algebras, in Proc. Conf. on Orders, Group Rings and Related Topics, Springer Lecture Notes 353 (1973), p.8-71.

(2) M.Auslander and I.Reiten, Representation Theory of Artin Algebras III, Commun. Algebra, 3 (1975), p.239-294.

(3) M.Auslander and I.Reiten, Representation Theory of Artin-Algebras IV, Commun. Algebra, 5 (1977), p.443-518.

(4) M.Auslander and I.Reiten, Representation Theory of Artin-Algebras VI, Commun. Algebra, 6 (1978), p.257-300.

(5) E.C.Dade, Blocks with cyclic defect groups, Ann. of Math. 84 (1966), p.20-48.

(6) P.W.Donovan and M.R.Freislich, The indecomposable modular repre-

sentations of certain groups with dihedral sylow-subgroup, preprint.

(7) K.R.Fuller, Weakly symmetric rings of distributive module type, commun. Algebra 5 (1977), 997-1008.

(8) P.Gabriel, Indecomposable representations II, Symp. Math. Ist. Naz. Alta Math. XI (1973, 81-101.

(9) P.Gabriel and Chr.Riedtmann, Group representations without groups, Commentarii Math. Helvet. 54 (1979), 240-287.

(10) G.J.Janusz, Indecomposable representations for finite groups, Ann. of Math. 89 (1969), 209-241.

(11) H.Kupisch, Projective Moduln endlicher Gruppen mit zyklischer p-Sylow-Gruppe, J.Algebra 10 (1968), 1-7.

(12) H.Kupisch, Quasi-Frobenius-Algebras of finite representation type, Lecture Notes in Mathematics No. 488, 184-200, Springer-Verlag, New York/Berlin, 1975.

(13) I.Reiten, Stable equivalence of self-injective algebras, J.Algebra 40 (1976), 63-74.

Eberhard Scherzler
Fachbereich 6, Mathematik
der Gesamthochschule Essen
Universitätsstr. 2
D-4300 Essen

Josef Waschbüsch
II. Mathematisches Institut
Freie Universität Berlin
Königin-Luise-Str. 24/26
D-1000 Berlin 33

RIGHT PURE SEMISIMPLE HEREDITARY RINGS

Daniel Simson

A ring R is called <u>right</u> (resp. <u>left</u>) <u>pure semisimple</u> if every right (resp. left) R-module is a direct sum of finitely presented modules (see [12,15-17]). It is well known that every ring of finite representation type is both left and right pure semisimple (see [1], [14,15] . Moreover by [1,9,11] the converse is also true. However, it is still an open question if every right pure semisimple ring is of finite representation type.

Auslander [2] answers the question in the affirmative for artin algebras. A discussion of this problem can be also found in [9-11], [17,18]. Unfortunately the proof of Theorem 2.3 in [18] is not correct. In this note we discuss the above question for hereditary rings. Our main results are the following three theorems.

<u>Theorem 1</u>. <u>Suppose that</u> R <u>is either a hereditary ring or</u> $J(R)^2 = 0$, <u>where</u> $J(R)$ <u>is the Jacobson radical of</u> R . <u>If</u> R <u>is right pure semisimple and</u> $R/J(R)$ <u>is an artin algebra then</u> R <u>is of finite representation type</u>.

<u>Theorem 2. The following statements are equivalent:</u>

(a) <u>Every right pure semisimple hereditary ring is of finite representation type</u>.

(b) <u>Every right pure semisimple hereditary ring is left artinian</u>.

(c) <u>If</u> F,G <u>are division rings and</u> ${}_FM_G$ <u>is an F-G-bimodule such that the ring</u> $\begin{pmatrix} F, & {}_FM_G \\ 0, & G \end{pmatrix}$ <u>is right pure semisimple then</u> $\dim_F M$ <u>is finite</u>.

<u>Theorem 3</u>. <u>A hereditary ring</u> R <u>is of finite representation type if and only if</u> R <u>is right pure semisimple, left artinian, and for any pair</u> $X \to Y$ <u>of indecomposable preprojective finitely generated</u>

right R-*modules connected by an irreducible map* [3] *the ring* $\mathrm{End}(X \oplus Y)/J^2\mathrm{End}(X \oplus Y)$ *is left artinian*.

Here, a finitely generated right module P over a hereditary right artinian ring R is called *preprojective* if there is only finitely many of indecomposable finitely generated R-modules N with the property $\mathrm{Hom}_R(N,P) \neq 0$.

In view of the Propositions 1 and 2 below it is enough to prove Theorems 1-3 for tensor rings of species. We recall that a *species* $\underline{M} = (F_i, {}_iM_j)_{i,j \in I}$ is a finite set of division rings F_i together with a set of F_i-F_j-bimodules ${}_iM_j$. Throughout we will suppose that $\dim_{F_i}({}_iM_j)$ and $\dim({}_iM_j)_{F_j}$ are both finite. The *tensor ring* of $\underline{M}$ is the ring

$$T = F \oplus M \oplus M^2 \oplus \ldots \oplus M^i \oplus \ldots$$

with $F = \prod_{i \in I} F_i$, $M = \oplus_i M_j$ and $M^{i+1} = M \otimes_F M^i$. Given a species $\underline{M}$ the category $r(\underline{M})$ of right finite dimensional representations of $\underline{M}$ is defined in such a way that $r(\underline{M})$ is equivalent with the category mod-T of all finitely generated right T-modules [6].

The following two propositions follow from [19,Theorem 4.5] and [18,Proposition 2.1].

Proposition 1. *If* R *is a right pure semisimple hereditary ring and* R *is left artinian then* R *is Morita equivalent with a tensor ring of a species*.

Proposition 2. *Suppose* R *is a right pure semisimple ring and* $S = \mathrm{End}\, I$ *where* I *is a minimal injective cogenerator in* mod-R . *Then* S *is left artinian, right pure semisimple and there exists an equivalence of categories* $(\text{mod-}R)^{op} \cong \text{mod-}S^{op}$.

The method we use in the proof of our main theorems are almost split sequences [3] and partial Coxeter functors [4,6,8] .

We say that the species $\underline{M}$ admits a *right* (resp. *left*) *sequence* of partial Coxeter functors if there exist species $\underline{M}^{(i)}$, $i \geq 1$, (resp. $\underline{M}^{(j)}$, $j \leq -1$) and a sequence of pairs of partial Coxeter functors (in the sense of [4.8])

$$r(\underline{M}) \underset{S_1^-}{\overset{S_1^+}{\rightleftarrows}} r(\underline{M}^{(1)}) \underset{S_2^-}{\overset{S_2^+}{\rightleftarrows}} r(\underline{M}^{(2)}) \rightleftarrows \cdots$$

$$(\text{resp. } \cdots \rightleftarrows r(\underline{M}^{(-2)}) \underset{S_{-2}^-}{\overset{S_{-2}^+}{\rightleftarrows}} r(\underline{M}^{(-1)}) \underset{S_{-1}^-}{\overset{S_{-1}^+}{\rightleftarrows}} r(\underline{M})) .$$

For our purpose the following result is very useful.

Proposition 3. Let $\underline{M}$ be a species and let T be the tensor ring of $\underline{M}$. The following conditions are equivalent:

(1) $\underline{M}$ admits a right (resp. left) sequence of partial Coxeter functors.

(2) If X and Y are indecomposable preprojective right (resp. left) finitely generated T-modules connected by an irreducible map $X \to Y$, then the ring $\mathrm{End}(X \oplus Y)/J^2\mathrm{End}(X \oplus Y)$ is left artinian.

(3) For every indecomposable right (resp. left) finitely generated preprojective T-module X there exists an almost split sequence

$$0 \to X \to Y \to Z \to 0$$

where Y and Z are preprojective modules.

The main fact we use in the proof of Proposition 3 is the following simple lemma.

Lemma. Let F and G be division rings and let ${}_F M_G$ be an F-G-bimodule with $\dim M_G$ finite. Let P be the simple projective right module over the ring $\begin{pmatrix} F, & {}_F M_G \\ 0, & G \end{pmatrix}$. There exists an almost split sequence

$$0 \to P \to Q \to N \to 0$$

if and only if $\dim_F M$ is finite.

In view of properties of partial Coxeter functors [4.8] the implications $(2) \longleftarrow (1) \longrightarrow (3)$ in Proposition 3 can be proved by applying the lemma. The converse implications can be proved by using arguments applied in [8].

Now by Proposition 3, [16,Theorem 6.3], [1,Theorem 3.1] and the main result in [8] we get the following proposition which allows us to prove our Theorems 1-3.

Proposition 4. Let $\underline{M}$ be a species. If the tensor ring T of $\underline{M}$ is right pure semisimple then

(a) $\underline{M}$ admits a left sequence of partial Coxeter functors.

(b) If $\underline{M}$ admits a right sequence of partial Coxeter functors then T is of finite representation type.

Note that Theorem 1 follows immediately from (b) because we know from [7] that every quasi-Artin species admits both left and right sequence of partial Coxeter functors.

We end the paper by some observations concerning the existence of almost split sequences over arbitrary artinian hereditary rings. We know from [5] that there exists a pair of division rings $G \subset F$ such that $\dim F_G = 2$ and $\dim_G F$ is infinite. Note that the hereditary ring $T = \begin{pmatrix} F, & {}_F F_G \\ 0, & G \end{pmatrix}$ is both left and right artinian. T is of infinite representation type, the corresponding valued graph of T is the Dynkin diagram B_2, and if P is the nonsimple projective right T-module then there is no almost split sequence of the form

$$0 \to P \to X \to Y \to 0 .$$

P.M. Cohn has pointed out to me that the division ring G can be chosen in such a way that G is isomorphic with F (apply methods from [5]). If we take such G then the trivial extension $S = F \ltimes {}_F F_G$ is a local ring of infinite representation type. Moreover, S has a Morita duality which is not a self duality (see [13,Theorem 10]) and the separated valued graph of S is the Dynkin diagram B_2.

References

1. Auslander, M., Representation theory of artin algebras, Comm. in Algebra, 1(1974), 269-310.
2. Auslander, M., Large modules over artin algebras, Algebra, Topology and Category Theory, Academic Press, 1976, pp. 3-17.
3. Auslander, M., Reiten, I., Representation theory of artin algebras III, Almost split sequences, Comm. in Algebra, 3(1975), 239-294.
4. Auslander, M., Platzek, M.I., Reiten, I., Coxeter functors without diagrams, Trans. Amer. Math. Soc., to appear.
5. Cohn, P.M., On a class of binomial extensions, Illinois J. Math. 10(1966), 418-424.
6. Dlab, V. Ringel, C.M., Representations of graphs and algebras, Memoirs Amer. Math. Soc., 173, 1976.
7. Dowbor, P., Simson, D., Quasi-Artin species and rings of finite representation type, J. Algebra, to appear.
8. Dowbor, P., Simson, D., A characterization of hereditary rings of finite representation type, to appear.
9. Fuller, K.R., On rings whose left modules are direct sums of finitely generated modules, Proc. Amer. Math. Soc., 54(1976), 39-44.
10. Gruson, L., Simple coherent functors, Lecture Notes in Math., 488, Springer-Verlag, 1975, pp. 156-159.
11. Gruson, L., Jensen, C.U., L-dimensions of rings and modules, in preparation.
12. Kiełpiński, R., Simson, D., On pure homological dimension, Bull. Acad. Polon. Sci., 23(1975), 1-6.
13. Müller, B.J., On Morita duality, Can. J. Math., 21(1969), 1338-1347.

14. Ringel, C.M., Tachikawa, H., QF-3 rings, J. Reine Angew. Math., 272(1975), 49-72.

15. Simson, D., Functor categories in which every flat object is projective, Bull. Acad. Polon. Sci., 22(1974), 375-380.

16. Simson, D., On pure global dimension of locally finitely presented Grothendieck categories, Fund. Math., 96(1977), 91-116.

17. Simson, D., On pure semi-simple Grothendieck categories, Fund. Math., 100(1978), 211-222.

18. Simson, D., Pure semisimple categories and rings of finite representation type, J. Algebra, 48(1977), 290-296.

19. Simson, D., Categories of representations of species, J. Pure Appl. Algebra, 14(1979), 101-114.

Institute of Mathematics
Nicholas Copernicus University
ul. Chopina 12/18
87-100 Toruń, Poland

REPRESENTATIONS OF TRIVIAL EXTENSIONS OF HEREDITARY ALGEBRAS

Hiroyuki Tachikawa

Introduction. In the study of representations of algebras there are splendid results on tensor algebras, by Gabrial [7] and Dlab-Ringel [3], [4] and [5], which are free from a restriction "algebras with square-zero radicals". The purpose of this paper is to present resuls on indecomposable representations over another type of algebras which are similarly free from the same restriction and are closely connected with the above quoted algebras.

Let A be a basic artin algebra over a center C and Q an A-bimodule $\mathrm{Hom}_C(A, I)$, where I is the minimal injective cogenerator C-module. Then the trivial extension R of A by Q is a weakly symmetric algebra, i.c. a quasi-Frobenius algebra and A is ring-isomorphic to R/Q. In these situation we have the following problem : What relations are there between representation types of A and $R(=A \ltimes Q)$.

This problem was considered already by Müller [9] and Green-Reiten [8], but rings A treated by them had always square-zero radicals.

For the case of A being hereditary our main result answers to the question that the representation types of A and R are almostly same. More precisely, the cardinal number of isomorphism-classes of indecomposable R-modules is twice of one of indecomposable A-modules. Further we shall determine the strict relation between Auslander-Reiten functors on mod-R and mod-A and using this relation we shall show that the set of indecomposable projective A-modules (and indecomposable injective A-modules) are important role in the study of Auslander-Reiten qiver in mod-R, not only for the case R being of finite representation type but also of infinite representation type.

Throughout this paper unless otherwise specified A-modules and R-modules are unital right A-modules and right R-modules, and homomorphisms operate from the right hand.

1. Indecomposable R-modules and Indecomposable A-modules

Let A be a basic artin algebra with a center C, I the injective envelope of C/Rad C and Q the A-bimodule $Hom_C(A, I)$. Then we can construct a new algebra $R = A \ltimes Q$; i.e. $R = A \oplus Q$ as an additive group and the multiplication is defined by the following equation

$$(a, q)(a', q') = (aa', aq'+qa') \quad \text{for} \quad (a, q), (a',q') \in R.$$

Clearly Rad R = Rad A $\oplus$ Q and for primitive idempotents $e_i \in A$, $i = 1,2,\ldots, n$, we have primitive idempotents $(e_i, 0) \in R$. Identifying e_i with $(e_i, 0)$, $Re_i/Rad(Re_i) \cong Soc\ Re_i$, because $Soc\ Re_i \cong Soc\ Qe_i \cong Re_i/Rad(Re_i)$. This implies R is weakly symmetric and hence R is quasi-Frobenius.

At first we have

PROPOSITION 1.1. Let A be a basic artin (not necessarily hereditary) algebra. Then it holds that

(i) End $(Q_A) = A^{\circ}$, where the opoeration on Q with an element of A° is same with one induced by the left multiplication with the corresponding element of A.

(ii) By functors $- \otimes_A Q$ and $Hom_A(Q, -)$ the full subcategories of all projective A-modules and all injective A-modules are equivalent.

Proof. (i) Let C be the center of A. Then A is

a finitely generated C-module and C is basic and artinian. Hence, for $I = E(C/\mathrm{Rad}C)$ $\mathrm{End}(C_C) = C$ and the conclusion follows immediately.

(ii) The proof is straight-forward.

Since A is a subring of R, for a given R-module X the following exact sequence of R-modules

$$0 \longrightarrow XQ \longrightarrow X \rightarrow X/XQ \longrightarrow 0$$

can be naturally considered as an exact sequence of right A-modules. Here Q is considered as an ideal of R. On the other hand, Q is an injective right A-module and XQ is a homomorphic image of a direct sum of copies of Q_A. Hence we have a decomposition

$$X = XQ \oplus X/XQ$$

of X, because A is a hereditary algebra.

From now on we shall denote X/XQ and XQ by U and V respectively.

Let $(X_A \xrightarrow{\phi} [{}_AQ_A, X_A]_A)$ be an expression of X as an object of $[Q_A, -]_A \ltimes$ mod-A. Cf. [6]. Then it holds

$$\phi|V = 0 \quad \text{and} \quad \mathrm{Im}\,\phi \subset [Q, V].$$

In fact $u\phi = (\sum_i x_iq_i)\phi = \sum_i((q_i)[(x_i)\phi])\phi$ for $u = \Sigma x_iq_i \in XQ = V$, where $x_i \in X$ and $q_i \in Q$. But the definition of trivial extension $((q_i)[(x_i)\phi])\phi = x_i(\phi\cdot[Q, \phi]) = 0$ and hence $\phi|V = 0$

On the other hand, $\mathrm{Im}\,\phi \subset [Q, V]$ follows from $(q)[(x)\phi] = x\,q \in XQ = V$ for $q \in Q$ and $x \in X$.

Thus we may put

$X_R = (X_A \xrightarrow{\phi} [{}_AQ_A, X_A]_A) = (U_A \xrightarrow{\phi} [{}_AQ_A, V_A]_A)$ and we call the last term the canonical expression of X.

<u>PROPOSITION 1.2</u> Let A be a hereditary algebra, $Q = \mathrm{Hom}_C(A, I)$ and $R = A \times Q$. Let $(U \xrightarrow{\phi} [{}_AQ_A, V_A])$ be the canonical expression of R-module X and assume $\phi \neq 0$.

Then X is indecomposable if and only if either one of the following conditions (i) and (ii) is satisfied :

(i) ϕ is an isomorphism and U_A is indecomposable projective

(ii) ϕ is a monomorphism (but not an epimorphism), U_A is projective, $\mathrm{Im}\ \phi$ is a small submodule of $[{}_AQ_A, X_A]_A$ and $\mathrm{Cok}\ \phi$ is an indecomposable A-module.

In case (i) X is a projective and injective R-module.

Proof: Since Q_A is an injective cogenerator, $\mathrm{End}(A_A) = A^{\circ}$ and V_A is injective, $[{}_AQ_A, V_A]_A$ is projective. Since A is hereditary, $\mathrm{Im}\ \alpha_A$ is projective and $\mathrm{Ker}\ \phi$ is a direct summand of U_A. Therefore $(\mathrm{Ker}\ \phi \xrightarrow{o} [Q, 0])$ is a direct summand of $(U \xrightarrow{\phi} [Q, V])$. Thus, if $(U \xrightarrow{\phi} [Q, V])$ is indecomposable, $U_A(=[Q, V]_A)$ is projective and ϕ is a monomorphism.

Now we shall divide our consideration into two cases :

(i) ϕ is surjective : Since $\mathrm{End}(Q_A) = A^{\circ}$ we have an isomorphism $\theta : U \otimes Q \longrightarrow V_A$ which is a composition of maps $U \otimes Q \quad u \otimes q \longrightarrow u\,\phi \otimes q \in [Q, V]$ and $[Q, V] \otimes Q \in u\,\phi \otimes q \longrightarrow (q)[u\,\phi] \in V$.

Suppose $U_A = U_1 \oplus U_2$ and put $V_1 = \theta(U_1 \otimes Q), V_2 = \theta(U_2 \otimes Q)$. Then it can be proved easily $(U_1 \xrightarrow{(\phi|U_1)} [Q, V_1])$ is a direct summand of $(U \xrightarrow{\phi} [Q, V])$.

Conversely, from a decomposition $X_R = X_1 \oplus X_2$ we have decompositions $U_A = U_1 \oplus U_2$, $V_A = V_1 \oplus V_2$ and $\phi = \phi|U_1 \oplus \phi|U_2$, where $V_1 = X_1Q$ and $V_2 = X_2Q$. Since Q_A is an injective cogenerator, $U_i(\cong [Q_A, V_i]) \neq 0$, $i = 1,2$, if and only if $V_i \neq 0$, and consequently if and only if $X_i \neq 0$.

(ii) ϕ is not surjective : Suppose $\operatorname{Im}\phi$ is not small as a submodule of $[{}_AQ_A, V_A]$. Then there is a non-zero projective A-module P_0 such that $[{}_AQ_A, V_A]_A = P_0 \oplus P_1$ and $\operatorname{Im}\phi \supset P_0$.

Take U_0 and V_0 such that $U_0 = \phi^{-1}P_0$ and $[Q, V_0] = P_0$ by using a similar isomorphism as θ in (i). Then $(U_0 \xrightarrow{\phi|U_1} [Q, V_0])$ is a direct summand of $(U \xrightarrow{\phi} [Q, V])$. Further, under the assumption that $\operatorname{Im}\phi$ is small in $[{}_AQ_A, V_A]_A$ a direct sum decomposition of $(\operatorname{cok}\phi)_A$ induces a direct sum decomposition of projective resolusion $U_A \xrightarrow{\phi} [Q, V]_A \to \operatorname{cok}\phi \to 0$. Then by PROPOSITION 1.1 we have a drect sum decomposition of $(U \xrightarrow{\phi} [Q, V])$.

Conversely, assume ϕ is a monomorphism and $\operatorname{Im}\phi$ is small. Then

$$0 \to (U \xrightarrow{\phi} [Q,V]) \xrightarrow{(\phi, 1_V)} ([Q,V] \xrightarrow{1_{[Q,V]}} [Q,V]) \xrightarrow{(\operatorname{cok}\phi, 0)} (\operatorname{cok}\phi \xrightarrow{o} [Q,0]) \to 0$$

is exact as a sequence of R-modules, because R-homomorphisms $(\phi, 1_V)$ and $(\operatorname{cok}\phi, 0)$ are clearly compatible with the following diagram :

$$\begin{array}{ccc}
U & \xrightarrow{\phi} & [Q, V] \\
\downarrow \phi & & \downarrow [Q, 1_V] \\
[Q, V] & \xrightarrow{1_{[Q, V]}} & [Q, V] \\
\downarrow & & \downarrow 0 \\
\operatorname{cok}\phi & \xrightarrow{0} & [Q, 0]
\end{array},$$

where each square is commutative. By PROPOSITION 1.1 an R-module $P = ([Q, V] \xrightarrow{1_{[Q,V]}} [Q, V])$ is projective and injective. Further $\operatorname{Ann}_P Q = \operatorname{Ann}_X Q = V$, and consequently $\operatorname{Soc}(P_R) = \operatorname{Soc}(X_R)$. Hence $\operatorname{Im}(\phi, 1_V) (\cong X)$ is essential in P and X_R is indecomposable if and only if $(\operatorname{cok}\phi \to [Q, 0])$, equivalently cok ϕ_A is indecomposable.

The last statement is also clear from PROPOSITION 1.1,(i).

A right R-module X has another expression $(X_A \otimes {}_AQ_A \xrightarrow{\psi} X_A)$ as an object of mod-A $\ltimes$ $(- \otimes Q)$ which corresponds to in the adjoint relation $\mathrm{Hom}_A(U \otimes_A Q_A, V_A) = \mathrm{Hom}_A(U_A, \mathrm{Hom}_A({}_AQ_A, V_A))$.

It is known ψ is a composition of $\phi \otimes Q$ and $[Q, X] \otimes Q \ni f \otimes q \longrightarrow (q)f \in X$. Therefore we may put $X_R = (X_A \otimes_A Q_A \xrightarrow{\psi} X_A) = (U_A \otimes_A Q_A \xrightarrow{\psi} V_A)$, where $U = X/XQ$ and $V = XQ$, and we say the last term the canonical expression of X. $\psi \neq 0$ if and only if $\phi \neq 0$, and we have the following dual PROPOSITION which was independently obtained by T.Wakamatsu [14].

PROPOSITION 1.3. Let A be a hereditary artin algebra, $Q = \mathrm{Hom}_C(A, I)$ and $R = A \times Q$. Let $(U_A \otimes_A Q_A \xrightarrow{\psi} V_A) \in$ mod-A $\times (-\otimes_A Q_A)$ be a canonical expression of an R-module X with $\psi \neq 0$.

Then X is an indecomposable R-module if and only if either one of the following conditions (i) and (ii) is satisfied:

(i) ψ is an isomorphism and U_A is indecomposable and projective

(ii) ψ is an epimorphism (but not monomorphism), U_A is projective, Ker ψ_A is a large submodule of $U \otimes Q_A$ and is indecomposable.

In case (i) X is a projective and injective R-module.

From now on we shall call an indecomposable R-module $X = (U \xrightarrow{\phi} [Q, V]) = (U \otimes Q \xrightarrow{\psi} V)$ is of 2nd kind (resp. 1st kind) if $\phi \neq 0$ (resp. $\phi = 0$), same with $\psi \neq 0$ (resp. $\psi = 0$).

Further, we shall call Cok ϕ (resp. Ker ψ) the cokernel (resp. Kernel) of X.

THEOREM 1.4. Let A be a hereditary artin algebra and ${}_AQ_A = \mathrm{Hom}_C(A, I)$. Let R be a trivial extension of A by Q. Then the cardinal number of the set of isomorphism-classes of all indecomposable R-modules is twice of one of all indecomposable A-modules. Especially R is of finite representation type if and only if so is A.

Proof: Let $X = (U_A \xrightarrow{\phi} [{}_AQ_A, V_A]_A) \in$ mod-R. If $\phi = 0$, then $V = XQ = 0$ and $X = U$. Hence X is indecomposable as an R-module if and only if X is indecomposable as an A-module.

On the other hand, if $\phi \neq 0$ and ϕ is not an isomorphism, by PROPOSITION 1.2, (ii) X is indecomposable as an R-module if and only if $U_A \xrightarrow{\phi} [{}_AQ_A, V_A]_A \rightarrow \mathrm{Cok}\ \phi \rightarrow 0$ is the projective cover of $\mathrm{Cok}\ \phi$ and $\mathrm{Cok}\ \phi_A$ is indecomposable. Therefore there exists a one-to one correspondence between isomorphism-classes of these indecomposable R-modules X and one of all non-projective, indecomposable A-modules.

Further the remaining indecomposable R-modules correspond uniquely to indecomposable projective R-modules by PROPOSITION 1.2,(i) and the last statement. This completes the proof.

2. Auslander-Reiten functors DT_r^R and DT_r^A

To begin with we shall consider the desocription of morphisms between R-modules.

Let $(U \otimes Q \xrightarrow{\psi} V)$ and $(U' \otimes Q \xrightarrow{\psi'} V')$ be canonical expression of R-modules X and X'. Then $X = U \oplus V$ and $X' = U' \oplus V'$ are decompositions as A-modules, and any R-homomorphism from X to X' has the matrix expression $\begin{pmatrix} \alpha & \delta \\ \gamma & \beta \end{pmatrix}$, where $\alpha : U \rightarrow U'$, $\beta : V \rightarrow V'$, $\gamma : V \rightarrow U'$ and $\delta :$ $U \rightarrow V'$ are A-homomorphisms. In this case α and β are compatible with the following commutative diagram :

$$(1)\quad \begin{array}{ccc} U\otimes Q & \xrightarrow{\psi} & V \\ \downarrow{\scriptstyle \alpha\otimes Q} & & \downarrow{\scriptstyle\beta} \\ U'\otimes Q & \xrightarrow{\psi'} & V' \end{array}$$

and $\gamma = 0$ because $V = UQ \ni \sum_i u_i q_i \gamma = \sum_i (u_i\gamma) q_i \in V' \cap U'$ $= 0$.

Conversely it can be easily checked that a matrix $\begin{pmatrix}\alpha & \delta \\ 0 & \beta\end{pmatrix}$ with A-homomorphisms δ and α, β, of which the last two homomorphisms are compatible with the commutative diagram (1), determines a R-homomorphism of X to X'.

If $V = 0$ or $V' = 0$, we shall abbreviate it (α, δ) or $\begin{pmatrix}\alpha \\ 0\end{pmatrix}$ respectively.

PROPOSITION 2.1.

(i) For an indecomposable A-module X, let $\rho : P_A$ X_A be the projective cover. Then

$$(P \otimes Q \xrightarrow{1_P \otimes 1_Q} P \otimes Q) \xrightarrow{\binom{\rho}{0}} (X \otimes Q \xrightarrow{0} 0)$$

is the projective cover of an R-module $X = (X \otimes Q \xrightarrow{0} 0)$.

(ii) For an indecomposable R-module $X = (U \otimes Q \xrightarrow{\psi} V)$

$$(U \otimes Q_A \xrightarrow{1_U \otimes 1_Q} U \otimes Q_A) \xrightarrow{\begin{pmatrix}1_U & 0 \\ 0 & \psi\end{pmatrix}} (U \otimes Q_A \xrightarrow{\psi} V_A)$$

is the projective cover of X_R.

Proof: (i) Denote $(P \otimes Q_A \xrightarrow{1_P \otimes 1_Q} P \otimes Q_A)$ by Y. Then $YQ = P \otimes Q$ and $Y/YQ = P$. If N is the radical of A, the radical of R is $(N, Q) = \{(a, q) \in A \ltimes Q \mid a \in N$ and $q \in Q\}$. Therefore $Y/Y(N,Q) = X/XN$, where the isomorphism is both an A- and R-isomophism. So $\binom{\rho}{0}$ is the projective cover of X_R.

(ii) Similarly as in the proof of (ii) $X(N,Q) = UN$ and $Y(N,Q)) = UN$ for $X = (U \otimes Q \xrightarrow{\psi} V)$ and $Y = (U \otimes Q \xrightarrow{1_U \otimes 1_Q} U \otimes Q)$, and hence $X/X(N,Q) = U/UN$ $= Y/Y(N,Q)$ as both an A-and R-isomorphisms. On the other hand, ψ is an epimorphism by PROPOSITION 1.3,(ii) and $(U \otimes Q \xrightarrow{1_{U\otimes Q}} U \otimes Q)$ is a projective R-module. Hence $\begin{pmatrix} 1_U & 0 \\ 0 & \psi \end{pmatrix}$ is the projective cover of X_R.

In order to prove Brauer-Thrall conjecture for Artin algebras Auslander-Reiten [1] introduced the Endofunctor DT_r of a category of modules over an Artin algebra. Now in our context we can consider the following two kinds of DT_r, one of which is the endofunctor on mod-A and the other the endofunctor on mod-R, and Emphaszing their difference we shall denote them DT_r^A and DT_r^R respectively.

Then the next Theorem states the strict relations between DT_r^A and DT_r^R

<u>THEOREM 2.2.</u>

(i) Let X be a non-projective indecomposable A-module. Then $DT_r^R(X) = DT_r^A(X)$, where in the left term X is considered as an R-module.

(ii) Let X be a projective indecomposable A-module, and $DT_r^R(X) = (U \otimes Q \xrightarrow{\lambda} V)$. Then

$$\mathrm{Ker}\ \lambda \cong DT_r^A(X \otimes Q) \quad .$$

(iii) Let $X = (U \otimes Q \xrightarrow{\psi} V)$ be an indecomposable R-module and $DT_r^R(X) = (U' \otimes Q \xrightarrow{\lambda} V')$. If $\mathrm{Ker}\ \psi_A$ is not projective, then $\ker \lambda = DT_r^A(\ker \psi)$.

(iv) Let $X = (U \otimes Q \xrightarrow{\psi} V)$ be an indecomposable R-module. If $\mathrm{Ker}\ \psi_A$ is projective, then it holds that

$$DT_r^R(X) \cong (\mathrm{Ker}\psi \otimes Q \otimes Q \xrightarrow{0} 0),$$

i.e. $DT_r^R(X)$ is isomorphic to an A-module $\mathrm{Ker}\ \psi \otimes Q$.

In our proof of THEOREM 2.2 the following LEMMA is necessary.

LEMMA 2.3. Let A be an artin algebra, i.e. an artin ring which is finitely generated over its center C.

Let $\cdots \longrightarrow P_1 \xrightarrow{\rho_1} P_0 \xrightarrow{\rho_0} X \longrightarrow 0$ be a projective resolution of an A-module X. Then $DT_r(X) \cong \mathrm{Ker}(\rho_1 \otimes Q)$, where $Q = \mathrm{Hom}_C(A_C, I_C)$ for the injective cogenerator C-module I, D is the dual functor $\mathrm{Hom}_A(- \ , {}_AQ)$, and T_r is the transpose.

Especially, in case A being a quasi-Frobenius algebra, $DT_r(X) = \ker \rho_1$.

Proof of LEMMA 2.3 : Denote $\mathrm{Hom}_A(- \ , A_A)$ by $*$. Then by the definitions $DT_r(X) = D(\mathrm{cok}\ \rho_1^*)$, i e.

$$0 \quad D(\mathrm{Cok}\ \rho_1^*)(=[{}_A\mathrm{Cok}\ \rho_1^*,\ {}_AQ]) \hookrightarrow [{}_AP_1^*,\ {}_AQ] \xrightarrow{[\rho_1^*, {}_AQ]} [{}_AP_0^*,\ {}_AQ]$$

is exact. However

$$\begin{aligned}[{}_AP_i^*,\ {}_AQ] &= [{}_A[P_{i\ A},\ A_A],\ {}_A[{}_CA,\ {}_CI]] \\ &\cong [{}_CA\ {}_A\otimes\ {}_A[P_{i\ A},\ A_A],\ {}_CI] \\ &= [{}_C[P_{i\ A},\ A_A],\ {}_CI] \qquad \text{for } i = 0,\ 1.\end{aligned}$$

So it is sufficient to prove the maps θ_i :

$$P_1 \otimes Q = P_i \otimes [{}_CA, {}_CI] \quad p \otimes f \longrightarrow (\phi \longrightarrow f(\phi(p))) \varepsilon [{}_C[P_{i\ A}, A_A], {}_CI]$$

are A-isomorphisms and their naturality. But the proofs are both immediate.

Proof of THOREM 2.2: (i). Let $P \xrightarrow{\rho} X$ be the projective cover of X. Since A is hereditary, $0 \quad \ker \rho \overset{i}{\hookrightarrow} P \xrightarrow{\rho} X \longrightarrow 0$ becomes a projective resolution of X. Then

$$0 \longrightarrow (\mathrm{Ker}\,\rho \otimes Q \xrightarrow{i\otimes 1_Q} P \otimes Q) \xrightarrow{\begin{pmatrix} i & 0 \\ 0 & 1_{P\otimes Q}\end{pmatrix}} (P \otimes Q \xrightarrow{1_P \otimes 1_Q} P \otimes Q) \xrightarrow{\binom{\rho}{0}} (X\otimes Q \xrightarrow{0} 0) \rightarrow 0$$

and

$$(\mathrm{Ker}(i\otimes Q)\otimes Q \xrightarrow{0} 0) \xrightarrow{(0,j)} (\mathrm{Ker}\rho\otimes Q \xrightarrow{1_{\mathrm{Ker}\rho}\otimes Q} \mathrm{Ker}\rho\otimes Q) \xrightarrow{\begin{pmatrix} 1_{\mathrm{Ker}\rho} & 0 \\ 0 & i\otimes Q\end{pmatrix}} (\mathrm{Ker}\rho\otimes Q \xrightarrow{i\otimes Q} P\otimes Q) \rightarrow 0$$

are exact sequences of R-modules, where j is the injection $\mathrm{Ker}(i \otimes Q) \hookrightarrow \mathrm{Ker}\,\rho \otimes P$, and their middle terms are both projective R-modules.

As R is weakly symmetric (quasi-Frobenius), $(\mathrm{Ker}(i \otimes Q) \otimes Q \xrightarrow{0} 0) \cong DT_r^R\ (X \xrightarrow{0} 0)$. But from LEMMA 2.3 it follows that $\mathrm{Ker}(i \otimes Q)_A \cong DT_r^A\ (X)$.

(ii) Assume X_A is projective and let

$$0 \quad \mathrm{Ker}\,\xi \xhookrightarrow{i} P \xrightarrow{\xi} X \otimes Q \quad 0$$

be a projective resolution of $X \otimes Q_A$. Then by LEMMA 2.3 $DT_r^A\ (X \otimes Q) \cong \mathrm{Ker}(i \otimes Q)$. On ther other hand

$$0 \rightarrow (X\otimes Q\otimes Q \xrightarrow{0} 0) \xrightarrow{(0,\ 1_{X\otimes Q})} (X\otimes Q \xrightarrow{1_{X\otimes Q}} X\otimes Q) \xrightarrow{\binom{1_X}{0}} (X\otimes Q \xrightarrow{0} 0) \rightarrow 0$$

and

$$0 \rightarrow (\mathrm{Ker}\,\xi\otimes Q \xrightarrow{i\otimes Q} P\otimes Q) \xrightarrow{\begin{pmatrix} i\otimes Q & 0 \\ 0 & 1_{P\otimes Q}\end{pmatrix}} (P\otimes Q \rightarrow P\otimes Q) \xrightarrow{\binom{\xi}{0}} (X\otimes Q\otimes Q \rightarrow 0) \rightarrow 0$$

are exact as R-modules, and the middle terms are both projective R-modules. Since R is weakly symmetric

$$DT_r^R(X) = (\mathrm{Ker}\,\xi\otimes Q \xrightarrow{i\otimes Q} P\otimes Q)$$

It follows $\lambda = i\otimes Q$ and $\mathrm{Ker}\,\lambda \cong DT_r^A(X\otimes Q)$.

(iii) Let $0 \rightarrow \mathrm{Ker}\,\rho \xhookrightarrow{i} P \rightarrow \mathrm{Ker}\,\psi$ be an exact sequence of A-modules with the projective cover $\rho : P_A \rightarrow \mathrm{Ker}\,\psi_A$ of $\mathrm{Ker}\,\psi$. Then, since A is hereditary $(\mathrm{Ker}\,\rho)_A$ is projective, $\mathrm{Ker}(i \otimes Q) = DT_r^A\ (\mathrm{Ker}\,\psi)$. On the other hand

$$0 \quad (\mathrm{Ker}\,\psi \otimes Q \xrightarrow{0} 0) \xrightarrow{(0,\ j)} (U \otimes Q \rightarrow U \otimes Q) \xrightarrow{\begin{pmatrix} 1_U & 0 \\ 0 & \psi\end{pmatrix}} (U \otimes Q \xrightarrow{\psi} V) \rightarrow 0$$

and

$$0 \longrightarrow (\mathrm{Ker}\rho \otimes Q \xrightarrow{i\otimes Q} P\otimes Q) \xrightarrow{\begin{pmatrix} i\otimes Q & 0 \\ 0 & 1_{P\otimes Q}\end{pmatrix}} (P\otimes Q \xrightarrow{1_P \otimes Q} P\otimes Q) \xrightarrow{\begin{pmatrix}\rho \\ 0\end{pmatrix}} (\mathrm{Ker}\psi \otimes Q \xrightarrow{0} 0) \rightarrow 0^*$$

are exact sequence of R-modules, where j denotes the injection of Ker $\psi \hookrightarrow U \otimes Q$, and their middle terms are both projective R-modules.

Hence, similarly as in cases (i) and (ii) we have

$$(\mathrm{Ker}\rho \otimes Q \xrightarrow{i\otimes Q} P \otimes Q) \cong DT_r^R(X)$$

This implies $U' = \mathrm{Ker}\ \rho$, $V' = P \otimes Q$ and $\lambda = i \otimes Q$.

(iv) Similarly as in the preceding proofs the following two exact sequences of R-modules assures the conclusion :

$$0 \rightarrow (\mathrm{Ker}\ \psi\otimes Q \xrightarrow{0} 0) \xrightarrow{(0,i)} (U\otimes Q \xrightarrow{1_{U\otimes Q}} U\otimes Q) \xrightarrow[\psi]{1_{U\otimes Q}} (U\otimes Q \xrightarrow{\psi} V) \rightarrow 0,$$

$$0 \rightarrow (\mathrm{Ker}\psi\otimes Q\otimes Q \xrightarrow{0} 0) \xrightarrow{(0,1_{\mathrm{Ker}\psi\otimes Q})} (\mathrm{Ker}\psi\otimes Q \rightarrow \mathrm{Ker}\psi\otimes Q) \xrightarrow{\begin{pmatrix}1_{\mathrm{Ker}\ \psi} \\ 0\end{pmatrix}} (\mathrm{Ker}\psi\otimes Q \xrightarrow{0} 0) \rightarrow 0.$$

Dually we have

<u>THEOREM 2.4.</u>

(i) Let X be a non-injective indecomposable A-module. Then

$$T_r D^R(X) = T_r D^A(X),$$

where in the left term X is considered as an R-module.

(ii) Let X be an injective indecomposable A-module and $T_r D^R(X) = (U \xrightarrow{\zeta} [Q, V])$.

Then

$$\mathrm{Cok}\ \zeta_A = T_r D^A([Q, X]).$$

(iii) Let $X = (U \xrightarrow{\phi} [Q, V])$ be an indecomposable R-module and $T_r D^R(X) = (U' \xrightarrow{\zeta} [Q, V'])$.

Then

$$\mathrm{Cok}\ \zeta_A = T_r D^A(\mathrm{Cok}\ \phi_A).$$

(iv) Let $X = (U \xrightarrow{\phi} [Q, V])$ be an indecomposable R-module. If $\mathrm{Cok}\,\phi_A$ is injective, then it holds that

$$T_rD^R(X) \cong ([Q, \mathrm{Cok}\,\phi] \xrightarrow{0} [Q, 0]),$$

i.e. $T_rD^R(X)$ is isomorphic to an A-module [Q, Cok].

THEOREM 2.5. Let A be a hereditary artin algebra and R a trivial extension of A by Q, where Q is ${}_A\mathrm{Hom}_C(A,I)_A$. If R is of finite representation type, then non-isomorphic indecomposable injective (resp. Projective) A-modules is the DT_r-bases (resp. T_rD-bases) as R-modules, i.e. any non-injective (resp. non-projective) indecomposable R-module can be obtained by taking DT_r^R (resp. T_rD^R) repeatedly from indecomposable injective (resp. projective) A-modules.

Proof : Dlab-Ringel [4] and Platzek-Auslander [10] proved that if a hereditary artin algebra A is of finite representation type, then every indecomposable A-module can be obtained by applying Coxeter functor C^+ repeatedly to indecomposable injective A-modules. Further in this case Brenner-Butler [2] proved C^+ is equivalent to DT_r^A.

At first, take an indecomposable injective A-module X and, considering X as an R-module, apply DT_r^R to X. Then by THEOREM 2.2,(i) we obtain $DT_r^A(X)$. Applying DT_r^R,i.e. DT_r^A several times we obtain at last a projective A-module X' by the above quoted result. Now, take $X'\otimes Q$ which is an indecomposable injective A-module, and put $Y = DT_r^A(X'\otimes Q)$ and $DT_r^R(X') = (U\otimes Q \xrightarrow{\psi} V)$. Then by THEOREM 2.2,(iii) $\mathrm{Ker}\,\psi \cong Y$. Further Theorem 2.2,(ii) implies that $\mathrm{Ker}\,\psi' \cong (DT_r^A)^i\, Y$ for $i = 1,2,\ldots$, where $(DT_r^R)^{i+1}(X') = (U'\otimes Q \xrightarrow{\psi'} V')$.

Hence by the above quoted result we know all non-projective indecomposable R-modules of 2nd kind and also all indecomposable R-modules of 1st kind are obtained by applying

DT_r^R repeatedly to indecomposable injective A-modules X's.

The proof for dual statement will be easy from THEOREM 2.4.

Quite recently T.Wakamatsu [14] proved the converse of THEOREM 1.4.

THEOREM 2.6. Let A be an artin algebra and $R = A \times Q$. If each indecomposable R-module is isomorphic to either an indecomposable A-module or $(U \otimes Q \xrightarrow{\alpha} V)$ such that α is an isomorphism or a monomorphism, U_A is projective and V_A is injective, and for a non-isomorphic monomorphism α $\mathrm{Ker}\,\alpha$ is essential in $U \otimes Q_A$ and indecomposable, then A is hereditary.

3. Auslander-Reitern quivers in mod-R and mod-A

Though his situation is more general than ours, K.Yamagata studied in a recent article [15] Auslander-Reiten quivers in mod R for R of finite representation type. He has proved that the subdiagram consist of all indecomposable projective A-modules appears as a tree of stable Auslander-Reiten quiver in mod-R. Since the subdiagram is identified with the quiver of hereditary algebra A, his result clarifies in our case the reason why Dynkin diagrams play an important role in the classification of self-injective algebras $R = A \ltimes Q$. Further it suggest us every Dynkin diagram, not only A_n, D_n, E_6, E_7, E_8 but also B_n, C_n, F_4, G_2, appears as a tree of Auslander-Reiten quiver in Mod-R.

In this section we shall give a survey on Auslander-Reiten quiver in Mod R for R of infinite representation type.

LEMMA 3.1. Let $(U \otimes Q \xrightarrow{\psi} V)$ and $(U' \otimes Q \xrightarrow{\psi'} V')$ be canonical expressions of indecomposable R-modules of 2nd kind and $X = (X \otimes Q \xrightarrow{0} 0)$ an A-module. If

$$\begin{array}{ccc}
(U \otimes Q \xrightarrow{\psi} V) & \xrightarrow{\begin{pmatrix}\alpha & \delta \\ 0 & \beta\end{pmatrix}} & (U' \otimes Q \xrightarrow{\psi'} V') \\
& \begin{pmatrix}\sigma_1 \\ \tau_1\end{pmatrix} \searrow \qquad \nearrow (\sigma_2, \tau_2) & \\
& (Y \otimes Q \xrightarrow{0} 0) &
\end{array}$$

is commutative for R-homomorphisms $\begin{pmatrix}\alpha & \delta \\ 0 & \beta\end{pmatrix}$, $\begin{pmatrix}\sigma_1 \\ \tau_1\end{pmatrix}$, (σ_2, τ_2), then $\alpha, \beta = 0$ and $\tau_1, \sigma_2 = 0$

Proof : From the commutativity we have $\psi\, \beta = 0$. then $\beta = 0$ because ψ is an epimorphism. $\alpha = 0$ follows from the following commuative diagram :

$$\begin{array}{ccc}
U \cong [Q, U \otimes Q] & \xrightarrow{[Q,\ V]} & [Q,\ V] \\
\alpha \downarrow & & \downarrow [Q,\ \beta] \\
U' \cong [Q, U' \otimes Q] & \xrightarrow{[Q, \psi']} & [Q,\ V']
\end{array}$$

with a monomorphism $[Q, \psi']$.

As similarly as β, we obtain $\tau_1 = 0$. The commutative diagram

$$\begin{array}{ccc}
Y & \longrightarrow & [Q,\ 0] \\
\sigma_2 \downarrow & & \downarrow \\
U' \cong [Q, U' \otimes Q] & \xrightarrow{[Q, \psi']} & [Q,\ V']
\end{array}$$

with a monomorphism $[Q, \psi']$ induces also $\sigma_2 = 0$.

<u>LEMMA</u> 3.2. Let $X = (U \otimes Q \to V)$ and $X' = (U' \otimes Q \xrightarrow{\psi'} V')$ be indecomposable R-modules of 2nd kind and $\Theta : X \to X'$ an R-homomorphism. Let θ : Ker $\psi \to$ Ker ψ' be the A-homomorphism induced by Θ. Then Θ is irreducible if and only if θ is irreducible.

Proof. Suppose we have a factorization of Θ such as in LEMMA 3.1 :

$$\Theta : X \xrightarrow{\binom{\sigma_1}{\tau_1}} Y \xrightarrow{(\sigma_2,\tau_2)} X' ,$$

then $\alpha = 0$ and $\beta = 0$ for $\theta = \begin{pmatrix} \alpha & \delta \\ 0 & \beta \end{pmatrix}$.

Hence both Θ and θ are not irreducible, for Θ is neither monomorphism nor epimorphism and $\theta = \alpha \otimes Q|\mathrm{Ker}\ \psi$. Therefore any factorization of $\Theta = \begin{pmatrix} \alpha & \delta \\ 0 & \beta \end{pmatrix}$ may be assumed to be the following one :

$$\Theta : X \xrightarrow{\Theta_1} (U''\otimes Q \xrightarrow{\psi''} V'') \xrightarrow{\Theta_2} X' \quad \text{with} \quad \psi'' \neq 0 ,$$

and $\begin{pmatrix} \alpha & \delta \\ 0 & \beta \end{pmatrix} = \begin{pmatrix} \alpha_1 & \delta_1 \\ 0 & \beta_1 \end{pmatrix}\begin{pmatrix} \alpha_2 & \delta_2 \\ 0 & \rho_2 \end{pmatrix}$ for $\Theta_i = \begin{pmatrix} \alpha_i & \delta_i \\ 0 & \beta_i \end{pmatrix}$, $i = 1,2.$

Then the above factorization induces a factorization of θ : $\mathrm{Ker}\ \psi \xrightarrow{\theta_1} \mathrm{Ker}\ \psi'' \xrightarrow{\theta_2} \mathrm{Ker}\ \psi'$ such that

$\theta_1 = \alpha_1 \otimes Q \mid \mathrm{Ker}\ \psi$ and $\theta_2 = \alpha_2 \otimes Q \mid \mathrm{Ker}\ \psi''$.

Assume Θ is irreducible, then either Θ_1 is a splittable monomorphism or Θ_2 is a splittable epimorphism. It results obviously either θ_1 is a splittable monomorphism or θ_2 is a splittable epimorphism.

Conversely, assume θ is irreducible, then either θ_1 is a splittable monomorphism or θ_2 is a splittable epimorphism. If θ_1 is a splittable epimorphism, then at the following commutative diagram :

$$\begin{array}{ccccccccc}
0 & \longrightarrow & \mathrm{Ker}\ \psi & \longrightarrow & U \otimes Q & \xrightarrow{\psi} & V & \longrightarrow & 0 \\
 & & \big\downarrow \theta_1 & & \big\downarrow \alpha_1 \otimes Q & & \big\downarrow \beta_1 & & \\
0 & \longrightarrow & \mathrm{Ker}\ \psi'' & \longrightarrow & U''\otimes Q & \longrightarrow & V'' & \longrightarrow & 0
\end{array}$$

$\alpha_1 \otimes Q$ is a splittable monomorphism, because $\mathrm{ker}\ \psi$ is essential in an injective A-module $U \otimes Q$. Hence by PROPOSITION 1.1,(ii) α_1 is a splittable monomorphism and β_1 is also a splittable monomorphism.

Thus there are A-homomorphisms $\alpha_1' : U'' \to U$ and $\beta_1' : V'' \to V$ such that $\alpha_1\alpha_1' = 1_U$, $\beta_1\beta_1' = 1_V$ and they are compatible with

$$\begin{array}{ccc} U''\otimes Q & \xrightarrow{\psi''} & V'' \\ \Big\downarrow \alpha_1'\otimes Q & \psi & \Big\downarrow \beta_1' \\ U \otimes Q & \longrightarrow & V \end{array} .$$

Now putting $\Theta' = \begin{pmatrix} \alpha_1' & \delta_1' \\ 0 & \beta_1' \end{pmatrix}$ for $\delta_1' = -\alpha_1'\delta_1\beta_1'$

we have an R-homomorphism Θ' and $\Theta \cdot \Theta' = \begin{pmatrix} 1_U & 0 \\ 0 & 1_V \end{pmatrix}$ implies Θ is a splittable monomorphism.

In case of θ_2 being a splittable epimorphism we can prove similarly that Θ_2 is a splittable epimorphism.

LEMMA 3.3. Let $\Theta : X \quad Y$ be an irreducible R-homomorphism.

(i) If X is of 2nd kind and Y is of 1st kind, then Y is a projective A-module.

(ii) If X is of 1st kind and Y is of 2nd kind, then X is an injective A-module.

Proof : For the case (i) we may put $\Theta = (\alpha, 0)$ for $\alpha : U \to Y$, where $X = (U \otimes Q \xrightarrow{\psi} V)$ and $Y = (Y \otimes Q \xrightarrow{0} 0)$. Suppose Y_A is not projective. Then there is an epimorphism $\alpha_2 : P \to Y$ such that α_2 is not splittable and P_A is projective. Further we have a factorization of $\alpha : U \xrightarrow{\alpha_1} P \xrightarrow{\alpha_2} Y$. It follows a factorization of Θ :

$$(U \otimes Q \xrightarrow{\psi} V) \xrightarrow{(\alpha_1,\ 0)} (P \otimes Q \to 0) \xrightarrow{\alpha_2} (Y \otimes Q \to 0).$$

But since $(\alpha_1, 0)$ is not a (splittable) monomorphism,

$_2$ is a splitable epimorphism. This contradicts Y_A is not projective.

The proof for the case (ii) is obtained dually.

Now assume R is of infinite representation type and let X be an indecomposable injective A-module. Then X is not projective, for otherwise R is of finite representation type. And therefore by THEOREM 2.2 $DT_r^R(X) \cong DT_r^A(X)$. Let

$$0 \longrightarrow DT_r^R(X) \longrightarrow \underset{i}{\oplus} X_i^{(1)} \longrightarrow X \longrightarrow 0$$

be an almost splittable R-sequence and $X_i^{(1)}$, i = 1,2,..., indecomposable R-modules. Then all $X_i^{(1)}$ are of 1st kind, i.e. A-modules and hence pre-injective A-modules, for if some $X_i^{(1)}$ is of 2nd kind, $DT_r^R(X)$ must be an injective A-module by LEMMA 3.3, but $DT_r^A(X)$ is clearly not injective.

Next, let

$$0 \longrightarrow X \rightarrow \underset{i}{\oplus} X_j^{(-1)} \longrightarrow T_rD^R(X) \longrightarrow 0$$

be an almost splittable R-sequence and $X_j^{(-1)}$, j=1,2,..., indecomposable R-modules. Then by THEOREM 2.4 $T_rD^R(X)$ is of 2nd kind, and the canonical kernel of $T_rD^R(X)$ is isomorphic to $T_rD^A([Q, X])$ which is a pre-projective A-module. If $X_j^{(-1)}$ is of 1st kind, by LEMMA 3.3 it is an injective A-module. If $X_j^{(-1)}$ is of 1st kind, by LEMMA 3.3, it is an injective A-module. If $X_i^{(-1)}$ is of 2nd kind, by LEMMA 3.2 the cononical kernel of $X_j^{(-1)}$ is Pre-projective.

So, if we denote one of $X_i^{(1)}$, $X_j^{(-1)}$ by Y, then Y may be assumed to be an R-module of the following types :

(i) Y is a pre-injective A-module or

(ii) Y is an R-module of 2nd kind,

of which the canonical kernel is a pre-projective A-module.

Next, replacing X by Y, as similarly as above we can

construct modules $Y_i^{(1)}$ and $Y_j^{(-1)}$.

In case Y being of type (i), Y_A is not projective and by a similar argument as above we know $Y_i^{(1)}$ and $Y_j^{(-1)}$ are again of type (i) and (ii).

If Y is of type (ii), then by LEMMAS 3.2 and 3.3 $Y_i^{(1)}$ is either an injective A-module or an R-module of 2nd kind with a pre-projective canonical kernel, and $Y_j^{(-1)}$ is of 2nd kind with a pre-projective canonical kernel, because the canonical kernel of Y is not injective.

Hence repeating the above discussions and taking their duals we can conclude.

<u>THEOREM 3.4</u>. Assume that a hereditary artin algebra A is of infinite representation type and denote by Λ(resp. Γ) the connected component of stable Auslander-Reiten quiver in Mod-A consist of all pre-injective (resp. pre-projective) A-modules. Then the connected component of stable Auslander-Reiten quiver Σ(resp. T) in Mod-R containing an indecomposable injective (resp. projective) A-module is obtained by indentifying injective X in Λ(resp. projectives Y in Γ) with projectives $[Q, X]$ in Γ(resp. injectives $Y \otimes Q$ in Λ). And each indecomposable R-module in Σ(resp. T) which corresponds to a vertex of Γ(resp. Λ) is of 2nd kind such that its canonical kernel is isomorphic the same vertex of Γ(resp. Λ).

Further by LEMMAS 3.2 and 3.3 we have

<u>THEOREM 3.5</u>. Under the same assumption with THEOREM 3.4 it holds that

(i) each connected component of stable Auslander-Reiten quiver in Mod-A consist of regular modules becomes itself one in Mod-R.

(ii) Let S be a set of indecomposable R-modules such that the canonical kernel of at least one R-module belonging to S is a regular A-module. Then S becomes a set of vertices of a stable Auslander-Reiten quiver in mod-R if and only if the set of canonical kernels becomes one of vertices of stable Auslander-Reiten quiver in Mod-A.

Now according to THEOREMS 3.5 and 3.6. It seems to us the following definitions are reasonable : An indecomposable R-module of 2nd kind is said to be pre-projective (resp. pre-injective, regular) if its canonical kernel is pre-injective (resp. pre-projective, regular as an A-module).

There are several interesting results on the structure and the construction of regular modules over hereditary algebras [4], [5], [11] and [12]. At the end of this section we would like mension the following THEOREM 3.7 showing that the structure of regular R-modules of 2nd kind is similar to one of regular A-modules.

THEOREM 3.7. Let $X(=U \otimes Q \xrightarrow{\psi} V)$ and $X'(=U' \otimes Q \xrightarrow{\psi'} V')$ be regular R-modules of 2nd kind. Then

$$\mathrm{Ext}^1_R(X, X') \cong \mathrm{Ext}^1_A(\mathrm{Ker}\ \psi, \mathrm{Ker}\ \psi').$$

So defining quasi-simple R-modules of 2nd kind, as regular R-modules with quasi-simple A-modules as their canonical kernels, we have similar theorems concerning regular R-modules as ones of hereditary algebras A. C.f.[12].

References

[1] Auslander, M. and Reiten, I. : Representation theory of artin algebras III ; almost split sequences. Comm. in Algebra 3, 239-294 (1975).

[2] Brenner, S. and Butler, M. C. R. : The equivalence of certain functors occuring in the representation theory of artin algebras and species, J. London Math. Soc. (2) 14, 183-187 (1976).

[3] Dlab, V. and Ringel, C. M. : On algebras of finite representation type, J. Algebra 33, 306-394(1975).

[4] Dlab, V. and Ringel, C. M. : Representations of graphs and algebras, Memoirs Amer. Math. Soc. 173, Providence (1976).

[5] Dlab, V. and Ringel, C. M. : The representations of tame hereditary algebras, Lecture notes in pure and applied math. 37 (Proc. Philadelphia conf.),329-353 (1976).

[6] Fossum, R. M., Griffith, P. A. and Reiten, I : Trivial extensions of abelian categories, Lecture notes in math. 456, Berlin-Heidelberg-New York : Springer (1975).

[7] Gabriel, P. : Indecomposable representations II, Symposia Mathematica, Vol. XI, Academic Press, New york/ San Francisco/London, 81-104 (1973).

[8] Green, E. L. and Reiten, I : On the construction of ring extensions, Glasgow Math. J. 17, 1-11 (1976).

[9] Müller, W. : Unzerlegbare Moduln über Artinschen Ringen, Math. Z. 137, 197-226 (1974).

[10] Platzeck, M. T. and Auslander, M. : Representation theory of hereditary artin algebras, Lecture notes in pure and applied math. 37 (Proc. Phildelphia conf.), 389-353 (1976).

[11] Ringel, C. M. : Representations of K-species and bimodules, J. Algebra 41, 269-302(1976).

[12] Ringel, C. M. : Finite dimensional hereditary algebras of wild representation type (to appear).

[13] Tachikawa, H. : Trivial extensions of finite representation type, Proc. Symp. on Representations of groups and algebras, University of Tsukuba, Ibaraki, 69-80 (1978) (In Japanese).

[14] Wakamatsu, T. : Trivial extensions of Artin algebras (to apear)

[15] Yamagata, K : Extensions over hereditary artinian rings with self-dualities 1 (to appear).

ALMOST SPLIT SEQUENCES FOR TrD-PERIODIC MODULES

Gordana Todorov

Brandeis University, Waltham, Massachusetts
University of Georgia, Athens, Georgia

Abstract

In this paper we associate to each TrD-periodic module, over an artin algebra, a diagram and show that the diagram is one of the Dynkin diagrams or one of the

A_∞

${}_\infty A_\infty$

B_∞

C_∞

D_∞

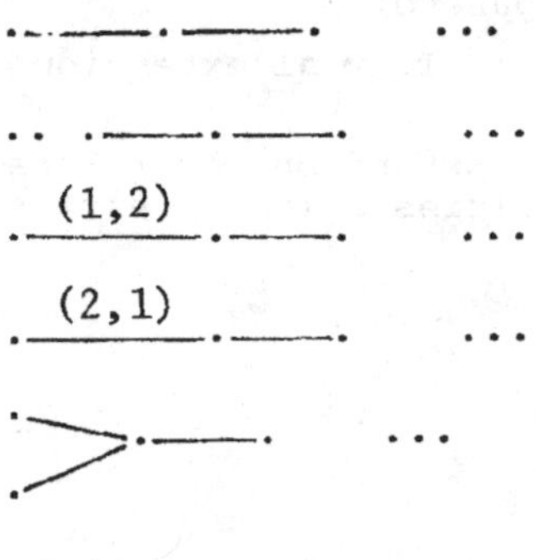

If the algebra is of finite representation type we show that the diagram is a Dynkin diagram.

Introduction

Throughout this paper we assume that Λ is an artin algebra, that is an artin ring that is a finitely generated module over its center C, which is also an artin ring. We denote by mod Λ the category of finitely generated Λ-modules. Let D : mod $\Lambda \to$ mod Λ^{op} be the usual duality given by $X \to \mathrm{Hom}_C(X,I)$, where I is the injective envelope over C of C/rad C and let Tr : $\underline{\mathrm{mod}}\ \Lambda \to \underline{\mathrm{mod}}\ \Lambda^{op}$ be the duality between the category $\underline{\mathrm{mod}}\ \Lambda$ of finitely generated modules modulo projectives over Λ and $\underline{\mathrm{mod}}\ \Lambda^{op}$, the category of finitely generated modules modulo projectives over Λ^{op} given by the transpose (see [3]). Let M be an indecomposable Λ-module. We say M is TrD-periodic if there exists an integer $k \neq 0$ such that $\mathrm{TrD}^k M \simeq M$.

In this paper we associate to each TrD-periodic module M a diagram $\mathcal{D}_M$ in the following way:
To the module M we associate the point $\cdot$m. Let

$$0 \to M \to E \to \mathrm{TrDM} \to 0$$

be an almost split sequence (for the definition and properties see [3]). Let E_1 be an indecomposable summand of E_1 which is TrD-periodic. Let d be the multiplicity of E_1 in E. Let

$$0 \to DTrE_1 \to F \to E_1 \to 0$$

be an almost split sequence and d′ the multiplicity of M in F. Then we put an arrow $m \cdot \overset{(d,d')}{} \cdot 1$. We define the rest of the diagram by induction. Suppose $\cdot j$ is an end point of an arrow. Let

$$0 \to E_j \to E \to TrDE_j \to 0$$

be an almost split sequence. Let E_{j+1} be an indecomposable summand of E, which is TrD-periodic, but not isomorphic to $TrDE_{j-1}$. We define the multiplicities in the same way as we did in the first step.

The Dynkin diagrams which were used to give a classification of hereditary artin algebras of finite representation type [8], appear now in a study of TrD-periodic modules over arbitrary artin algebras in the following result.

<u>THEOREM</u>: Let Λ be an artin algebra and M a TrD-periodic module. Then the diagram $\mathcal{D}_M$ is either a Dynkin diagram or one of the:

A_∞

${}_\infty A_\infty$

B_∞

C_∞

D_∞

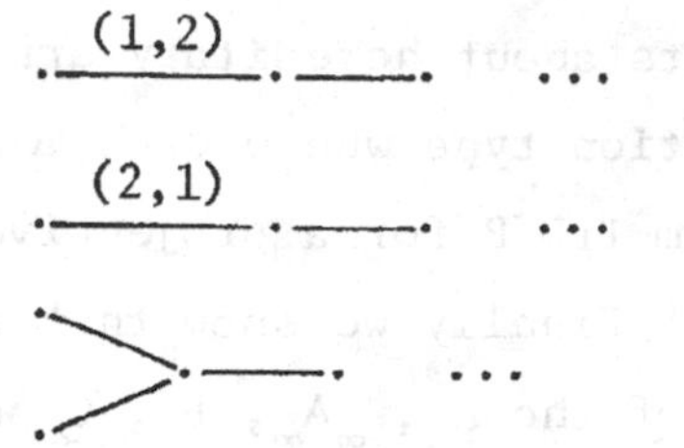

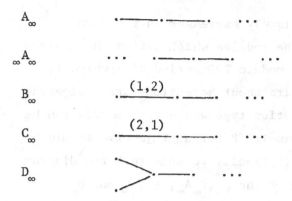

As a consequence we obtain that if Λ is of finite representation type, and if we consider only non-isomorphic modules then the diagram is a Dynkin diagram.

In sections 2 and 3 we consider more general definition (i.e. without the restriction to non-isomorphic modules). We show that if Λ is a selfinjective artin algebra of finite representation type then the diagram is a Dynkin diagram. This way our results from section 1 specialize to some of the results that Christine Riedtmann obtained for selfinjective artin algebras of finite representation type over algebraically closed field [9].

Section 3 is about TrD-periodic modules with the property that there are no projective, and therefore no injective modules in the class [M] ([M] denotes the class of all modules which are connected to M by chains of irreducible maps)(for the definition and properties of irreducible maps see [4]). We say that two modules X and Y are TrD-isomorphic if there exists an integer k such that $Y \simeq TrD^k X$. We show that if two modules correspond to two different points in the diagram then they are not TrD-isomorphic. So we obtain that every module in the class [M] can be uniquely expressed in the form TrD^kE, where E is one of the modules which define the diagram $\mathcal{D}_M$, and k is unique modulo TrD-period of E, which is similar to the results about hereditary artin algebras of finite representation type where each module can be expressed in the form TrD^kP for a projective module P (see [8] and [2]). Finally we show that the diagram in this case is one of the A_∞, ${}_\infty A_\infty$, B_∞, C_∞ or D_∞.

I want to thank María Inés Platzeck for many helpful conversations and suggestions.

1. General Case

In this section we will define certain chains of irreducible maps for a given TrD-periodic module M and give a description of almost split sequences for TrD-periodic modules which are in the class [M], in terms of the chains of irreducible maps. We will also associate a diagram to the module and show that the diagram is one of the Dynkin diagrams or A_∞, ${}_\infty A_\infty$, B_∞, C_∞ or D_∞ with no restriction on the algebra or the module. And if algebra is of finite representation type we will show that the diagram is a Dynkin diagram.

DEFINITION 1.1 Let M be a TrD-periodic module and let

$$0 \to M \to \coprod_{j=1}^{s} E_j \coprod X \to \mathrm{TrDM} \to 0$$

be an almost split sequence, where E_j's are TrD-periodic and X has no TrD-periodic summands. Then we define:

$$\delta(M) = s \text{ and } \delta(\Lambda) = \sup\{\delta(M) \mid M \text{ TrD-perodic } \Lambda\text{-module}\}$$

Let $\mathcal{C}$ be a collection of finitely generated modules. Then we define

$$\delta(\mathcal{C}) = \sup\{\delta(M) \mid M \text{ in } \mathcal{C}\}$$

To each TrD-periodic module M we associate chains of irreducible maps in the following way. Let

$$0 \to M \to E \to \mathrm{TrD}M \to 0$$

be an almost split sequence. Let E_1 be a TrD-periodic summand of E. We define E_{j+1} by induction: let

$$0 \to E_j \to E' \to \mathrm{TrD}E_j \to 0$$

be an almost split sequence. Define E_{j+1} to be a TrD-periodic summand of E′, not isomorphic to $\mathrm{TrD}E_{j-1}$, if it exists. This way we obtain chains of irreducible maps between TrD-periodic modules. Let

$$\mathscr{A}_M = \{M \to E_1^i \to E_2^i \to \dots\}_{i \in I}$$

be all such chains, where two chains are considered to be the same if $E_j^i \simeq E_j^{i'}$ for all j. If a chain is finite we will denote the last module by $E_{s_i}^i$.

We will give information about these chains: possible numbers of them and lengths.

<u>DEFINITION 1.2</u> Let $M \to \coprod_{i=1}^{s} E_i$ be a minimal left almost split map where E_i are indecomposable modules. We define $\alpha_L(M) = s$ and $\beta_L(M)$ to be the number of non-projective summands. Similarly, if $\coprod_{i=1}^{t} E_i \to M$ is a minimal right almost split map, we define $\alpha_R(M) = t$ and $\beta_R(M)$ to be the number of non-projective summands. (for the definitions see [4])

<u>LEMMA 1.3</u> Let Λ be an artim algebra and M a non-injec-

tive, non-projective indecomposable Λ-module with $\alpha_L(M) \geq 4$. Then:

a) $\ell(\mathrm{TrDM}) - \ell(M) \geq \ell(M) - \ell(\mathrm{DTrM})$

b) The equality holds if and only if $\alpha_R(M) = \beta_L(M)$ and if E is an indecomposable module and $M \to E$ is irreducible map, then $\alpha_R(E) = 1$.

PROOF: Let $\alpha_L(M) = t \geq 4$. Since M is non-injective, there is an almost split sequence:

$$0 \to M \to \coprod_{i=1}^{s} E_i \coprod (\coprod_{i=1}^{s'} E_i') \to \mathrm{TrDM} \to 0$$

with E_i non-projective and all E_i' projective modules and $s + s' = t$. Then:

$$(1) \quad \ell(\mathrm{TrDM}) - \ell(M) = \sum_{i=1}^{s} \ell(E_i) + \sum_{i=1}^{s'} \ell(E_i') - 2\ell(M) =$$

$$= \sum_{i=1}^{s} [\ell(E_i) - \ell(M)] + \sum_{i=1}^{s'} [\ell(E_i') - \ell(M)] + (t-2)\ell(M).$$

Since E_i is not projective there exists an almost split sequence:

$$0 \to \mathrm{DTrE}_i \to F_i \coprod M \to E_i \to 0$$

where F_i might be zero. From this we obtain:

$$\ell(E_i) - \ell(M) = \ell(F_i) - \ell(\mathrm{DTrE}_i).$$

So from (1) we have:

$$(2)\quad \ell(\mathrm{TrDM}) - \ell(M) = \sum_{i=1}^{s} [\ell(F_i) - \ell(\mathrm{DTrE}_i)] + \sum_{i=1}^{s'} [\ell(E_i') - \ell(M)] +$$

$$+ (t - 2)\ell(M)$$

Since M is not projective there exists an almost split sequence:

$$0 \to \mathrm{DTrM} \to I \coprod (\coprod_{i=1}^{s} \mathrm{DTrE}_i) \to M \to 0$$

where I is either zero or injective. Then

$$\ell(M) - \sum_{i=1}^{s} \ell(\mathrm{DTrE}_i) = \ell(I) - \ell(\mathrm{DTrM}).$$

This combined with (2) gives:

$$(3)\quad \ell(\mathrm{TrDM}) - \ell(M) = \sum_{i=1}^{s} \ell(F_i) + \ell(I) - \ell(\mathrm{DTrM}) - \ell(M) +$$

$$+ \sum_{i=1}^{s'} [\ell(E_i') - \ell(M)] + (t - 2)\ell(M) =$$

$$= \ell(M) - \ell(\mathrm{DTrM}) + (t - 4)\ell(M) + \sum_{i=1}^{s'} [\ell(E_i') - \ell(M)] +$$

$$+ \sum_{i=1}^{s} (F_i) + \ell(I).$$

We are assuming $t \geq 4$ and E_i' are projective so that $\ell(E_i') - \ell(M) > 0$. So we have:

$$\ell(\mathrm{TrDM}) - \ell(M) \geq \ell(M) - \ell(\mathrm{DTrM}).$$

From (3) it follows that the equaility holds if and only if $t = 4$, $s = 0$, $F_i = 0$ and $I = 0$. This implies that

$\alpha_L(M) = \beta_L(M) = 4 = \alpha_R(M)$. From $F_i = 0$ it follows that $\alpha_R(E_i) = 1$ for all i.

As a corollary we have the following :

PROPOSITION 1.4 Let Λ be an artin algebra and M an indecomposable TrD-periodic module. Then there exists an integer k such that $\alpha_L(\mathrm{TrD}^k M) \leq 3$.

PROOF: Let $\mathrm{TrD}^m M \simeq M$. If $\alpha_L(\mathrm{TrD}^i M) \geq 4$ for all i, then by lemma 1.3 $\ell(\mathrm{TrDM}) - \ell(M) = \ell(\mathrm{TrD}^{m+1}M) - \ell(\mathrm{TrD}^m M) \geq \ldots \geq$ $\geq \ell(\mathrm{TrDM}) - \ell(M)$. Thus equality must hold and by lemma 1.3 b) the only modules in the class of [M] are M, $\mathrm{TrD}^i M$ for i = 1, ..., m-1 and E_j and $\mathrm{TrD}^i E_j$, where $0 \to M \to \coprod E_j \to \mathrm{TrDM} \to 0$ is an almost split sequence. None of this modules are projective. Hence there is a finite number of non-isomorphic modules in the class [M] , and there are no projectives in the class, which is impossible by [6].

COROLLARY 1.5 Let be an artin algebra. Then $\delta(\Lambda) \leq 3$.

PROOF: Let M be a module in mod Λ. If M is TrD-periodic then $\delta(M) \leq 3$ by proposition 1.4.

Suppose M is not TrD-periodic and suppose $\delta(M) \geq 4$. Let

$$0 \to M \to \coprod_{i=1}^{s} E_i \coprod Y \to \mathrm{TrDM} \to 0$$

be an almost split sequence where E_i are TrD-periodic, Y is not periodic and $s \geq 4$. Let

$$0 \to DTrE_i \to Z \coprod M \to E_i \to 0$$

be an almost split sequence. Now by a result from [3] it follows that there is an integer k such that $DTR^k \cong P$, a projective module. Since $\delta(M) \geq 4$ it follows that $\alpha_L(TrD^k P) \geq 4$ for all k. Then by lemma 4.1.1 [10] it follows that

$$\ell(TrDP) - \ell(P) > 0 \text{ and by lemma 4.2.3 [10]}$$

$$\ell(TrD^k P) - \ell(TrD^{k-1} P) \geq \ell(TrD^{k-1} P) - \ell(TrD^{k-2} P) \geq \ldots > \ell(TrDP) - \ell(P).$$

By the result from [3] there is also an integer k' such that $TrD^{k'} P \cong I$, an injective module. But then

$$\ell(TrD^{k'} P) - \ell(TrD^{k'-1} P) > 0.$$

This is the same as

$$\ell(DTrI) - \ell(I) < 0$$

Which is impossible by lemma 4.1.1 b) [10].

Let M be a TrD-periodic module with $\delta(M) = 3$. Let $\mathscr{A}_M = \{M \to E_1^i \to E_2^i \to \ldots\}_{i \in I}$ be the chains of irreducible maps. For finite chains denote the last module by E_{s_i} and let $s_1 \leq s_2 \leq \ldots$. The following lemmas will show that $s_1 = 1$, $s_2 \leq 2$ and if $s_2 = 2$ then $s_3 \leq 4$.

<u>LEMMA 1.6</u> Let M be a TrD-periodic module with $\delta(M) = 3$.

Let

$$0 \to M \to \coprod_{i=1}^{3} E_i \coprod X \to \mathrm{TrD}M \to 0$$

be an almost split sequence, with E_i TrD-periodic and X has no TrD-periodic summands. Then there exists E_i such that $\delta(E_i) = 1$.

PROOF: Suppose not and consider the following graph of irreducible maps:

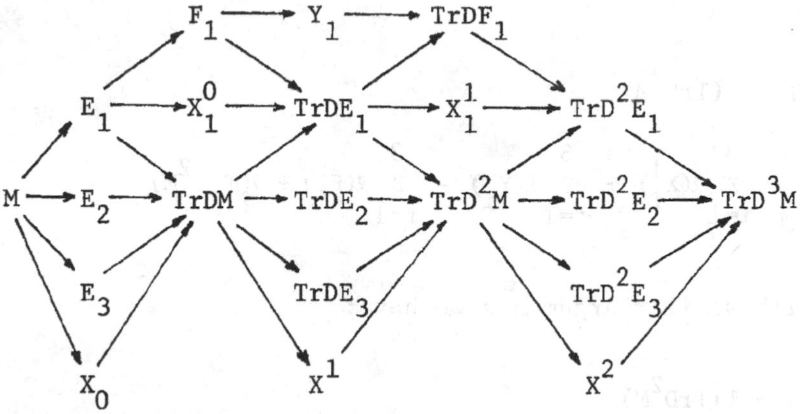

with similar almost split sequences for E_2 and E_3. Then:

(1) $\ell(\mathrm{TrD}^3M) - \ell(\mathrm{TrD}^2M) =$

$$= \ell(X^2) + \sum_{i=1}^{3} [\ell(\mathrm{TrD}^2E_i) - \ell(\mathrm{TrD}^2M)] + \ell(\mathrm{TrD}^2M)$$

Considering almost split sequences

$$0 \to \mathrm{TrD}E_i \to \mathrm{TrDF}_i \coprod X_i^1 \coprod \mathrm{TrD}^2M \to \mathrm{TrD}^2E_i \to 0.$$

From (1) we obtain:

$$(2)\quad \ell(\mathrm{TrD}^3M) - \ell(\mathrm{TrD}^2M) = \ell(X^2) + \sum_{i=1}^{3} \ell(X_i^1) + \sum_{i=1}^{3} \ell(\mathrm{TrDF}_i)$$

$$- \sum_{i=1}^{3} \ell(\mathrm{TrDE}_i) + \ell(\mathrm{TrD}^2M).$$

Now from (2) and almost split sequences

$$0 \to F_i \to Y_i \coprod \mathrm{TrDE}_i \to \mathrm{TrDF}_i \to 0$$

we have

$$(3)\quad \ell(\mathrm{TrD}^3M) - \ell(\mathrm{TrD}^2M)$$

$$= \ell(X^2) + \sum_{i=1}^{3} \ell(X_i^1) + \sum_{i=1}^{3} \ell(Y_i) - \sum_{i=1}^{3} \ell(F_i) + \ell(\mathrm{TrD}^2M)$$

Continuing with similar arguments we have:

$$(4)\quad \ell(\mathrm{TrD}^3M) - \ell(\mathrm{TrD}^2M) =$$

$$= \ell(X^2) + \sum_{i=1}^{3} \ell(X_1^i) + \sum_{i=1}^{3} \ell(Y_i) - \sum_{i=1}^{3} \ell(F_i) +$$

$$+ \ell(X^1) + \sum_{i=1}^{3} \ell(\mathrm{TrDE}_i) - \ell(\mathrm{TrDM}) + 2\ell(\mathrm{TrDM}) =$$

$$= \ell(X^2) + \sum_{i=1}^{3} \ell(X_1^i) + \sum_{i=1}^{3} \ell(Y_i) - \sum_{i=1}^{3} \ell(F_i) +$$

$$+ \ell(X^1) + \sum_{i=1}^{3} \ell(X_i) + \sum_{i=1}^{3} \ell(F_i) - \sum_{i=1}^{3} \ell(E_i) + 2\ell(\mathrm{TrDM}) =$$

$$= \ell(\mathrm{TrDM}) - \ell(M) + \ell(X^0) + \ell(X^1) + \ell(X^2) +$$

$$+ \sum_{i=1}^{3} [\ell(X_i^0) + \ell(X_i^1) + \ell(Y_i)].$$

Hence $\ell(\mathrm{TrD}^3 M) - \ell(\mathrm{TrD}^2 M) \geq \ell(\mathrm{TrDM}) - \ell(M)$ and the equality holds if and only if all the modules X^j, X_i^j, Y_i are zero. Similarly $\ell(\mathrm{TrD}^{j+3}M) - \ell(\mathrm{TrD}^{j+2}M) \geq \ell(\mathrm{TrD}^{j+1}M) - \ell(\mathrm{TrD}^j M)$ and the equality holds if and only if the corresponding modules in the almost split sequences are zero. Since M is TrD-periodic the equalities must hold and therefore there is only a finite number of modules in [M] and all of them are non-projective and non-injective. Contradiction by [6]. |

From Lemma 1.6 and with the notation introduced before the lemma we know that $s_1 = 1$.

<u>LEMMA 1.7</u> Let $\mathscr{A}_M = \{M \to E_1^i \to E_2^i \to \ldots\}_{i \in I}$ be the chains of irreducible maps and suppose $\delta(M) = 3$. Then $s_2 \leq 2$.

<u>PROOF:</u> Suppose not and consider the following graph of irreducible maps, where ΔX denotes TrDX.

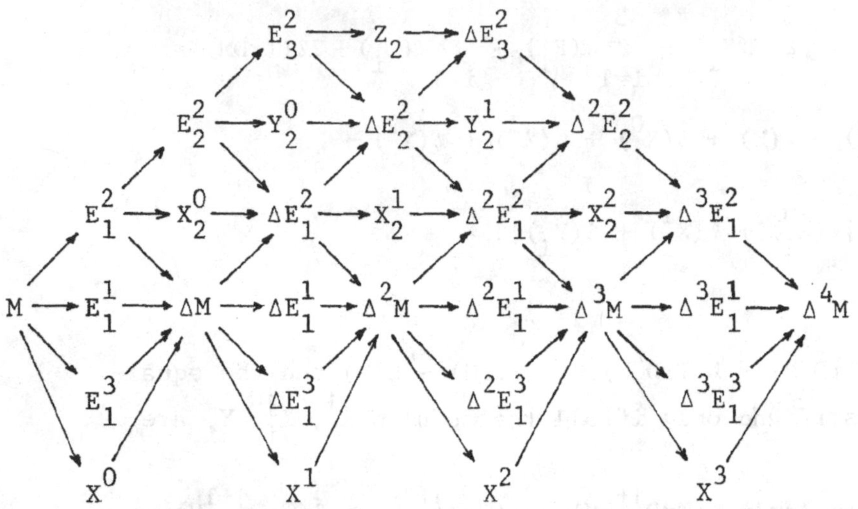

There are similar almost split sequences for E_1^3, E_2^3, E_3^3 and from the previous lemma we know that $\delta(E_1^1) = 1$. From the graph we have that

$$\ell(\mathrm{TrD}^4 M) - \ell(\mathrm{TrD}^3 M) = k + \ell(\mathrm{TrD}M) - \ell(M)$$

where k is a positive constant, which has as a summand

$$\ell(X^0) + \ell(X_1^0) + \ell(X_2^0) + \ell(X_3^0) + \ell(Y_2^0) + \ell(Y_3^0) + \ell(Z_2) + \ell(Z_3).$$

The same argument as in the previous lemma shows that this implies that the only modules in [M] are the TrD^i images of the modules that appear in the chains in $\mathscr{A}_M$. Contradiction by [6].

<u>LEMMA 1.8</u> Let $\mathscr{A}_M = \{M \to E_1^i \to E_2^i \to \ldots\}_{i \in I}$ be the chains of irreducible maps and suppose $\delta(M) = 3$, $s_1 = 1$ and $s_2 = 2$. Then $s_3 \leq 4$.

PROOF: If we consider again graph of irreducible maps assuming that $s_3 \geq 5$, similar length argument will show that $\ell(TrD^6M) - \ell(TrD^5M) \geq \ell(TrDM) - \ell(M)$ unless the only modules in the class [M] are the TrD^i images of the modules in $\mathscr{A}_M$, which gives a contradiction. |

REMARK We will talk about a module in $\mathscr{A}_M$, meaning a module that appears in a chain which is in $\mathscr{A}_M$.

The following lemma shows that among the modules in $\mathscr{A}_M$ at most one of them has $\delta(X) = 3$.

LEMMA 1.9 Let $\mathscr{A}_M = \{M \to E_1^i \to E_2^i \to \ldots\}_{i \in I}$ and suppose $\delta(M) = 3$. Then $\delta(E_j^i) \leq 2$.

PROOF: Let $M \to F_1$, $M \to F_2$, $M \to E_1 \to \ldots \to E_j \to N$ be chains of irreducible maps which are parts of the chains in $\mathscr{A}_M$, where $\delta(E_i) \leq 2$ for all $i \leq j$ and $\delta(N) = 3$. By considering almost split sequences we show that

$$\ell(TrD^{j+4}M) - \ell(TrD^{j+3}M) \geq \ell(TrDM) - \ell(M)$$

and the equality holds if the only modules in [M] are TrD^i images of the moduels in $\mathscr{A}_M$. Contradiction. |

LEMMA 1.10 Let $\mathscr{A}_M = \{M \to E_1^i \to E_2^i \to \ldots\}_{i \in I}$ be the chains of irreducible maps and suppose $\delta(M) = 3$. Then $\delta(E_{s_i}) = 1$ for every finite chain $M \to E_1^i \to \ldots \to E_{s_i}^i$.

PROOF: If $\delta(E_{s_i}^i) = 2$ then an almost split sequence for

$E^i_{s_i}$ looks like $0 \to E^i_{s_i} \to TrDE^i_{s_i-1} \coprod TrDE^i_{s_i-1} \coprod Y \to TrDE^i_{s_i} \to 0$.

Consider the chain of irreducible maps

$$M \to E^i_1 \to \ldots \to E^i_{s_i} \to TrDE^i_{s_i-1} \to TrD^2E^i_{s_i-2} \to \ldots \to TrD^{s_i}M.$$

Then $\delta(M) = \delta(TrD^{s_i}M) = 3$ and $\delta(X) = 2$ for all other modules X in the chain. Using the same length argument as in the previous lemma we get a contradiction. |

<u>LEMMA 1.11</u> Let $\mathscr{A}_M = \{M \to E^i_1 \to E^i_2 \to \ldots\}_{i \in I}$ be the chains of irreducible maps. Suppose $\delta(\mathscr{A}_M) = 2$ and

$$0 \to M \to E \coprod E \coprod X \to TrDM \to 0$$

is an almost split sequence with E TrD-periodic. Then:

a) There is only one chain $M \to E_1 \to E_2 \ \ldots$.

b) If it is finite, then $\delta(E_s) = 1$.

<u>PROOF:</u> a) This is clear since $\delta(\mathscr{A}_M) = 2$.

b) Let $E_s = N$ and suppose $\delta(N) = 2$. Then an almost split sequence for N looks like:

$$0 \to N \to TrDE_{s-1} \coprod TrDE_{s-1} \coprod Y \to TrDN \to 0.$$

From the graph of irreducible maps we can see that $\ell(TrD^{s+1}M) - \ell(TrD^sM) \geq \ell(TrDM) - \ell(M)$ and if equality holds the only modules in the class [M] are TrD^i images of the modules in $\mathscr{A}_M$. Contradiction. |

To each TrD-periodic module M we will associate a diagram $\mathcal{D}_M$ in a similar way that we constructed the chains $\mathcal{A}_M$.

To the module M we associate the point .m. Let

$$0 \to M \to E \to TrDM \to 0$$

be an almost split sequence. Let E_1 be an indecomposable summand of E, which is TrD-periodic. Let d be the multiplicity of E_1 in E. Let

$$0 \to DTrE_1 \to F \to E_1 \to 0$$

be an almost split sequence and d′ the multiplicity of M in F. Then we put on arrow $m\cdot \xrightarrow{(d,d')} \cdot 1$. We define the rest of the diagram by induction. Suppose $\cdot_j$ is an end point of an arrow. Let

$$0 \to E_j \to E \to TrDE_j \to 0$$

be an almost split sequence. Let E_{j+1} be an indecomposable summand of E, which is TrD-periodic, but not isomorphic to $TrDE_{j-1}$. Let ℓ be the multiplicity of E_{j+1} in E and ℓ' be the multiplicity of E_j in the middle term of an almost split sequence

$$0 \to DTrE_{j+1} \to F \to E_{j+1} \to 0$$

Then we put an arrow $_j\cdot \xrightarrow{(\ell,\ell')} \cdot_{j+1}$

With the definition and from previous lemmas we have the following theorem.

<u>THEOREM 1.12</u> Let Λ be an artin algebra and let M be a TrD-periodic module. Then the diagram $\mathcal{D}_M$ is either one of the Dynkin diagrams or one of the diagrams A_∞, ${}_\infty A_\infty$, B_∞, C_∞ or D_∞.

<u>COROLLARY 1.13</u> Let Λ be an artin algebra of finite representation type and let M be a TrD-periodic module. If in the definition of the diagram $\mathcal{D}_M$ we consider only non-isomorphic modules the diagram is a Dynkin diagram.

<u>PROOF:</u> This is true since there is only a finite number of non-isomorphic Λ-modules. So if the diagram is one of A_∞, ${}_\infty A_\infty$, B_∞, C_∞ or D_∞ there must be repetition in the finite chains. |

It would be interesting to know if two modules in the definition of $\mathcal{D}_M$ can be isomorphic or even more if they can be TrD-isomorphic.

2. Self-injective artin algebras of finite representation type

In section 1 we defined the chains $\mathscr{A}_M$ for a given TrD-periodic module M. We also associated a diagram $\mathscr{D}_M$ to the module M and showed that the diagram is either a Dynkin diagram or one of the A_∞, ${}_\infty A_\infty$, B_∞, C_∞ or D_∞. From corollary 1.13 it follows that if we consider only nonisomorphic modules in the definition of $\mathscr{D}_M$ then if Λ is of finite representation type the diagram is a Dynkin diagram.

In this section we will show that if Λ is a self-injective artin algebra of finite representation type then we obtain a Dynkin diagram even without the above restriction.

LEMMA 2.1 Let M be a TrD-periodic module and let $\mathscr{A}_M = \{M \to E_1^i \to E_2^i \to \ldots\}_{i \in I}$ be the chains of irreducible maps. Suppose the only modules in the class [M] are either TrD-periodic modules or projective-injective modules. Then, if N is a TrD-periodic module in [M] there exist i, j, k such that $N \simeq \mathrm{TrD}^k E_j^i$ or $N \simeq \mathrm{TrD}^k M$.

PROOF: Since N is in the class [M] there exists a chain of irreducible maps:

$$M - X_1 - X_2 - \ldots - X_n = N$$

Where — means that there is either an irreducible map like $\to$ or $\leftarrow$. The proof will be by induction on the length of the chains. If n = 1 then N is isomorphic to one of the

E_1^i's or $DTrE_1^i$'s. Now suppose $X_{n-1} \to N$ is irreducible. Sine N is TrD-periodic there exists an almost split sequence:

$$0 \to DTrN \to E \amalg X_{n-1} \to N \to 0.$$

If X_{n-1} is projective then $DTrN \to X_{n-1}$ is the only irreducible map to X_{n-1} so $X_{n-2} \simeq DTrN$ and therefore TrD-periodic. Otherwise X_{n-1} is TrD-periodic. So by induction we may assume that there esixt k, i, j such that $X_{n-1} \simeq TrD^k E_j^i$ or $X_{n-1} \simeq TrD^k M$. Let

$$0 \to X_{n-1} \to F \amalg N \to TrDX_{n-1} \to 0$$

be an almost split sequence. We will prove lemma in the case $X_{n-1} \simeq TrD^k E_j^i$ and the proof is similar if $X_{n-1} \simeq TrD^k M$. By applying DTr^k to the above sequence we obtain (see [5]).

$$0 \to E_j^i \to DTr^k F' \amalg DTr^k N \amalg P \to TrDE_j^i \to 0$$

where F′ is non-projective summand of F and P is either projective or zero. But by the definition of E_j^i's we know that the summands of the middle term of an almost split sequence for E_j^i are $TrDE_{j-1}^i$ (or TrDM), E_{j+1}^i or projective. Hence $DTr^k N \simeq TrDE_{j-1}^i$ (or TrDM) or $DTr^k N \simeq E_{j+1}^i$ for some i and j. |

LEMMA 2.2 The same assumption as in the lemma 2.1. Then the following are equivalent:

a) There exists a module X in [M] with $\delta(X) = \alpha$.

b) There exists a module X in the $\mathscr{A}_M$ with $\delta(X) = \alpha$.

PROOF: It follows from lemma 2.1 |

LEMMA 2.3 Let Λ be an artin algebra and M TrD-periodic module. Suppose there is a module X in $\mathscr{A}_M$ with $\delta(X) \neq 2$ or a module X with $\delta(X) = 2$, which has two isomorphic summands in the middle term of its almost split sequence. If there is an infinite chain of irreducible maps:

$$M \to E_1 \to E_2 \to \dots$$

then $E_j \not\simeq \mathrm{TrD}^k E_i$ for all $i \neq j$ and all k.

PROOF: Since we are proving $E_j \not\simeq \mathrm{TrD}^k E_i$ for any k and modules in $\mathscr{A}_X$ are just TrD-images of modules in $\mathscr{A}_M$, we may assume M = X.

Suppose now that $E_j \simeq \mathrm{TrD}^k E_i$. Let $i < j$. Then by lemma 1.9 $\delta(E_i) = \delta(E_j) \neq 3$ and clearly $\delta(E_i) \neq 1$. Therefore $\delta(E_i) = \delta(E_j) = 2$ and the summands in the middle term of almost split sequence for E_i (and also for E_j) are not isomorphic. So either $E_{j+1} \simeq \mathrm{TrD}^k E_{i+1}$ or $E_{j+1} \simeq \mathrm{TrD}^{k+1} E_{i-1}$. In the first case we have $E_{j-1} \simeq \mathrm{TrD}^k E_{i-1}$ and using the same argument as above we have that $\delta(E_{j-1}) = \dots = \delta(E_{j-2}) = 2$ and $E_{j-i} \simeq \mathrm{TrD}^k M$. This gives a contradiction since we assumed that $\delta(M) \neq 2$ or that the summands of the middle term of its almost split sequence are isomorphic. Similarly in the second case we show that $\delta(E_{j+i}) = 2$ and $E_{j+i} \simeq \mathrm{TrD}^{k+i} M$. Contradiction. |

PROPOSITION 2.4 Let Λ be a self-injective artin algebra of finite representation type and let M be a TrD-periodic module. Then the diagram $\mathcal{D}_M$ is a Dynkin diagram.

PROOF: Since Λ is self-injective of finite representation type all Λ-modules are either TrD-periodic or projective-injective. Then by [4] we know that there is a TrD-periodic module X with $\alpha(X) = 1$. If $\delta(X) = 0$ then the only TrD-periodic modules in the class [M] are TrD-images of M and therefore the diagram is just A_1. Otherwise $\delta(X) = 1$ and by lemma 2.2 there is a module Y in $\mathcal{A}_M$ with $\delta(Y) = 1$. Now, if the diagram $\mathcal{D}_M$ is not Dynkin than it has an infinite chain of irreducible maps and since Λ is of finite representation type there must be two modules in the chain which are isomorphic. By lemma 2.3 we know that this is impossible. Hence the diagram is Dynkin. |

This way Theorem 1.12 specializes to a result of Christine Riedtmann [9].

3. TrD-periodic modules M with no projectives in the class [M]

In this section we assume that M is a TrD-periodic module and that there are no projective, and therefore no injective modules in the class [M]. We will show that if two modules X and Y correspond to two different points in the diagram $\mathcal{D}_M$ then $X \not\cong Y$. Even more, we show that $Y \not\cong TrD^k X$ for any k. This enables us to give a simpler definition of the diagram $\mathcal{D}_M$ which coincides with the definition from section 1 in this case, and we show that the diagram is one of A_∞, ${}_\infty A_\infty$, B_∞, C_∞ or D_∞.

REMARK: For the simplicity of notation, whenever $\delta[M] = 3$ we will assume $\delta(M) = 3$. Now the following two problems are equivalent:

A. $Y \not\cong TrD^k X$ for any k and any two modules X and Y which correspond to two different points in the diagram $\mathcal{D}_M$.

B. $Y \not\cong TrD^k X$ for any k and any two modules X and Y which appear in $\mathcal{A}_M$.

LEMMA 3.1 Suppose M is a TrD-periodic module with no projectives in the class [M]. Then there is at least one infinite chain of irreducible maps in $\mathcal{A}_M$.

PROOF: If all chains were finite, and since we know that there is only a finite number of the chains, by lemma 2.1 there would be only a finite number of non-isomorphic modules in the class [M] which is impossible [6]. |

DEFINITION 3.2 We will say that two modules X and Y are

TrD-isomorphic if there exists an integer k such that $Y \simeq TrD^k X$. We denote $TrD^k X$ by X^k.

From lemma 3.1 it follows that the diagram $\mathcal{D}_M$ is one of the A_∞, ${}_\infty A_\infty$, B_∞, C_∞ or D_∞ and from lemma 2.3 it follows that in the case A_∞ and B_∞ none of the modules in $\mathcal{A}_M$ are TrD-isomorphic. The following lemmas will be used to show that the same is true for ${}_\infty A_\infty$.

LEMMA 3.3 Suppose that M and N are TrD-periodic modules and if

$$0 \to N \to M^1 \coprod M^i \coprod X \to N^1 \to 0 \qquad (*)$$

$$0 \to M \to N \coprod Y \to M^1 \to 0 \qquad (**)$$

are almost split sequences than X does not have a summand isomorphic to M^j for any j, and Y does not have a summand isomorphic to N^j for any j. Then

a) $N^{i-1} \simeq N$.

b) $M^{2i-2} \simeq M$.

PROOF: a) Since there is an irreducible map $M \to N$ there must be an irreducible map $M^i \to N^i$. But from the almost split sequence (*) there is an irreducible map $M^i \to N^1$. From the assumption on the almost split sequence (**) it follows that $N^i \simeq N^1$ and hence $N^{i-1} \simeq N$.

b) From the almost split sequence (*) it follows that there are irreducible maps $N \to M^1$ and $N \to M^i$ and therefore $M^{i-1} \to M^i$ and $N^{i-1} \to M^{2i-1}$. From a) we know that

$N^{i-1} \simeq N$ and since X does not have any summands isomorphic to M^j for any j, either $M^1 \simeq M^i$ or $M^1 \simeq M^{2i-1}$. Therefore $M \simeq M^{i-1}$ or $M \simeq M^{2i-2}$. |

LEMMA 3.4 Suppose there are no projectives in the class [M], and suppose $\delta([M]) = 2$. Then it is impossible to have an irreducible map $M \to M^i$.

PROOF: Let $M^m \simeq M$. Then we have the following almost split sequences:

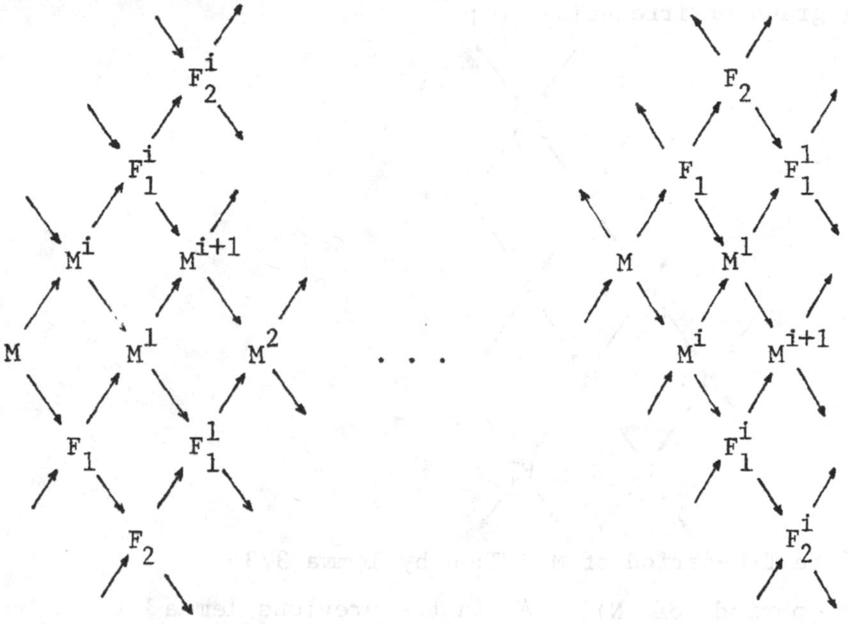

So we have that:

$$2[\ell(M) + \ell(M^1) + \ldots + \ell(M^{m-1})] =$$

$$= \ell(M^i) + \ell(F_1) + \ell(M^{i+1}) + \ell(F_1^1) + \ldots + \ell(M^{i-1}) + \ell(F_1^{i-1}) =$$

$$= \ell(M) + \ell(M^1) + \ldots + \ell(M^{m-1}) + \ell(F_1) + \ell(F_1^1) + \ldots + \ell(F_1^{m-1})$$

Let $M = M^0$ and $F_i = F_i^0$ and $\underline{m} = \sum_{i=0}^{m-1} \ell(M^i)$, $\underline{f}_j = \sum_{i=0}^{m-1} \ell(F_j^i)$.

So we have from above that $\underline{f}_1 = \underline{m}$. Using the same argument we show that $2\underline{f}_1 = \underline{m} + \underline{f}_2$ which implies $\underline{f}_2 = \underline{m}$ and also that $\underline{f}_j = \underline{m}$ for all j. Therefore the lengths of the modules in [M] are bounded which is impossible by [1].

LEMMA 3.5 Suppose there are no projectives in the class [M] and suppose $\delta([M]) = 2$. Then it is impossible to have irreducible maps $M \to N \to M^i$ for $i \neq 1$.

PROOF: Suppose there is such a chain. Then we have the following graph of irreducible maps

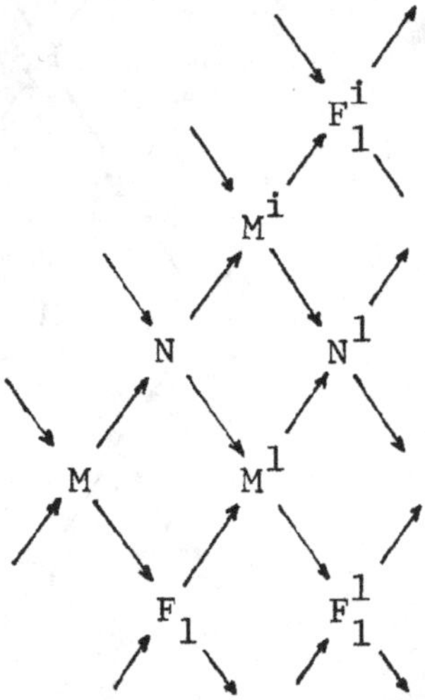

Let m be the TrD-period of M. Then by lemma 3.3 $m = 2\cdot$(TrD-period of N). As in the previous lemma let

$$\underline{m} = \sum_{i=0}^{m-1} \ell(M^i),\ \underline{n} = \sum_{i=0}^{m-1} \ell(N^i)\ \underline{f}_j = \sum_{i=0}^{m-1} \ell(F_j^i).$$ Then we can

show that $2\underline{n} = 2\underline{m}$, $2\underline{m} = \underline{n} + \underline{f}_1$, $2\underline{f}_1 = \underline{m} + \underline{f}_2$, $2\underline{f}_j = \underline{f}_{j-1} + \underline{f}_{j+1}$ and therefore $\underline{n} = \underline{m} = \underline{f}_j$ for all j. Hence the lengths of the modules in [M] are bounded, which gives a contradiction. (by proposition 6.3,[1]). |

<u>PROPOSITION 3.6</u> Suppose there are no projective modules in the class M and suppose $\delta([M]) = 2$. Let X and Y be two modules in $\mathcal{A}_M$. Then $Y \not\simeq X^k$ for any k.

<u>PROOF:</u> Since $\delta([M]) = 2$ we know that there is either one or two chains in $\mathcal{A}_M$. Suppose there are two modules X and Y in $\mathcal{A}_M$ such that $Y \simeq X^k$. Then by lemma 2.3 we know that there are two infinite chains in $\mathcal{A}_M$. Let

$$M \to F_1 \to F_2 \to \dots.$$

$$M \to G_1 \to G_2 \to \dots$$

be the chains and suppose $X = F_i$ and $Y = G_j$. If we denote DTr^kZ by Z^{-k} then we have the following chain of irreducible maps:

$$X^{k-j} = Y^{-j} \to G_{j-1}^{-(\ -1)} \to \dots \to G_1^{-1} \to M \to F_1 \to \dots \to F_i = X$$

So we reduced to the case when X and X^k appear in the same chain. Let:

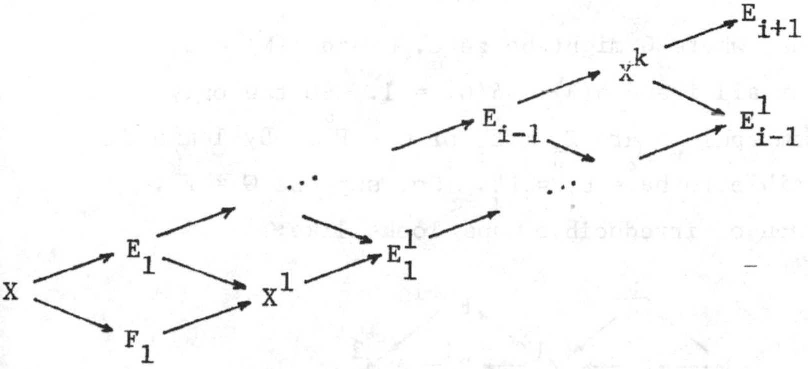

be the graph of irreducible maps. Then either $E_{i+1} \simeq E_1^k$ or $E_{i+1} \simeq F_1^k$. If $E_{i+1} \simeq E_1^k$ then $E_{i-1}^1 \simeq F_1^k$ and this way we obtain that all F_j's are TrD-images of E_j's, hence there is only a finite number of modules in [M]. So we may assume $E_{i+1} \simeq F_1^k$ and $E_{i-1}^1 \simeq E_1^k$. Therefore $E_{i-1} \simeq E_1^{k-1}$. So we obtained a chain which is of length two less than the chain $X \to E_1 \to \ldots \to X^k$. By Lemmas 3.4 and 3.5 we know that it is impossible to have such chains of lengths one or two, so we may apply induction. |

<u>PROPOSITION 3.7</u> Suppose there are no projectives in the class [M] and suppose $\delta([M]) = 3$. Let X and Y be two modules in $\mathscr{A}_M$. Then $Y \not\simeq X^k$ for any k.

<u>PROOF:</u> From the remark at the beginning of this section we may assume that $\delta(M) = 3$. Then there are either two or three chains in $\mathscr{A}_M$. Let

$$M \to E_1 \to E_2 \to \ldots$$

$$M \to F \quad \text{and} \quad M \to G$$

be the chains, where G might be zero. Then $\delta(M) = 3$, $\delta(E_i) = 2$ for all i and $\delta(F) = \delta(G) = 1$. So the only possible isomorphisms are $E_j \simeq E_i^k$ or $G \simeq F^k$. By lemma 2.3 it is impossible to have $E_j \simeq E_i^k$. So, suppose $G \simeq F^k$. Then the graph of irreducible maps looks like:

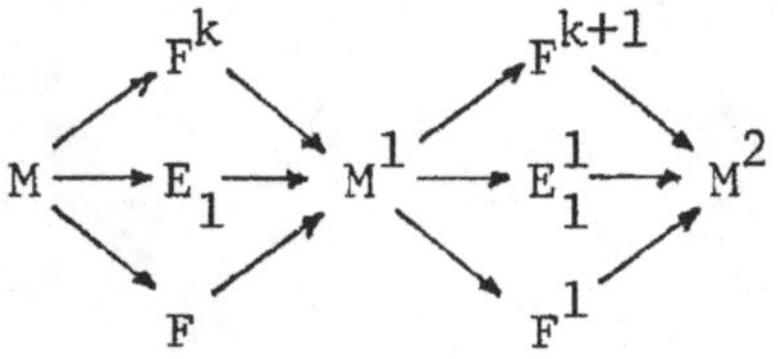

Then by lemma 3.3 $M^k \simeq M$ and $F^{2k} \simeq F$.
Let

$$\underline{m} = \sum_{i=o}^{2k-1} \ell(M^i),\ \underline{f} = \sum_{i=0}^{2k-1} \ell(F^i),\ \underline{e}_j = \sum_{i=0}^{2k-1} \ell(E_j^i).$$

Then one can show that $2\underline{f} = \underline{m}$, $2\underline{m} = 2\underline{f} + \underline{e}_1$, $2\underline{e}_1 = \underline{m} + \underline{e}_2$ and $2\underline{e}_{j=1} = \underline{e}_{j+2} + \underline{e}_j$. Therefore

$$\underline{m} = \underline{e}_1 = \underline{e}_2 = \dots = \underline{e}_j$$

Hence all modules have bounded lengths. Contradiction. |

From the last two propositions it follows that for a TrD-periodic module M, with no projectives in the class [M], none of the modules in $\mathcal{A}_M$ are TrD-isomorphic. This enables us to give a new definition for the diagram $\mathcal{D}_M$. To each module in $\mathcal{A}_M$ we associate a point and the arrows and multiplicities are defined in the same way as at the end of section 1. Then we have the following theorem.

THEOREM 3.8 Let Λ be an artin algebra and M a TrD-periodic module with no projective modules in the class M. Then the diagram $\mathcal{D}_M$ is one of the A_∞, ${}_\infty A_\infty$, B_∞, C_∞ or D_∞.

Since none of the modules in $\mathcal{A}_M$ are TrD-isomorphic, the class of the modules in $\mathcal{A}_M$ forms a section in the sense of Raymundo Bautista [7] (see also [9]). We also have the following result.

COROLLARY 3.9 Let M be A TrD-periodic module with no projectives in the class [M]. Let X be a module in [M].

Then there is a unique module E in $\mathcal{A}_M$ and an integer (unique module TrD-period of E) such that $X \simeq TrD^k E$.

REFERENCES

[1] AUSLANDER, M.: Applications of Morphisms Determined by Objects, (Proc. Conf. Temple University, Philadelphia, PA, 1976, 245-327), Lecture Notes in Pure and Applied Math., Vol. 37, Dekker, New York, (1978).

[2] AUSLANDER, M., PLATZECK, M. I.: Representation Theory of Hereditary Artin Algebras, (Proc. Conf. Temple University, Philadelphia, PA, 1976, 389-424), Lecture Notes in Pure and Applied Math., Vol. 37, Dekker, New York, (1978).

[3] AUSLANDER, M., REITEN, I.: Representation Theory of Artin Algebras III: Almost Split Sequences. Communications in Algebra, 3 (3), 239-294, (1975).

[4] AUSLANDER, M., REITEN, I.: Representation Theory of Artin Algebras IV: Invariants given by Almost Split Sequences. Communications in Algebra, 5 (5), 443-518, (1977).

[5] AUSLANDER, M., REITEN, I.: Representation Theory of Artin Algebras V: Methods for Computing Almost Split Sequences and Irreducible Morphisms, Communications in Algebra,

[6] Representation Theory of Artin Algebras VI: A functorial Approach to Almost Split sequences, Communications in Algebra, 6 (3), 257-300, (1978).

[7] BAUTISTA, R.:

[8] DLAB, V., RINGEL, C.M.: Indecomposable Representations of graphs and Algebras, Memoirs of the

A.M.S., No 173, (1976)

[9] RIEDTMANN, Ch.: Algebren, Darstellungsköcher, Ueberlangerungen und Zurück, Thesis, (1979) (Zürich)

[10] TODOROV, G.: Almost Split Sequences in the Representation Theory of Certain Classes of Artin Algebras, Thesis, Brandeis University, (1978)

Gordana Todorov
Department of Mathematics
University of Georgia
Athens, Georgia 30602

A CLASS OF SELF-INJECTIVE ALGEBRAS AND THEIR INDECOMPOSABLE MODULES

Josef Waschbüsch

Let K be an algebraically closed field. We are concerned with finite dimensional self-injective K-algebras which have the property that for each indecomposable projective module P the subfactormodule rad P/soc P is a direct sum of n(P) uniserial (nontrivial) modules. If n is the maximum of the n(P)'s we call such an algebra n-serial.

In our joint work with E. Scherzler (7) we classified the 2-serial self-injective K-algebras of finite representation type. The indecomposable modules of these algebras are determined by P. W. Donovan and M. R. Freislich (2). Here we classify the 3-serial self-injective K-algebras of finite representation type and describe their indecomposable modules. We point out that if A is an n-serial self-injective K-algebra of finite representation type then n must be smaller or equal to 3.

In order to state the main results we assume that the reader is familiar with the terminology and the basic definitions of (7).

The invariants for the Morita-equivalence classes of 2-serial self-injective algebras of finite representation type are the 2-si-systems T = (G,f,n) (see also (7)) consisting in a Brauer-tree G (i.e. a finite connected tree with cyclic ordering on each maximal set of edges with a common vertex), a natural number n and a map f from G to the nonnegative integers such that the following conditions are satisfied:

1) $f(S) \neq 1$ and $f(\beta) \neq 1$ for at most one vertex S and for at most one edge β of G.
2) If $f(S) \geq 1$ for all vertices then $f(\beta) = 1$ for all edges and $f(S)$ and n are relatively prime for all vortices S.
3) If $f(S_o) = 0$ for the vertex S_o then $f(\beta_o) = 0$ for an edge β_o ending at S_o and n is even, $n = 2m$.
4) If G consists in just one vertex S, then $f(S) = 1 = n$.

Two 2-si-systems T and $T' = (G', f', n')$ are <u>equivalent</u>, if $n = n'$ and if there exists an isomorphism $g : G \xrightarrow{\sim} G'$ of Brauer-trees such that $f'g = f$.

2-si-systems are an essential part of the invariants for 3-serial self-injective algebras of finite representation type. To describe the latter we need the following

<u>Definition</u>. Let G be a Brauer-tree, $|G|$ the number of edges of G and $|S|$ the number of edges ending in the vertex S of G. A sequence

$$w = \beta_{11}, \beta_{12}, \ldots, \beta_{1r_1}, \beta_{21}, \ldots, \beta_{n1}, \beta_{n2}, \ldots \beta_{nr_n}$$

of edges of G is called a <u>walk</u> in G, if it satisfies the conditions

1) $\beta_{i1}, \ldots \beta_{ir_i}$ have a common vertex S_i and are alternatively consecutive edges either in the cyclic ordering or in the reversed cyclic ordering of the edges ending in S_i for $1 \leq i \leq n$.
2) $1 < r_i \leq |S_i|$ for $1 \leq i \leq n$.
3) $S_i \neq S_{i+1}$ and $\beta_{ir_i} = \beta_{i+1,1}$ for $i = 1, 2, \ldots, n-1$

We say that the walk w <u>starts</u> with β_{11} in S_1. For an edge $\beta = (S, S')$ between the vertices S and S' let $\ell_s(\beta)$ be the number of walks starting with β in S; by p(β) we denote the rational number $\frac{1}{\ell_S(\beta)+1}+\frac{1}{\ell_{S'}(\beta)+1}$

Now a pair R = (T,x), where T = (G,f,n) is a 2-si-system and where x is an element of G, is called a <u>3-si-system</u>, if it satisfies the conditions

1. $|f| := \prod_{\text{vertices } S} f(S) \leq 1$, $|G| \geq 1$
2. If $|f| = 0$ then $|G| = 2$ and $x = S_0$ = vertex of G with $f(S_0) = 0$
3. If $|f| = 1$ then x = (S,S') is and edge with $p(x) > 1/2$

Two 3-si-systems R and R'= (T' = (G',f',n'),x') are said to be <u>equivalent</u> if T and T' are under an isomorphism $g : G \xrightarrow{\sim} G'$ with g(x) = x'.

To each connected 3-serial self-injective algebra A of finite representation type we construct a 3-si-system $R(A) = (T_A, x_A)$, $T_A = (G,f,n)$, where to each vertex S of G there corresponds an "elementary" self-injective algebra (see (7), 1.1. and (1.1.) below) $e_S A e_S$, $e_S^2 = e_S \in A$. Their type is given by n, f(S) and the position of S in G. Also there corresponds a further elementary self-injective algebra $\hat{e} A \hat{e}$, $\hat{e}^2 = \hat{e} \in A$, to x and x describes the position of the indecomposable projective module P with n(P) = 3 in A.

Reversely we construct to each 3-si-system R a connected 3-serial self-injective algebra A(R) of finite representation type as the bounden quiver algebra of a bounden quiver (Q_R, I_R) determined by R.

For the compositions of the maps $A \longmapsto R(A)$ and $R \longmapsto A(R)$ we show that A(R(A)) is Morita-equivalent to A and that R(A(R)) is equivalent to R. Thus we get

<u>Theorem 1</u>: The maps $A \longmapsto R(A)$ and $R \longmapsto A(R)$ induce a bijection between the Morita-equivalence classes of connected 3-serial self-injective algebras of finite representation type and the

equivalence classes of 3-si-systems.

For a 3-si-system $R = (T,x)$, $T = (G,f,n)$ let Δ_R be the ordered set

$$\Delta_R = \{1, 1' < 2' < \ldots < r', 1'' < 2'' < \ldots < s''\},$$

where $r = s = 2$ if $|f| = 0$, $r = \mathrm{Max}(\ell_S(x), 1)$, $s = \mathrm{Max}(\ell_{S'}(x), 1)$ if $|f| = 1$ and $x = (S,S')$. By $\mathfrak{U}_R$ we denote a set of representatives for the types of indecomposable representations V of the ordered set Δ_R satisfying $V_1 \neq 0$ if $|f| = 0$ or if $|f| = 1$ and $\ell_S(x) \cdot \ell_{S'}(x) \neq 0$, $V_1 + V_{1'} \neq 0$ if $|f| = 1$ and $\ell_S(x) \neq 0$, $\ell_{S'}(x) = 0$ and $V_1 + V_{1'} + V_{1''} \neq 0$ if $|f| = 1$ and $\ell_S(x) = \ell_{S'}(x) = 0$. Also we put $\tilde{n} = n$ if $|f| = 1$ and $\tilde{n} = {}^n/2$ if $|f| = 0$ and denote by $(\tilde{n})$ the set of the integers $-\tilde{n}, -\tilde{n}+1, \ldots, -1, 1, 2, \ldots, \tilde{n}$.

To each $p \in (\tilde{n})$ we construct a functor $\Psi_p : \mathrm{mod}_K\, \Delta_R \longrightarrow \mathrm{mod}\, A(R)$ from the category of finite dimensional K-representations of Δ_R to the category of finitely generated left A(R)-modules such that the following theorem holds:

<u>Theorem 2</u>: Let $R = (T,x)$ be a 3-si-system, $A = A(R)$ and let $A_o = e_o A e_o$ be the algebra A(T) (see (<u>7</u>)). Then the set $\Psi_p(V)$, $p \in (\tilde{n})$, $V \in \mathfrak{U}_R$, is a family of nonisomorphic indecomposable nonprojective A-modules and for any indecomposable nonprojective A-module M just one of the following assertions holds:

1) M is simple and $e_o M = 0$
2) $M = e_o M$ is an indecomposable nonprojective A_o-module
3) M is isomorphic to $\Psi_p(V)$ for properly choosen $p \in (\tilde{n})$, $V \in \mathfrak{U}_R$.

We recall first some of the basic definitions in (<u>7</u>) and give some auxiliar lemmas concerning quivers with relations. In the second section we show that a 3-serial self injective algebra A of finite representation type is determined by a 2-serial self-injective algebra $A_o = e_o A e_o$, $e_o^2 = e_o \in \Lambda$, and by an elementary algebra $\hat{A} = \hat{e} A \hat{e}$, $\hat{e}^2 = \hat{e} \in A$, which together lead us to the system $R(A) = (T(A_o), x)$. In section 3 we describe the indecomposable A-modules and get precise conditions which R(A) has to satisfy.

We shall present the main ideas and the course of the proofs without giving any detailed calculations. The complete proofs are given in the authors "Habilitationsschrift" and will be published elsewhere.

1. For the definitions of a quiver Q, a path w in Q, a bounden quiver (Q;I), the (bounden) quiver algebra K(Q) (K(Q,I)) and related definitions and terminology we refer to (7) (see also (5)). As in (7) certain bounden quivers S = (Q,I) rising in this paper have additional structure, given by a cyclic ordering of the arrows of Q such that they form a closed path in their cyclic ordering. We call such a bounden quiver S a track.

Examples for tracks are the bounden quivers $Z_e^h = (Z_e; I_e^h)$ and $S_{m,k} = (Q_{m,k_i} \, I_{m,k})$ in (7). Further tracks S_m rising in this paper are given by quivers Q_m (in figure 1) with the indecated cyclic ordering of the 6 m arrows of Q_m and by the ideal I_m which is generated by all elements of the forms

1) paths of length ≥ 3,
2) products $\alpha_j \alpha_i$ of arrows α_i, α_j which are not consecutive elements in the cyclic ordering of the arrows of Q,
3) differences of parallel paths of length 2.

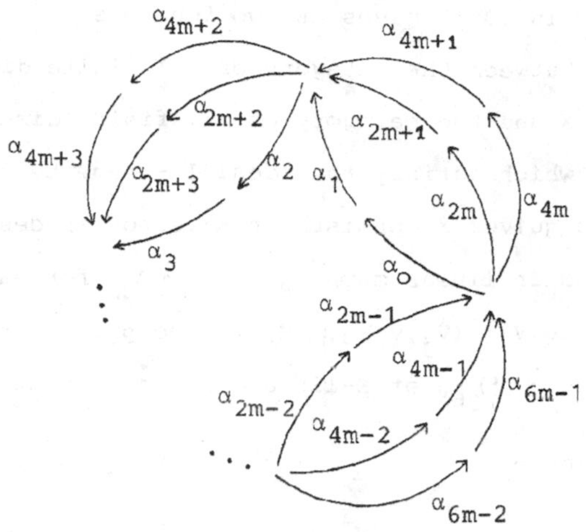

figure 1

If (Q,I) is a bounden quiver each closed path $w = \alpha_{t-1} \cdots \alpha_2 \alpha_1 \alpha_o$ in Q with $\alpha_i \neq \alpha_j$ for $i \neq j, i,j = 0,1,\ldots t-1$, furnishes a track $S_w = (Q_w, I_w)$: Q_w is the underlying quiver of w, $I_w = I \cap K(Q_w)$ and the cycling ordering of the arrows of Q_w is induced by w.

For the proofs of the theorems we use the concept of a X-sequence in a bounden quiver (Q,I), where X is a quiver (without relations): we define it to be a quiver homomorphism $F : X \to Q$ satisfying the condition

(1.1) If $\lambda \cdot F\omega + u \in I$ for $\lambda \in K$, w a path in X, u a path in Q, then $\lambda = 0$.

We say that the X-sequence F is locally faithful if for each vertex j of X F is injective both on the set of all arrows starting at j and on the set of all arrows ending in j.

A X-sequence F in (Q,I) gives us two functors $\mathrm{mod}_K X \underset{\tilde{F}}{\overset{\hat{F}}{\rightleftarrows}} \mathrm{mod}_K(Q,I)$ between the category of all finite dimensional K-representations of X and the category of all finite dimensional representations of Q which satisfy the ideal I - remember that a representation V of a quiver X consists in K-vector spaces V_j for each vertex j of X and in linear maps $V_\beta : V_j \to V_k$ for each arrow $\beta = j \to k$ of X. We write $V = (V_j, V_\beta; j, \beta \in X)$. A morphism f from V to V' is a family $f = (f_j : V_j \to V_j')_{j \in X}$ of K-linear maps such that all squares

$$\begin{array}{ccc} V_j & \xrightarrow{f_j} & V_j' \\ V_\beta \downarrow & & \downarrow V_\beta' \\ V_k & \xrightarrow{f_k} & V_k' \end{array}$$ commute.

Define $\hat{F}$ by $\hat{F}(V_j, V_\beta; j, \beta \in X) = (W_i, W_\alpha; i, \alpha \in Q)$ where $W_i = \bigoplus_{Fj=i} V_j$ and where $W_\alpha : \bigoplus_{Fj_1=i_1} V_{j_1} \longrightarrow \bigoplus_{Fj_2=i_2} V_{j_2}$ is given by the matrix $(g_{j_1 j_2})$ with $g_{j_1 j_2} = \sum_{\beta = j_1 \to j_2 \in X} V_\beta$ ($g_{j_1 j_2} = 0$ if there is no arrow form j_1 to j_2 in X); $\hat{F}(f_j)_{j \in X} = (h_i)_{i \in Q}$ where $h_i = \bigoplus_{Fj=i} f_i$.

$\tilde{F}$ is defined by the formulas $\tilde{F}(W_i, W_\alpha ; i, \alpha \in Q) = (V_j, V_\beta; j, \beta \in X)$ with $V_j = W_{Fj}$ and $V_\beta = W_{F\beta}$ and $\tilde{F}(h_i)_{i \in Q} = (f_j)_{j \in X}$ with $f_j = h_{Fj}$.

We need these functors only in situations where neither X nor Q has double arrows (= subquivers of the form $\cdot \rightrightarrows \cdot$). Then it is clear from (1.1) that the representation $\hat{F}(V_j, V_3 ; i, \beta \in X)$ satisfies the ideal I, and we get the following

<u>Lemma</u> 1.2 Let F be a locally faithful X-sequence in (Q,I) and suppose that neither X nor Y have double arrows. Then we have $\tilde{F}\hat{F} V = V \oplus *$ for each representation V of X.

<u>Corollary</u> 1.3. Under the assumption of Lemma we get: If (Q,I) is of finite representation type then the underlying graph of X is a

Dynkin-Diagramm (i.e. a diagramm of type A_n, D_n, E_6, E_7 or E_8).

2. In this section we look at the (ordinary) quiver of a connected 3-serial self-injective algebra A of finite representation type in order to prove Theorem 1. Thus we may assume that A is basic. If N is the radical of A and $1 = e_1 + e_2 + \dots + e_q$ is a decomposition of the unit element of A into a sum of orthogonal primitive idempotents $e_i \in A$ the quiver Q_A of A has the vertices $1,2,\dots,q$ and $\dim_K e_i N/N^2 e_j$ arrows from j to i. We fix elements $a_{ij} \in e_i Ne_j \setminus e_i N^2 e_j$ representing a K-base of $e_i N/N^2 e_j$ and get an algebra-epimorphism $K(Q_A) \longrightarrow A$ from $K(Q_A)$ onto A sending i to e_i and an arrow $\alpha_{ij} = j \to i$ to a_{ij}.

Thus we have

(2.1) A is isomorphic to an algebra $K(Q_A, I_A)$ for a suitable ideal I_A of Q_A.

The ideal I_A in (2.1) depends on the choice of the elements a_{ij} above. Later we will see that for properly chosen a_{ij} I_A has a "canonical" form.

We denote by π the Nakayama-permutation of A and by ord i the number of arrows starting at the vertex i of Q_A. The first lemma is the base of all what follows, for the proof see also (7), Lemma 2.2.

Lemma 2.2 1) π can be extended to an automorphism of Q_A.

2) Each arrow α determines a unique closed path $\omega = \alpha_r \dots \alpha_1 \alpha_0$ in Q_A such that $\alpha_0 = \alpha$, $\alpha_i \neq \alpha_j$ for $i \neq j$, $i,j = 0,1,\dots,r$ and $\alpha_{i+1} \alpha_i \notin I_A$ for $i = 0,1,\dots,r$.

3) By 2) each arrow α determines a unique track

$S_\alpha = (Q_\alpha, I_\alpha) \subseteq (Q_A, I_A)$ such that

a) $S_\alpha = S_{\alpha'}$ if and only if $\alpha' \in S_\alpha$

b) S_α is <u>nondegenerate</u> (i.e. for consecutive arrows α_i, α_{i+2} in the cyclic ordering in S_α we have $\alpha_{i+1}\alpha_i \notin I_\alpha$

c) If S is any nondegenerate track with $Q_S \subseteq Q_A$, $I_S = I_A \cap K(Q_S)$ then $S = S_\alpha$ for each $\alpha \in Q_S$

In the following we refer to the tracks S_α as <u>the tracks of</u> Q_α .

To determine the structure of the tracks of Q_A we use X-sequences in Q_A for quivers X of type $A_n, D_n, E_6, E_7, E_8, \tilde{A}_n, \tilde{D}_n, \tilde{E}_6, \tilde{E}_7, \tilde{E}_8$. We call such sequences also (A)-<u>sequences</u>, (D)-<u>sequences</u> ... . The number of vertices of X is the <u>length</u> of F.

<u>Lemma 2.3</u> 1) The lengths of (A)-sequences in (Q_A, I_A) are bounded.

2) If $S_1, S_2, \ldots, S_r$ is a sequence of tracks of Q_A such that $S_i \neq S_{i+1}$, $S_i \cap S_{i+1} \neq \emptyset$ for $i = 1,2, \ldots ,r-1$, there exists an (A)-sequence F of length $\geq r-1$ starting at S_1 and ending at S_r (i.e. the extreme vertices of X are mapped onto vertices of S_1 and S_r respectively).

The first assertion can be proved observing that the (A)-sequences give rise to indecomposable modules (see (2) and (7)); the second assertion is proved as in (7), Lemma 2.7.

<u>Corollary 2.4.</u> 1) Each track S of Q_A is of the form Z_e^h, $S_{m,k}$ or S_m for properly chosen natural numbers e,h,m,k (depending on S; also I_S may be different from $I_{m,k}$ or I_m if S is of the form $S_{m,k}$ or

S_m - see the remark after proposition 2.11 in (7)).

2) The vertices i of order 3 are just one orbit under π and are vertices of a track $\hat{S}$ of type $S_m, S_{m,2}$ or Z_c^2 such that all the other vertices of $\hat{S}$ have order 1.

3) There is no track S of type Z_e^h for $h > t(s)$ = number of orbits of vertices $i \in S$ under π

4) If S_o is a track of the form $S_{m,k}$, the double points of S_o are just the vertices of order 3 of Q_A.

5) If there is a track S of type S_m, we have $Q_A = Q_s$ and $A \cong K(S_m, I_m)$.

Let e_o be the sum of all primitive idempotents $e_1, e_2, \ldots, e_q$ but those corresponding to vertices $i \in \hat{S}$ of order 1. Then by proposition 2.4 $A_o = e_o A e_o$ is a connected 2-serial self-injective algebra of finite representation type and the 2-si-system $T = T(A_o) = (G, f, n)$ satisfies $|f| \leq 1$, $|G| \geq 1$. The track $\hat{S}$ is completely determined by T and the set of vertices of order 3 which is described either by the vertex S_o of G with $f(S_o) = 0$ (if $|f| = 0$) or by an edge $x = (S, S')$ (if $|f| = 1$): In the first case we have $\hat{S} \cong Z_n^2$, in the latter case we have $\hat{S} \cong S_n$, if $p(x) = 2$, $\hat{S} \cong S_{n,2}$ if $1 < p(x) < 2$ and $\hat{S} \cong Z_{2n}^2$ if $p(x) \leq 1$.

So far we have seen that the quiver Q_A is uniquely determined by the system $R = (T, x)$ above. The next proposition shows that the same is true for I_A and such also for A itself.

Proposition 2.5. For properly chosen elements a_{ij} the ideals I_s of tracks S which are of type $S_{m,k}$ or S_m coincide with the ideals $I_{m,k}$ and I_m respectively and the ideal I_A in (2.1) is generated by all elements of the forms

1) $\alpha'\alpha$ for $\alpha \in S, \alpha' \in S'$, $S \neq S'$

2) $w(i,S)-w(i,S')$, $i \in S \cap S'$, $S \neq S'$.

3) $u \in I_s$.

Here S,S' are tracks of Q_A and $w(i,S)$ is a maximal path in Q_s but not in I_s starting at i and following the cyclic ordering of the arrows of S.

3. To finish the proof of theorem 1 we must prove that the system $R = (T,x)$ satisfies the conditions in the definition of 3-si-systems. But this will follow from the description of the indecomposable modules of algebras $A(R)$ determined by a system $R = (T,x)$, $T = (G,f,n)$ with $|G| \geq 1$, $|f| \leq 1$ and where $x = S_o$ if $f(S_o) = 0$, while $x = (S,S')$ if $|f| = 1$: We get $A(R)$ by extending the quiver Q_T of $A_o = A(T)$ by the track $\hat{S}$ (determined by T and x) to a quiver Q_R and by the ideal I_R described in proposition 2.5: $A = A(R) := K(Q_R,I_R)$; $A_o = K(Q_T,I_T) = e_o A e_o$.

In order to simplify the formulations we make no difference between A-modules and representations of the bounden quiver (Q_R,I_R). If V is an A-module let supp V be the full subquiver of Q_A given by all vertices i of Q_R with V_i $(= e_i \cdot V) \neq 0$.

A subquiver F of Q_R is said to be a (A)-sequence in (Q_R,I_R), if the imbedding of F in Q_A is a (A)-sequence in (Q_R,I_R). Later we will use the fact that each (A)-sequence F in (Q_T,I_T) is just of this form, because of $|f| \leq 1$.

If we denote the idempotent $1-e_o$ be $\hat{e}$ we know by the results of P.W. Donovan and M.R. Freislich (3) (see also (7), 2.6) and by construction that for each indecomposable nonprojective A-module M just one of the following conditions holds:

(3.1) 1) $M = e_o M$ and supp M is a (A)-sequence in (Q_T,I_T).

2) $M = \hat{e}M$ and $\operatorname{supp} M = \{i\}$ where $i \in \hat{S} \setminus Q_T$.

3) $e_o M \neq 0$ and $\hat{e}M \neq 0$.

Because the modules in 1) and 2) are well known (and there are only finitely many of this type!) it remains to determine the modules M with $e_o M \neq 0$ and $\hat{e}M \neq 0$.

An (A)-sequence F in (Q_R, I_R) can be written down in the form

$$F = i_o \overset{\alpha_1}{\text{---}} i_1 \overset{\alpha_2}{\text{---}} \cdots \overset{\alpha_r}{\text{---}} i_r ,$$

where $i_{k-1} \overset{\alpha_k}{\text{---}} i_k$ is an arrow with arbitrary direction between i_{k-1} and i_k. If we fix one end of F, we say that F <u>starts</u> with α_1 at i_o and <u>ends</u> with α_r in i_r. α_k is called a <u>direct arrow</u> of F if $\alpha_k = i_{k-1} \longleftarrow i_k$ and an <u>inverse arrow</u> of F if $\alpha_k = i_{k-1} \longrightarrow i_k$; $r+1$ is called the <u>length</u> of F.

The crucial step of the proof of Theorem 2 now is

<u>Lemma</u> 3.2 Let M be an indecomposable nonprojective A-module with $e_o M \neq 0 \neq \hat{e}M$ and let $e_o M \cong \bigoplus_{F \in \mathfrak{F}} M(F)$ be a decomposition of $e_o M$ in a direct sum of indecomposable A_o-modules M(F), where each F is an (A)-sequence in (Q_T, I_T). Then we have

1) If $\hat{S}$ is of type $S_{n,2}$ there exists an arrow $\alpha = i \text{---} i_1$, $\operatorname{ord} i = 3$, such that each $F \in \mathfrak{F}$ of length ≥ 2 starts with α at i.

2) If $\hat{S}$ is of type Z_n^2 or Z_{2n}^2 there are two arrows $\alpha = i \text{---} i_1$ $\beta = i \text{---} i_2$, $\operatorname{ord} i = 3$, both starting at i or both ending at i such that each $F \in \mathfrak{F}$ of length ≥ 2 contains either α or β as an arrow.

3) If $\hat{S}$ is of type S_n there is a subquiver Y of the form

$$\begin{array}{ccccc} i_1 & \overset{\alpha}{\text{---}} & i & \overset{\beta}{\text{---}} & i_2 \\ & & \mid{\scriptstyle\gamma} & & \\ & & i_3 & & \end{array}$$

in Q_R, where all arrows start at i or all end at i, such that M is an indecomposable representation of Y.

Using ideas of C. M. Ringel (6) we attach to each (A)-sequence $F = i_o \xrightarrow{\alpha_1} i_1 \xrightarrow{\alpha_2} \ldots\ldots \xrightarrow{\alpha_r} i_r$ a functor $\phi_F : \mathrm{mod}\, A \longrightarrow \mathrm{mod}\, K$: Let $t_k = t(\alpha_k) = 1$ if α_k is direct and $t_k = -1$ if α_k is inverse. Define $\phi_F(M) = (M_{\alpha_1})^{t_1} \ldots (M_{\alpha_r})^{t_r} (M^{t_r})$ where $M^1 = M$ and $M^{-1} = 0$.

If we look at the set $\mathfrak{F}\, i_o,\alpha_1$ of all (A)-sequences F in (Q_R, I_R) starting with α_1 at i_o we get the

<u>Lemma 3.3</u> $\mathfrak{F} i_o,\alpha_1$ is totally ordered by inclusion according to the relation

$$F_1 = i_o \xrightarrow{\alpha_1} i_1 \xrightarrow{\alpha_2} \ldots \xrightarrow{\alpha_r} i_r < F_2 = i_o \xrightarrow{\alpha_1} i_1 \xrightarrow{\beta_2} \ldots \xrightarrow{\beta_s} i_s$$

if and only if either

1) there exists $j_o = \mathrm{Min}(j : \alpha_j \neq \beta_j)$ and $t(\alpha_j) = 1$

or 2) $s > r$, $\alpha_j = \beta_j$ for $j = 1,\ldots,r$ and $t(\alpha_r) = 1$

or 3) $s < r$, $\alpha_j = \beta_j$ for $j = 1,\ldots,s$ and $t(\alpha_s) = -1$

Let us fix the vertices $i_1, i_2, \ldots, i_{\tilde{n}}$ of order 3 which are numbered in such a way that $i_{s+1} = \pi^s i_1$ for $s = 1,2,\ldots \tilde{n}$, where $\tilde{n} = n$ for $|f| = 1$ and $\tilde{n} = n/2$ for $|f| = 0$. These vertices have neighbourhoods of the form

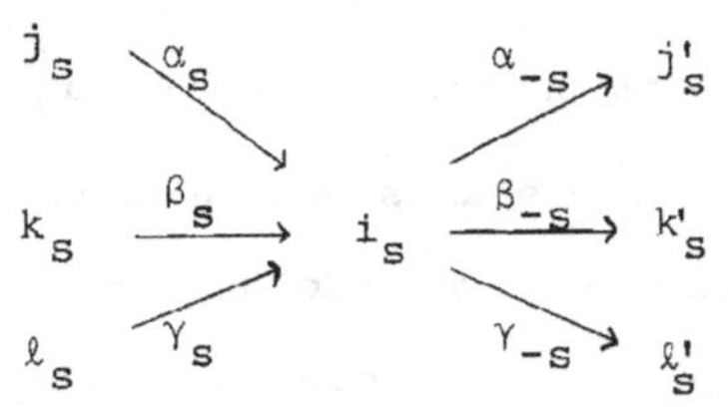

$s = 1,2,\ldots,\tilde{n}$

in Q_R. We can assume that $\gamma_s \in \hat{S}$. Put

$$\mathfrak{F}_s = \mathfrak{F} i_s,\alpha_s \sqcup \mathfrak{F} i_s,\beta_s \sqcup \mathfrak{F} i_s,\gamma_s$$

and $$\mathfrak{F}_{-s} = \mathfrak{F} i_s,\alpha_{-s} \sqcup \mathfrak{F} i_s,\beta_{-s} \sqcup \mathfrak{F} i_s,\gamma_{-s} ,$$

where we denote by $\Delta \sqcup \Delta'$ the disjoint union of two ordered sets Δ

and Δ'. Then all ordered sets $\mathcal{F}_s$ are isomorphic (indeed π^s maps $\mathcal{F}_1$ isomorphically onto $\mathcal{F}_{s+1}$) to $\Delta_{1,r,t}$ with $1 = |\mathcal{F}i_s,\gamma_s|$, $r = |\mathcal{F}i_s,\beta_s|$, $t = |\mathcal{F}i_s\alpha_s|$ while each $\mathcal{F}$-s is anti-isomorphic (and thus also isomorphic) to $\mathcal{F}_s$. Let $(\tilde{n})$ be the set $\{-\tilde{n},\tilde{n}+2,..,-1,1,2,...,\tilde{n}\}$ and let $\lambda(F)$ and $p(F)$ denote the arrow and the vertex of Q_R respectively such that the (A)-sequence F (of length ≥ 2) ends with $\lambda(F)$ in $p(F)$. We now get the following description of the quiver supp M for M as in Lemma 3.2.:

Corollary 3.4 Let M be an A-module as in Lemma 3.2. Then there exists a unique $s \in (\tilde{n})$ with $M_{i_{|s|}} \neq 0$ and such that:

1) To each vertex $j \in$ supp M there exists a unique $F \in \mathcal{F}_s$ with $p(F) = j$ and to each arrow $\alpha \in$ supp M there exists a unique $F \in \mathcal{F}_s$ with $\lambda(F) = \alpha$.

2) If $\lambda(F) = \alpha = j \longrightarrow j' \in$ supp M the linear map $M_\alpha : M_j \longrightarrow M'_j$ is a monomorphism if α is a direct arrow of F and an epimorphism if α is an inverse arrow of F.

With the aid of Lemma 3.3 and Corollary 3.4 we can construct functors

$$\phi_s : \text{mod } A \longrightarrow \text{mod}_K \mathcal{F}_s \qquad \text{for each } s \in (\tilde{n})$$
$$\psi_s : \text{mod}_k \mathcal{F}_s \longrightarrow \text{mod } A$$

as follows:

Define $\phi_s(M)$ to be the K-space $M_{i_{|s|}}$ with the subspaces $\phi_F(M)$, $F \in \mathcal{F}_s$; for a given homomorphism $g : M \longrightarrow M'$ let $\phi_F(g)$ be the K-linear map $g_{i_{|s|}} : M_{i_{|s|}} \longrightarrow M'_{i_{|s|}}$.

Reversely each K-space V with subspaces $V_F, F \in \mathcal{F}_s$ defines a K-representation M of (Q_k,I_k), $M = (M_i,M_\alpha\ ;\ i,\alpha \in Q_k)$ where $M_{p(F)}$ and $M_{\lambda(F)}$ is given by induction on the length of F:

$M_{i_{|s|}} = V$, for $F_1 = i_{|s|} \xrightarrow{\lambda(F)} p(F_1)$ let

$$M_{p(F_1)} = \begin{cases} V_F & \text{if } \lambda(F) \text{ is direct} \\ V/V_F & \text{if } \lambda(F) \text{ is direct} \end{cases}$$

and so on. Put $M_i = 0$ and $M_\alpha = 0$ for $i \notin p(\mathfrak{F}_s)$ and $\alpha \notin \lambda(\mathfrak{F}_s)$. These two functors satisfy

Proposition 3.5 Let D be the duality functor Hom(-,K) on $\mathrm{mod}_K \mathfrak{F}_s$ and let

$$\overline{\phi}_s = \begin{cases} \phi_s & \text{for } s > 0 \\ D\phi_s & \text{for } s < 0 \end{cases}, \qquad \overline{\psi}_s = \begin{cases} \psi_s & \text{for } s > 0 \\ \psi_s D & \text{for } s < 0 \end{cases}.$$

Then we have:

1) If M is an indecomposable nonprojective A-module with $e_o M \neq 0 \neq \hat{e} M$ then $(\overline{\phi}_s M)_F \neq 0$ for an (A)-sequence $F \in \mathfrak{F}_s$ of length ≥ 2, $F \subseteq \hat{S}$, and $\overline{\psi}_s \overline{\phi}_s M \cong M$

2) If V is an indecomposable representation of $\mathfrak{F}_s$ with $V_F \neq 0$ for an (A)-sequence $F \in \mathfrak{F}_s$ of length ≥ 2, $F \subseteq \hat{S}$, then $\overline{\phi}_s \overline{\psi}_s V \cong V$.

Using the fact that each $\mathfrak{F}_s$ is isomorphic to Δ_R we get the

Corollary 3.6 The functors ϕ, ψ induce a bijection between the types of indecomposable nonprojective A-modules M with $e_o M \neq 0 \neq \hat{e} M$ and the elements of $\bigcup_{s \in (\tilde{n})} \mathfrak{U}_s$ (each $\mathfrak{U}_s \cong \mathfrak{U}_R$). Thus A is of finite representation type if and only if Δ_R is of finite representation type.

Now Theorem 2 is proved. To finish the proof of Theorem 1 we use the well-known result that $\Delta_{1,r,t}$ is of finite representation type if and only if $\frac{1}{(r+1)} + \frac{1}{(t+1)} > \frac{1}{2}$.

References

(1) M.Auslander and I.Reiten: Representation Theory of Artin Algebras VI, Comm. in Algebra, 6 (1978), 257-300

(2) P.W. Donovan and M.R. Freislich: The indecomposable modular representations of certain groups with dihedral sylow-subgroup, preprint.

(3) P.Gabriel: Unzerlegbare Darstellungen I, Manuscripta Math. 6 (1972), 71-103.

(4) -------- : Indecomposable Representations II, Symp. Math. Ist. Naz. Alta Math. 11 (1973), 81-104.

(5) P.Gabriel and Ch. Riedtmann: Group representations without groups, Com. Math. Helv. 54 (1979) 240-287.

(6) C.M.Ringel: The indecomposable representations of the dihedral 2-groups, Math. Ann. 214 (1975), 19-34.

(7) E.Scherzler and J.Waschbüsch: A class of self-injective algebras of finite representation type (this volume).

Josef Waschbüsch
II. Mathematisches Institut
der Freien Universität Berlin
Königin-Luise-Str. 24/26
D 1000 Berlin 33

HEREDITARY ARTINIAN RINGS OF RIGHT LOCAL REPRESENTATION TYPE

Kunio Yamagata

Algebras of right local representation type (i.e., every indecomposable right module has the simple top) or local-colocal representation type (i.e., every indecomposable module has the simple top or the simple socle) are first studied by Tachikawa and the ideal-theoretical structure is completely determined [5], [6]. However, for Artinian rings of these representation types, the structure theorem is not known yet. As a related work, Auslander, Green and Reiten introduced the concept "waist", and they characterized Artinian rings of local-colocal represetation type as those rings over which every indecomposable module is simple or has a waist [1], [2]. On the other hand, the representation theory for hereditary algebras is deeply studied and well-known. Further, recently Tachikawa studied a representation theory for trivial extensions of hereditary Artin algebras. Such extensions are always quasi-Frobenius (in fact, they are weakly symmetric). In this paper we study some class of Artinian rings with self-dualities by applying the concept of waist to indecomposable modules over extensions. We are mainly concerned with hereditary Artinian rings of right local representation type, and it will be proved the structure theorem for these Artinian rings.

Let A be an Artinian ring with a self-duality defined by an A-bimodule Q and T an extension over A with kernel Q. Then, in the section 1, some results and definitions will be recalled from [7], [8] and [9]. In the section 2, we will consider the condition such that for an extension T over A with kernel Q, MQ is a waist in M for every indecomposable T-module M with MQ $\neq$ 0 (this will be called the W-condition for right T-modules). Then it will be proved that A is a biserial whenever there is an extension T with the W-condition for right T-modules. In the section 3, hereditary Artinian rings will be considered and it will be proved that for a hereditary Artinian ring A and an extension T, A is a direct product of Artinian rings of right or left local representation type if and only if T is of local-colocal representation type if and only if T satisfies the W-condition for right T-modules. This and the result proved in the section 2 will imply the structure theorem for hereditary Artinian rings of right local representation type. In the last section 4, we will consider arbitrary Artinian rings with self-dualities and, as a generalization of the theorem in the section 3, it will be proved that A is a direct product of hereditary Artinian rings of right or left local representation type and a serial Artinian ring with a serial extension if and only if there is an extension T such that it is of local-colocal representation type and every indecomposable projective T-module P has a waist PQ if and only if there is an extension T satisfying the W-condition for right T-modules.

1. PRELIMINARIES.

Throughout this paper, A will be a (left and right) Artinian ring with a bimodule Q such that $Hom_A(\ , Q)$ defines a Morita duality. Such a bimodule Q will be called a QF-module (cf. [7], [9]). All modules will be finitely generated right modules, unless otherwise stated. Following [3] a ring T is called an extension over A with kernel Q if there is a ring epimorphism $\rho: T \to A$ whose kernel is isomorphic to Q as additive groups and is squared zero, that is, Q as a T-bimodule canonically induced by ρ is isomorphic to the ideal Ker ρ in T. In this case we may identify Q with the Ker ρ and every A-module may be regarded as a T-module annihilated by Q via ρ. Here it should be noted that there always exist extensions over A with kernel Q, for example, consider the trivial extension $A \ltimes Q$. Since Q is nilpotent in T, every primitive idempotent e in A is lifted to a primitive idempotent e in T such that $\rho(e) = e$. Conversely, it is clear that $\rho(e)$ is a primitive idempotent in A for every primitive idempotent e in T. Hence, if there is no confusion, primitive idempotents in A will be identified with those in T.

For a study of extensions the following is the most essential.

<u>PROPOSITION 1.1</u> [7] Every extension over A with kernel Q is quasi-Frobenius.

Let M be a nonprojective T-module, $f_1 : P_1 \to M$ a projective cover and $f_2 : M \to P_2$ an injective hull in mod T. Then we put $\Omega(M) = \text{Ker } f_1$ and $\Omega^{-1}(M) = \text{Coker } f_2$. If M is indecomposable, by (1.1) both $\Omega(M)$ and $\Omega^{-1}(M)$ are indecomposable and $\Omega^{-1}\Omega(M) \simeq M \simeq \Omega\Omega^{-1}(M)$. Let M be a module and W a nonzero proper submodule of M. Then W is called a <u>waist</u> in M provided for every sub-module X of M it holds that $X \subseteq W$ or $W \subseteq X$ [1]. As for the properties of waists over an extension we will recall some results from [8] and [9], which are easily proved by the definition and (1.1).

<u>LEMMA 1.2</u> [9] Let T be an extension and let M and N be T-modules. If $f : M \to N$ be a morphism with $f(M)Q \neq 0$ and MQ a waist in M, then $M/MQ \simeq f(M)/f(M)Q$ and $f(M)Q$ is a waist in f(M). In particular, if M is projective in mod T, then $f(M)/f(M)Q$ is projective in mod A.

<u>LEMMA 1.3</u> [9] If A is hereditary, then every indecomposable projective T-module P has a waist PQ.

<u>LEMMA 1.4</u> [8] Let P be an indecomposable projective T-module with a waist PQ and M its submodule such that $PQ \subsetneq M$. Then if M/PQ contatins a projective submodule in mod A, it holds that MQ = PQ.

If every indecomposable T-module M with $MQ \neq 0$ has a waist MQ, then T is of finite representation type (this will be a consequence of (4.6), and so the direct proof is omitted). This fact is one of the motives for our study. In this paper, we say that a T-module M with $MQ \neq 0$ satisfies the W-condition provided MQ is a waist in M, and T satisfies the W-condition for right T-modules if every indecomposable right module M with $MQ \neq 0$ has a waist MQ.

LEMMA 1.5 If A is hereditary and T an extension, then T satisfies the W-condition for right modules if and only if it does the condition for left modules.

This will be still valid for an arbitrary Artinian ring A, which will be an easy consequence of (4.6), however the lemma 1.5 will be used to prove the theorem (4.6). The (1.5) is essentially proved in [9].

2. THE STRUCTURE THEOREM.

In this section we prove that a ring A is biserial if A has an extension T satisfying the W-condition for right modules.

LEMMA 2.1 Let M be an indecomposable torsionless A-module. If $\Omega(M)Q$ is a waist in $\Omega(M)$, then $\mathrm{top}(M)$ is simple.

PROOF. If M is projective in mod A, then the assertion is clear. Let M be nonprojective in mod A and $0 \to \Omega(M) \to \bigoplus_{i=1}^{n} P_i \to M \to 0$ a projective cover of M in mod T, where each P_i is indecomposable. Then we must show that $n = 1$. Since M is an A-module, $MQ = 0$ in mod T, so that we have that $\bigoplus_{i=1}^{n} P_iQ \subseteq \Omega(M)$. Hence each P_iQ is contained in $\Omega(M)$. On the other hand, it follows from [7,2.6] that $P_iQ \not\subseteq \Omega(M)Q$ for all i. Hence $\Omega(M)Q \subsetneq P_iQ$, because $\Omega(M)Q$ is a waist in $\Omega(M)$ by assumption. Furthermore $\bigoplus_{i=1}^{n} \mathrm{soc}(P_i) = \mathrm{soc}(\Omega(M)) \subseteq \Omega(M)Q$ by the same reason. Hence it holds that $\bigoplus_{i=1}^{n} \mathrm{soc}(P_i) \subseteq P_i$ for all

i. Therefore we know that $n = 1$.

For a right T-module M and a simple module S, $(M : S)$ denotes the number of composition factors of M which are isomorphic to S, and $\#_{\alpha+1}(M : S) = (\mathrm{soc}^{\alpha+1}(M)/\mathrm{soc}^{\alpha}(M) : S)$. Then it is clear that $(M : S) = \sum_{\alpha \geq 0} \#_{\alpha+1}(M : S)$. For a subset I of the ring T, $\ell_M(I)$ denotes the left annihilator of I in M. $|M|$ denotes the composition length of M.

<u>LEMMA 2.2</u> Let M be an indecomposable nonprojective A-module with the simple socle. If $\ell_{\Omega(M)}(Q)$ is a waist in $\Omega(M)$, then $|\mathrm{top}(M)| \leq 2$.

<u>PROOF</u> Let $S = \mathrm{soc}(M)$, $0 \to \Omega(M) \to P \xrightarrow{f} M \to 0$ a projective cover of M in mod T, and $P = \bigoplus_{i=1}^{n} P_i$, where each P_i is indecomposable projective in mod T. Since M is nonprojective in mod A and $MQ = 0$ in mod T, it holds that $\Omega(M)Q \neq 0$ and $\ell_{\Omega(M)}(Q) = PQ$ by [6,2.1]. Let $X_i = f^{-1}(S) \cap P_i$. Then $P_iQ \subsetneq X_i$ and $(X_i/P_iQ : S) \geq 1$, because S is the simple socle of M and hence $S \subseteq f(P_i)$. Let $X = \bigoplus_{i=1}^{n} X_i$. Then

$$(X/PQ : S) = \sum_{i=1}^{n} (X_i/P_iQ : S) \geq n.$$

Now, since $n = |\mathrm{top}(M)|$, it suffices to show that $n = 2$ whenever $n \geq 2$. Thus we assume that $n \geq 2$ in the rest. Let $Y = X \cap \Omega(M)$. Then $PQ \subseteq Y$ and $f(X) = S$. Hence the canonical sequence $0 \to Y/PQ \to X/PQ \to S \to 0$ is exact, so that

$$(\overline{Y} : S) = (\overline{X} : S) - 1 = \sum_{i=1}^{n} (\overline{X}_i : S) - 1, \qquad (1)$$

where $\overline{Y} = Y/PQ$, $\overline{X} = X/PQ$ and $\overline{X}_i = X_i/P_iQ$. Hence, if we can prove that $(\overline{X}_i : S) \geq (\overline{Y} : S)$ for each i, we will have that $n = 2$. Because, (1) and the inequality imply that

$$0 \leq (\overline{X}_1 : S) - (\overline{Y} : S) = 1 - \sum_{i=2}^{n} (\overline{X}_i : S) \leq 2 - n,$$

i.e., $n \le 2$ and so $n = 2$. Thus we have only to prove that $(\overline{X}_i : S) \ge (\overline{Y} : S)$ for any i. Let e be an idempotent in A such that $S \simeq \mathrm{top}(eA)$, and choose

$$0 \ne m_k = m_k e \in \mathrm{soc}^{\alpha+1}(\overline{Y})\setminus\mathrm{soc}^{\alpha}(\overline{Y})$$

such that

$$\mathrm{soc}^{\alpha+1}(\overline{Y})/\mathrm{soc}^{\alpha}(\overline{Y}) \supseteq \sum_{1\le k\le \#_{\alpha+1}(\overline{Y}:S)} \oplus ((\overline{m}_k T + \mathrm{soc}^{\alpha}(\overline{Y})/\mathrm{soc}^{\alpha}(\overline{Y})).$$

Since $(\overline{Y} : S) = \sum_{\alpha>0} \#_{\alpha+1}(\overline{Y} : S)$ and $(\overline{X}_i : S) = \sum_{\alpha>0} \#_{\alpha+1}(\overline{X}_i : S)$, it then suffices to show that for the canonical projection $p_i : P \to P_i$,

i) $\overline{p_i(m_k)} \in \mathrm{soc}^{\alpha+1}(\overline{X}_i)\setminus \mathrm{soc}^{\alpha}(\overline{X}_i)$ for any i, k,

ii) $\sum_{1\le k\ \#_{\alpha+1}(\overline{Y}:S)} (\overline{p_i(m_k)}T + \mathrm{soc}^{\alpha}(\overline{X}_i)/\mathrm{soc}^{\alpha}(\overline{X}_i))$ is a direct sum.

<u>PROOF OF (i)</u> : Let $J = \mathrm{rad}(T)$. Since $\overline{Y} \subseteq \overline{X}$, it is clear that $\overline{p_i(m_k)} \in \mathrm{soc}^{\alpha+1}(\overline{X}_i)$. Now suppose that there are i and j such that $\overline{p_i(m_j)} \in \mathrm{soc}^{\alpha}(\overline{X}_i)$, i.e, $p_i(m_j)J^{\alpha} \subseteq P_iQ$. Since $m_j = \sum_{1\le k\le n} p_k(m_j)$, it holds that

$$m_jJ^{\alpha} \subseteq (\sum_{k\ne i} p_k(m_j))J^{\alpha} \oplus P_iQ . \tag{2}$$

Moreover, $\overline{m}_j \in \mathrm{soc}^{\alpha+1}(\overline{Y})\setminus \mathrm{soc}^{\alpha}(\overline{Y})$ by the choice of m_j. Hence $m_jJ^{\alpha} \subsetneq PQ$, which implies that $PQ \subsetneq m_jJ^{\alpha}$ by the assumption that PQ is a waist in $\Omega(M)$. On the other hand, from (2), $m_jJ^{\alpha+1} \subseteq \bigoplus_{k\ne 1} P_k \oplus P_iQJ$. Hence $PQ \subseteq \bigoplus_{k\ne i} P_k \oplus P_iQJ$, so that $P_jQ \subsetneq P_jQJ$, which is a contradiction. Therefore we know that $\overline{p_i(m_k)} \notin \mathrm{soc}^{\alpha}(\overline{X}_i)$ for any i and k.

<u>PROOF OF (ii)</u> : Assume that there is i such that

$$\sum_{1\le k\le \#_{\alpha+1}(\overline{Y};S)} (\overline{p_i(m_k)}T + \mathrm{soc}^{\alpha}(\overline{X}_i)/\mathrm{soc}^{\alpha}(\overline{X}_i))$$

is not a direct sum. It should be noted that each term of the above sum is simple, which is isomorphic to top(eA). Now then, there is j such that

$$((\overline{p_i(m_j)}T + soc^{\alpha}(\overline{X}_i)/soc^{\alpha}(\overline{X}_i)) \cap \sum_{k\neq j}((\overline{p_i(m_k)}T + soc^{\alpha}(\overline{X}_i)/soc^{\alpha}(\overline{X}_i)) \neq 0.$$

Hence $\overline{p_i(m_j)}T \subseteq \sum_{k\neq j}(\overline{p_i(m_k)}T + soc^{\alpha}(\overline{X}_i))$, i.e., there is $t_k \in T$ such that $\overline{p_i(m_j)} + \sum_{k\neq j}\overline{p_i(m_k)t_k} \in soc^{\alpha}(\overline{X}_i)$.

Let $m = m_j + \sum_{k\neq j} m_k t_k$. Then $\overline{p_i(m)} \in soc^{\alpha}(\overline{X}_i)$. Since $soc^{\alpha}(\overline{X}_i)J^{\alpha} = 0$, $p_i(m)J^{\alpha} \subseteq P_iQ$ and hence

$$mJ^{\alpha+1} = (\sum_{1\leq \ell\leq n} P_{\ell}(m))J^{\alpha+1} \subseteq \bigoplus_{\substack{\ell\neq i\\ 1\leq\ell\leq n}} P_{\ell} \oplus P_kQJ. \tag{3}$$

Moreover, $\overline{m} \in soc^{\alpha+1}(\overline{Y})\setminus soc^{\alpha}(\overline{Y})$ by the choice of m_i. Hence $\overline{m}J^{\alpha} \neq 0$, i.e., $mJ^{\alpha} \not\subseteq PQ$. This means that $PQ \subsetneq mJ^{\alpha+1}$, because PQ is a waist in $\Omega(M)$ by assumption. Hence from (3) we have that $PQ \subsetneq \bigoplus_{1\leq\ell\neq i\leq n} P_{\ell} \oplus P_kQJ$, so that $P_kQ \subsetneq P_kQJ$, a contradiction.

<u>LEMMA 2.3</u>. Let M be an indecomposable submodule of the radical of a factor of some primitive right ideal in mod A, and let top(M) be simple. Then, if $\Omega^{-1}(M)Q$ is a waist in $\Omega^{-1}(M)$, soc(M) is simple.

<u>PROOF</u>. Let e be a primitive idempotent in A such that M is a submodule of $(eA/I)\,rad(A)$ for some submodule I of eA, and let $0 \to M \to P \xrightarrow{f} \Omega^{-1}(M) \to 0$ be an injective hull of M in mod T and $P = \bigoplus_{i=1}^{n} P_i$ with P_i indecomposable projective. Since $soc(M) \simeq soc(P) = \bigoplus_{i=1}^{n} soc(P_i)$, it suffices to show that $n = 1$.

First we show that $M \quad PQJ$, where $J = rad(T)$. (1)

Let E be an injective hull of eA/I in mod A. Then, since PQ is an injective hull of M in mod A (cf. [7,2.3]), we have that $E = PQ \oplus X$, and $E\,\mathrm{rad}(A) = PQ\,\mathrm{rad}(A) \oplus X\,\mathrm{rad}(A)$. Hence $M \subseteq PQ \cap E\,\mathrm{rad}(A) = PQ\,\mathrm{rad}(A)$, while it is clear taht $PQ\,\mathrm{rad}(A) = PQ\,\mathrm{rad}(T)$.

Next, since $MQ = 0$ in mod T, f induces an isomorphism $\overline{f} : P/PQ \tilde{\rightarrow} \Omega^{-1}(M)\Omega^{-1}(M)Q$ by [7,2.3]. Hence, for a nonzero element $\overline{x}_i = x_i + P_iQ \in \mathrm{soc}\,(P_i/P_iQ)$, $\overline{f}(\overline{x}_i)$ is a nonzero element in $\mathrm{soc}\,(\Omega^{-1}(M)/\Omega^{-1}(M)Q)$. This implies that $f(x_i)J = \Omega^{-1}(M)Q$, because $\Omega^{-1}(M)Q$ is a waist in $\Omega^{-1}(M)$. Therefore we have that

$$f^{-1}(\Omega^{-1}(M)Q) = x_iJ + M \qquad \text{for any } i. \qquad (2)$$

In particular, it holds that

$$x_1J + M = x_iJ + M \qquad \text{for any } i. \qquad (3)$$

Now then, we assume that $n \geq 2$, which will imply a contradiction. Let $p_i : P \rightarrow P_i$ be the canonical projection. Then $p_1(x_1J + M) = x_1J + p_1(M)$ and $p_1(x_iJ + M) = p_1(M)$.

Hence it follows from (3) and (1) that $x_1J \subseteq p_1(M)$ and $x_1J \subseteq p_1(PQJ) = p_1(P)QJ = P_1QJ$. Consequently we have from (1) and (2) that $f^{-1}(\Omega^{-1}(M)Q) \subseteq PQJ$. Thus we conclude that $\Omega^{-1}(M)Q \subseteq f(P)QJ = \Omega^{-1}(M)QJ$, a contradiction.

The following is obtained by a routine calculation.

<u>LEMMA 2.4</u> Let A be an arbitrary semi-perfect ring, and let an A-module M be a sum of two serial submodule. If M is not indecomposable, then M is a direct sum of two serial submodules.

<u>THEOREM 2.5</u> Let A be an Artinian ring with a QF-module Q. Assume that there is an extension over A with kernel Q which satisfies the W-condition for right modules. Then, for any indecomposable projective A-module P, either P is serial or $P\mathrm{rad}(A)$ is a direct sum of two serial submodules.

<u>PROOF</u> Let $P = eA$ for some idmpotent e in A. Since

$Hom_A(\ ,Q)$ defines a duality between mod A and mod A° (where A° is an opposite ring of A), it holds that $_A Hom_A(eA, Q) \approx {}_A Qe$ and $Hom_A(soc^2(Qe), Q)_A \approx eA/erad(A)^2_A$. Then $|top(soc^2(Qe))| \leq 2$ by [9] and (2.2), so that $|soc(eA/e\ rad(A)^2)| \leq 2$. Since $soc(eA/e\ rad(A)^2) = e\ rad(A)/e\ rad(A)^2$, it herefore holds that $e\ rad(A) = M_1 + M_2$, where each M_i is a factor of some primitive right ideal in A. If M_i is nonzero, $M_i/M_i rad(A)^m \subseteq (eA/M_i rad(A)^m) rad(A)$ for every $m \geq 1$, and $top(M_i/M_i rad(A)^m)$ is simple. It therefore follows from (2.3) that $soc(M_i/M_i rad(A)^m)$ is simple for $m \geq 1$. This means that m_i is serial. Thus, in case $e\ rad(A)$ is decomposable, we know that $e\ rad(A)$ is a direct sum of two serial submodules in view of (2.4). On the other hand, in case $e\ rad(A)$ is indecomposable, we know that $top(e\ rad(A))$ is simple by (2.1). From this and the fact that $e\ rad(A)$ is a sum of serial submodules, it is easy concluded that $e\ rad(A)$ is serial, that is, eA is serial.

Next we will consider a hereditary Artinian ring with an extension satisfying the W-condition.

<u>LEMMA 2.6</u> Let M be an indecomposable A-module such that top(M) is nonsimple and $0 \to \Omega(M) \to \bigoplus_{i=1}^{n} P_i \to M \to 0$ a projective cover of M in mod T, where each P_i is indecomposable. If there is a waist W in $\Omega(M)$ with $WQ = 0$, then $W = \bigoplus_{i=1}^{n} P_i Q$ and $P_i Q = \Omega(M) \cap P_i$. In particular, M is a sum of projective A-modules.

<u>PROOF</u> Let $X_i = \Omega(M) \cap P_i$. Then $P_i Q \subseteq X_i$, because $f(P_i)Q = 0$ in mod T. Since Q is a waist in $\Omega(M)$, $\bigoplus_{i=1}^{n} soc(P_i Q) \subseteq W$ because of the fact that $soc(\Omega(M)) = \bigoplus_{i=1}^{n} soc(P_i Q)$. Hence, in particular, $W \subseteq X_i$ in view of the assumption

"n > 1". It follows that $\bigoplus_{i=1}^{n} P_i Q \subseteq \bigoplus_{i=1}^{n} X_i \subseteq W$. On the other hand, we know that $W \subseteq \ell_{\Omega(M)}(Q) = \bigoplus_{i=1}^{n} P_i Q$, because $MQ = 0$ in mod T. Hence we have that $W = \bigoplus_{i=1}^{n} X_i = \bigoplus_{i=1}^{n} P_i Q$, i.e., $X_i = P_i Q$. The other assertions are easy consequences of the fact that $P_i Q = \Omega(M) \cap P_i$.

Let A be an Artinian ring. The right quiver $Q(A_A)$ of A is an oriented graph such that the set of vertices are the set of isomorphism classes of indecomposable projective right A-modules, and for two vertices $[P_1]$ and $[P_2]$ there are n arrows from $[P_1]$ to $[P_2]$ iff $\mathrm{top}(P_2\mathrm{rad}(A))$ has a submodule which is isomorphic to $\mathrm{top}(P_1)^{(n)}$ (a direct sum of n many copies of $\mathrm{top}(P_1)$). The left quiver $Q({}_A A)$ of A is similarly defined. The <u>quivers</u> of A is the set of the left and the right quivers of A.

<u>THEOREM 2.7</u> Let A be a hereditary Artinian ring with a QF-module Q such that there is an extension over A with kernel Q which satisfies the W-condition for right modules. Then the quivers of A is a disjoint sum of the following.

1) 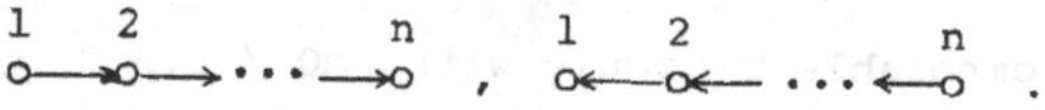

2) 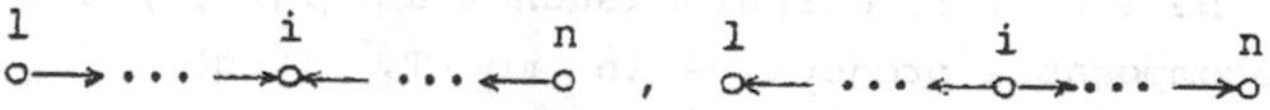

3)  $(n \geq 2)$

<u>PROOF</u> In the following proof it should be remenbered the lemma 1.5. Let e be an primitive idempotent in A. Then both e rad(A) and rad(A)e are serial or direct sums of two serial submodules by (2.5). Hence, considering the left quiver, we know that the right quiver $Q(A_A)$ does not contain the following subquivers

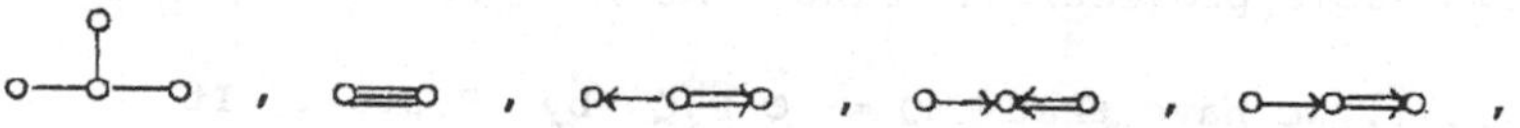

where o—o means an arrow with arbitrary orientation. Moreover, from (2.6) $Q(A_A)$ does not contain the subqunivers of the following types :

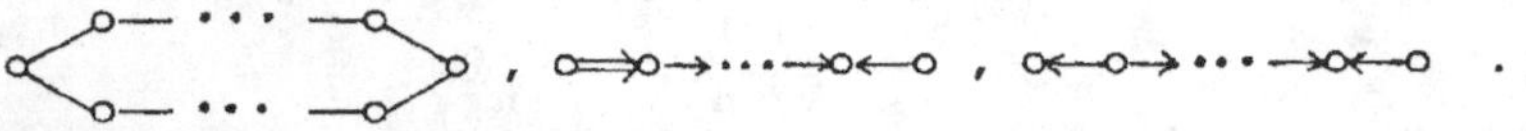

Thus the desired result will be easily obtained from those observations and (2.6).

3. HEREDITARY ARTINIAN RINGS OF RIGHT LOCAL REPRESENTATION TYPE

In this section we study hereditary Artinian rings with extensions satisfying the W-condition. Particularly it will be seen that the theorem (2.7) shows the structure of a hereditary Artinian ring with a self-duality and of right local representation type. Throughtout this section, we assume that the ring A is hereditary.

LEMMA 3.1 Assume that there is an extension T satisfying the W-condition for right T-modules. Then for every indecomposable T-module M, it holds that $|soc(M)| \leq 2$ and $|top(M)| \leq 2$.

PROOF Since T satisfies the W-condition for left modules by (1.5), by the duality we have only to show that $|soc(M)| \leq 2$ for every indecomposable T-module M.

i) Let M be an indecomposable T-module with $MQ \neq 0$. Let $f : P \to M$ be a nonzero morphism such that $f(P)Q \neq 0$ and P is indecomposable projective in mod T. Then $\ell_{f(P)}(Q) = \ell_M(Q)$ by [9], so that $f(P)Q = MQ$ by [7,2.7]. On the other hand, we know that $|top(PQ)| \leq 2$, because PQ is indecomposable injective in mod A. Since ${}_A Hom_A(PQ,Q)$ is projective, $|soc(Hom_A(PQ,Q))| \leq 2$ by (2,5). It follows from these facts that $|top(MQ)| \leq 2$. Now let $\bigoplus_{i=1}^{n} P_i$ be an injective hull of M in mod T, where each P_i is indecomposable projective. Since $MQ \neq 0$ and $MQ = \ell_M(Q)$ by [7,2.7], we have that $MQ = \bigoplus_{i=1}^{n} P_i Q$ by [7,2.3]. It therefore holds that

$$n \leq \sum_{i=1}^{n} |\mathrm{top}(P_i Q)| = |\mathrm{top}(MQ)| \leq 2,$$

while $\mathrm{soc}(MQ) = \bigoplus_{i=1}^{n} \mathrm{soc}(P_i Q)$ and each $\mathrm{soc}(P_i Q)$ is simple. Thus we know that $|\mathrm{soc}(M)| \leq 2$, because $|\mathrm{soc}(MQ)| = n \leq 2$ and $\mathrm{soc}(M) = \mathrm{soc}(MQ)$ by [7,2.7].

ii) Let M be an indecomposable T-module with $MQ = 0$. In this case we may assume that M is not injective in mod A, because if M is injective in mod A, it is clear that $|\mathrm{soc}(M)| = 1$. Then $\Omega^{-1}(M)Q \neq 0$ by [7,2.4]. Hence $|\mathrm{top}(\Omega^{-1}(M)Q| \leq 2$ by the above (i), which is clearly equivalent to that $|\mathrm{soc}(M)| \leq 2$. Thus we are done.

PROPOSITION 3.2 Assume that there is an extension satisfying the W-condition for right modules, and let A be a direct product of two rings A_1 and A_2 such that A_1 is not right serial. Then eQ is serial for every primitive idempotent e in A_1.

PROOF We may assume that A_1 is indecomposable, and suppose that there is a primitive idempotent e in A_1 such that eQ is not serial. Since Ae is isomorphic to ${}_A\mathrm{Hom}_A(eQ,Q)$, Ae is then nonserial. Hence $\mathrm{rad}(A)e$ is a direct sum of two serial submodules by (2.5), and so is $eQ/\mathrm{soc}(eQ)$. Because, $eQ/\mathrm{soc}(eQ) \simeq \mathrm{Hom}_A(\mathrm{rad}(A)e, Q)$ in mod A. This implies that $|\mathrm{top}(eQ)| = 2$, say $\mathrm{top}(eQ) = S_1 \oplus S_2$, where S_i denotes a simple submodule. Since A is hereditary and eQ is injective in mod A, each S_i is injective in mod A. There is then indecomposable projective T-modules P_i such that $P_i Q \simeq S_i$. On the other hand, since A_1 is indecomposable, it follows from(2.7) that there is an indecomposable projective right A_1-module, say $e_1 A$, such that every indecomposable projective A_1-module is isomorphic to a submodule of $e_1 A$. Such a module cleary has the injective top. (Particularly, there is a monomorphism $K : eA \to e_1 A$. Let $f : eT \to e_1 T$ be a morphism such that the diagram

$$\begin{array}{ccccc} & & eT & \xrightarrow{f} & e_1T \\ & & \downarrow & & \downarrow \\ 0 & \longrightarrow & eA & \xrightarrow{k} & e_1A \end{array}$$

is commutative, where the vertical morphisms are canonical. Then $0 \neq f(eQ) \subseteq e_1Q$ and, since $\mathrm{top}(e_1A)$ is injective in mod A, e_1Q is simple from [7,4.3]. Hence it must hold that $f(eQ) = e_1Q$. By the definition, $\mathrm{top}(eQ) \simeq P_1Q \oplus P_2Q$. Thus e_1Q is isomorphic to P_1Q. We may assume that $e_1Q \simeq P_1Q$, without loss of generality. In this case it holds that $P_1/P_1Q \simeq e_1A$.

Next, we will show that $P_1Q \not\simeq P_2Q$. (*)

Suppose that $P_1Q \simeq P_2Q$ to the centrary, and let $M = eT/eQ\mathrm{rad}(T)$. Then M is clearly indecomposable, and $P_1 \oplus P_2$ is an injective hull of M in mod T. Because, by the definition it is easily shown that $M/MQ \simeq eA$ and $MQ \simeq P_1Q \oplus P_2Q$, while we have known that P_iQ is simple socle of P_i, so that $\mathrm{soc}(M) \simeq P_1Q \oplus P_2Q$. Furthermore, since A is hereditary, $MQ = \ell_M(Q)$ and $\Omega^{-1}(M)Q = 0$ by [7,2.7,2.9]. Hence from [7,2.3] we have the canonical exact sequence

$$0 \to M/MQ \to P_1/P_1Q \oplus P_2/P_2Q \to \Omega^{-1}(M) \to 0.$$

On the other hand, we know that $\mathrm{soc}(eQ)$ is projective in mod A by (2.7), because eQ is indecomposable injective and nonserial. Hence eA is simple projective by [7,4.3], and so is M/MQ. It follows from these facts that

$$|\mathrm{soc}(\Omega^{-1}(M))| \geq \left(\sum_{i=1}^{2} |\mathrm{soc}(P_i/P_iQ)|\right) - 1.$$

However $|\mathrm{soc}(e_1A)| = |\mathrm{soc}(P_1/P_1Q)| = |\mathrm{soc}(P_2/P_2Q)|$, because

$P_1 \simeq P_2$ by assumption. Hence we have that $|soc(\Omega^{-1}(M))| \geq 3$, because $|soc(e_1A)| = 2$ by (2.5). But this contradicts (3.1).

Finally we will have a contradiction by showing that there is a nonzero morphism $f_1 : P_2/P_2Q \to e_1A$. Since $top(eQ) \to P_1Q \oplus P_2Q$, there is a nonzero morphism from eQ to P_2Q. By extending this morphism, we have a nonzero morphism $g : eT \to P_2$ such that $g(eQ) = P_2Q$, because P_2Q is simple. Hence g canonically induces a morphism $\overline{g} : eT/eQ \to P_2/P_2Q$, which is cleary nonzero. This means that P_2/P_2Q is an A_1-module, because $eT/eQ \simeq eA = eA_1$. Hence by the choice of e_1A there is a monomorphism $f_1 : P_2/P_2Q \to e_1A$, because P_2/P_2Q is projective in mod A_1. Now let $f : P_2 \to e_1T$ be an extension of f_1. Then $f(P_2Q) \subseteq e_1Q$. Since e_1Q and P_2Q are simple, this implies that $P_2Q \simeq e_1Q$. Therefore we have that $P_1Q \simeq P_2Q$, which contradicts the above fact (*). Thus we conclude the proof.

<u>LEMMA 3.3</u> If there is an extension satisfying the W-condition, then the ring A is a direct product of hereditary Artinian rings of right or left local representation type.

<u>PROOF</u>. Let T be an extension with the W-condition and B an indecomposable subring of A which is a direct summand as a ring. It suffices to show that B is of right or left local representation type.

i) First we consider the case that B is not right serial. We will show that every indecomposable right B-module has the simple top.

Let M be an indecomposable right B-module. Here we may assume that M is nonprojective in mod B. Let

$$0 \to N \to P = \bigoplus_{i=1}^{n} P_i \to M \to 0$$

be a projective cover of M in mod T, where each P_i is indecomposable. Since T is quasi-Frobenius, N is indecomposable and it holds that $PQ = \ell_N(Q) \subsetneq N$, because $MQ = 0$ in mod T and M is nonprojective in Mod A. (cf. [7,2.3].) Further, since A is hereditary by assumption, $NQ = \ell_N(Q)$, so that $NQ = PQ$ (see [7,2.7]). Hence there is an element $x = xe$ such that $x \in N \backslash PQ$, where e is a primitive idempotent in T. Let $p_i : P \to P_i$ be the projection. Then $x = \sum_{i=1}^{n} p_i(x)$ and hence there is $p_j(x)$ which is not in P_jQ. Let $f : eT \to P_j$ be a morphism defined by $f(et) = p_j(x)et$ for any $t \in T$. Then, since $p_j(x) = p_j(x)e$, $f(eQ) \neq 0$. Hence this induces a nonzero morphism from eT/eQ to P_j/P_jQ, while P_j/P_jQ is a B-module, because $\mathrm{Hom}_A(P_j/P_jQ, M) \neq 0$. It follows from this fact that eT/eQ is also a B-module. Hence we know that eQ is serial by (3.2). On the other hand, PQ is contained in xT, because $xT \not\subset NQ$ (=PQ) and NQ is a waist in N by assumption. Hence, by [7,2.1, 2.7] it holds that, $PQ = xQ$. However we have already known that xQ is serial and $PQ = \bigoplus_{i=1}^{n} P_iQ$. Therefore it must be that $n = 1$, which shows that top(M) is simple.

ii) Next assume that B is right serial. In this case, if B is also left serial, then B is serial. Hence it is well known [4] that every indecomposable left B-module is serial. On the other hand, if B is not left serial, B is of left local representation type by (i) and the symmetry of the assumptions in (3.3). Thus we know that B is of left local representation type.

PROPOSITION 3.4 If A is of right local or left local representation type and T an extension over A with kernel Q, then for every indecomposable right T-module M with $MQ \neq 0$, the soc(M) or top(M) is simple, respectively. Particularly T is of local-colocal representation type.

PROOF It suffices to show that, for a hereditary Artinian ring A of right local representation type, the soc(M) is simple. In this case, $\Omega^{-1}(M)$ is in mod A by [7,2.9]. Hence top($\Omega^{-1}(M)$) is simple by assumtion. This shows that soc(M) is simple, because every indecomposable projective T-module has the simple socle.

LEMMA 3.5 Let T be an extension of local-colocal representation type. Then T satisfies the W-condition.

PROOF Since A is hereditary, the W-condition holds for all indecomposable projctive T-modules by (1.3) and hence it holds for indecomposable T-modules with simple top. Hence it suffices to show that the W-condition holds for modules with simple socle. Let M be an indecomposable. T-module such that $MQ \neq 0$ and soc(M) is simple. Let $0 \to M \to P \to \Omega^{-1}(M) \to 0$ be an injective hull of M. Since soc(M) is simple, P is indecomposable and so PQ is a waist in P. Moreover it follows from [7,2.1, 2.7] that $PQ = \ell_M(Q) = MQ$. Hence MQ is a waist in M, because PQ is a waist in P and $PQ \subset M \subset P$.

In the following, it should be noted that, if A is of right local representation type, then every indecomposable injective right A-module is serial.

LEMMA 3.6. Let A be a hereditary Artinian ring of right local representation type. Let M be an indecomposable right A-module and E an injective hull of M in mod A. Then, if M is noninjective in mod A, E/M is serial. In particular, if soc(M) is not simple, then E/M is simple.

PROOF If soc(M) is simple, then E is indecomposable injective in mod A. Hence E is serial by assumption, so that E/M is clearly serial. Now assume that soc(M) is nonsimple and let $\underline{r}$ be the radical of A. Then top(M) is simple from the assumption, and it follows from (3.4), (3.5) and (3.1) that $|soc(M)| = 2$. Let $soc(M) = S_1 \oplus S_2$ and $S_i \simeq top(e_iA)$ for some primitive idempotents e_i in A.

Let top(M) ≃ top(eA) for some idempotent e in A. Then top(eA) is injective in mod A because of (2.7). Let E_i be an injective hull of S_i in mod A. Then each E_i is serial and $E_1 \oplus E_2$ may be considered as the E. Let ρ_i be the projection of $E = E_1 \oplus E_2$ onto E_i (i=1,2), $E_i\underline{r}^{\alpha_i} = E_i \cap M$ and $M_0 = E_1\underline{r}^{\alpha_1} \oplus E_2\underline{r}^{\alpha_2}$. Then, by making use of the facts that top(eA) is injective in mod A and E_i is serial, it is easily shown that each $\rho_i(M) = E_i$ for i = 1, 2. Moreover, for the canonical morphism $f : M/M_0 \to E/M_0$, it is seen that $f(M/M_0) \cap E_i/E_i\underline{r}^{\alpha_i} = 0$ (i=1,2). Thus we have the isomorphisms $E_1/E_1\underline{r}^{\alpha_1} \simeq M/M_0 \simeq E_2/E_2\underline{r}^{\alpha_2}$, i.e., $E_1/E_1\underline{r}^{\alpha_1} \simeq E_2/E_2\underline{r}^{\alpha_2}$. Hence $\alpha_1 = \alpha_2$. Let $\alpha = \alpha_1 = \alpha_2$. Next we will show that $\alpha = 1$. For this, suppose that $\alpha > 1$ to the contrary. Then $M_0 \subseteq M\underline{r}^2$ and hence $M/M\underline{r}^2$ is serial, because it is a factor of M/M_0. Particularly, $M\underline{r}/M\underline{r}^2$ is simple, say $M\underline{r}/M\underline{r}^2 \simeq$ top(e'A) for an idempotent e' in A. Since soc(M) $\subseteq M\underline{r}$, $M\underline{r}$ and so e'A are nonserial. Hence top(e'A) is injective by (2.7). But this implies that $M\underline{r}/M\underline{r}^2$ is a direst summand of $M/M\underline{r}^2$, a contradiction. Thus we have that $\alpha = 1$. Using this fact, it is easily shown that $M\underline{r} = E\underline{r}$. Therefore we have the isomorphism $E/M \simeq (E/E\underline{r})/(M/M\underline{r})$. Since $E/E\underline{r} \simeq E_1/E_1\underline{r} \oplus E_2/E_2\underline{r}$ and E_i is serial, $|E/E\underline{r}| = 2$, while $|M/M\underline{r}| = 1$. Hence $|E/M| = 1$, i.e., E/M is simple. Thus we conclude that soc(E/M) is simple.

Now we can chracterize the hereditary Artinian ring with an extension satisfying the W-condition.

<u>THEOREM 3.7.</u> Let A be a hereditary Artinian ring with a QF-module Q. Then the following are equivalent.

1) A is a direct product of an Artinian ring of right local representation type and an Artinian ring of left local representation type.

2) Every extension over A with kernel Q is of local-colocal representation type.

3) There is an extension over A with kernel Q which is of local-colocal representationtype.

4) There is an extension over A with kernel Q which satisfies the W-condition for right modules.

5) Every extension over A with kernel Q satisfies the W-condition for right modules.

PROOF $2 \Rightarrow 3$ This is clear, if we note that the trivial extension $A \ltimes Q$ does always exist. The $2 \Rightarrow 5$ and $3 \Rightarrow 4$ are proved in (3.5). The implication $4 \Rightarrow 1$ is also proved in (3.3), and $5 \Rightarrow 4$ is clear.

$1 \Rightarrow 2$: Let $A = A_1 \times A_2$, where A_1 and A_2 are Artinian rings of left local representation type and of right local representation type, respectively. Let T be an arbitrary extension and M an indecomposable right T-module. If $MQ = 0$, then M is in $\mathrm{mod}\, A$. Hence M is regarded as an A_1-module or an A_2-module. In any case, we know from the assumption for A that $\mathrm{top}(M)$ or $\mathrm{soc}(M)$ is simple. Now assume that $MQ \neq 0$, and let $0 \to \Omega(M) \to P \to M \to 0$ be a projective cover of M in $\mathrm{mod}\, T$. Since A is hereditary, $\Omega(M)Q = 0$ by [7,2.7, 2.9], i.e., $\Omega(M)$ is in $\mathrm{mod}\, A$. Hence $\Omega(M)$ may be considered as an A_1- or A_2-module. If it is an A_1-module, then $\mathrm{soc}(\Omega(M))$ is simple. Hence $\mathrm{soc}(P)$ is simple. This cleary shows that $\mathrm{top}(M)$ is simple, because $\mathrm{top}(M) \simeq \mathrm{top}(P)$ and P is indecomposable projective. Next cosider the case $\Omega(M)$ is an A_2-module. By [7,2.3] the canonical sequence $0 \to \Omega(M) \to PQ \to MQ \to 0$ is exact and an injective hull of $\Omega(M)$ in $\mathrm{mod}\, A$. Further $\Omega(M)$ is noninjective in $\mathrm{mod}\, A$ and so in $\mathrm{mod}\, A_2$ by [7,2.4]. Hence it follows from (3.6) that $\mathrm{soc}(PQ/\Omega(M))$ is simple, so that soc(MQ) is simple from the above sequence. On the other hand, $MQ = \ell_M(Q)$ by [7,2.7] and $soc(M) \subsetneq \ell_M(Q)$ becaus $Q \subsetneq rad(T)$. Therefore we know that $soc(M) \subsetneq MQ$, in particular, soc(M) = soc(MQ). Thus we have proved that soc(M) is simple.

EXAMPLE 3.8 As a conclusion of this section we will note that there is a decomposable ring A with an indecomposable extension satisfying the W-condition for right modules.

Let T be a serial quasi-Frobenius algebra with identity $1_T = e_1 + e_2 + e_3 + e_4$, where $\{e_i\}$ is a set of orthogonal primitive idempotents in T, such that the Loewy length is there, i.e., $J^2 \neq 0$ and $J^3 = 0$, where J = rad(T). Assume that $top(e_iJ) \simeq top(e_{i+1}T)$ for $1 \leq i \leq 3$ and $top(e_4J) \simeq top(e_1T)$. Let $Q_1 = e_1J^2 \oplus e_2J$, $Q_2 = e_3J^2 \oplus e_4J$, and $A_1 = (e_1T \oplus e_2T)/Q_1$, $A_2 = (e_3T \oplus e_4T)/Q_2$. Then each I_i is an ideal in T and so each A_i is a ring. Hence it is easily seen that A_i is hereditary. Let $A = A_1 \times A_2$ and $Q = Q_1 \times Q_2$. Then it is not difficult to show that ${}_AQ_A$ is a QF-module. T is therefore an extension over A with kenel Q. Now then, it is clear that A is decomposable but T is indecomposable.

4. ARTINIAN RINGS WITH EXTENSIONS SATISFYING THE W-CONDITION

In this final section we will determine the Artinian ring with an extension satisfying the W-condition for right modules. The theorem will show that the Artinian ring is a direct product of hereditary Artinian rings of right or left local representation type and some serial Artinian ring without simple projectives. By making use of the results in the section two, the proof will be reduced to the theorem proved in [8]. In this section the ring A will not be assumed that it is hereditary, and as usual, Q is a QF-bimodule and T an extension over A with kernel Q.

The following lemma is an easy consequence of the

fact that $Q \subset \mathrm{rad}(T)$.

<u>LEMMA 4.1</u> Let P be an indecomposable projective right T-moudle. If $\mathrm{top}(P)$ is projective in mod A, then $PQ = P\,\mathrm{rad}(T)$.

<u>LEMMA 4.2</u> Let T be an extension such that for any indecomposable T-module M with $MQ \neq 0$, MQ is a waist in M whenever M is projective in mod T or $\mathrm{soc}(M)$ is nonsimple. Then $\mathrm{soc}(P/PQ)$ is projective in mod A for any indecomposable projective T-module P such that $\mathrm{soc}(P/PQ)$ contains a projective submodule in mod A.

<u>PROOF</u> If $\mathrm{soc}(P/PQ)$ is simple, there is nothing to prove. Hence we may assume that $\mathrm{soc}(P/PQ)$ is not simple. Let M be a submodule of P such that $PQ \subseteq M$ and $M/PQ = \mathrm{soc}(P/PQ)$, and let $0 \to \Omega(M) \to \bigoplus_{i=1}^{n} P_i \xrightarrow{f} M \to 0$ be a projective cover of M in mod T with indecomposable projectives P_i. Then $n \geq 2$, i.e., both $\mathrm{top}(M)$ and $\mathrm{soc}(\Omega(M))$ are nonsimple. Here we may assume that $\mathrm{top}(P_1)$ is projective in mod A, because $\mathrm{top}(M) \simeq \bigoplus_{i=1}^{n} \mathrm{top}(P_i)$ and $\mathrm{top}(M)$ has a projective submodule by assumption. Then $P_1Q = P_1\,\mathrm{rad}(T)$ by (4.1). Next we will show that $\Omega(M)Q = 0$. Let $x \in \Omega(M)\ \Omega(M)Q$ and $x = \sum_{i=1}^{n} x_i$, where $x_i \in P_i$. Since $P_1Q = P_1\mathrm{rad}(T)$, it is then clear that $x_1 \in P_1Q$, i.e., $x_1Q = 0$. Hence $xQ = (\sum_{i=2}^{n} x_i)Q \subseteq \bigoplus_{i=2}^{n} P_iQ$, which shows that $\Omega(M)Q \cap P_1Q = 0$. However, if $\Omega(M)Q \neq 0$, $\Omega(M)Q$ is a waist in $\Omega(M)$, because $\mathrm{soc}(\Omega(M))$ is nonsimple. Hence $\Omega(M)Q \supseteq \mathrm{soc}(P_1Q)$, i.e., $\Omega(M)Q \cap P_1Q \neq 0$, a contradiction. Thus we know that $\Omega(M)Q = 0$. Hence M/MQ is projective in mod A by [7,2.4]. On the other hand, it follows from (1.4) that $MQ = PQ$, because M/PQ contains a projective submodule in mod A by assumption. Hence M/PQ is projective in mod A.

LEMMA 4.3 Let T be the extension given in (4.2). Then, for any indecomposable projective T-module P, if $\mathrm{soc}(P/PQ)$ is nonsimple, $\mathrm{soc}(P/PQ)$ is projective in mod A.

PROOF Let M_i $(i=1,2)$ be any submoudles of P containing PQ auch that

$$\mathrm{soc}(P/PQ) \supseteq M_1/PQ \oplus M_2/PQ$$

and M_i/PQ is simple. Then it suffices to show that each M_i/PQ is projective in mod A. Since M_i/PQ is simple and PQ is a waist in P, it is seen that $PQ = M_iJ$, where $J = \mathrm{rad}(T)$. Now let N be a maximal submodule of PQ and $M = (M_1 + M_2)/N$. Then $MJ = PQ/N$, so that MJ is the simple socle of M. Moreover $M/MJ \simeq M_1/PQ \oplus M_2/PQ$ and hence $|\mathrm{top}(M)| = 2$. Next, to show that $MQ \neq 0$, suppose the contrary. Since $|\mathrm{top}(M)| = 2$, there is a projective cover of M in mod T such that $0 \to \Omega(M) \to P_1 \oplus P_2 \xrightarrow{f} M \to 0$, where P_i is indecomposable. Since M is nonprojective in mod A, it follows from [7,2.4] that $\Omega(M)Q \neq 0$. Hence $\Omega(M)Q$ is a waist in $\Omega(M)$, because $\mathrm{soc}(\Omega(M))$ is nonsimple. Hence $\Omega(M)Q = P_1Q \oplus P_2Q$ by (2.6) and $P_iQ = \Omega(M) \cap P_i$. This shows that f induces mononorphisms $f_i : P_i/P_iQ \to M$. On the other hand, it is easily seen that there are $x_i \in P_i$ such that $0 \neq f(x_i) \in \mathrm{soc}(M)$ and $f(x_1 + x_2) = 0$. Let $x = x_1 + x_2$. Then $x \in \Omega(M)$ and $x \notin \Omega(M)Q$, and so $\Omega(M)Q \subsetneq xT$, because $\Omega(M)Q$ is a waist in $\Omega(M)$. Hence $\mathrm{soc}(P_iQ) \subsetneq xT$ and so $\mathrm{soc}(P_iQ) \subset xQ$, because xQ is a waist in xT by assumption. Moreover, since $P_1Q \oplus P_2Q = \Omega(M)Q \subsetneq xT$, it holds that $P_iQ \subsetneq xQ$ by the same reason. Of course, this means that $xQ = P_1Q \oplus P_2Q$, i.e., $x_iQ = P_iQ$. Hence $x_iT/x_iQ \subseteq P_i/P_iQ$, while x_iT/x_iQ is projective in mod A by (1.2). Since $x_iT/P_iQ \simeq \mathrm{soc}(M)$ via f_i, $\mathrm{soc}(M)$ is therefore simple projective in mod A. Hence N is a direct summand of PQ i.e., $N = 0$, because $PQ/N = \mathrm{soc}(M)$

and PQ is indecomposable. Hence $M_iQ = 0$. But this contadicts that $PQ \subsetneq M_i$. Thus we proved that $MQ \neq 0$. Hence we may assume that $(M_1/N)Q \neq 0$. Then we have that $(M_1/N)Q = PQ/N$, because $|M_1/N| = 2$ and $PQ/N \subsetneq M_1/N$. Since $M_1Q \not\subset N$ and M_1Q is a waist in M_1, it therefore holds that $M_1Q = PQ$. Hence $M_1/M_1Q \subseteq P/PQ$. As a consequence, since M_1/M_1Q is projective in mod A, soc(P/PQ) has a projective submodule in mod A. Thus we know form (4.2) that soc(P/PQ) is projective. Thus we are done.

<u>LEMMA 4.4</u> Assume that T satisfies the W-condition for right modules, and let P be any indecomposable projective T-module. Then, for any indecomposable submodule M of P with $MQ \neq 0$, it holds that M/MQ is projective in mod A.

<u>PROOF</u> We first note that $PQ \subsetneq M$, because $M \not\subset PQ$ and PQ is a waist in P. It follows from (2.5) that either P/PQ is serial or PJ/PQ is a direct sum of two serial submodules, where J = rad(T).

i) Assume that P/PQ is serial. Since $PQ \subsetneq M$ and PQ is a waist in M, we have that $PQ \subseteq MJ$. Hence M/MJ is a factor of M/PQ. Since M/PQ is serial by assumption, it therefore follows that top(M/MJ) is simple, and so is top(M/MQ), because $MQ \subseteq MJ$. Hence M/MQ is projective in mod A by (1.2).

ii) Next assume that PJ/PQ is a direct sum of two serial submodules. It then follows from (4.3) that soc(P/PQ) is projective in mod A. Now we can put $M/PQ = M_1/PQ \oplus M_2/PQ$, where $PQ \subseteq M_i$ and M_i/PQ is indecomposable with simple top if it is not zero. Since soc(P/PQ) contains soc(M/PQ) and soc(M_i/PQ), soc(M_i/PQ) is projective in mod A if it is not zero. Hence it follows from (1.4) that MQ = PQ and $M_iQ = PQ$ if $M_i/PQ \neq 0$. On the other hand, we know from (1.2) that M_i/M_iQ is projective in mod A (if it is nonzero), because top(M_i) is simple. Thus we have that M/PQ is projective in mod A, so that M/MQ is projective.

From those results we can easily obtain the following main theorem.

THEOREM 4.5 I) For an Artinian ring A with a QF-module Q, the following are equivalent.

1) There is an extension T over A with kernel Q such that MQ is a waist in M for every indecomposable T-module M with $MQ \neq 0$.

2) There is an extension T over A with kernel Q such that T is of local-colocal representation type and PQ is a waist in P for every indecomposable projective T-module P.

3) A is a direct product of two Artinian rings A_1 and A_2 such that

i) A_1 is a direct product of hereditary Artinian rings of right or left local representation type and it has a QF-module Q_1.

ii) A_2 is a serial Artinian ring with a QF-module Q_2 and without simple projectives, and there is a serial extension over A_2 with kernel Q_2.

II) If the property (3) in (I) holds, then for any extensions $\mathbb{E}_i : 0 \to Q_i \to T_i \to A_i \to 0$ such that T_2 is serial, the extension $\mathbb{E}_1 \oplus \mathbb{E}_2$ satisfies the properties in (1) and (2). Conversely, if $\mathbb{E} : 0 \to Q \to T \to A \to 0$ is an extension given in (1) or (2), then there are extensions $\mathbb{E}_i : 0 \to Q_i \to T_i \to A_i \to 0$ such that $\mathbb{E} \simeq \mathbb{E}_1 \oplus \mathbb{E}_2$, where A_i and Q_i satisfy the properties in (3).

PROOF (I) $1 \Rightarrow 3$: This is an easy consequence of (4.4) and the theorem in [8]. $3 \Rightarrow 2$: Let $\mathbb{E}_i : 0 \to Q_i \to T_i \to A_i \to 0$ be any extensions over A_i with kernel Q_i such that T_2 is serial, and let $\mathbb{E} = \mathbb{E}_1 \oplus \mathbb{E}_2$. Then the (2) is an immediate consequence of (3.7). $2 \Rightarrow 1$: Let T be an extension of local-colocal represntation type such that every indecomposable projective right T-modules satisfies the W-condition.

We will show that T satisfies the W-condition for right modules. For this, in view of (1.2) it suffies to consider indecomposable modules with nonsimple tops. Now let M be an indecomposable module such that $MQ \neq 0$ and $\text{top}(M)$ is nonsimple, and let P be an injective hull of M in mod T. Since T is of local-colocal representation type, $\text{soc}(M)$ is simple. Hence P is indecomposable and so PQ is a waist in P. Hence $PQ \subsetneq M$.

i) If $\text{soc}(P/PQ)$ is not simple, $\text{soc}(P/PQ)$ is projective in mod A by (4.3), and so is $\text{soc}(M/PQ)$. It then follows from (1.4) that $MQ = PQ$. Since PQ is a waist in P, MQ is therefore a waist in M.

ii) Assume that $\text{soc}(P/PQ)$ is simple. Let $N = M/PQ$. Then $\text{soc}(N)$ is simple and $\text{top}(N)$ is nonsimple. Hence $\Omega(N)$ is an indecomposable T-module such that $\Omega(N)Q \neq 0$ and $\text{soc}(\Omega(N))$ is nonsimple. Hence $\text{top}(\Omega(N))$ is simple by assumption, so that $\Omega(N)Q$ is a waist in $\Omega(N)$ by (1.2). It therefore follows from (2.6) that N is a sum of projective A-modules. Hence $MQ = PQ$ by (1.4). Thus we have that MQ is a waist in M.

(II) This is easily obtained from the proof of (I).

The following is an immediate consequence of [7,2.11,2.9].

PROPOSITION 4.5 Let A be a serial Artinian ring with a QF-module Q. Then A is hereditary if and only if the trivial extension $A \ltimes Q$ is serial.

PROPOSITION 4.7 Let T be a trivial extension of A by Q. Then the following are equivalent.

1) A is a direct product of hereditary Artinian rings of right or left local representation type.

2) T is of local-colocal representation type and every indecomposable projective right T-module P has a waist PQ.

PROOF $1 \Rightarrow 2$: Since A is hereditary, the (2) is a consequence of (3.7) and (1.3). $2 \Rightarrow 1$: Let $\mathbb{E} : 0 \to Q \to T \to A \to 0$ be the given trivial extension. Then it follows from (4.6) that there are extensions $\mathbb{E}_i : 0 \to Q_i \to T_i \to A_i \to 0$

such that $\mathbb{E} \simeq \mathbb{E}_1 \oplus \mathbb{E}_2$, A_2 has no simple projectives and T_2 is serial. It is easily seen that $\mathbb{E}_i$ is a trivial extension. Hence A_2 is hereditary by (4.6). But A_2 has no simple projective modules, it must be that $A_2 = 0$, i.e., $A = A_1$.

<u>EXAMPLE 4.8</u> In the conclusion of this paper, we remark that the waist condition for indecomposable projective modules in (4.5-2) or (4.6-2) cannot be removed. This is seen in the following example.

Let K be an algebraically closed field and let A be the subring of the matrix algebra $(K)_4$:

$$A = \left\{ \begin{pmatrix} x_1 & & & \\ x_4 & x_2 & & \\ & & x_3 & \\ & & x_5 & x_1 \end{pmatrix} : x_i \in K \right\} .$$

Let $T = A \ltimes Q$, where $Q = \mathrm{Hom}_k(A,K)$. Then it is seen from [6] that T is of local-colocal representation type, and it is clear that A is not hereditary but serial with simple projectives, i.e., A does not satisfy the condition (4.5-3) nor (4.7-1). Moreover, P_1Q is not a waist in P_1, where P_1 is an indecomposable projective T-module such that

$P_1/P_1Q \simeq e_1A$ (e_1 denotes $\begin{pmatrix} 1 & & & \\ & 0 & & \\ & & 0 & \\ & & & 1 \end{pmatrix}$) .

REFERENCES

[1] Auslander, M., Green, E.L. and Reiten, I. : Modules having waist. "Representations of Algebras", Lecture Notes in Math. 488, Springer-Verlag, 20-28 (1975).

[2] Gordon, R. and Green, E.L. : Modules with cores and amalgamations of indecomposable modules. Memoirs of Amer. Math. Soc. 187 (1977).

[3] Cartan, H. and Eilenberg, S. : Homological algebra. Princeton, New Jersey Princeton Unicersity press 1956.

[4] Nakayama, T. : On Frobenius Algebras II. Ann, Math. 42, 1-21 (1941).

[5] Tachikawa, H. : On rings of which every indecomposable right modules has a unique maximal submodule. Math. Zeit. 71, 200-222 (1959).

[6] Tachikawa, H. : On algebras of which every indecomposable representation has an irreducible one as the top or the bottom Loewy constituent. Math. Zeit. 75, 215-227 (1961).

[7] Yamagata, K. : Extensions over hereditary Artinian rings with self-dualities I, to appear.

[8] Yamagata, K. : Extensions over herediatry Artinian rings with self-dualities II, to appear.

[9] Yamagata, K. : On extensions over Artinian rings with self-dualities, to appear.

Instiute of Mathematics,
University of Tsukuba,
Sakuramura Niiharigun Ibaraki, 300-31
Japan.